Outdoor Lighting: Physics, Vision and Perception

Duco Schreuder

Outdoor Lighting: Physics, Vision and Perception

Author
Duco Schreuder
Spechtlaan 303
2271 BH Leidschendam
The Netherlands

ISBN: 978-1-4020-8601-4   e-ISBN: 978-1-4020-8602-1

Library of Congress Control Number: 2008931487

© 2008 Springer Science + Business Media B.V.
No part of this work may be reproduced, stored in a retrieval system, or transmitted in any form or by any means, electronic, mechanical, photocopying, microfilming, recording or otherwise, without written permission from the Publisher, with the exception of any material supplied specifically for the purpose of being entered and executed on a computer system, for exclusive use by the purchaser of the work.

Printed on acid-free paper

9 8 7 6 5 4 3 2 1

springer.com

Contents

	Preface	xiii
1	**Introduction: The function of outdoor lighting**	**1**
1.1	Why lighting the outdoors?	1
1.2	Lighting engineering	3
1.3	The function of outdoor lighting	4

1.3.1 Road lighting, street lighting and public lighting 4
1.3.2 The advancement of human well-being 4
1.3.3 The function of road lighting 4
1.3.4 The driving task 6
  (a) Driving task analysis 6 • (b) Task elements 7 • (c) Manoeuvres 7

1.4	Cognitive aspects of vision	9
1.5	Tools and methods	10

1.5.1 Models 10
1.5.2 Quick-and-dirty statistics 11
1.5.3 Scales in psycho-physiology 11

1.6	Conclusions	13
2	**Physical aspects of light production**	**17**
2.1	The physics of light	17

2.1.1 Definitions of light 17
2.1.2 Light rays 19
2.1.3 Waves and particles 20
  (a) Waves 20 • (b) Particles 25 • (c) Photons 26

2.2	General aspects of light production	28

2.2.1 Principles of light generation 28
2.2.2 The efficacy of light sources 29

2.3	Incandescence	30

2.3.1 Thermal radiation 30
  (a) The laws of black-body radiation 30 • (b) Grey bodies 33 • (c) Non-electric incandescent lamps 34 • (d) Electric incandescent filament lamps 35
2.3.2 Characteristics of electric incandescent lamps 38
  (a) The filament 38 • (b) Filament evaporation and bulb blackening 38 • (c) Lamp life and design values 39 • (d) Balance between the efficacy and the lamp life 39

    2.3.3 *Halogen incandescent lamps* 40
        (a) The role of gas pressure in the bulb 40 • (b) The halogen cycle 41 •
        (c) Characteristics of halogen incandescent lamps 41 • (d) Why use incandescent lamps? 42
2.4   Gas-discharge lamps   42
    2.4.1 *Quantum aspects of light* 42
        (a) Bosons, baryons, and fermions 42 • (b) The physics of metals 43 • (c) Quantum aspects of light; gas discharges 45 • (d) The construction of gas-discharge lamps 46 • (e) The influence of the vapour pressure 48 • (f) The main families of gas-discharge lamps 51
    2.4.2 *Fluorescence* 51
        (a) Fluorescence in gas-discharge lamps 51 • (b) The conversion from UV radiation into light 53 • (c) The efficiency of the fluorescent process 55 • (d) Fluorescent materials 57
    2.4.3 *Types of gas-discharge lamps* 59
        (a) Four families of lamps 59 • (b) Low-pressure mercury lamps 61 • (c) Low-pressure sodium lamps 62 • (d) High-pressure gas-discharge lamps 63 • (e) High-pressure mercury lamps 64 • (f) Metal-halide lamps 64 • (g) High-pressure sodium lamps 65
2.5   Semiconductor light   65
    2.5.1 *The physics of semiconductors* 65
        (a) Intrinsic and extrinsic semiconductors 65 • (b) Semiconductor diodes 67 • (c) Semiconductor light-emitting diodes 69
    2.5.2 *Anorganic LEDs* 70
        (a) The construction of anorganic LEDs 70 • (b) The colour of anorganic LEDs 71 • (c) The performance of anorganic LEDs 71 • (d) Use of anorganic LEDs 74
    2.5.3 *Organic LEDs* 75
2.6   Conclusions   78

3   Radiometry and photometry   85
3.1   Radiometry   85
    3.1.1 *Principles of radiometry* 85
        (a) The difference and the similarity between radiometry and photometry 85 • (b) Radiant power 86 • (c) The basic formula of radiometry 86 • (d) Terminology 88
    3.1.2 *The solid angle* 88
3.2   Basic photometric concepts   90
    3.2.1 *The SI-units* 90
        (a) Seven basic units 90 • (b) The photometric units 91 • (c) Units of the ISO-photometry 91
    3.2.2 *The luminous flux* 92
        (a) Definition 92 • (b) Reflectance, transmittance 92
    3.2.3 *The luminous intensity* 93
    3.2.4 *The illuminance* 94
        (a) Definition 94 • (b) Horizontal, vertical and semicylindrical illuminance 95 • (c) The average illuminance and the non-uniformity 96 • (d) The inverse square law 97 • (e) The distance law for large sources 98 • (f) The distance law for bundled light 100 • (g) The cosine law 101 • (h) The cosine to the third law 102

Contents   vii

    3.2.5 *The luminance* 103
        (a) General definition 103 • (b) The luminance of light-reflecting objects 105 •
        (c) The luminance of light emitting objects 105
3.3   Conclusions   106

4   **The mathematics of luminance**   109
4.1   The field concept   109
    4.1.1 *Field theory* 109
        (a) Light fluid and light vectors 109 • (b) Fields 110 • (c) Forces and potentials 110 •
        (d) Morphic fields 111
    4.1.2 *The light field* 111
        (a) The photic field 111 • (b) Fields of light rays 112 • (c) The speed of light 112 •
        (d) The basic formula of photometry 114
4.2   Some aspects of hydrodynamics   118
    4.2.1 *The continuity principle* 118
    4.2.2 *Bernoulli-fluids* 119
    4.2.3 *The equation of continuity* 120
    4.2.4 *Friction and diffraction* 121
        (a) The width of a light ray 121 • (b) Diffraction 121 • (c) The minimum separable
        123
4.3   The luminance of real and virtual objects   124
    4.3.1 *The need for a proper definition of luminance* 124
    4.3.2 *The general definition of luminance* 125
        (a) The direction aspects of the luminance 125 • (b) Light tubes 125 • (c) The
        physics of light tubes 127 • (d) The definition of the geometric flux 127 • (e) The
        throughput of light 129
4.4   The luminance of reflecting surfaces   131
    4.4.1 *The use of surface luminances* 131
    4.4.2 *Definition of reflection* 132
    4.4.3 *The luminance factor* 134
        (a) The reflection factor and the luminance factor 134 • (b) A description of the
        luminance factor 134 • (c) The luminance factor of a perfect diffuser 135 • (d) The
        definition of the luminance factor 135
    4.4.4 *The luminance factor of practical materials* 137
4.5   Conclusions   138

5   **Practical photometry**   141
5.1   General aspects of photometry   141
    5.1.1 *Five stages in the history of photometry* 141
    5.1.2 *Definition of measurement* 143
        (a) A general description of measurement 143 • (b) Measuring in nominal scales 143
        • (c) Measuring in ordinal scales 144 • (d) Measuring in quantitative scales 144 •
        (e) The accuracy when measuring in different scales 145
    5.1.3 *The relation between radiometry and photometry reconsidered* 145
    5.1.4 *Calibration* 146
        (a) Gauging and calibration 146 • (b) Standards 146
5.2   Traditional subjective photometry   148

    5.2.1 *Brightness estimation* 148
    5.2.2 *Visual photometry* 149
        (a) The sensitivity of the eye 149 • (b) The photopic V_λ-curve 151 • (c) The determination of the V_λ-curve 154 • (d) Heterochromatic and isochromatic photometry 155 • (e) The contrast method of photometry 156 • (f) Flicker effects 157 • (g) Flicker photometry 160
5.3   Traditional objective photometry   162
    5.3.1 *Instrumental photometry* 162
        (a) Counting photons 162 • (b) Sensors; Control and decision-making systems 162 • (c) The S/N ratio 164
    5.3.2 *Detectors* 165
        (a) Photocells 165 • (b) Barrier-layer photo-effect 165 • (c) Photocells for internal photo-effects 167 • (d) Photocells for the external photo-effect 167 • (e) Photomultipliers 168
    5.3.3 *Measuring photometric quantities* 170
        (a) Basic considerations 170 • (b) Luxmeters 170 • (c) Luminance meters 171 • (d) Measuring the luminous flux, and the light distribution 172 • (e) Accuracy 173 • (f) Examples of cosine and colour corrections for lux-meters 175
5.4   Modern objective photometry   177
    5.4.1 *CCDs* 177
        (a) CCDs for taking pictures 177 • (b) The properties of CCDs 178 • (c) The performance of CCDs 179 • (d) CCD data extraction and data processing 180
    5.4.2 *CCDs in photometry* 181
5.5   Conclusions   181

6   The human observer; physical and anatomical aspects of vision   187
6.1   The ability to see   187
6.2   The nervous system   188
    6.2.1 *The structure of the nerve cells* 188
        (a) Neurones 188 • (b) Synapses 189
    6.2.2 *The central nervous system* 191
6.3   The anatomy of the human visual system   191
    6.3.1 *The overall anatomy* 191
    6.3.2 *The optical elements, the cornea* 195
    6.3.3 *The optical elements, the eye lens* 196
        (a) The anatomy of the eye lens 196 • (b) Accommodation 197 • (c) Fourier optics 198 • (d) The point spread function or PSF 199 • (e) Optical aberrations; monochromatic aberrations 200 • (f) Optical aberrations; heterochromatic aberrations 200
    6.3.4 *The optical elements, the iris* 202
        (a) The anatomy of the iris 202 • (b) The Stiles-Crawford effect 203
    6.3.5 *The optical elements, the retina* 204
        (a) The anatomy of the retina 204 • (b) The photoreceptors 206 • (c) Cones and rods 207 • (d) The spatial distribution of rods and cones 209 • (e) Retinal ganglion cells 210
6.4   The optical nerve tracts   211
    6.4.1 *Image forming and non-image forming effects of light*   *211*

Contents ix

    6.4.2 *The visual neural pathways*   212
        (a) The organization of the retinal visual system 212 • (b) The optical nerve 213 • (c) Pathways for rod vision 215 • (d) Pathways for daylight levels 216 • (e) Movie tracks 218
    6.4.3 *The anatomy of the brain*   219
        (a) The main structure of the brain 219 • (b) Brain anatomy and brain functions 219 • (c) The cerebrum 220 • (d) The cortex 221
6.5   Conclusions   223

7   **The human observer; visual performance aspects**   229
7.1   The functions of the human visual system   229
    7.1.1 *The sensitivity of the eye* 229
        (a) Standard observers 229 • (b) Assessing the sensitivity of the eye 231 • (c) Two-degree and ten-degree photometry 232
7.2   The sensitivity of the human visual system   233
    7.2.1 *The duplicity theorem* 233
    7.2.2 *Photopic vision* 233
        (a) Three families of cones 233 • (b) The V_λ-curve 234 • (c) V_λ as a function 235 • (d) Tabulated values of V_λ 235
    7.2.3 *The scotopic spectral sensitivity curve* 236
    7.2.4 *Mesopic vision* 237
        (a) The limits of mesopic vision 237 • (b) The transition between photopic and scotopic vision 238 • (c) The high-mesopic region 239 • (d) The Purkinje-shift 240 • (e) Mesopic spectral sensitivity curves 240 • (f) Mesopic photometry 242 • (g) Mesopic metrics 242 • (h) Mesopic brightness impression 243
7.3   Visual performance   244
    7.3.1 *Human performance and visual performance* 244
        (a) Ergonomic aspects 244 • (b) Task aspects 245
    7.3.2 *Visual performance and Weber's Law* 246
        (a) The concept of visual performance 246 • (b) The law of Weber 246 • (c) Limits of Weber's Law 247 • (d) Validity of Weber's Law 249
7.4   The primary visual functions   250
    7.4.1 *Introducing the primary visual functions* 250
    7.4.2 *Adaptation* 251
    7.4.3 *Luminance discrimination* 253
        (a) The contrast 253 • (b) The minimum detectable contrast 254 • (c) Neural aspects of achromatic contrast phenomena 254 • (d) The laws of Ricco and Piper 255 • (e) The relation between size and threshold contrast 258 • (f) Lights with periodic brightness 259 • (g) The sensitivity to changes in the contrast 259 • (h) The RSC-curve 260
    7.4.4 *The visual acuity* 262
        (a) Visus 262 • (b) Measurement of the visual acuity 263 • (c) Visual acuity in relation to colour 265
7.5   Conclusions   266

8   **The human observer; visual perception**   273
8.1   Derived visual functions   273

8.1.1 *Field of view* 273
    (a) The field of vision 273 • (b) The functional visual field 275 • (c) Binocular vision 275 • (d) Stereopsis 276 • (e) The stereoscopic range 277
8.1.2 *The speed of observation; flicker-effects* 278
    (a) The discrimination in time 278 • (b) Flicker effects 279 • (c) Discomfort by flicker effects 280
8.1.3 *Subjective brightness* 281
8.1.4 *Detection of movement* 284
    (a) Constancy 284 • (b) Movement detection 285
8.2 Blinding glare   286
8.3 Disability glare   287
8.3.1 *Sources of disability glare* 287
    (a) Glare sources, the glare angle 287 • (b) Stray light in the eye 288 • (c) The equivalent veiling luminance 289 • (d) The nature of disability glare 289
8.3.2 *Characteristics of disability glare* 289
    (a) The effect of the light veil 289 • (b) Colour effects of disability glare 290 • (c) Practical implications of the colour effects 291 • (d) The four-component model 294 • (e) The θ-dependence 295 • (f) Age effects of disability glare 296 • (g) The limits for the θ-dependence 297 • (h) The CIE Standard Glare Observer 301
8.4 Discomfort glare   303
8.5 Conclusions   305

9   The human observer; colour vision   313
9.1 Colour aspects   313
9.1.1 *A description of colour* 313
9.1.2 *The importance of colour* 314
9.1.3 *Experiencing colours* 315
9.2 Colour vision physiology   315
9.2.1 *Three cone families* 315
    (a) The relative spectral sensitivity 315 • (b) Colour defective vision 318
    (c) The absolute spectral sensitivity 320
9.2.2 *The neural circuity in cone vision* 322
    (a) Colour vision theories 322 • (b) The opponent-process theory 323 • (c) The opponent-process theory circuitry 324 • (d) The construction of V_λ of self-luminous objects 327
9.3 Colour metrics and colorimetry   328
9.3.1 *Terminology* 328
9.3.2 *Colorimetry* 329
    (a) Additive and subtractive processes 329 • (b) The CIE system of colorimetry 329 • (c) The 1931 CIE Standard Chromaticity Diagram 330 • (d) Standard conditions for colour vision 333 • (e) Colour names 335 • (f) Colour points 335 • (g) The Munsell system 336 • (h) Metamerism 337
9.4 The colour characteristics of light sources   338
9.4.1 *Chromatic adaptation effects* 338
    (a) The colour impression 338 • (b) Chromatic adaptation 338
9.4.2 *The colour temperature* 339
    (a) The definition of the colour temperature 339 • (b) Incandescent light sources 340

Contents   xi

         • (c) The locus of the black-body radiators 340 • (d) Near-white light sources 341 • (e) Colour differences 342
  9.4.3 *The colour rendering* 344
         (a) The colour rendering in illuminating engineering 344 • (b) The colour rendition 345 • (c) The standard colours 345 • (d) The selection of the standard light source 346 • (e) The colour rendering of light sources 347
9.5   Conclusions   349

10   Road lighting applications   357
10.1 Geometric optics   358
  10.1.1 *Definitions of light* 358
         (a) Four models for the description of light 358 • (b) Light rays 358
  10.1.2 *Design of optical devices* 359
         (a) Principles of image-forming and lighting equipment 359 • (b) Image-forming equipment 360 • (c) Non-image forming equipment 363
10.2 Luminaire design   364
  10.2.1 *Optical elements* 364
         (a) Lamp and road axis 364 • (b) Road lighting luminaire light distributions 364 • (c) Road lighting luminaire classification 366 • (d) The proposal of Narisada and Schreuder 368
  10.2.2 *The optics of road lighting luminaires* 370
  10.2.3 *Ingress protection* 372
10.3 Light pollution   373
  10.3.1 *Description of light pollution* 373
         (a) Light pollution and sky glow 373 • (b) Victims of light pollution 374 • (c) Description of light pollution effects 375 • (d) Conspicuity of point sources 376 • (e) Application of flat luminaire covers 376
  10.3.2 *Limits of light pollution* 378
         (a) The natural background radiation 378 • (b) Artificial sky glow 379 • (c) CIE Limits 379 • (d) IAU limits 382
  10.3.3 *Remedial measures* 382
         (a) Limiting sky glow 382 • (b) Switching off the lights 383 • (c) Gated viewing 383 • (d) Light control 384 • (e) Reduction of reflection 384 • (f) Using monochromatic light 385 • (g) Filtering the light 385
10.4 Reflection properties of road surfaces   385
  10.4.1 *Road reflection as a road lighting design characteristic* 385
         (a) The luminance technique in road lighting 385 • (b) Reflection characteristics 386 • (c) Documentation of reflection characteristics 388
  10.4.2 *The classification of road surface reflection* 388
  10.4.3 *Standard reflection tables* 390
  10.4.4 *Field measurements of the road reflection* 390
10.5 Conclusions   393

11   Road lighting design   401
11.1 Design methods for road lighting installations   401
  11.1.1 *Principles of lighting design* 401

11.1.2 Design methods based on the road luminance 403
    (a) Systems of road lighting quality assessment 403 • (b) The method using E-P diagrams 403 • (c) The basis of computer-assisted luminance-design methods 406 • (d) LUCIE 407 • (e) Shortcomings of LUCIE 408

11.1.3 Alternative design parameters 410
    (a) Illuminance 410 • (b) Road lighting design methods based on revealing power 411 • (c) Visibility Level 411 • (d) Small Target Visibility 411 • (e) The Narisada visibility-based design method for road lighting 412 • (f) Guidance lighting 414 • (g) Visual comfort and city beatification 415 • (h) Fear for crime and subjective safety 417 • (j) Cost-benefit considerations 418

11.2   Road lighting for developing countries   418

11.2.1 Recommendations 418
    (a) The visibility approach 418 • (b) The traffic engineering approach 419

11.2.2 Low-maintenance lighting installations 420
    (a) The maintenance of outdoor lighting installations 420 • (b) Open luminaire options 421 • (c) Comparing open and closed luminaires 422

11.3   Simplified design methods   425

11.3.1 Characteristics of simplified design methods 425
    (a) The need for simplified design methods 425 • (b) Road classification 425 • (c) Lighted and unlit roads 426

11.3.2 A practical method for simplified lighting design 426
    (a) The design approach 426 • (b) The effective road width 427 • (c) The mounting height 428 • (d) The spacing 428 • (e) The light level 428 • (f) The utilization factor 428 • (g) The tabulated lighting design 429

11.4   Conclusions   430

Preface

The present book is based on the experience of the author. The experience is mainly the result of years of research, of consulting work, and in participation in policy decision making in many fields, most, but not all, related to outdoor lighting. To some degree, the book represents the preference of the author. The selection of the subjects is based on more than 50 years of experience of what is desirable to know for persons engaged in scientific research or practical application in the fields of lighting and vision. The subjects deal with a number of fundamental aspects. The theorists must have them at their fingertips, whereas the practical engineers may assume them as known in their daily work. The selection of subjects is based in part by the questions that came to the author over the years, but even more by the preference of the author himself. In this respect, it is a personal book. Thus, it should be stressed that the book is not a 'handbook' or even a 'textbook'; many subjects that commonly are treated in such books are not included here. Not because they lack importance, but because the author feels that they are adequately treated elsewhere. Some relevant works are mentioned in the References.

Over the years, the author has been engaged in giving courses on vision and lighting, lately more in particular on Masterclasses on a post-graduate or post-doctorate level. These courses were attended by small numbers of experienced lighting and vision engineers. The present book did profit from these courses.

There are many fundamental aspects that are essential for understanding why lighting is installed and why it works the way it does. However, most practical problems can be satisfactorily solved by leaving out all fundamental considerations. This may be regarded as a typical engineering approach. Engineers have to take care that things work, and do not always have to bother why they work. Typically, engineers use rules, rules-of-thumb even, that prove to work by experience. They are likely to use 'gut-feeling' to select the rules; in a nicer wording, they use 'engineering judgement'.

There are several rather condescending terms for this attitude, like 'lets duck it', or even 'lets dodge it'. What they mean is that it is tempting to avoid unpleasant or difficult things. Usually, unpleasant or difficult things go away when we wait for a while – usually, but not always. If you ignore the mechanics of vibration, bridges may collapse in storms, as they do quite often. If you ignore statical mechanics, halls may collapse under snow loads, as they do quite often. If you ignore geriatric psychology, the elderly get far too much light, which bothers them without helping them. This is what this book is all about: avoiding pitfalls that may be disastrous. The fundamental issues that were mentioned above are discussed in considerable detail in another book that is under preparation (Schreuder, 2008). The two books relate to each other like Sherlock Holmes and Dr. Watson. The theories of Holmes are essential, but without Dr Watson's down-to-earth pragmatism nothing would come out of it – no problems would be solved.

The book covers several of these points that are essential for the insight in public lighting. Many of these points tend to be overlooked in the day-to-day practice of lighting engineers and lighting designers. The book focusses on insight – backgrounds – coherence. There is another point. In a book where the physics of lighting has a prominent place, several chemical elements have a place. In order to make the book more accessible for an international readership, we did decide to use the more general terms instead of the correct English terminology. So the reader will meet wolfram instead of tungsten, and aluminium instead of aluminum. When needed, both terms are used side-by-side.

Finally about the language. As the book aims at an international readership for whom English is often not the mother language, and as the author is not a native speaker of English, compromises between purism and understanding had sometimes to be made. In this respect, the old 'Odham' often proved more helpful than the traditional dictionaries of Oxford and Webster (Smith, A.H. & O'Loughlin, J.L.N., eds., 1948. Odhams Dictionary of the English Language, reprinted. London, Odhams Press, 1948). As was done in the past, my good friend Stephen Harris took the trouble to read through the text. We mostly agreed, particularly on the controversial point of using 'commas' on a wider scale than is usually done, particularly in US texts. Many thanks for this! In spelling we tried to follow British-English; so the reader will meet colour and not color; metre and not meter. However, one may find organization, etc., with a 'z'.

The present book leans on several other books by the present author (Schreuder, 1964, 1967, 1967a, 1998; Narisada & Schreuder, 2004). And it goes without saying, that a large number of textbooks, monographs, and journal publications have been used in an attempt to select and discuss the subjects that were considered essential. Ample reference has been made throughout the book. In line with the lecture-character of the book, almost everything covered in this book stems from published material. It might be helpful to quote a few of the major sources that have been used, not only as

Preface

a tribute to these great works, but also as an attempt to provide the reader with more detailed textbooks, if further study is desired. Most books are quite recent; however, there are a few 'classics' that seem not to be surpassed. Also, in these works, many subjects are discussed that have not been included in the present book. As is mentioned in the Foreword, there is an element of personal preference in the subjects that have been treated.

- Mathematics: Bronstein et. al., 1997.
- Physics: Feynman et al., 1977; Joos, 1947; Van Heel, 1950. Von Weizsäcker, 2006; Wachter & Hoeber, 2006.
- Physiology: Augustin, 2007; Chalupa & Werner, eds., 2004; Gregory, ed., 2004.
- Lighting engineering: Baer, ed., 2006; De Boer, ed., 1967; Hentschel, ed., 2002; Moon, 1961; Van Bommel & De Boer, 1980.

Over the decades, the author made good use of the experience of those who acted as his teachers. At Delft University, the great Professor Van Heel taught that the work of an engineer is to look for optimal solutions, not for perfection. At Philips, the equally famous Professor De Boer did show that theoretical studies are nice, but that they serve only a purpose if they result in some sort of product. And at the Institute of Road Safety Research SWOV, Professor Asmussen stressed the need for a functional approach, realising that all products or measures must serve a purpose, and that this purpose must be made explicit. Also he taught me how to deal with public or semi-public authorities. It is the author's wish that these wise lessons and pragmatic teachings will be reflected in the approach the lectures on visual science and outdoor lighting engineering.

The author wishes to thank
- Mr Ian Mulvany and Mrs Mieke van der Fluit of Springer for their support and advice;
- Mr Paul de Kiefte for his thoughts on the didactic aspects of the book;
- Mr Jan Faber of LINE UP for his excellent work on the production of the book;
- Mr Stephen Harris for his invaluable help in English, and for the many stimulating discussions we had in the process.

Most of all I would like to thank my wife Fanny who supported me all the time, and who never complained although I have declared several times that I would 'never again' embark on writing a book.

In this book, a number of abbreviations are used. Most of them are listed here, although many are explained in the text where they show up. The translations given here are indicative only. Most are not authorised; they are given only for matters of information.

- ALCoR: Astronomical Light Control Region
- Anon: Anonymous: author not known, or not indicated
- ASP: Astronomical Society of the Pacific, San Francisco, USA
- CBS: Centraal Bureau voor de Statistiek (Central Statistical Bureau). The Hague, the Netherlands
- CCD: Charge-coupled Device
- CGPM: Conference Générale des Poids et Mesures (General Meeting on Weights and Measures)
- CEN: Comité Européen de Normalisation. Brussels, Belgium
- CROW: Centrum voor Regelgeving en Onderzoek in de Grond-, Water- en Wegenbouw en de Verkeerstechniek (Centre for regulation and research in soil, water and road engineering and traffic engineering)
- DIN: Deutsches Institut für Normung e.V. (German Standards Institution)
- EN: European Norm
- GLS-lamp: General Lighting Service lamp
- HID-lamp: High Intensity Discharge lamp
- IAU: International Astronomical Union
- IDA: International Dark-Sky Association
- IEC: International Electrotechnical Commission
- IES: The Illuminating Engineering Society of North America
- IEIJ: Illuminating Engineering Institute of Japan
- ILE: The Institution of Lighting Engineers
- ISO: International Standards Organization
- IZF/TNO: Instituut voor Zintuigphysiologie TNO (Institute for Perception TNO); now TM/TNO
- JCIE: Japanese National Committee of CIE
- LITG: Lichttechnische Gesellschaft (Association for illuminating engineering)
- LNV: Ministerie van Landbouw, Natuurbeheer en Visserij (Ministry of Agriculture, Nature Policy and Fishery)
- NSVV: Nederlandse Stichting voor Verlichtingskunde (Illuminating Engineering Society of The Netherlands)
- NOVEM: Nederlandse Maatschappij voor Energie en Milieu bv (Society for energy and the environment of the Netherlands)
- OECD: Organization for Economic Cooperation and Development
- PAOVV: Orgaan voor postacademisch onderwijs in de vervoerswetenschappen en de verkeerskunde. (Organization for post-graduate studies in transportation sciences and traffic engineering)
- SANCI: South African National Committee on Illumination

Preface

- SCW: Studiecentrum Wegenbouw (Study Centre for Road Construction SCW); now CROW
- SVT: Stichting Studiecentrum Verkeerstechniek (Study Centre for Traffic Engineering); now CROW
- SWOV: Stichting Wetenschappelijk Onderzoek Verkeersveiligheid (Institute of Road Safety Research SWOV)
- TRB: Transportation Research Board
- UNESCO: United Nations Educational, Scientific and Cultural Organization
- V&W: Ministerie van Verkeer en Waterstaat (Ministry of Transport, Public Works and Water Management)
- VROM: Ministerie van Volkshuisvesting, Ruimtelijke Ordening en Milieubeheer (Ministry of Public Housing, Land Planology and Environmental Policy)
- WHO: World Health Organization
- WVC: Ministerie van Welzijn, Volksgezondheid en Cultuur (Ministry of Well-being, Public Health and Culture)

References

Augustin, A. (2007). Augenheilkunde. 3., komplett überarbeitete und erweiterte Auflage (Ophthalmology. 3rd completely revised and extended edition). Berlin, Springer, 2007.

Baer, R., ed. (2006). Beleuchtungstechnik; Grundlagen. 3., vollständig überarbeitete Auflage (Essentials of illuminating engineering, 3rd., completely new edition). Berlin, Huss-Media, GmbH, 2006.

Bronstein, I.N.; Semendjajew, K.A.; Musiol, G. & Mühlig, H. (1997). Taschenbuch der Mathematik. 3. Auflage (Manual of mathematics. 3rd edition). Frankfurt am Main, Verlag Harri Deutsch, 1997.

Chalupa, L.M. & Werner, J.S., eds. (2004). The visual neurosciences (Two volumes). Cambridge (Mass). MIT Press, 2004.

De Boer, J.B., ed. (1967). Public lighting. Eindhoven, Centrex, 1967.

Feynman, R.P.; Leighton, R.B. & Sands, M. (1977). The Feynman lectures on physics. Three volumes. 1963; 6th printing 1977. Reading (Mass.), Addison-Wesley Publishing Company, 1977.

Gregory, R.L., ed. (2004). The Oxford companion to the mind. Second edition. Oxford, Oxford University Press, 1987.

Hentschel, H.-J. ed. (2002). Licht und Beleuchtung; Grundlagen und Anwendungen der Lichttechnik; 5. neu bearbeitete und erweiterte Auflage (Light and illumination; Theory and applications of lighting engineering; 5th new and extended edition). Heidelberg, Hüthig, 2002.

Joos, G. (1947). Theoretical physics (First edition in German 1932). London, Blackie & Sons Limited, 1947.

Moon, P. (1961). The scientific basis of illuminating engineering (revised edition). New York, Dover Publications, Inc., 1961

Narisada, K. & Schreuder, D.A. (2004). Light pollution handbook. Dordrecht, Springer, 2004.

Schreuder, D.A. (1964). The lighting of vehicular traffic tunnels. Eindhoven, Centrex, 1964.

Schreuder, D.A. (1967). The theoretical basis for road lighting design. Chapter 3. In: De Boer, ed., 1967.

Schreuder, D.A. (1967a). Measurements. Chapter 8. In: De Boer, ed., 1967.

Schreuder, D.A. (1998). Road lighting for safety. London, Thomas Telford, 1998. (Translation of: Schreuder, D.A., Openbare verlichting voor verkeer en veiligheid. Deventer, Kluwer Techniek, 1996).

Schreuder, D.A. (2008). Looking and seeing; A holistic approach to vision. Dordrecht, Springer, 2008 (in preparation)
Van Bommel, W.J.M. & De Boer, J.B. (1980). Road lighting. Deventer, Kluwer, 1980.
Van Heel, A.C.S. (1950). Inleiding in de optica; derde druk (Introduction into optics, third edition). Den Haag, Martinus Nijhoff, 1950.
Von Weizsäcker, C. F. (2006). The structure of Physics (Edited, revised and enlarged by Görnitz, T., and Lyre, H.). Dordrecht, Springer, 2006.
Wachter, A. & Hoeber, H. (2006). Compendium of theoretical physics (Translated from the German edition). New York, Springer Science+Business Media, Inc., 2006.

1 Introduction: The function of outdoor lighting

In this chapter, several aspects of the subject matter of this book are discussed. First we will consider the question why the outdoor areas are lit. In the 20th century, the technology allowed it, and the 24/7-economy required it. Its roots are in the idea in many people that Light is Good, and that Dark is Evil. However, removing darkness from the world did mean at the same time that the starry night became invisible, with unknown but probably serious consequences for people's self-esteem. Detail of these considerations are given in a separate book that is in some sense is parallel to this one.

A second item of this chapter is the function of outdoor lighting, more in particular of road lighting. Two ways to consider the function of road lighting results in two different priorities of the lighting quality criteria. The first is the detection of small objects on the road, the other is the driving task that describes what is necessary to reach the destination of a trip.

A third item is a brief description of some tools that are frequently used in the design and the assessment of outdoor lighting, viz.: models; 'quick-and-dirty' statistics, and scales.

1.1 Why lighting the outdoors?

Humanity did welcome the availability of artificial light. By means of what is one of the major technological achievements of the second half of the 20th century, the Curse of Darkness was banned once and for all. The 24/7 economy was born – 24 hours a day, 7 days a week. The fact that the starry night was lost seemed to be a small price to pay. It is not right to blame the lighting industry for this: it was the social pressure of a culture that believes in rationality as the ultimate level of human consciousness that caused the loss.

In our world today, we have the custom to light the outdoors. Why? At a first glance the answer seems to be obvious. Because you are able to see, you may experience a more pleasant outdoor scene ('amenity'), and you may avoid being a victim of accidents and crime. This seems to be enough for most decision makers and politicians, and, indeed, it may solve the majority of the practical problems. It is based on the idea that Light is Good, an idea that is maintained by almost all religions. However, most outdoor lighting is ugly and sometimes even repulsive, and accidents an crime still occur. Furthermore, light is bad for the environment. It requires energy and it causes light pollution.

In many religions, as well as in many philosophical systems, Light is the incarnation of God, of Good, of Virtue, whereas Dark is the incarnation of the Devil, of Evil, of Sin. Light is considered as a synonym for safety, comfort, and beauty, whereas darkness is considered as a synonym for danger, fear, and ugliness. From this idea comes the emotional conviction that Light is Good, so it is a moral responsibility for the governing bodies to provide light to the people. Although all this seems a little far-fetched, the idea is made clear in many policy papers where road authorities argue the need of road and street lighting. And because Light is Good, it is a small step only to maintain that More Light is Better. It is generally assumed that city authorities have the responsibility, even the moral responsibility, towards their citizens to provide road and street lighting, more in particularly for urban residential areas. The corresponding lighting requirements are often quite high as regards light level and light colour, not only to avoid accidents and crime, but to ensure an agreeable surrounding: the well-known amenity lighting. After all, in the 21st century, amenity and comfort are the ultimate Good!

This is not all there is to say about light and dark, good and evil. Darkness is not only the absence of light. There is a completely different way to consider darkness. We will coin this way the starry night. As the word suggests, it deals with the night sky, made luminous by stars. It seems more in particular that the starry night may offer to contemplate and to meditate. In this, it differs fundamentally from the concepts of darkness and lack of light that have been discussed earlier. The faint sources may contribute to understanding the evolution of the early Universe. Philosophers and poets can feel the darkness of the night sky, and children can marvel at the night sky, and comprehend their place in the Universe (Narisada & Schreuder, 2004, chapter 1).

There seems to be little to say against these standpoints, apart from the fact that they are non-rational. Delving deeper into this matter opens up a very wide vista of philosophical and religious questions and their answers. Precisely this is the subject matter of another book by the same author – a book that is, to a certain extent, parallel to the present book (Schreuder, 2008).

1.2 Lighting engineering

Although these aspects are essential for understanding why lighting is installed and why it works the way it does, most practical problems can be satisfactorily solved by leaving out all fundamental considerations. This may be regarded as a typical engineering approach. Engineers have to take care that things work, and do not always have to bother why they work. Typically, engineers use rules, rules-of-thumb even, that prove to work by experience. They are likely to use 'gut-feeling' to select the rules; in a nicer wording, they use 'engineering judgement'.

Rules almost always work. That's why they became rules in the first place. However, rules are based on certain assumptions. Usually, these assumptions are hidden. The users do not know them. Nowadays, technology evolves very rapidly, and in many cases the new methods, the new products, and the new applications are based on other assumptions. The rapid evolution does not allow the new assumptions to condense in new rules. So there is a real danger that old, out-of-date rules are applied in new fields. Pitfalls abound!

One of the golden rules is the functional approach of outdoor lighting. There are three classes of outdoor lighting, viz.:
1. The purely utilitarian lighting like e.g. road traffic lighting, lighting of industrial complexes or sports facilities, etc. Its main function is to improve the task performance; boundary conditions lie in the promotion of safety and security.
2. Amenity lighting like e.g. the lighting of pedestrian malls, residential streets, floodlighting of public buildings, etc. Visibility aspects are important, but the main function is to promote of the feeling of well-being. Boundary conditions lie in the reduction of the number and severity of criminal acts and in the promotion of feeling secure.
3. Decorative lighting like e.g. illumination of Christmas trees, laser beam displays, floodlighting of fountains and trees. Their function is exclusively to enliven the scene.

Clearly, all three types have a function in society. The function may be, as suggested above, rather different, but the basic idea is that lighting is always put up with some purpose, some goal, some function in mind. Lighting is not put up at random. This is, of course, just a rule-of-thumb, that easily can be wrong. A major pitfall looms: if the designer assumes that the lighting serves some purpose, and if the authorities only put it up out of habit, severe misunderstandings are likely to arise.

1.3 The function of outdoor lighting

1.3.1 Road lighting, street lighting and public lighting

To clarify the points that were raised in the preceding part of this section, we will discuss briefly the function of lighting, using public lighting as an example. First the terminology. For lighting designers and traffic engineers, road lighting, street lighting and public lighting are considered as synonyms. Many politicians think otherwise, and so do accounting authorities.

1.3.2 The advancement of human well-being

Safety, health, and happiness have a special place in human society, being aspects of human well-being. The relation to lighting is clear. The advancement of human well-being is usually considered as the main function of lighting.

Many believe that functionality is a typical Western-European consideration (Schreuder, 2001). It is closely related to the idea of utility, and thus to ideas of purpose and profit. People are supposed to do things with profit – material profit – in mind; if there are no advantages, people do not do it. This is a rather limited approach, which may describe only limited aspects of human behaviour.

Lighting is basically functional. The functions depend on the area application of the lighting. Over the years, most work has been done in the area of public lighting (Schreuder 1970, 1974, 1997, 1998; De Boer, 1967; Van Bommel & De Boer, 1980).

1.3.3 The function of road lighting

As regards road lighting, its main function is to make road traffic possible at night with an acceptable degree of safety and comfort. This holds not only for car drivers, but for other road users as well. Road lighting is an effective and efficient accident and crime counter-measure. Apart from these functions, road lighting may also contribute to amenity and the quality of life, more in particular to the aesthetic aspects of the night-time scenery. In this respect, road lighting has a lot in common with decorative lighting and illumination ('flood lighting'; Narisada & Schreuder, 2004, chapter 2; Forcolini, 1993; Cohu, 1967).

In history, the primary function of road lighting was to ensure the security of people and goods – in modern terms the crime-reduction function. Later, the enhancement of the status of cities and countries has been added – in modern terms the economic

function. More recent functions relate to the safety and throughput of motorised traffic. And still more recently, amenity and the security of pedestrians were added. In history, these functions developed in the following sequence:
- crime prevention;
- economic and aesthetic enhancement of cities;
- provide orientation;
- safety for traffic (pedestrian and carriage);
- enhance traffic flow of motor transport;
- enhance safety of motor transport;
- enhance social safety of residents, pedestrians etc.;
- enhance amenity for residents.

Further inspection of these criteria suggests that they are of a different nature. Some are more directly of an engineering nature, whereas some others show a more aesthetic – or architectural – nature. Safety and traffic throughput belong to the first; they are the main area of lighting engineers. Others, like the economic and aesthetic aspects, and the amenity for residents, are the area of lighting designers. The provision of the orientation has aspects of both approaches. It is an important aspect of safety, but also – in a more emotional sense – it determines the 'feel' of the area in question. It should be added that environmental protection and cost management play an important role as well (Narisada & Schreuder, 2004).

For each type of road or street, there is another combination of the predominant functions, and anther set of priorities. These depend on:
1. The location (urban or rural roads);
2. The land use (industrial, commercial, residential);
3. The network function (flow; access; local);
4. The infrastructure (dual carriageway; motor traffic only; mixed traffic).

In order to classify all roads and streets, a matrix approach would be desirable. In practice, most national and international standards and recommendations do not go so far. Usually, they are restricted to a number of frequently occurring combinations. For each combination, the requirements are given in terms of photometric characteristics. In most cases, the photometric requirements for traffic routes are expressed in luminance values, whereas the photometric requirements for local and residential streets are expressed in illuminance terms. Some examples of such standards are: CEN, 2002; CIE, 1965, 1992, 1995; NSVV, 2002.

For traffic route lighting, there are four main criteria for the quality of road lighting:
1. The average road surface luminance;

2. The degree of glare restriction;
3. The uniformity of the luminance pattern;
4. The optical – visual – guidance.

For the lighting of local and residential streets, there are five main criteria for the quality:
1. The average illuminance on the road surface;
2. The illuminance on areas adjoining the carriage-way;
3. The uniformity of the illuminance pattern;
4. The degree of glare restriction;
5. The colour of the light.

1.3.4 The driving task

(a) Driving task analysis

For each of these functions, quantitative criteria, expressed in photometric terms, can be established. They differ as to whether the detection of small obstacles or the fulfillment of the driving task is given priority. The driving task can be described as the collection of observations and decisions a car driver has to make in order to reach the goal of the trip. The visual task is usually understood as being a part of the driving task (Griep, 1971; Narisada & Schreuder, 2004, sec. 10.3; Schreuder, 1988, 1991).

The driving task requires a considerable amount of information, for the greater part visual information. Drivers need to look ahead while driving. The term coined for this is foresight. The concept of 'foresight' can be used, because there is only one driving task, in spite of the fact that the visual field is complex.

Contrary to other modes of transport, like e.g. commercial air traffic and rail traffic, in road traffic conflicts are accepted. It is assumed that the road traffic system will operate in such a way that, although conflicts may be frequent, accidents will be avoided – or rather, will be limited to a socially acceptable level of severity.

Obviously, the propose of taking part in traffic is to reach the trip destination. It is convenient to subdivide this purpose in three distinct goals, although in reality they are very much intertwined. The goals each have their sub-tasks. The three sub-tasks are:
1. Reaching the destination by selecting and maintaining the correct route;
2. Avoiding obstacles while travelling towards the destination;
3. Coping with emergencies while performing the two other sub-tasks.

(b) Task elements

The sub-tasks are quite different in nature. The first sub-task – the selection and maintenance of the route – involves decisions that are made for a large part even before the beginning of the trip. When the decisions are incorrect, the result is that the destination will not be reached, or not reached in time, resulting in loss of time and/or money. The loss may be called 'economic'.

The second sub-task originates while driving. It refers to discontinuities in the run of the road, and to the presence of other traffic participants. The decisions relate to the avoidance of the hazards that are presented by the discontinuities and the other participants. When the decisions are incorrect, collisions may result. Apart from not reaching the destination with its economic loss, the consequences are further the losses of goods and properties, and maybe even injuries or even fatalities. The consequences are of a road safety nature.

In both cases the essential feature is that there is adequate time to acquire and process the necessary information, to make the decision, and to execute the manoeuvre. In many instances, this is not the case. Unexpected and unwanted emergencies may arise that require a fast reaction of the driver in order to avoid collisions. This is the third sub-task: coping with emergencies. Usually, the information on which the decision must be made is inadequate, incomplete or even wrong, the time for making the decision is very short, and the time to execute the manoeuvre is often simply not sufficient.

(c) Manoeuvres

Manoeuvres can always be considered as a means to reach a goal, in traffic terms to reach the destination (Schreuder, 1991). Means and goals can be listed in an hierarchical system in such a way that the means at a certain level is a sub-goal on a lower level. Means and goals are only relative concepts. Manoeuvres are usually listed in an hierarchical system (Asmussen, 1972; Schreuder, 1972, 1973, 1974). In broad terms, there are three levels: the strategic level, where the higher decisions are made about the selection of the trip destination and the route. These decisions are usually made before the start of the actual trip; the tactical level, where the decisions concerning the manoeuvres are made, and the operational level, where the decisions concerning the vehicle handling are made.

As has been explained earlier in this section, the driving task refers to the manoeuvres. The main interest is therefore on the tactical level as indicated above. Within that level, three levels, or sub-levels, are of particular interest for the description of the driving task. In each level the relevant manoeuvres can be listed. This has been explained in the following scheme.

(1) The level of complex manoeuvres
The major complex manoeuvres are:
- just going on (as a result of the relevant decision);
- negotiating a curve;
- overtaking and passing a preceding vehicle without opposing traffic;
- overtaking and passing a preceding vehicle when opposing traffic is present;
- passing a (priority) intersection;
- passing an intersection with traffic signals;
- passing a roundabout;
- coming to a stop for a T-junction, or for a traffic signal.

(2) The level of elementary manoeuvres
The elementary manoeuvres are:
- just going on;
- adjusting speed;
- swerving around;
- leaving the traffic lane;
- coming to a stop.

(3) The level of manoeuvre-parts
The manoeuvre-parts are:
- just going on;
- adjusting the speed (to the desired speed or to the speed of the preceding vehicle);
- adjusting the lateral position within the driving lane.

The hierarchical structure of the manoeuvres results from the fact that each manoeuvre represents the outcome of a decision. Each decision has a 'way' and a 'goal'. The way must be followed in order that the goal can be reached.

When priority is given to the driving task and not to the detection of small obstacles as is given in the preceding part of this section, for traffic route lighting, the four main criteria for the quality of road lighting are the same but grouped differently:
1. The optical guidance;
2. The degree of glare restriction;
3. The uniformity of the luminance pattern;
4. The average road surface luminance.

1.4 Cognitive aspects of vision

Visual perception has to do in the last instance with transfer of information. This aspect of human activities is the subject of cognitive psychology. After many years where psychology was dominated by behaviourists, gradually the cognitive psychology became a subject worth while to study. The reason was that practice required data about cognition, more in particular the practice of the advertising business. It became clear that the theories of the behaviourists could not explain why people bought one product and disregarded another. This is similar to the aspects of 'Gestalt'. Furthermore, the information technology required data on how patterns were recognized, mainly for computers that could recognize handwriting, like e.g. the signatures on bank cheques, or for computers that could recognize spoken words and convert them into data files.

When we discuss perception, we assume that there are objects outside us to be perceived. Particularly in visual perception, we assume that this process gives us relevant information about the things, the objects, that are perceived. If we have no interest in the objects, there would be no point in trying to make them perceivable by means of lighting. But at the other hand, when discussing the physical aspects of light generation, and the basics of a concept like luminance, we assume that mathematics and logic give us information that gives us access into the real outside world, in spite of the fact that mathematics and logic are purely mental exercises.

The relation between the certainties that are found in mathematics differ in essential aspects from the probabilities that come out of physical studies are investigated. However, the combined efforts can lead to better results.

One aspect that needs to be mentioned is reductionism as opposed to holism. In simple terms, reductionism means that it is enough to describe one set of areas – e.g. the parts – in order to be able to describe another area – e.g. the total. In still more simple terms, reductionism means that 'the total equals the sum of the parts'. There are reasons to believe that reductionism hardly does give any insight in the real world. It is valid only for algebra, and possibly only for a small part of it. Still, it is a model that is used throughout mathematics, physics, and most branches of applied physics. However, looking at nature without prejudices makes it perfectly clear that the assumption of 'the total equals the sum of the parts' cannot to be made to fit any living organism. One should be very careful in dismissing a holistic approach, and rely completely on reductionism – as well in science as in life. Holism is understood to be any doctrine emphasizing the priority of a whole over its parts.

Another 'credo' that must be taken into account is determinism. It really is a 'credo', an 'act of faith' that cannot be proved or disproved, a thing you may believe or not. It is the cornerstone of positivism, which in its turn is accepted as the basis for all theoretical and quantifiable considerations about nature.

In positivism, amongst other things, it is generally understood that mathematics present the only valid model of reality. It was Galilei who began to suspect that mathematics could describe nature itself. Since then, many believe that mathematics is the true nature of physics. The nature of things seems to be mathematical in essence. True or not, this belief presents a solid basis for all practical aspects of science, such as engineering. "If the mathematical structures are not themselves the reality of the physical world, they are the only key we possess to that reality" (Kline, 1981, p. 465).

1.5 Tools and methods

1.5.1 Models

Models are a useful tool to make an approximation or a similarity to reality. A model is a projection of reality; a model can also be used as a paradigm. A paradigm is a thematic programme for a research project. Conventionally, models are split in two categories:
1. Descriptive models;
2. Predictive models.

As the term says, descriptive models just describe the reality, whereas predictive models allow to make predictions about the result of manipulations of the reality.

We will list here some of the more important characteristics of models:
1. Models do not give any information in how far the concepts of reality are true. There are no 'true' or 'false' models; there are only 'clever' and 'dumb' models. Clever models 'work' and dumb models don't.
2. The most suitable model should be used. Sometimes this is interpreted, although somewhat sloppily, as a variation of Ockham's razor.
3. At any moment in time, the need may arise to change the model, to adapt it, to extend it or to throw it over board altogether. Thus, we are completely free to select and use any model we like.
4. And finally, they are called 'models', because they all miss something essential: they depict reality, but they are not reality themselves!

Introduction

Models, their uses and their restrictions are described in many publications, e.g. Narisada & Schreuder (2004, sec. 10.1.4); Nauta (1970); Schreuder (1998, sec. 7.1.4); Verschuren (1991).

1.5.2 Quick-and-dirty statistics

Humans are biological systems. They are not constant. Their variability is both of a physical and a biological nature (Dixon & Massey, 1957; Moroney, 1990; Schreuder, 1998, sec. 7.1.2; Stevens, ed., 1951). To find out something useful about them, such as things related to vision, means that all measurements must be repeated often. The scores collected in experiments often show a bell shape. The bell-shape curves are described by two characteristics: the mean and the width, or the spread in the observations, characterised by the standard deviation, and usually designated by σ.

From such simplified normal distributions of all scores, 34,13% fall between $x = 0$ and $x = \sigma$; 13,59% between $x = \sigma$ and $x = 2 \times \sigma$, and 2,15% between $x = 2 \times \sigma$ and $x = 3 \times \sigma$. Because in real life measurements never are exact, we may round off these percentages to 33,5%; 13% and 2%. This means that in the simplified normal distribution:

- 67% (or 2/3!) of the scores is between σ and $-\sigma$;
- 93% between $2 \times \sigma$ and $-2 \times \sigma$;
- 97% between $3 \times \sigma$ and $-3 \times \sigma$, and
- 3% is at either more that $3 \times \sigma$ or less than $-3 \times \sigma$.

This brings us back to the concept of 'normal', but now in a different context. Earlier, we dealt with the statistical normal distribution; now we discuss what is normal in society, and what is exceptional. We will repeat the ranges mentioned earlier, but we will add to them in how far they may represent 'normal' or 'exceptional'. The range between σ and $-\sigma$, comprising 2/3 of the scores, one might call 'normal'. The range between σ and $2 \times \sigma$, comprising 13% of the sores, may be called 'large'. Conversely, the range between $-\sigma$ and $-2 \times \sigma$, comprising also 13% of the sores, is called 'small'. The range between $2 \times \sigma$ and $3 \times \sigma$, comprising 2% of the sores, is called 'very large', and finally the range greater than $3 \times \sigma$, comprising 1,5% of the scores, may be called 'exceptional large'. Conversely, all ranges also apply to 'small', 'very small' etc.

1.5.3 Scales in psycho-physiology

In many cases, the results of psycho-physiological studies are presented in the form of scales. A scale is a collection of groups of parts of the results, grouped along one specific characteristic. In terms of the set-theory all results together form a 'set'. A set A is defined as a collection of any kind of objects. The objects are called the elements.

When the results are grouped in one way or another, each group is a subset. A set B is a subset of A if and only if whatever is a member of B is also a member of A. Any group can therefore be designated as B_i. Further, scaling research results makes sense only if all data are included, and if the groups have no overlap. In terms of set theory this means that the set of members not belonging to A is an empty set, and that all possible cross-sections between the subsets $B_1...B_n$ are empty sets as well. In other words, this means that no results are left out, and that any result falls in one and only one group. The definitions used here are based on Daintith & Nelson, 1989 (p. 293; 311).

In applied psychology, as well as in applied engineering, four different types of scale are in common use. All meet the requirements given earlier. The first is the nominal scale. A nominal scale is a scale of items that does not include any quantitative measure. An example is the alphabetical order. It does not make any sense at all to maintain that 'a' is better than 'b', etc..

The next, more specific, scale is the ordinal scale. An ordinal scale is a scale where items are scaled according to their magnitude, irrespective of how much their quantitative difference is. All one can say is that a higher class represent larger items, but it is not possible to say how much larger. An example of an ordinal scale is the distribution of medals in sports championships. All one can say that the gold medalist was faster/ higher/stronger that the silver medalist, and so forth, but not how much. Therefore, many think that championship competitions are somewhat dull. Another example is the well-known nine-point scale of the Glare Control Mark, that is used to quantify discomfort glare in road lighting. The origin of this scale is explained in Narisada & Schreuder (2004, sec. 9.2.3), its use in Schreuder (1967); Adrian & Schreuder (1971); De Boer & Schreuder (1967), and Van Bommel & De Boer (1980). The scale runs from G = 1: Unbearable glare to G = 9: Unnoticeable glare, according to CIE, 1994, Table 1. In this scale, one may say that level 3 represents more glare than level 5, but not how much more. There is no way to ascertain whether the difference between the levels 5 and 3 is equal, larger or smaller than the difference between the levels 7 and 5. In mathematical terms this means, arithmical actions with the items are forbidden. More to the point, is it not permitted to take the average of several items. If, as an example, one installation is appraised by two observes as '6' one as '7' and one as '8' is it fundamentally wrong to say: "the installation scores as 27/4 = 6,75". In spite of these objections, this is precisely what always has been done.

The next step in scales is the interval scale. An interval scale is an ordinal scale where the differences between the levels are equal. An example is the temperature scale of Celsius. The difference between 5° and 7° is the same as between 8° and 10°. However,

Introduction

it is not correct to say that 40° is twice as warm as 20°. Additions and subtractions are permitted in an interval scale but multiplications are not.

And finally the metric scale. A metric scale is an interval scale with a real 'zero' point. An example is the temperature-scale of Kelvin. The zero point of the Celsius scale is just a convention, whereas the zero of the Kelvin-scale is 'really' zero. In 1960, the International System of Unities or SI was adopted. The kelvin (K) is one of the seven basic entities of the SI. It is the unity for thermodynamic temperature (Illingworth, ed., 1991, p. 484). The Kelvin scale is, contrary to the Celsius scale a metric scale. Another example is the length, another of the seven basic entities of the SI. A piece of string of 2 metre is twice as long as a piece of string of 1 metre. In a metric scale, additions and subtractions, but also multiplications are permitted. Details on the different scales and their underlying theories are given in Blackburn (1996, p. 236).

There are a number of pitfalls when expressing research results in scales. The mathematical restrictions when using ordinal or interval scales have been mentioned already. A very common mistake, the use of the arithmical average of the sores in an ordinal scale, has also been mentioned. Another warning should be made, particularly to those who might want to use scale values in order to justify policy decisions: expressing items in a nominal scale is a qualitative action, whereas expressing them in ordinal scale, in an interval scale, or in a metric scale is a quantitative action.

1.6   Conclusions

Considering Light as Good may easily lead to a lighting design that is based on amenity rather than on traffic engineering or traffic safety considerations, particularly in residential areas. There is a real danger that amenity requirements will push up the photometric requirement, resulting in over-lighting. This may not only result in waste of energy but also to blot out the starry night from the world of people. Taking into account that Dark is Evil will lead to economically based cost-benefit considerations that often may disregard amenity and the quality of life.

In the functional approach these different aspects may be taken into account, leading to a rational basis for road lighting standards and recommendations that include the quantitative aspects of costs and safety, as well as the qualitative aspects of amenity.

Often, experimental results are represented in scales. Four types are described: nominal scales, ordinal scales, interval scales, and metric scales. A common mistake is to use the arithmical average of the sores in an ordinal scale. Also, one must take into account that expressing items in a nominal scale is a qualitative action, whereas expressing them in an ordinal scale, in an interval scale, or in a metric scale is a quantitative action.

References

Asmussen, E. (1972). Transportation research in general and travellers decision making in particular as a tool for transportation management. In: OECD, 1972.
CEN (2002). Road lighting. European Standard. EN 13201-1..4. Brussels, Central Sectretariat CEN, 2002 (year estimated).
CIE (1965). International recommendations for the lighting of public thoroughfares. Publication No. 12. Paris, CIE, 1965.
CIE (1992). Guide for the lighting of urban areas. Publication No. 92. Paris, CIE, 1992.
CIE (1995). Recommendations for the lighting of roads for motor and pedestrian traffic. Technical Report. Publication No. 115-1995. Vienna, CIE, 1995.
Cohu, M. (1967). Floodlighting of buildings and monuments. Chapter 10. In: De Boer, ed., 1967.
De Boer, J.B. (1967). Visual perception in road traffic and the field of vision of the motorist. Chapter 2. In: De Boer, ed., 1967.
De Boer, J.B., ed. (1967). Public lighting. Eindhoven, Centrex, 1967.
Forcolini, G. (1993). Illuminazione di esterni (Exterior lighting). Milano, Editore Ulrico Hoepli, 1993.
Griep, D.J. (1971). Analyse van de rijtaak (Analysis of the driving task). Verkeerstechniek, 22 (1971) 303-306; 370-378; 423-427; 539-542.
Kline, M. (1981). Mathematics and the physical world (Original edition 1959). New York, Dover Publications, Inc., 1981.
Narisada, K. & Schreuder, D.A. (2004). Light pollution handbook. Dordrecht, Springer, 2004.
Nauta, D. (1970). Logica en model (Logic and model). Bussum, de Haan, 1970.
NSVV (2002). Richtlijnen voor openbare verlichting; Deel 1: Prestatie-eisen. Nederlandse Praktijkrichtlijn 13201-1 (Guidelines for public lighting; Part 1: Performance requirements. Practical Guidelines for the Netherlands 13201-1). Arnhem, NSVV, 2002.
OECD (1972). Symposium on road user perception and decision making. Rome, OECD, 1972.
Schreuder, D.A. (1970). A functional approach to lighting research. In: Tenth International Study Week in Traffic and Safety Engineering. OTA, Rotterdam, 1970.
Schreuder, D.A. (1972). The coding and transmission of information by means of road lighting. In: SWOV, 1972.
Schreuder, D.A. (1973). De motivatie tot voertuiggebruik (The motivation for vehicle usage). Haarlem, Internationale Faculteit, 1973. Leidschendam, Duco Schreuder Consultancies, 1973/1998.
Schreuder, D.A. (1974). De rol van functionele eisen bij de wegverlichting (The rol of functional requirements in road lighting). In: Wegontwerp en verlichting tegen de achtergrond van de verkeersveiligheid (Road design and lighting in view of road safety); Preadviezen Congresdag 1974, blz. 111 t/m 137. Vereniging Het Nederlandsche Wegencongres, 's-Gravenhage, 1974.
Schreuder, D.A. (1987). Visual aspects of the driving task on lighted roads. CIE Journal. 7 (1988) 1 : 15-20.
Schreuder, D.A. (1991). Visibility aspects of the driving task: Foresight in driving. A theoretical note. R-91-71. Leidschendam, SWOV, 1991.
Schreuder, D.A. (1997). The functional characteristics of road and tunnel lighting. Paper presented to the Israel National Committee on Illumination on Tuesday, 25 March 1997 at the Association of Engineers and Architects in Tel Aviv. Leidschendam, Duco Schreuder consultancies, 1997.
Schreuder, D.A. (1998). Road lighting for safety. London, Thomas Telford, 1998 (Translation of "Openbare verlichting voor verkeer en veiligheid", Deventer, Kluwer Techniek, 1996).

Schreuder, D.A. (2001). Principles of Cityscape Lighting applied to Europe and Asia. Paper presented at International Lightscape Conference ICiL 2001, 13 – 14 November 2001, Shanghai, P.R. China. Leidschendam, Duco Schreuder Consultancies, 2001.

SWOV (1972). Psychological Aspects of Driver Behaviour. Symposium Noordwijkerhout, 2-6 August 1971. Voorburg, SWOV, 1972.

Van Bommel, W.J.M. & De Boer, J.B. (1980). Road lighting. Deventer, Kluwer, 1980.

Verschuren, P.J.M. (1991). Sructurele modellen tussen theorie en praktijk (Structural models between theory and practice). Utrecht, Spectrum, Aula, 1991.

2 Physical aspects of light production

Light can be understood as the aspect of radiant energy which an observer perceives through visual sensation, but also as a physical phenomenon. For this, four different models may be used: Light as a collection of light rays; light as an electromagnetic wave; light as a stream of photons; light as fluid of power. Each model has its own use and its own applications. Light rays are straight and infinite narrow, and do not show any smaller detail. They are the basis of geometric optics. Electromagnetic waves and streams of photons appear in duplicity of light. Waves are essential to describe diffraction phenomena. The corpuscular theories of light are essential to describe photo-electric phenomena.

As regards the generation of light, three principles are useful in light applications:
- *Incandence – the basis for incandescent lamps;*
- *Recombination of electrons and ions in a plasma – the basis for gas-discharge lamps;*
- *Recombination of electrons and holes in a semiconductor – the basis for semiconductor lamps like LEDs.*

In this chapter these three major families and their applications will be discussed in considerable detail.

2.1 The physics of light

2.1.1 Definitions of light

We must begin with the statement that light is not defined in a consistent way. At the one hand, light is the aspect of radiant energy which an observer perceives through visual sensation. The visual system is the detector that transforms radiant power into luminous sensation. Hence, one may say that a burning candle emits radiation, but that the radiation does not become light unless a human perceives it (Sterken & Manfroid, 1992, p. 2). This poetic description is not a proper definition, but it suggests that, at the other hand, there is radiation of many kinds; one of the kinds is light. We must add that, when calling an item 'radiation', we mean 'electromagnetic radiation'. In sec. 3.1.1a the relation between radiation and light is explained when discussing their measurement, that is to say the relation between radiometry and photometry. In this chapter we will concentrate on the physical aspects of light. It must be stressed that, light being 'just' one of the many sorts of – electromagnetic – radiation, most if what is discussed in this chapter, applies to all other sorts of radiation was well. In most of this section, we will use only the term 'light'.

There are four ways to look at light as a physical phenomenon. One might call it, as is explained in sec. 1.5.1, as four different 'models':
1. Light is a collection of light rays. This is called the phenomenological approach to light;
2. Light is an electromagnetic wave;
3. Light is a stream of rapidly moving particles, or photons;
4. Light is fluid of power (wattage).

The first model is the model that is used in optical imaging. The light rays are straight as long as there is no reflection or refraction. They have no speed or propagation; they are just 'there'. The models (2) and (3), that shows the duplicity of light, are mutually exclusive. Model (2) is described by classical physics, and model (3) by relativistic quantum mechanics. Both are characterised by a large, but finite speed of light. In the classical model, the speed of light depends on the speed of sender and receiver, whereas in the relativistic model of the quantum mechanics the speed is constant ('a constant of nature'). It not usual that the fourth model is made explicit. The fluid of power is, however, the basis of radiometry and, consequently, of photometry. The fluid follows most of the rule of hydrodynamics, although the question of propagation is never raised. The fourth model, the power fluid model, is explained in sec. 4.1.2, when the 'light field' concept is discussed. Before that, the relation between light and power must be explained. This is done in sec. 3.1.1a, where the relation between radiometry and photometry is discussed. However, just as in the light ray model, it is assumed that the fluid is just 'there'. It all four cases, the speed of light is mentioned. In models (1) and (4) is in fact assumed to be infinite, whereas in models (2) and (3) the speed is finite. However, in all lighting engineering applications, the speed is so high that for all practical purposes it can be considered as infinite, and thus can be disregarded in almost all considerations. The four models will be discussed briefly. Details can be found in Narisada & Schreuder (2004, sec. 11.1).

2.1.2   Light rays

The first model, the model of light rays that are infinite small in width and that do not show any smaller detail, is the basis of optical imaging, which is studied in geometric optics. Light rays are straight as long as there is no reflection or refraction. They have no speed or propagation; they are just 'there'. In every lighting design method, in every calculation that is used in illuminating engineering, this approach is used, both in luminaire design as in the design of lighting installations. Geometric optics are used in graphical design methods, as well as in computer-aided luminaire design. The same methods are used in the design of optical instruments, optical telescopes included. In sec. 10.1 further details are given, more in particular in relation to the design of road lighting equipment.

Most of geometric optics is based on the principle of Fermat (Feynman et al., 1977, Vol. I, p. 26-4). This principle states that a light ray always takes the shortest possible path; more precisely, when media of different refraction indexes are involved, the path that takes the least time. Here, the speed of light is apparently introduced, but it has no physical meaning. One of the consequences of Fermat's principle is that light rays can be reversed: all results of geometric optics are invariant with respect to the direction of the light. Fermat's principle is a special case of the Principle of Least Effort, one of the basic principles in Nature. The principle as such has a long history of at least 2000 years. See for this the section on the history of optics in Blüh & Elder (1955, p. 351-352).

The light ray concept is often combined with the wave theory of Huygens. According to Huygens, light is to be considered as spherical waves that originate from the source and that propagate through space. Each wave has a wave front. The wave front is the geometric envelope of the wave. At any moment in time, each point of the wave front acts as a secondary light source, emitting new elementary waves and creating a new wave front. The new wave front is the sum of the different elementary waves (Prins, 1945, p. 49-50). This is the Huygens Principle that is depicted in Figure 2.1.1 (Illingworth, ed., 1991, p. 221-222). The light rays can be considered as being in the direction of the normal to the wave front (Blüh & Elder, 1955, secs. 16.4; 16.11).

It must be stressed here, right from the beginning, that the wave concept of Huygens is completely different from the wave concept of Maxwell that are discussed in sec. 4.2.1. Huygens describes space, whereas Maxwell describes electromagnetic phenomena that are essentially a field that exists in vacuum. The field concept is discussed in sec. 4.1.

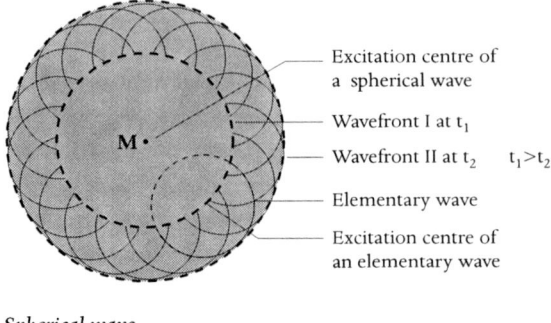
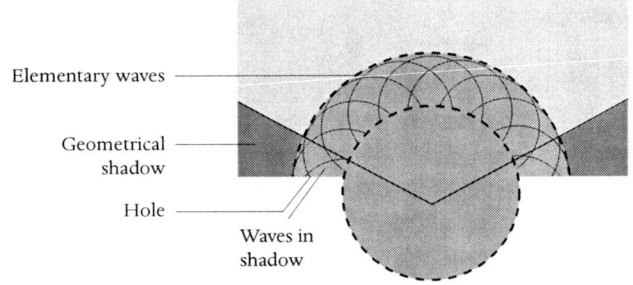

Figure 2.1.1 The Huygens Principle (After Breuer, 1994, p. 88).

2.1.3 Waves and particles

(a) *Waves*

In the preceding parts of this section, there are four models to describe light: as light rays; as an electromagnetic wave; as a stream of photons; as a fluid of power. The duplicity of light refers to the second and third models. It has been found that phenomena involving the propagation of light can be interpreted adequately on the wave basis but when interactions of light e.g. the photoelectric effect are considered the quantum theory must be employed (Illingworth, ed., 1991, p. 266). The electromagnetic theory and the wave theory are regarded as complementary, or rather, as mutually exclusive.

In classical physics, light is characterized by its wavelength or the frequency. Because the speed of light is supposed to be a constant, these two are exchangeable:

$$\nu = \lambda / c \qquad [2.1\text{-}1]$$

Physical aspects of light production

with:
- ν: the frequency (Hz);
- λ: the wavelength (m);
- c: the speed of the light (m/s).

In the years around 1700 it seemed that the wave theory was the correct one. There are three phenomena that can be explained only by assuming that light is waves: interference, diffraction, and polarization. At that time, one was convinced that aether was the medium for the waves. Apparently at the time people were not overly concerned about the fact that the aether was not observed, and that it did show a number of strange characteristics. That seemed to be better when Maxwell postulated no medium but just a field for the waves to proceed in.

The wave character of light is immediately clear when one observes the colours of thin layers like soap bubbles, oil films etc. They are very easy to observe, even with the naked eye. The underlying phenomena are called interference. They are described already by Hooke, Newton, and Young. See, as mentioned earlier, the section on the history of optics in Blüh & Elder (1955, p. 351-352). These phenomena can only be explained by waves that are reflected at the top and the bottom of the thin layer. If the travelling distance of the two beams differs a half wavelength they cancel each other out. If, however, the travelling distance of the two beams differs a full wavelength, they reinforce each other. The result is that, when the angle of incidence the light changes, the reflected light becomes zero, or maximum, or anything in between. It is immaterial if a number of full wavelength is added to the travelling distance. Only the difference in the travelling distance matters. This is depicted in Figure 2.1.2.

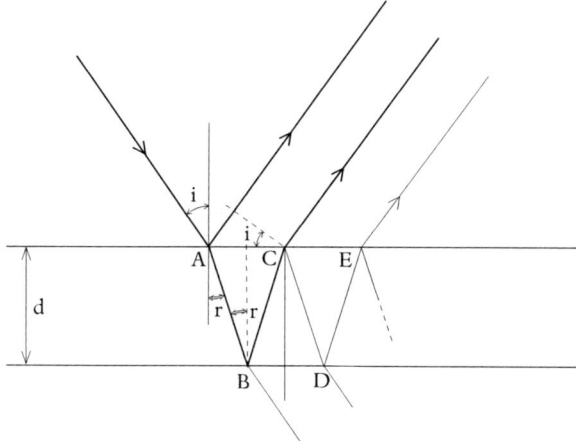

Figure 2.1.2 Diffraction of monochromatic light at a thin film.

If the incident light is white, which according to Newton means that it contains all wavelengths, some colours will be extinguished whereas other colours are reinforced. That is precisely what makes soap bubbles so beautiful to look at (Minnaert, 1942). It may be added that the sequence of interference colours is quite different from the colour sequence from refraction. In refraction the different spectral colours show up side by side, whereas in interference the colours one may see result from particular wavelength areas that are missing. The theory of the origin of interference colours is explained clearly in Van Heel (1959). The theory and practice of colour vision and colorimetry is discussed in some detail in another chapter of this book. See also Narisada & Schreuder, 2004, sec. 8.3, and Schreuder, 1998, sec. 7.4.

The wave character of light is also closely related to the phenomenon of diffraction. These phenomena are described already by Young and Fresnel. See again the section on the history of optics in Blüh & Elder (1955, p. 351-352). When light passes through any opening, with or without a lens in it, diffraction plays a role. The rim of the opening causes the light to bend, so that it does not follow precisely a straight line. In terms of the Huygens Principle one may state that each point of the rim acts as a secondary light source that emits light waves, which in turn combine to a new wave front.

When the light hits a screen after passing through a small aperture, a ring-shaped pattern will be seen. This pattern is called the diffraction pattern. For an arrangement with circular symmetry, the illuminance at a point P on the screen is given by the relation:

$$E = E_0 \cdot \frac{J_1 (2m)^2}{m} \qquad [2.1\text{-}2]$$

with

$$m = \frac{\pi \cdot \delta r_0}{2 \lambda z_0} \qquad [2.1\text{-}3]$$

in which
    δ: diameter of light source;
    r_0: displacement of P in the screen from the axis;
    λ: wavelength of the light;
    z_0: the distance between the source and the screen;
    $J_1 (2m)$: the Bessel function of the first order;
    E: illuminance in P;
    E_0: illuminance for $r_0 = 0$;

Physical aspects of light production

The relation is given here in the form as presented by Moon (1961, equation 12.11, p. 425). See also Van Heel (1950) and Longhurst (1964). Measurements of the retinal distribution are described by Vos et al. (1976).

The mathematical derivation is given by Feynman et al. (1977, Vol. I. sec. 30-1, p. 30-2). See also Wachter & Hoebel (2006, Appendix A5). In essence it is based on the Huygens Principle. For the calculation of the pattern it is assumed that a large number (n) of oscillators of equal strength are spread evenly over the aperture. Their phase (ϕ) differs for different positions on the screen. This, of course, is the cause for the diffraction pattern to arise. The resulting intensity on the screen shows a pronounced maximum at phase zero – straight ahead. For increasing phase, the intensity on the screen passes through a series of equally spaced minima with maxima in between. The intensity in the minima is zero; the intensity of the maxima decreases rapidly with increasing phase. The intensity at the first maximum is less than 5% of that of the maximum at phase zero – more exact, 0,047. See Figure 2.1.3.

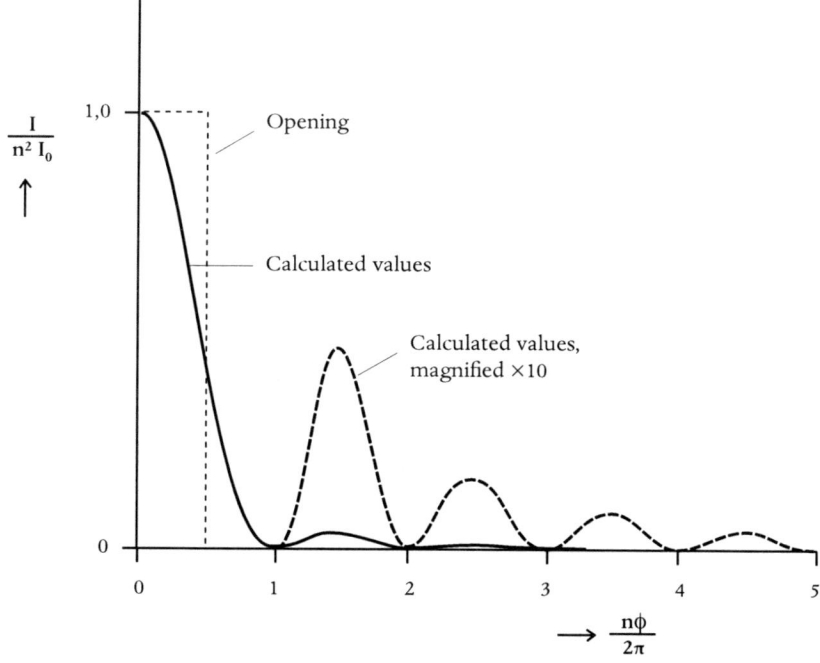

Figure 2.1.3 *The calculated refraction pattern for a circular opening (After Narisada & Schreuder, 2004, Fig. 9.1.12).*

For a two-dimensional case, the pattern is depicted in Figure 2.1.4.

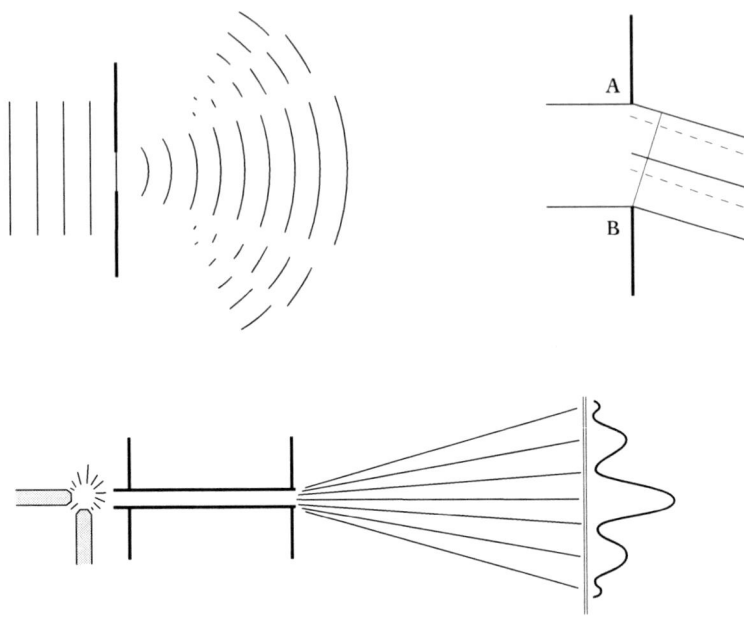

Figure 2.1.4 Diffraction at one split (After Prins, 1945, Fig. 33).

In many cases of experimental research and of applied engineering it is important to know under which conditions as regards the angular distance two point sources, e.g. two stars, can be seen separately. This is called the minimum separable. The importance for astronomical observation is described in Narisada & Schreuder (2004, sec. 9.1.4). Its use for the practice of determining visual acuity is described in sec. 7.4.4. Crudely speaking, the apparent size of a point source is said as being equal to the width of the first spread-out maximum, that is the place where the first minimum occurs, or, more precisely, the diameter of the first diffraction ring as is explained earlier in this section. The minimum distance is determined by the refraction in the eye in such a way that the two objects seem to be situated at consecutive maxima, that are, as indicated earlier, equidistant in angular measure. Using the pattern as given in Figure 2.1.3, the minimum separable can be calculated. See Figure 2.1.5.

When in an optical instrument all other errors are eliminated, one may speak of diffraction limited optics. In lighting engineering diffraction effects can be disregarded.

Physical aspects of light production

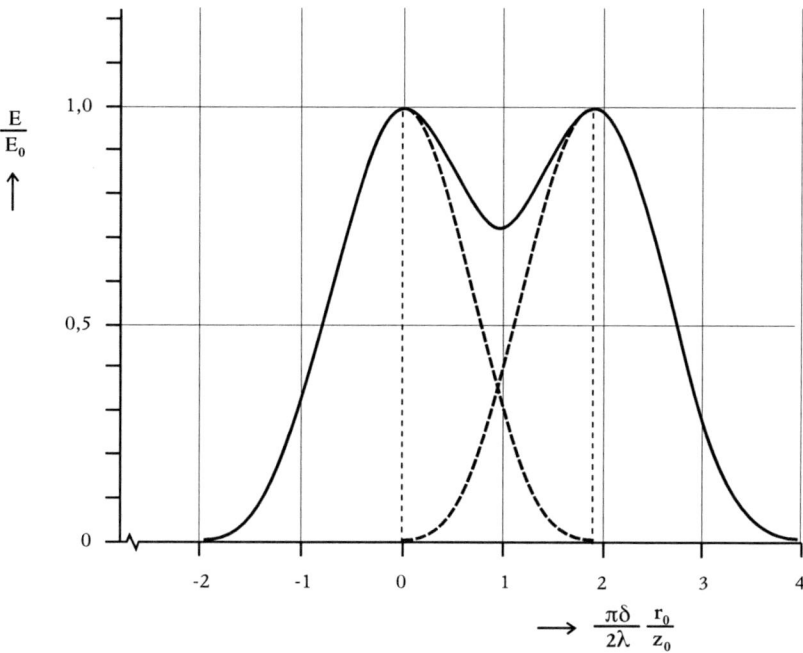

Figure 2.1.5 *The calculated retinal illumination caused by two light sources at minimum distance (After Narisada & Schreuder, 2004, Fig. 9.1.13).*

(b) Particles

The corpuscular theories of light have, just as the wave theories, a long history. It started in classical times when it was sometimes assumed that the eye did send out particles. If these particles did strike an object, the sensation of vision did arise. This idea does not conform to modern physics and it is almost completely forgotten. However, 'sight rays' may seem to have some meaning in non-sensory perception. The corpuscular theory of light, that assumes that light consists of particles, is a hypothesis about the nature of light that has been championed by Newton (Lafferty & Rowe, eds., 1994, p. 154). In its modern form it is the quantum theory of light that is explained further on in this section.

According to the theories of Newton, every colour is already present in white light. They all traverse space at the same speed along straight lines. White light is a mixture of corpuscles of different kinds, belonging to different colours. A glass prism may separate them due to forces acting on the particles of light and originating from the particles of glass (Einstein & Infeld, 1960, p. 97). The substance theory of light seems to work splendidly in all cases. However, as Einstein and Infeld state: "The necessity for introducing as many substances as colors make us somewhat uneasy" (Einstein & Infeld, 1960, p. 99).

Here we come to the point of the 'stories' that are hard to swallow. The validity of stories, or hypotheses, is explained in Schreuder (2008), where the propositions of Kuhn and the involvement of paradigms are discussed.

In a very different kind of experiments, it had been found around the beginning of the 20th century that electrons show interference patterns similar to those caused by waves, light waves or any other type of wave. Electrons are thought of as small particles that have a certain mass, also when at rest (about $9 \cdot 10^{-31}$ kg; Illingworth, ed., 1991, p. 141). As long as photons are considered as particles as well, although with a rest mass of zero, they must show interference patterns as well. According to the principles of wave mechanics that had been developed by Schrödinger, the wave process is governed by a certain oscillating quantity ψ, whose square is equal to the electric charge density. In order to reconcile the wave and corpuscular concepts, Born made the assumption that ψ^2 represents the probability that an electron, thought of as a particle, will be found in a certain region of space. The waves then have the character of guiding waves, along whose maximum amplitudes the particles of mass or charge are propagated (Blüh & Elder, 1955, p. 654).

The theory of wave mechanics is discussed in detail in Feynman et al. (1977), Joos (1947), and Wachter & Hoeber (2006, sec. 2.9). A more easy to grasp description is given in Gribbin (1984); Breuer (1994), and Kuchling (1995).

(c) *Photons*

As mentioned earlier, the other model can be described in terms of the relativistic quantum physics, where light is characterised by the momentum carried by the photons.

Photons are particular objects. They carry energy, but they have no rest-mass, or mass in the more traditional sense of the word. The momentum of a photon is:

$$m = h \cdot \nu / c \qquad [2.1\text{-}4]$$

with:
 h: Planck-constant, equal to $6,626\,076 \cdot 10^{-34}$ J·s (Illingworth, ed., 1991, p. 354).
 ν: the frequency;
 c: the speed of light.

Physical aspects of light production

The energy of a photon depends on the wavelength:

$$E = H \cdot v \qquad [2.1-5]$$

and so does its mass:

$$M = E / c^2 \qquad [2.1-6]$$

For visible light this corresponds to about $3 \cdot 10^{-36}$ kg (Breuer, 1994, p. 181). It was mentioned earlier that the rest mass of a photon is zero. According to the relativity mechanics, the mass of a moving particle depends on its speed:

$$m = m_0 / \sqrt{(1 - v^2/c^2)} \qquad [2.1-7]$$

(Illingworth, ed., 1991, p. 405). Because the rest mass m_0 of a photon is zero, also $(1 - v^2/c^2)$ must be zero, or v must equal to c. All photons travel with the speed of light, independent of their other characteristics. That is, of course, the reason for dominant position of the speed of light in relativistic physics! This statement can, of course, also be reversed: because light travels by definition with the speed of light, photons cannot have a rest-mass.

There is an interesting consequence of this statement. For this, we will refer to the uncertainty principle introduced by Heisenberg. This principle states that the product of the uncertainty in the measured value of the momentum p and the uncertainty in the position x is about equal to the Planck constant that is mentioned earlier (Illingworth, ed., 1991, p. 503). In the most precise form:

$$\Delta p \cdot \Delta x \geq h / 4\pi \qquad [2.1-8]$$

As is well known, the momentum of a body is the product of the mass and the speed (Illingworth, ed., 1991, p. 309). The mass of a photon is zero and its speed is has a finite value, viz. the speed of light. So the momentum is zero and thus the uncertainty of measuring it. It follows that Δx must be infinitely large. The consequence is that it is not possible to predict where the photon is at any moment in time. One might say that it is at the source and at the detector and at any place in between at any time. Or, photons stretch over the full length of the distance it is supposed to travel.

2.2 General aspects of light production

2.2.1 Principles of light generation

In sec. 2.1.1, four ways have been introduced to look at light as a physical phenomenon:
- Light as a collection of light rays;
- Light as an electromagnetic wave;
- Light as a stream of photons;
- Light as a fluid of power.

In sec. 2.1.3 it is mentioned that phenomena involving the propagation of light can be interpreted adequately on the wave basis but when interactions of light e.g. the photoelectric effect are considered the quantum theory must be employed.

As regards the generation of light, there are several families of sources of light. Only three of the principles are useful in light applications:
1. Incandence – the basis for incandescent lamps;
2. Recombination of electrons and ions in a plasma – the basis for gas-discharge lamps;
3. Recombination of electrons and holes in a semiconductor – the basis for semiconductor lamps like LEDs.

In this section these three major families and their applications will be discussed in considerable detail. But first, we will briefly discuss the efficacy of light sources, because that is the final criterion whether a light generating principle will actually be used in lamps. This is of course an economic criterion: when we need a lot of energy to create only a little light, the generation process obviously is not efficient, and industry will not be interested. However, over time the picture may change. Gas discharges were known for many decades before gas-discharge lamps finally did conquer the world.

In the early 21st century, nearly the whole world is lit by electric light. It may for many be a sobering thought to realise that this is only very recently so. As we all know, combustion may generate flames, and flames were for millennia the only light sources available to humanity, even as recent as the beginning of the 20th century. The most famous of all engineering handbooks devoted in the first quarter of the 20th century more space to gas lighting than to electric lighting (Hütte, 1919). It might be argued that the safety aspects of fire and gas poisoning were the major incentive for the conversion from gas lighting to electric lighting, but certainly the higher efficacy and the greater consumers' convenience played a major role. Presently, fires and candles are used as light sources

Physical aspects of light production

only in the poorest parts of the developing world (Schreuder, 2001, 2004) and of course in religious and social rituals, also nowadays. And even there one may find changes that some people, but not all, might call progress: at Christmas time, fake candles with LED lighting are for sale! The next section on the efficacy of light sources will therefore be devoted almost completely to electric light sources.

2.2.2 The efficacy of light sources

Electric light sources convert electric energy in electromagnetic energy. The efficiency of this conversion is, however, not directly relevant to the visual efficiency of a lamp. Visual efficiency implies, as is explained in another chapter of this book, the weighing of the electromagnetic radiation according to the spectral sensitivity of the visual system. In practice, we need the efficacy of the light source. The efficacy of a light source is defined as the ratio between the luminous flux in lumens (lm) generated by the lamp. and the energy, or rather the work in watts (W) consumed by the lamp. The efficacy is therefore expressed in (lm/W), and not in (%). In energy terms lm/W is dimensionless, because both the lumen and the watt represent 'work'. As an example, in Table 2.2.1, examples of the efficacy of a number of traditional lamp types is given. The efficacy of compact fluorescent lamps is difficult to assess because they from an integrated construction together with the ballast. It is not possible to measure the lamp separately. Something similar is the case with LEDs. The actual light emitting part is a tiny chip of semiconductor material that cannot be used in the open air, leave alone be subjected to photometric measurements. The characteristics of LEDs as light sources are discussed in a further section of this chapter. The photometry of LEDs and some of the problems attached to it are discussed in CIE (1997). The corresponding values given in Table 2.2.1 are only approximations.

Lamp category	Type	watt	Efficacy [lm/W]
Incandescent lamp, Krypton filled	100 volt	90	17
Incandescent lamp, halogen	110 volt	130	18,5
Fluorescent tube	white	40	77,5
Fluorescent tube	tri-phosphor	32	105
Compact flu. lamp	PLE	11	54
High-pressure sodium		360	132
Metal halide lamp, Scandium*)		400	100
Low-pressure sodium		180	175
LED white	50°	0,07	21

*) usually, other metals are added as well.

Table 2.2.1 Examples of the efficacy of several common lamp types (After Narisada & Schreuder, 2004, Table 2.1.1; Anon., 1997, 2003; Nichia Company, 2000).

In Table 2.2.2, a comparison between non-electric and electric light sources may be made.

Lamp family	Type	watt	cd	lumen	Lm/W
candle	wax	55	0,1	1	0,02
candle	stearin	80	1	10	0,125
petrol	0,02 l/h	200		10	0,05
petrol	0,05 l/h	488		100	0,21
incand. lamp	signal	3		30	10
incand. lamp	GLS	60		730	12
fluorescent	PLE*)	11		600	54
LED white	50°	0,07		1,5	21

*Table 2.2.2 Examples of the efficacy of several lamp types. *) ballast included. [After Schreuder (2004). Wax candles and petrol lamps after Mills (1999). Stearin candle after Vermeulen (2000). Indecent and fluorescent lamps based on data from Philips, Eindhoven (Anon., 1997). White LED after Nichia Company (2000).]*

2.3 Incandescence

2.3.1 Thermal radiation

(a) *The laws of black-body radiation*

At all temperatures, any body will emit electromagnetic radiation, the characteristics of which depend exclusively on the temperature of the body – in theory at least. This is called thermal radiation. The amount of energy that is emitted is quantified by the emissivity. At the other hand, when radiation hits a body, part of the energy is absorbed. The amount of energy that is absorbed is quantified by the absorption.

The universal law of conservation of energy requires that the emissivity and the absorption are equal. This law is based on the equivalence of heat and mechanical energy (Blüh & Elder, 1955, p. 128-129). It is commonly known as the first law of thermodynamics. In essence it is nothing but a law of experience. In thermodynamics it is considered as an axiom, meaning that it is an assumption that is followed through consistently, and that has never been found to be in contradiction with observation (Blüh & Elder, 1955, p. 454).

Usually, a body absorbs and emits radiation simultaneously. In conditions of thermodynamic equilibrium, the two are equal. This is called Kirchhoff's Law. The

Physical aspects of light production

mathematical derivation of this law is given in Joos (1947, sec. XXXVI 3, p. 583-585). Using the rules of the specific heat of solids that is introduced by Debye and described by Joos (1947, sec. XXXV 3, p. 572-576) in combination with Kirchhoff's Law, it is possible to derive Planck's Law of radiation. "The Law of Planck contains all other laws of radiation" (Joos, 1947, p. 586). The mathematical derivation of Planck's Law is highly technical. It can be found in Joos (1947, chapters XXXV and XXXVI). In simplified form it is given in Illingworth, ed. (1991. p. 39-40).

The radiation of a black body can be fully described by two equations that can be derived from Planck's Law according to Joos (1947, p. 585-587). See also Moon (1961, sec. 5.03). These two equations are sufficient for the description of almost all practical applications of thermal radiators. We do not need to concern ourselves too much with the complications of Planck's Law.

The first equation that describes the black-body radiation is:

$$\lambda_{max} = C_2/T \qquad [2.3-1]$$

This equation is called Wien's Law, or also Wien's displacement law, because it describes the wavelength λ_{max} at which the maximum energy is emitted by a black body at a temperature T, with C_2 a constant. When the temperature increased, the wavelength of the maximum emission shifts from long to short. Because each wavelength corresponds with a specific colour, the colours shift from red to blue. This equation shows directly why light sources that are based on the principles of black-body radiation, like e.g. incandescent lamps, go from invisible, emitting only infrared radiation, to red to yellow and to blue-ish in aspect when they are heated up.

The displacement law of Wien is depicted in Figure 2.3.1

The second of the two equations that describe the black-body radiation is:

$$J = C_1 \cdot T^4 \qquad [2.3-2]$$

This equation is called the Stefan-Boltzmann's Law. The law of Stefan-Boltzmann may be regarded as an integration of the law of Wien. It describes the total energy J emitted by a black body at a temperature T, with C_1 a constant. This equation describes the maximum light emission and therefore also the maximum luminous efficacy that may be reached by any light source that is based on the principles of black-body radiation. More in particular, it proves directly that incandescent lamps never can be 'efficient' light sources.

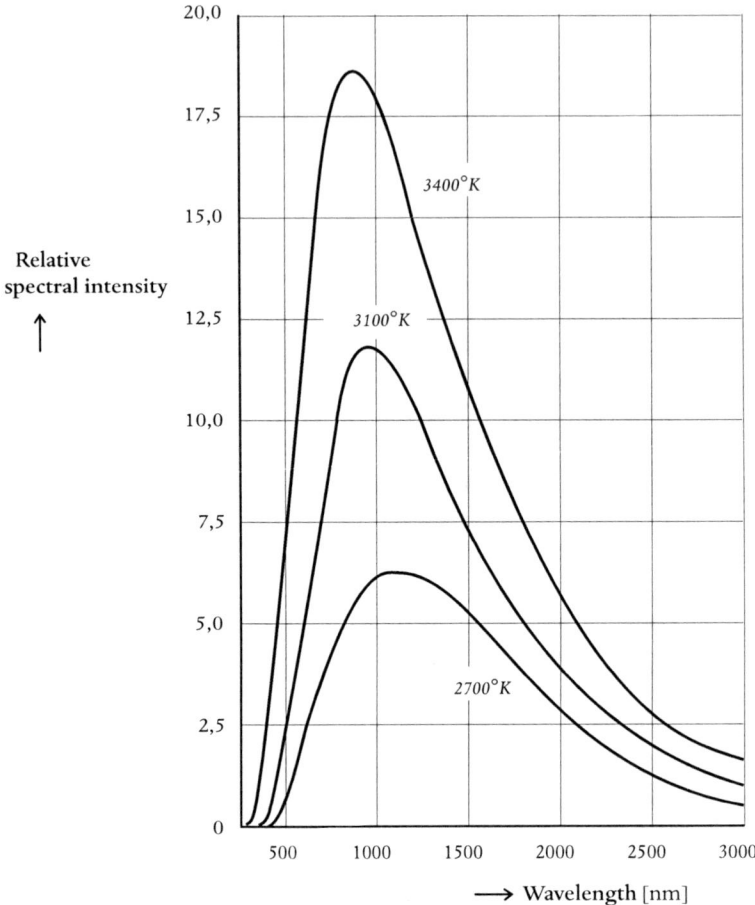

Figure 2.3.1 The maximum intensity moves up to shorter wavelengths at increasing temperature (After Narisada & Schreuder, 2004, Fig. 11.1.3).

As an example, in Figure 2.3.2 the thermal balance of a standard general lighting service lamp or GLS-lamp of 100 W is given.

Physical aspects of light production

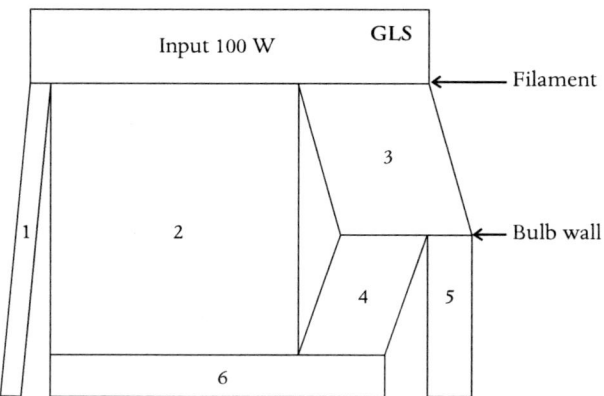

Figure 2.3.2 *The thermal balance of a 100 W GLS-lamp (After Anon., 2005).*

The numbers in the figure have the following meaning:
1. Visible radiation – 5%;
2. Infrared radiation of the filament – 61%;
3. Convection and conduction losses from the filament to the bulb – 34%;
4. Infrared radiation of the bulb – 22%;
5. Total convection and conduction losses from the bulb – 12%;
6. Total infrared radiation – 83%.

As we all know, incandescent lamps are poor light sources, but excellent heaters (Narisada & Schreuder, 2004, sec. 11.3.3).

For some bodies the emissivity and the absorption are equal to one. A body that has these characteristics is called full radiator. Because such a body absorbs all incident radiation, it will look completely black. As will be explained later on in this section, full radiators, and consequently black bodies are not to be found in nature. There are many materials, however, that come fairly close to a black body.

(b) *Grey bodies*

It should be noted that the relations [2.3-1] and [2.3-2] are valid only for a black body. For a black body, the emission equals the absorption. In practice, both the emissivity and the absorption usually are smaller than unity. One may call a body like that a grey body.

It must be made clear that a black body needs not to be a 'real' body. A hollow room gives the best approximation for a black body that can be found in practice. As a matter of fact, in the past, such a device was used when defining the candela – the unit of luminous intensity that is explained in sec. 3.2.3 when discussing objective photometry. Only after 1979, a more stable definition of the candela is introduced (Anon., 1979, as explained by Hentschel, ed., 2002, p. 39).

In Figure 2.3.3, a device is depicted that was used for defining the candela until 1979.

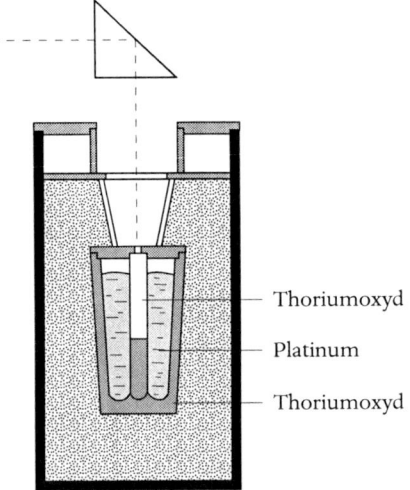

Figure 2.3.3 *A device for the creation of the standard candela (After Keitz, 1967, Fig. 9, p. 27).*

The reason to discuss this obsolete definition of the candela is not so much its historic interest, but rather the fact that little hollows may make the filament of an electric incandescent lamp look a little like a black body, thus enhancing its efficiency as an emitter of light. The 'coiled' filament is explained in sec. 2.3.3a.

(c) *Non-electric incandescent lamps*

Before dealing with electric incandescent lamps, it must be noted that most 'ancient' non-electric lamp types like wax lamps, candles, oil, and kerosine lamps are incandescent lamps as well. The fuel, which is usually a hydrocarbon, contains small carbon particles. The heat of the flame makes them glow and emit light, so they are temperature radiators and must be classified as incandescent lamps. The same is true

Physical aspects of light production

for carbon-arc lamps. And so are gas lamps. The envelope is heated by the gas flame, and becomes an incandescent body (Gladhill, 1981; Hütte, 1919). As is explained in sec. 2.2.2, it is their inherent danger and user's inconvenience that made them obsolete, as well as their low efficacy. At the other hand, one should realise that about 1200 million people do not have direct access to the main electricity grid; additionally, most of them do not have the money to afford electric lighting. So, notwithstanding the danger and the low efficacy, candles and oil lamps are still the main source of light for a large part of the people in this world.

(d) Electric incandescent filament lamps

When we deal with outdoor lighting, the main subject matter of this book, it is justified to concentrate on electric lamps. Almost all incandescent lamps are electric filament lamps, their main component being a metal filament. An electric current is passed through the filament. Because the electric resistance of the filament material, the filament heats up. Depending on the lighting goal of the lamp, and the actual construction that is derived from it, the operating temperature may be selected within a wide range, from low-temperature infrared and heat sources up to high-intensity light sources. In most modern lamps, the filament is made of tungsten – or wolfram – and sometimes of carbon. The melting point temperature of tungsten is about 3650 K (Hentschel, ed., 2002, Table 5.2). For practical reasons, the filament temperature in standard incandescent lamps is considerably lower. A value of 2500° Celsius (corresponding to 2773 K) is given by Baer, ed. (2006, sec. 2.1.2, p. 186).

It is well-established that most operation characteristics of electric incandescent lamps depend heavily on the temperature of the filament, more in particular the lamp life. As an example, a lamp with a filament temperature of 3000° Celsius would have a life only 0,2% of that of a lamp with a filament temperature of 2500° (Baer, 1990, p. 181).

Usually, it is not the filament temperature but the operating voltage that is used as the yard stick for the lamp performance. The operating voltage is usually understood as the value of the lamp voltage for which the lamp has been designed. The design voltage would also be an appropriate term for this voltage. The reasons are that differences in lamp construction and in lamp operation do not allow for a general rule between the operating voltage and the filament temperature, and that the most product information is based on the operating voltage as the main variable. That is done so because in almost all fields of lighting applications it is the operating voltage that is known, and usually not much more.

In Figure 2.3.4, the relation between the operating voltage of a GLS-lamp and a number of operating characteristics is given as a percentage of the nominal of design values. The dotted lines in the figure designate the limits of 'useful operation' of the lamps.

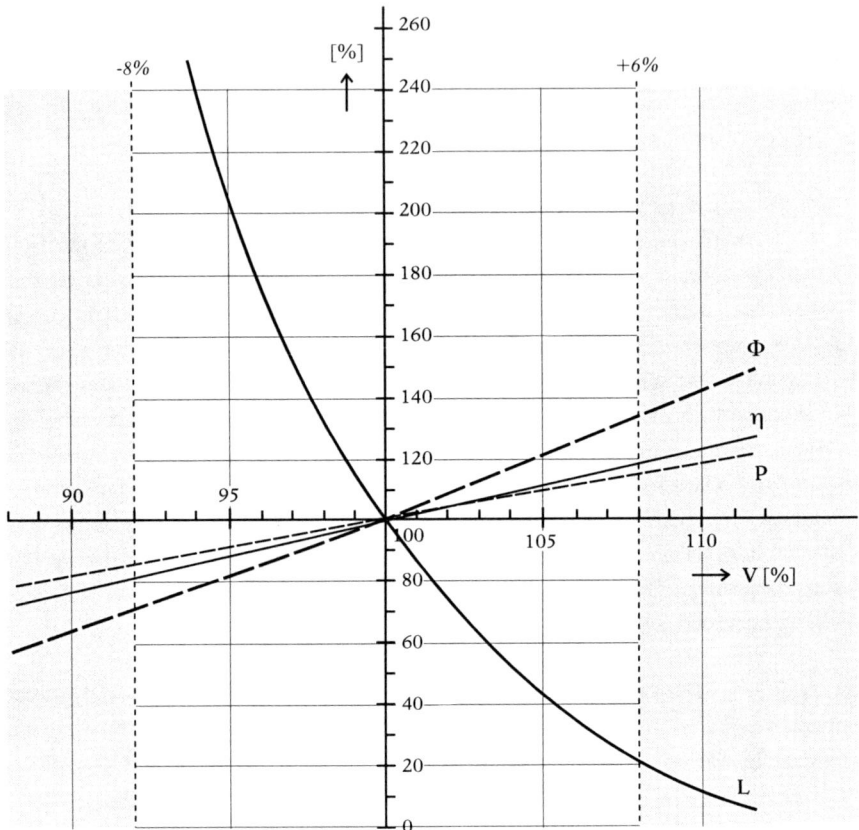

Figure 2.3.4 The relation between the operating voltage of a GLS-lamp and a the lamp life (L), the luminous flux (Φ), the luminous efficacy (η), and the dissipated power (P) as a percentage of the design values (After Van Ooyen, 2005, Fig. 1 [5-11])

All visual impressions, both for photopic vision as for scotopic vision, are restricted to a wavelength region between about 400 nm to about 700 nm. Radiation with a wavelength shorter than 400 nm or longer than 700 nm may have all sorts of physical or physiological effects; they, however, never will be 'light'. Many of these aspects are discussed in chapter 3 of this book. From Figure 2.3.1 it will be clear that a black body of about 2700 K does not emit much energy in total and the small amount of energy that

Physical aspects of light production

is emitted, falls mostly in a wavelength area well above 700 nm. It is sometimes stated that incandescent lamps are good heaters but poor light sources!

Finally, it should be mentioned that most practical materials hardly resemble black bodies. Metals are 'selective radiators', i.e. the radiation relative to that of a black body differs for different wavelengths. Some examples are given in Table 2.3.1.

Material	Temp [kelvin]	Relative emission at λ	
		665 nm	463 nm
Tantalum	2400	0,404	0,450
Platinum	1800	0,310	0,386
Nickel	1400	0,375	0,450
Gold	1275	0,140	0,632
Molybdenum	2400	0,341	0,371

Table 2.3.1 *Values of the spectral radiation factor of some metals (After Moon, 1961, Table XVII, based on data of Worthing, 1926).*

Strangely enough, the table as given by Moon does not include the two materials that are widely in use as incandescent materials in light sources; viz.: carbon and tungsten. As regards carbon it is only remarked that "it approximates a nonselective radiator – a grey body" (Moon, 1961, p. 124). No data on the total emission have been given. For tungsten, however, a graphical representation is given. See Figure 2.3.5.

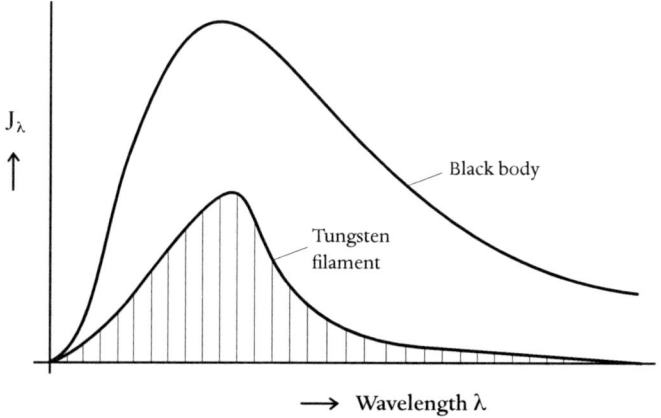

Figure 2.3.5 *The emission of tungsten compared to that of a black body for 3038 K (After Narisada & Schreuder, 2004, Fig. 11.1.4).*

The total emission of tungsten is between 0,46 and 0,42 for temperatures between 1200 K and 2800 K (Hentschel, ed., 2002, p. 117).

2.3.2   Characteristics of electric incandescent lamps

(a)   The filament

Electric incandescent lamps have a filament as the light-emitting element in an outer bulb. As mentioned earlier, in modern incandescent lamps the filament is almost always made of tungsten. The reason is its high melting point temperature and the fact that the mechanical properties do not change appreciably until a temperature close to the melting point temperature. A disadvantage is the fact that tungsten cannot be cast, so that filaments are always made by 'drawing' the sintered material. In order to enhance the efficiency of the light source, the filament is coiled into a spiral; sometimes, the process of coiling is repeated, so that there are 'coiled-coiled' lamps. When the filament is heated by the electric current up to a certain temperature, light is emitted. The primary goal of coiling the filament is to reduce its overall size, therewith making it easier to construct a lamp of limited size. In sec. 2.3.3c, when discussing halogen lamps, it is pointed out that there is a lower limit of the bulb size, dictated by the bulb temperature.

As is explained earlier, the light emission is considerably lower than that of a black body or Planckian radiator of the same temperature. In sec. 2.3.1a, when discussing black-body radiation, it is explained how a coil can create a sort of hollows that act as black bodies of the same temperature as the filament, and so may enhance the efficacy of the lamp. Actually, there is even more. Because of the coiled parts of the filament shield each other, and therefore restrict the radiation of the filament into the bulb, the temperature of the hollows is slightly higher than the average filament temperature, adding to the enhancement of the efficacy of the lamp.

Details of the construction and of the operation as well of the physical properties of incandescent lamps are described in detail in Barrows (1938), Hentschel (1994), Moon (1961), and Schreuder (1998).

(b)   Filament evaporation and bulb blackening

As mentioned earlier, incandescent lamps generate light because the hot filament emits radiation. Physically, the efficacy of the lamp improves as the temperature of the metal filament increases. A drawback of the increase in the filament temperature, however, is the increase in the speed of evaporation of the filament metal. The higher evaporation leads to a shorter the lamp life, because the filament gradually gets thinner. Its electric resistance increases, and so does its temperature. This increases the evaporation even more, and the result is a cascade effect that ends in lamp failure. In the process, the efficacy may increase a little, but as the resistance increases, the power

Physical aspects of light production

– the wattage – that is dissipated by the lamp decreases. Therefore, the overall luminous flux of the lamp diminishes as well.

Furthermore, the evaporated metal deposits onto the inner surface of the outer bulb of the lamp. This causes the outer bulb to appear black as the burning hours progress. For obvious reasons, this phenomenon is called the blackening of the outer envelope of the lamp. Blackening reduces the transmittance of the bulb and causes a rapid decrease in the luminous output of the lamp. The decreases in the luminous flux output by blackening deteriorate the efficacy of the lamp accordingly. Furthermore, the absorbed energy increases the bulb – and thus the lamp – temperature, reducing lamp life even further.

(c) Lamp life and design values

It is customary to define the life of a lamp as the time where 50% of a bulk of lamps has failed – the failure rate. More important for the lighting design is the economic life. For incandescent lamps, this is usually taken as the moment where the overall light output of lamps has decreased down to 90% of the original value that has been used for the design. The reason is that at that time the performance of the lighting installation also has decreased down to 90% of the value that has been used for the design. It might be mentioned that many believe that 90% is a much too severe criterion, because a difference of 10% usually is well below the threshold of visual detection.

In the actual design process, usually the value of the total depreciation is added to the design value. The total depreciation is a combination of the reduction of the lamp light output and other factors like soiling and corrosion. In chapter 11, when discussing the design procedures for outdoor lighting installations, it is explained how the depreciation is incorporated in the actual design process. The lighting design is over-dimensioned by a factor equal to the reciprocal value of the depreciation. Thus, in a new condition, the installation will provide more light than needed. In modern installations, this surplus light and the related extra costs can be avoided by selective dimming. It will be clear that this design procedure is relevant for all lamp types and not only for incandescent lamps.

(d) Balance between the efficacy and the lamp life

In almost all incandescent lamps, an inert gas (Nitrogen, Argon, Krypton etc.) or a mixture of inert gases is put into the bulb. The gas pressure acts as a counter force to the evaporation pressure of the filament and thus prevents, or at least reduces, the evaporation of the filament. For obvious reasons, these lamps are called gas-filled lamps, contrary to the presently more rare vacuum lamps. However, part of lamp energy is lost to the bulb and consequently to the outer world. The temperature of the filament will decrease, although the same electric energy is fed to it. This leads to a reduction in the efficacy of the lamp. For this reason, the gas pressure in the bulb is limited, in order to establish a balance between the efficacy and the blackening. The present state of lamp technology is such, that for the near future, dramatic improvements in the efficacy or in the life of the normal incandescent lamps cannot be expected. The optimum of the balance between the efficacy and the lamp life is chosen in relation to the field of application of the lamp type and the related design criteria.

In Table 2.3.2, examples of different characteristics for various incandescent lamp types for outdoor lighting are listed.

Type	Length [mm]	Cap type	Luminous flux [lumen]	Nominal life [hours]
J220V500W	118	R7S	8 750	2000
J220V1000W	208	R7S	21 000	2000
J220V1000W/I	138	wire	23 000	1000
J220V1500W	248	R7S	33 000	2000

Table 2.3.2 Characteristics of various incandescent lamps for outdoor lighting (After Narisada & Schreuder, 2004, Table 11.3.4. Based on data from Anon., 2003).

Because many other lamp types have a lamp life much superior to that of incandescent lamps, the use of the normal incandescent lamps is rapidly diminishing for outdoor lighting. They are superceded by halogen incandescent lamps or gas-discharge lamps.

2.3.3 Halogen incandescent lamps

(a) *The role of gas pressure in the bulb*

To improve the efficacy of the incandescent lamps without reducing the life of the lamps, the so-called halogen incandescent lamp was introduced. This method was invented in 1959 by G. Zebler (Narisada & Schreuder, 2004, p. 466). The construction and the operating principles of the halogen lamp are as follows. The metal filament is housed in a small bulb or a narrow tube, depending on the lamp type. Although most halogen lamps have a more or less tubular outer envelope, we will use the familiar term 'bulb' here. The bulb is usually made of quartz. Quartz is a transparent material that is able to withstand much higher bulb temperature than normal glass without losing its mechanical strength. The bulb can be filled with an inert gas at a pressure much higher than in normal incandescent lamps.

Physical aspects of light production

By increasing the gas pressure in the bulb the evaporation of the filament is reduced. A small bulb or a narrow tube brings about another advantage. Because there is only little space between the filament and the bulb wall, the gas convection, which carries energy from the filament to the bulb wall is reduced and consequently the convection losses as well. However, small bulbs or narrow tubes are more prone to blackening of their inner surface, due to the evaporation of filament material. For this, the halogen lamp that was mentioned earlier may offer a solution.

(b) The halogen cycle

To overcome the problem of bulb blackening, a halogen – such as iodine – is added to the filling of the lamp, allowing the so-called halogen cycle to develop. The halogen cycle takes place between the filament metal evaporated and halogen vapour in the bulb. The evaporated metal near the filament, due to very high temperature, diffuses toward the inner wall of the bulb where it is cooler than near the filament. Iodine forms a gaseous compound with tungsten metal around the inner wall. As the temperature around the inner wall of the bulb or tube is kept higher than about 250° C, the iodine compound is kept the gaseous state and, by convection, is moved towards the filament. Due to the very high temperature of the filament, the iodine compound is decomposed to the metal and the iodine. The metal deposits on the relatively cool spot on the filament and iodine repeats the process again. The principle is the same when other halogens are used. The process is described in Hentschel, ed. (2002, sec. 5.2.2) and Heinz (2006a, p. 35-41).

(c) Characteristics of halogen incandescent lamps

Some types of halogen incandescent lamps have an infrared reflection coating on the inner surface of the bulb. A part of the infrared radiation reflected back by the coating contributes to heat the metal filament and to increase the efficacy of the lamp.

Earlier in this section a brief explanation has been given of the halogen cycle in the halogen incandescent lamp. In this way, the blackening inside the bulb or tube of small dimension is prevented. Consequently, the halogen lamps may reach a high efficacy, long life and less decrement in the luminous output during the life while keeping the excellent colour rendering properties of regular incandescent lamps. Halogen lamps are widely in use as projector lamps, as motorcar headlamps, and in floodlighting installations outdoors, although gradually they are being replaced in many applications by discharge lamps, and more recently by LEDs. Gas-discharge lamps and LEDs are discussed in the following parts of this chapter.

(d) Why use incandescent lamps?

One might wonder why, in spite of the apparent disadvantages, the incandescent lamp is far out the most popular lamp type in the world. The reason is not to be found in their efficiency but in a number of other characteristics. The most important ones are:

1. Incandescent lamps show a continuous spectrum; therefore, they have an excellent colour rendition. In fact, the 100% value of the colour rendering index is defined as that belonging to incandescent lamps. The terms colour rendition and colour rendering index, and their implications for light application, are discussed in another chapter of this book;
2. Incandescent lamps are very versatile. One may find them in all sizes and shapes. Therefore, they are suitable in an excellent way to a great variety of lighting applications;
3. Incandescent lamps can be connected directly, without any ballast to the electricity grid;
4. And finally, incandescent lamps are very cheap.

It might be added that recently there is a political drive to outphase incandescent lamps in favour of lamp types that are more energy-efficient, like compact fluorescent lamps or LEDs.

2.4   Gas-discharge lamps

2.4.1   Quantum aspects of light

(a) Bosons, baryons, and fermions

As has been indicated earlier, light can be regarded not only as an electromagnetic wave, but also as a stream of rapidly moving photons. Photons belong to the family of elementary particles that are characterized by their 'spin' being either 0 or 1. They are called bosons. They follow the Bose-Einstein statistic (Illingworth, ed., 1991, p. 379). This means that any number of such particles may exist in the same quantum state. For all practical purposes, it means that their number is not limited. One may make or destroy as many photons as one likes. This is, of course, the basic characteristic of light. Light can be generated in lamps, and destroyed by absorption. A summary is given in Breuer (1994, p. 363). Contrary to photons, other familiar elementary particles, like e.g. electrons, neutrons, and protons, are fermions. They have a spin of $1/2$. They follow the Fermi-Dirac statistic (Illingworth, ed., 1991, p. 379; Breuer, 1994, p. 363). Their characteristic is that only a limited number – usually only one – can be in a specific quantum state. Almost all elementary particles are baryons. "Baryons are all

Physical aspects of light production

elementary particles that can undergo strong interactions. They all have a mass equal to or greater than that of a proton. Baryons are thus Fermions. An additive quantum number called the baryon number can be defined. The total baryon number is conserved in all particle interactions" (Illingworth, ed., 1991, p. 32). In a technical sense, one of the 'Great Conservation Principles' is that the baryon number is conserved (Feynman, 1990, p. 62). In non-technical terms, it means that one cannot create nor destroy baryons – or fermions – at will, contrary to bosons. So baryons are not suitable as 'light'.

As is explained in sec. 2.4.1, the quantum aspects of light are characterised by the momentum carried by the photons, which is expressed as follows:

$$m = h \nu / c \qquad [2.4\text{-}1; 2.1\text{-}4]$$

with
   h: 'Planck-constant', equal to $6,626\,076 \cdot 10^{-34}$ J·s (Illingworth, ed., 1991, p. 354)

(b) The physics of metals

In theory, all metals can be used in making a gas discharge. To explain that, a few words must be said about the nature of a metal – what makes a metal a metal. First we will need to explain the model of atoms, at least in its most simple way. As is well known, atoms consist of a nucleus with a positive electric charge, surrounded by a number of negatively charged electrons. The number of electrons equals the atom number, or the position of the atom in the periodic system of elements (Illingworth, ed., 1991, p. 342, 540). This table is usually called the Table of Menedeleyev (Clugston, ed., 1998, p. 583-584).

According to the atom models of Bohr, Dirac, and Sommerfeld, the electron orbits are fixed and separated by 'forbidding zones' (Illingworth, ed., p. 24). According to the Pauli exclusion principle, only two electrons can have a place in each orbit (Illingworth, ed., p. 340-341). This is because electrons are fermions as mentioned earlier. Each electron orbit corresponds with a specific energy level. In each orbit, the two electrons will have an opposite spin with slightly different energy levels (Clugston, ed., 1998, p. 569). For atoms with an average or higher than average atom number, several orbits are possible that are rather close as regards their energy levels, each forming an electron shell (Illingworth, ed., 1991, p. 146). As mentioned earlier, the atomic number equals the number of protons in the nucleus, and consequently the number of electrons, the atom as a whole being neutral as regards the electrical charge.

Now there are three possibilities regarding the way the shells are filled with electrons:
1. The shells may be almost, but not completely full. There are a few openings that easily can be filled by surplus electrons. This is the basis for their chemical nature.

These elements are called alkali metals (Clugston, ed., 1998, p. 17). Examples are Fluorine, Chlorine, and Bromine.
2. The shells are completely full. It follows from this fact that they do not react with atoms of the same element. Therefore they are always gaseous. They do not react with other elements either, so they are inert gasses. Usually they are called noble gasses (Clugston, ed., 1998, p. 538). Examples are Helium, Neon, Argon, and Krypton. Because they are inert, they are suitable as filling gasses in incandescent lamps (secs. 2.2.2 and 2.3.2).
3. Most shells are completely full. However, there are a few surplus electrons that may begin to fill the next shell. These electrons are not attached very tightly to the atom; they may be taken away rather easily. This is the basis for their chemical nature. These elements are called metals (Clugston, ed., 1998, p. 493). Examples are sodium (Natrium) and potassium (Kalium); but also the elements like mercury, iron, silver, gold and aluminium are metals.

The surplus electrons can easily detach themselves from the atoms they belong to, particularly in the cubic crystal grid that is characteristic for many metals (Clugston, ed., 1998, p. 493). They can form a 'sea' of loose electrons. When a voltage is put on the metal, an electric current will flow, feeling only a limited resistance. This makes the metals an electric conductor. Sometimes the resistance is for some reason quite considerable. The metal is called a semiconductor. Semiconductors are essential in almost all modern electronic devices. They are discussed in sec. 2.5.1. It might be added that if the electric resistance is really very large, the material is called an insulator. The sea of surplus electrons conducts not only very easily electricity but also heat. Thus, conductors of electricity usually are also heat conductors, and insulators of electricity usually are also heat insulators. The physics of these phenomena is rather complicated. We refer to the classical textbooks of theoretical physics like e.g. Feynman et al. (1977); Joos (1947); Prins (1945); Von Weizsäcker (2006); and Wachter & Hoeber (2006).

Metals have two characteristics that make them essential for industry. The first is the well-known 'metallic nature' of electric and heat conductivity, pliancy, etc. These characteristics all have to do with the sea of surplus electrons as is explained earlier. At the beginning of technology, metals are predominant, even so far that whole aeons of human prehistoric development are named after metals, like e.g. the copper age, the bronze age, and the iron age.

For lamp design, the second characteristic is essential. As mentioned earlier, in metals the outer shells contain only a few electrons that are loosely attached to their 'own' atom. These electrons can easily be knocked out of the shell altogether. When this happens, the result is an ionised atom or ion and a loose electron. The electrons can be

Physical aspects of light production 45

knocked out by thermal agitation, or by collisions with other elementary particles. And this last effect is the reason that gas discharges can exist.

In theory, all metals can be used in making a gas discharge. To explain that, a few words have been said about the nature of metals. All that is needed is that the outer nuclear electron orbit – or energy level – is almost empty of electrons. This is precisely what characterizes metals: the small number of electrons from the outer shell that can easily be freed. We will come back to this point in one of the following parts of this section.

(c) Quantum aspects of light; gas discharges

As will be explained in the next part of this section, all gas-discharge lamps look rather similar. They consist of a tube filled with the appropriate gaseous material. The tube contains two electrodes. One is the negatively charged cathode and the other is the positively charged anode. The gas discharge is caused by the electrons that flow from the cathode to the anode.

Gas discharges are physically and technically rather complicated. We will give a brief explanation, based on the 'classical' work of Orange (1942). Theoretical details are given in Elenbaas, ed. (1959, 1965). See also Heinz (2006a, chapters 4 and 5). Details about the construction and the application of lamps are given in De Groot & Van Vliet (1986) and in Meyer & Nienhuis (1988). See also Baer, ed. (2006); Breuer (1994); Hentschel (1994); Kuchling (1995); Narisada & Schreuder (2004), and Schreuder (1998, sec. 4.3).

When the lamp is in operation, negative electrons are emitted from the negatively charged cathode. Under the influence of the voltage difference between the two electrodes, they travel towards the positively charged anode. In the process, they may collide with the atoms of the gas in the tube. When this happens, one or more of three effects may take place (after Oranje, 1942, sec. 3; Meyer & Nienhuis, 1988, sec. 1.1). In all three cases, there is some form of energy transfer from the electron to the atom. The energy comes from the kinetic energy of the electrons that get that energy from the voltage difference between the cathode and the anode – thus, ultimately, from the electric supply of the lamp. The three effects are:

1. At low energy levels, the collision is a elastic one. Both electron and atom stay unchanged; only their speed and direction of movement – their momentum – may change. Elastic collisions do not contribute to the generation of light in the lamp. In the process, some of the energy may be converted into heat. As has been mentioned earlier, some heat generation is needed to bring to the lamp at the right temperature for optimal operation. Too much heat, however, is just a loss. It is one of the most difficult aspects of lamp design to keep the temperature in the right range;
2. At higher energy levels, the electrons that collide with the vapour atoms may cause one or more of the electrons of that atom to shift into an higher energy level. The

electrons come into an excited state (Illingworth, ed., 1991, p. 162). This effect is called excitation, and can be described as the electron circling the atom nucleus in a higher orbit. In order for this to happen, a precise amount of energy is required. Any surplus energy of the electron is transferred into heat. After a very short time, usually in the order of 10^{-7} seconds, the electron falls back into its ground state, emitting the same amount of energy that was absorbed earlier. Taking into account the equation [2.1-5], the energy corresponds to a very specific frequency or wavelength of the light. The effects of excitation and fall-back are not important for discharge lamps but they are essential for semiconductor lamps that are discussed in another chapter of this book;

3. At still higher energy levels, the electron may knock one or more of the electrons out of the atom. The electrons are free and join the electrons coming from the cathode; the atom with less electrons is an ion. The process is ionization. For knocking out a specific electron out of a specific atom, again a very precise amount of energy is needed. Any surplus energy of the electron is again transferred into heat. The ions have a positive charge, and begin to move towards the cathode. In moving, most ions meet a free electron and recombine again to a neutral atom, emitting again the same amount of absorbed ionization energy. It forms a spectral line. This, of course, is the spectral line that corresponds to that quantum transition. Such lines are called resonance lines (Meyer & Nienhuis, 1988, p. 23);

(d) The construction of gas-discharge lamps

The most essential component of a discharge lamp is obviously the gaseous material that provides the ions that have been introduced in the preceding part of this section. It has been suggested earlier that in principle, any metal could be used in a gas discharge. For practical reasons, only sodium and mercury are used on a large scale. As most metals that could theoretically be used in discharge lamps have many electrons, and each electron can be in many different states, the possible number of spectral lines is very large. The choice to use mercury and sodium in discharge lamps is governed by the fact that for these atoms, and for these atoms only, most strong spectral resonance lines are emitted in a wavelength range that is in, or very close to, the range of visible light (Meyer & Nienhuis, 1988, p. 23-24). Further details may be found in Schreuder (1998, sec. 4.3, Figures 4.3.1 and 4.3.2). See also Elenbaas, ed. (1965, Fig. 1.10) and De Groot & Van Vliet (1986, Fig. 3.6).

Another reason to use these materials is that at room temperature sodium is a solid and mercury a fluid. That allows to insert a surplus of the material in the lamp envelope. This surplus acts as a buffer. When there is not enough material in gas form in the envelope, some of the buffer material will evaporate, and when there it too much material, some

Physical aspects of light production

of it will condense. The implication is that the gas pressure in the envelope is determined only by the temperature (Meyer & Nienhuis, 1988, sec. 1.5.2, p. 29-30).

It must be stressed that the use of mercury is strongly disapproved by health authorities because of its very poisonous nature. It may be added that modern lamp design aims at a stringent reduction of the amount of mercury that is used in lamp manufacture. Furthermore, the disposal of broken lamps is strictly regulated, at least in most environment-conscious countries. In the process, up to 99% of the mercury is recycled (Anon., 2007).

As mentioned earlier, the actual construction of all gas-discharge lamps is basically very similar. They consist of a tube, usually made of glass, of quartz or of a transparent ceramic material. The tube can have any shape, but usually it is straight. It contains two electrodes that are, via a ballast, connected to the main electric supply. One is the negatively charged cathode and the other is the positively charged anode. In most cases, an alternate electric supply is used of 50 or 60 Hz. That means that, 50 or 60 times per second, the cathode and the anode change their function. Modern lamps often use a much higher frequency, up to the kHz region, for a number of technical and economic reasons. Ballasts can be smaller, and the efficacy of fluorescent tubes can be up to 10% higher (Van Ooyen, 2005, p. 25). Similar effects may be found in high-pressure sodium lamps. Further details are given in De Groot & Van Vliet (1986, p. 210-211). An example is given in Figure 2.4.1.

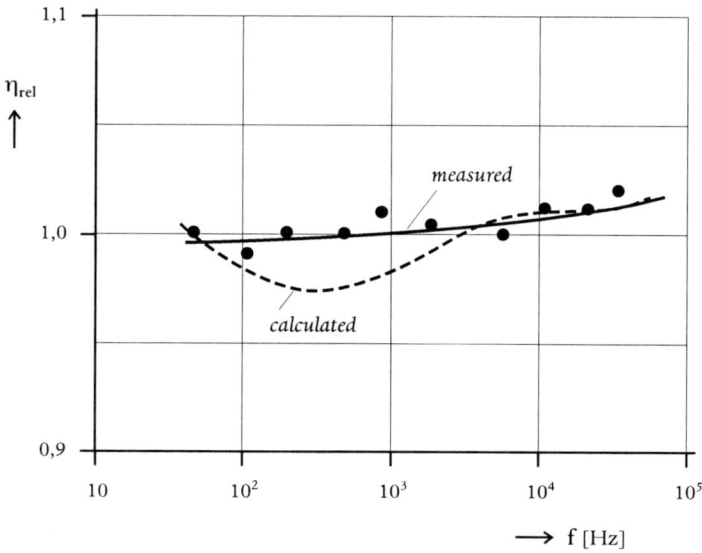

Figure 2.4.1 Relative luminous efficacy of a 70W high-pressure sodium lamp in relation to the supply current frequency (After De Groot & Van Vliet, 1986, Fig. 7.7).

Under the right conditions, that are discussed in detail in further parts of this section, a 'gas discharge' will take place. The main function of the tube is to contain the gas discharge and to help to maintain the conditions that are required in order to keep the discharge going. Therefore, the tube is called the 'discharge tube'. For short, sometimes the term burner is used, although it is, of course, not correct, because nothing actually 'burns'. The function of the ballast is explained in Narisada & Schreuder (2004, sec. 11.4.1d); Hentschel, ed. (2002, sec. 5.3), and Baer, ed. (2006, sec. 2.4 and 2.5).

As mentioned earlier, in modern outdoor lighting, only mercury and sodium are used in discharge lamps. Mercury and sodium are monovalent, so both terms 'atom' and 'molecule' may be used. Because, at room temperature, mercury is a fluid and sodium is a solid, it is really the vapour phase that is relevant here. In the past, this was sometimes expressed in the lamp nomenclature. In order to get a vapour pressure high enough to allow an effective operation of the lamp, its temperature must be rather high. Discharge lamps only operate in an optimum fashion within a narrow temperature range because for optimum lamp operation, the vapour pressure must be kept within narrow limits. It is well known that the vapour pressure depends heavily on the temperature. This is contrary to traditional incandescent lamps where the lamp temperature has little influence on lamp efficiency. Halogen lamps require a specific lamp – or bulb – temperature, not only for optimum operation but also for a long lamp life. Further details on the construction and the use of gas-discharge lamps are given in Narisada & Schreuder (2004, sec. 11.3.4).

>   (e)   *The influence of the vapour pressure*
>   When the vapour pressure is low, the spectral lines are very sharp, but their intensity is low. When the vapour pressure is higher, all sorts of additional elastic collisions occur, resulting in a broadening of the spectral lines. This effect is used in special types of high-pressure sodium lamps, where a further increase in the vapour pressure results in an improved colour rendering. Such lamps are in particular suitable for flower gardens (Akutsu et al., 1984).

When the vapour pressure is higher, the spectral lines are broadened, mainly as a result of elastic collisions between the particles in the gas discharge (De Groot & Van Vliet, 1986, p. 25-26). The effect of the broadening of the spectral lines is illustrated in Figures 2.4.2 and 2.4.3 for sodium discharges and Figures 2.4.4 and 2.4.5 for mercury discharges. When the pressure increases even further, another process gains importance, the self-absorption.

Physical aspects of light production

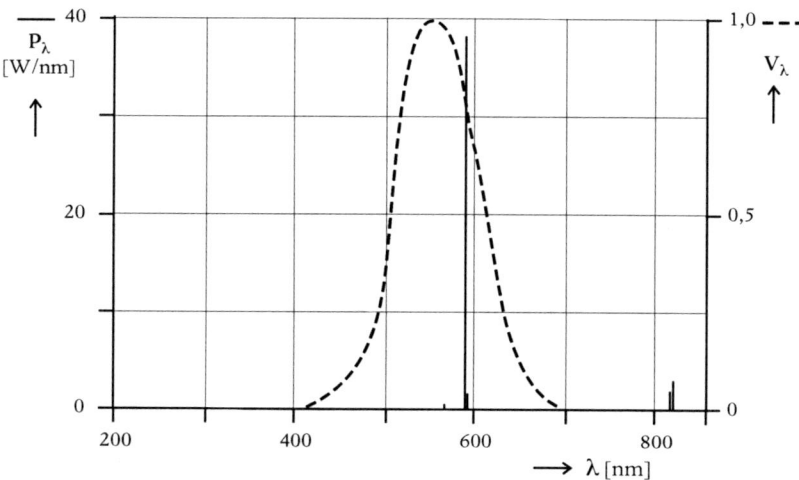

Figure 2.4.2 Full curve: the spectral power distribution curve of a low-pressure sodium lamp (SOX-E 131); dashed curve V_λ curve (After De Groot & Van Vliet, 1986, Fig. 1.9 a).

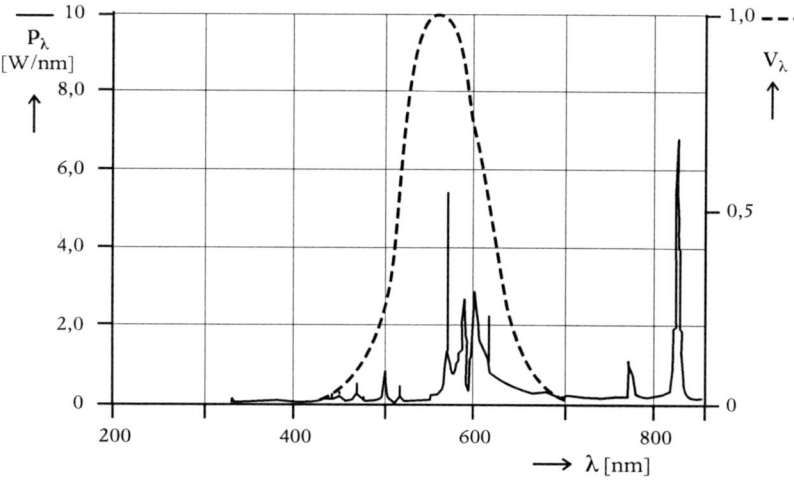

Figure 2.4.3 Full curve: the spectral power distribution curve of a high-pressure sodium lamp (SON 400 W); dashed curve V_λ curve (After De Groot & Van Vliet, 1986, Fig. 1.9 b).

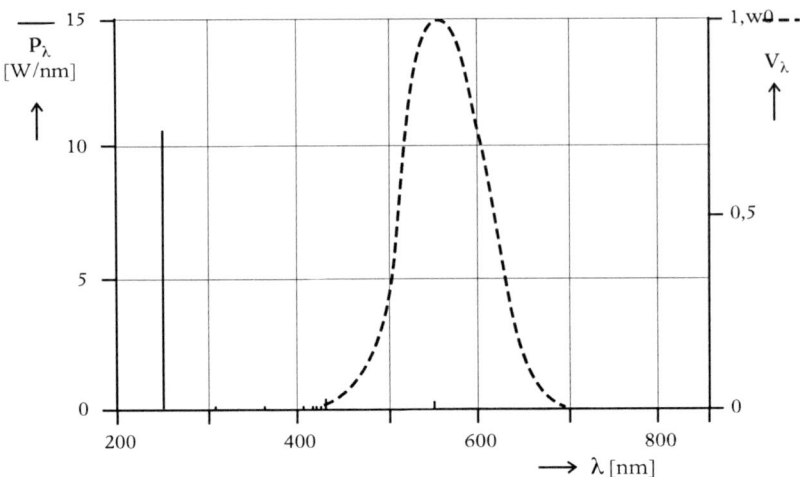

Figure 2.4.4 Full curve: the spectral power distribution curve of a germicidal low-pressure mercury lamp (TUV 40 W); dashed curve V_λ curve (After De Groot & Van Vliet, 1986, Fig. 1.10 a).

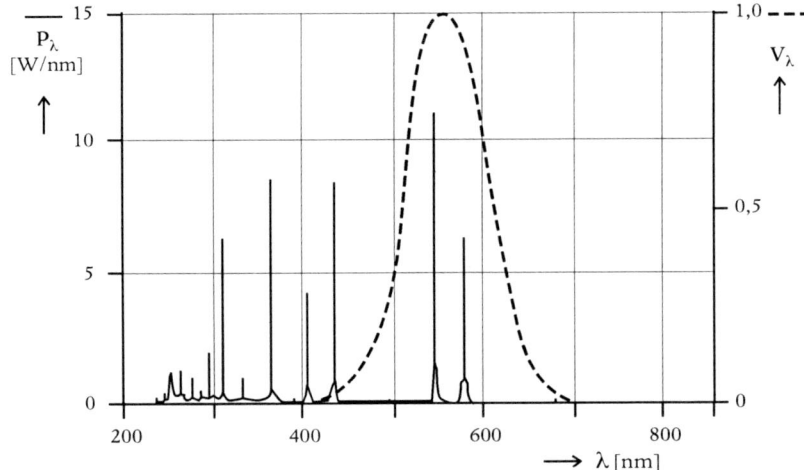

Figure 2.4.5 Full curve: the spectral power distribution curve of a clear high-pressure mercury lamp (400 W); dashed curve V_λ curve (After De Groot & Van Vliet, 1986, Fig. 1.10 b).

Using a high vapour pressure results in a smaller lamp with a corresponding increase in light source luminance, even if the total luminous flux may be lower. Thus, low-pressure sodium lamps emit an almost pure monochromatic light, whereas high-pressure sodium lamps emit a spectrum that contains, although not being exactly continuous,

Physical aspects of light production 51

most colours. As is explained in another chapter of this book, when light pollution is discussed, this makes low-pressure sodium lamps very suitable for outdoor lighting near astronomical observatories, and high-pressure sodium lamps particularly suitable for general outdoor lighting purposes.

(f) *The main families of gas-discharge lamps*
It follows directly from the foregoing that there are four main families of gas-discharge lamps:
1. Low-pressure mercury lamps;
2. Low-pressure sodium lamps;
3. High-pressure mercury lamps;
4. High-pressure sodium lamps.

We will discuss these four families in some detail in the following parts of this chapter. For other types of lamp, like e.g. neon lamps, glow-lamps, or super-high-pressure mercury lamps we refer to the literature, more in particular to Anon. (1997; 2005); De Groot & Van Vliet (1986); Meyer & Nienhuis (1988); Hentschel (1994); Narisada & Schreuder (2004), and Schreuder (1998).

Before discussing these four main families of discharge lamps, we will discuss in sec. 2.4.2 some detail the phenomena of fluorescence. Fluorescence allows us to use discharges that emit light in the ultra-violet range, because in that process light of short wavelengths is converted into light of longer wavelengths, i.e. from the ultra-violet to the visible parts of the electromagnetic spectrum.

2.4.2 Fluorescence

(a) *Fluorescence in gas-discharge lamps*
Luminescence is produced when atoms are excited, as by other radiation, electrons etc., and then decay to the ground state. The general term is luminescence where energy is converted into light, or, more specific, where short-wave light is converted into long-wave light. A typical time interval is between 10^{-8} and 10^{-6} seconds (Heinz, 2006a, p. 5). If the luminescence ceases as soon as the source of energy is removed, the phenomenon is called fluorescence. A typical time interval is between 10^{-8} and 10^{-6} seconds (Heinz, 2006a, p. 5). If it persists, the phenomenon is called phosphorescence (Illingworth, ed., 1991, p. 276, 346)

Most fluorescent materials consist of zinc sulfide and cadmium sulfide, and of all sorts of chemical combinations of silicium, tungsten, and – as mentioned earlier – phosphor, and additionally of mixtures of these materials.

It should be mentioned that there is a considerable amount of similarity between fluorescent materials and the semiconductors that are discussed in sec. 2.5.1. One is the fact that both materials only work properly if an activator, often called a doping material is added, be it in a very low concentration of some 10^{-4} tot 10^{-5}. In Figure 2.4.6 the role of the activator is clarified.

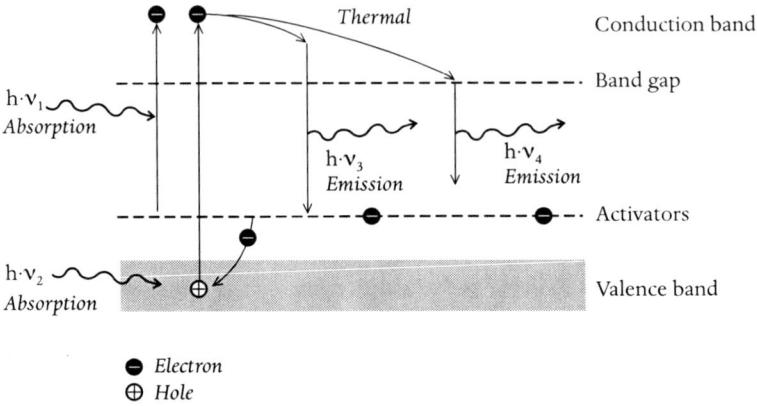

Figure 2.4.6 Fluorescence in a crystal, schematic (After Hentschel, ed., 2002, Fig 5.8, p. 139).

The conduction band is shown at the top of the figure, the valence band at the bottom. If the electrons would, after excitation, jump from the valence band to the conduction band and back again, nothing would have been gained. The absorbed energy would have been re-emitted at precisely the same wavelength, just as in a gas discharge.

The role of the activator is as follows. Their presence disturbs the pure crystalline grid. The activator atoms lay energetically speaking closely above the valence band. The same disturbance creates holes or open spaces closely below the conduction band. When an excitating quantum hits the crystal, one of the activator atoms may jump to the conduction band. Because they start higher, they will end higher up in the conduction band. They absorb the light quantum of the excitation. Its energy is

$$E = h \cdot v_L \qquad [2.4\text{-}2]$$

As a result of the thermal agitation the electrons drift down to the holes close below the conduction band. After that, they fall back in the well-known way to their original place closely above the valence band, in the process emitting a light quantum with an energy of

Physical aspects of light production

$$E = h \cdot \nu_E \qquad [2.4\text{-}3]$$

The end result is the emission of a light quantum with the energy equal to $h \cdot \nu_E$, and not of a light quantum with the energy equal to $h \cdot \nu_L$. Figure 2.4.6 shows that $h \cdot \nu_E < h \cdot \nu_L$. This is called Stokes' Law (Illingworth, ed., 1991, p. 177, Heinz, 2006a, p. 51).

Fluorescence is widely used in lamp production and application. In fluorescence, the emitted radiation is of longer wavelength than the incident electromagnetic radiation. This is Stokes Law that has been mentioned earlier. Although it might seem to be the case, Stokes' Law is not in conflict to the first law of thermodynamics. That law, that agrees to the law of energy conservation that is discussed in another chapter of this book, states that in any process no energy can be destroyed, nor generated. In the conversion from light of a short wavelength or a high energy state into light of a long wavelength or a low energy state, the surplus energy is converted into heat. Stokes' Law also indicates, on the ground of the same law of energy conservation, that it is not possible to convert light of a long wavelength into light of a short wavelength (Illingworth, ed., 1991, p. 461).

(b) The conversion from UV radiation into light

In low-pressure mercury discharges, almost all energy is emitted in the region of near-ultraviolet wavelengths. It may be helpful to recall that ultraviolet or UV radiation, in spite of the fact that it is often called 'UV light', is not light at all, because it does not provoke any sensation of light in the retina. Fluorescence is very helpful to convert this invisible and harmful radiation into visible light by coating the inner surface of the discharge tube by appropriate chemical substances. A suitable selection of materials allow to reach light with a high intensity, an almost continuous spectrum and a very good colour rendering. Additionally, coating the inner wall of the tube prevents the harmful ultraviolet being emitted as it is absorbed by the glass. The resulting lamp type, usually called the fluorescent tube did allow the lighting revolution of the 1960s to occur.

In fluorescence, the wavelength of the emitted light does not depend on the wavelength of the exciting – the incident – light; however, the efficiency of the transition does. As long as the wavelength of the exiting falls into absorption band, all is well. In Figure 2.4.7, a schematic picture is given of the absorption band and the emission band of a fluorescent material, as well as the spectral distribution of the exciting light. The two major resonance lines of a low-pressure mercury discharge both fall within the absorbtion band.

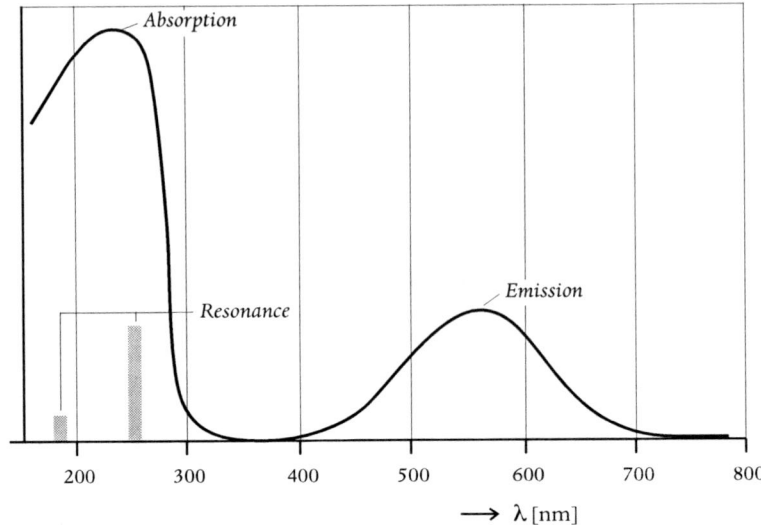

Figure 2.4.7 *A schematic picture of the fluorescence process (After Meyer & Nienhuis, 1988, Fig. 3.11, p. 78).*

The energy of the absorbed light quantum is

$$E_R = h \cdot f_R = (h \cdot c)/\lambda_R \qquad [2.4\text{-}4]$$

with:
    E_R: the energy of the absorbed light quantum;
    h: Planck's constant;
    f_R: the frequency of the absorbed light;
    c: the speed of light.
    λ_R: the wavelength of the absorbed light.

The energy of the emitted light quantum is:

$$E_L = h \cdot f_L = (h \cdot c)/\lambda_L \qquad [2.4\text{-}5]$$

with:
    E_L: the energy of the emitted light quantum;
    h: Planck's constant;
    f_L: the frequency of the emitted light;
    c: the speed of light;
    λ_L: the wavelength of the emitted light.

Of course, these relations are the same as those that have been given earlier, only a little more in detail.

Physical aspects of light production

From [2.4-4] and [2.4-5] follows the ratio between the absorbed and emitted energy:

$$E_R/E_L = \lambda_L/\lambda_L \qquad [2.4\text{-}6]$$

In this context, the quantum yield (QY) is defined:

$$QY = N(e_L)/N(e_R) \qquad [2.4\text{-}7]$$

with:
  $N(e_L)$: the number of emitted electrons;
  $N(e_R)$: the number of absorbed electrons.

The most important UV emission line of a low-pressure mercury gas discharge has a wavelength of 254 nm. The efficiency of the energy transformation cannot be higher than 62% for an emission at 400 nm and 31% at 800 nm. The rest of the absorbed energy is transformed into heat. Fluorescence is applied in other types of light sources as well, in order to improve their colour characteristics, such as in high-pressure mercury lamps and in LEDs.

(c) The efficiency of the fluorescent process

As mentioned earlier, the conversion from light of a short wavelength or a high energy state into light of a long wavelength or a low energy state will lead to loss of radiation energy (Meyer & Nienhuis, 1988, sec. 3.2.1, p. 77-79). To explain this effect in terms of the energy of quanta of radiation, we need again the equations from sec. 2.1.3,c that describe the energy and the mass of a photon as it depends on the wavelength:

$$E = H \cdot v \qquad [2.1\text{-}5]$$

$$M = E/c^2 \qquad [2.1\text{-}6]$$

For convenience we rewrite the equations:

$$E = (h \cdot c)/\lambda \qquad [2.4\text{-}8]$$

For the energy of the absorbed energy quantum E_R this is:

$$E_R = (h \cdot c)/\lambda_R \qquad [2.4\text{-}2a]$$

For the energy of the emitted energy quantum E_L this is:

$$E_L = (h \cdot c)/\lambda_L \qquad [2.4\text{-}2b]$$

This yields:

$$E_R/E_L = \lambda_L/\lambda_R \qquad [2.4\text{-}9]$$

This equation gives the maximum efficiency of the conversion of UV radiation into light. As an example we take the conversion of the energy of the strongest UV resonance line of a low-pressure mercury discharge into yellow light. The resonance line has a wavelength of 254 nm and the yellow light has a wavelength of 510 nm. The energy of the respective quanta is inversely proportional to the wavelength. So [2.4-6] becomes:

$$E_L = (\lambda_R/\lambda_L) \cdot E_R \qquad [2.4\text{-}10]$$

In this example, the maximum efficiency of the conversion is 254/510, or almost 50%.

The overall efficiency – or rather the overall lamp efficacy – of a fluorescent low-pressure mercury discharge lamp can be assessed in this way. There is another strong resonance line of UV radiation, viz. at 185 nm. The efficiency of the conversion into light of different wavelengths of the emitted light is given in Figure 2.4.8. In this figure, curve a is for the line of 254 nm, and curve b for the line of 185 nm. If one assumes that the lamp takes 90% of its light from the resonance line at 254 nm, and the other 10% from the line at 185 nm, the overall lamp efficacy is given by the curve designated by (a + b) in the figure.

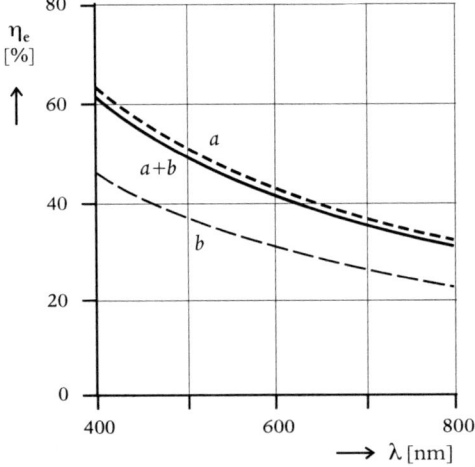

Figure 2.4.8 The theoretical efficiency of the conversion from UV radiation into visible light of different wavelengths (After Meyer & Nienhuis, 1988, Fig. 3.12).

(d) Fluorescent materials

Most fluorescent materials have been developed for fluorescent tubes. Usually they are called phosphors. However, this term is not correct. As is explained earlier, the general term is luminescence, and the short term effect is fluorescence. In this respect, the colloquial term flu-powder is more appropriate. One may assume that the term phosphorescence stems from the past, when the chemical element Phosphorus was an important ingredient for flu-powders. In the following we will adhere to the custom, and call the relevant materials also phosphors.

In current fluorescent lamps, there are three main groups of phosphors (Anon., 1993, p. 16-17; Van Ooyen, 2005).
1. Standard or traditional phosphors. They exhibit an emission that covers almost the whole visible spectrum. This results in a high lamp efficacy but in poor colour rendering. Most standard phosphors are based on halophosphates (Anon., 2006).
2. Tri-phosphors. The application of rare earths like e.g europium or terbium in flu-powders allows to create lamps that combine a high efficacy with good colour characteristics (Anon., 2006). The flu-powders show sharp radiation peaks in three well-defined wavelength areas, viz. blue, green, and red. These lamps are called three-phosphor lamps, in the Philips-coding the 'colour 80' series, referring to the colour rendering index R_a of 80 or more. The colour rendering index R_a is explained in sec. 9.4.3. Under some circumstances, the three sharp spectral bands of the lamps can be seen separately, creating peculiar colour effects at prismatic glass or plastic objects.
3. Multi-phosphors. The peculiar colour effects that were mentioned above can be avoided by using a mix of phosphors that cover the whole of the visible spectrum. These lamps show excellent colour characteristics. In the Philips-coding these lamps are called the 'colour 90' series, referring to the colour rendering index R_a of 90 or more.

As an example, in Figure 2.4.9 and Figure 2.4.10, the emission spectrum of two different types of mixed trichromatic phosphors for fluorescent lamps is depicted. Figure 2.4.9 is for warm-coloured lamps with a colour temperature of 2700 - 2900 K and a colour rendering index of 85. The chromaticity coordinates are x = 0,465; y = 0,445 (Stanford, 2006, LPTB 28).

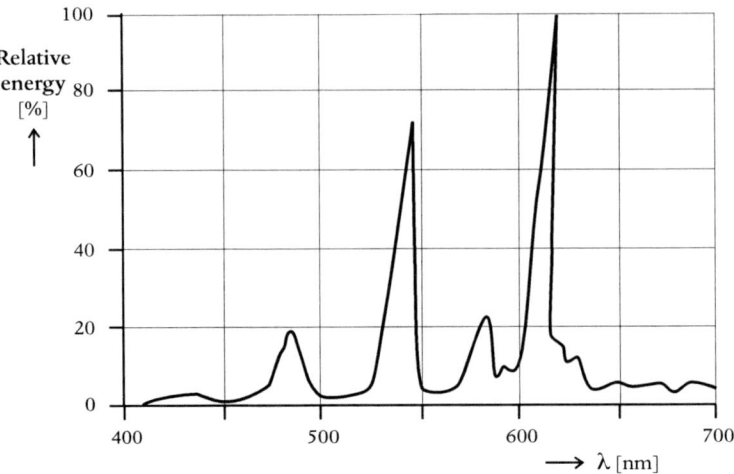

Figure 2.4.9 The emission spectrum of a mixed trichromatic phosphor for warm-coloured lamps (After Stanford, 2006, LPTB 28).

Figure 2.4.10 is for cool-coloured lamps with a colour temperature of about 5600 K and a colour rendering index of 90. The chromaticity coordinates are x = 0,314; y = 0,334 (Stanford, 2006, LPTB 65).

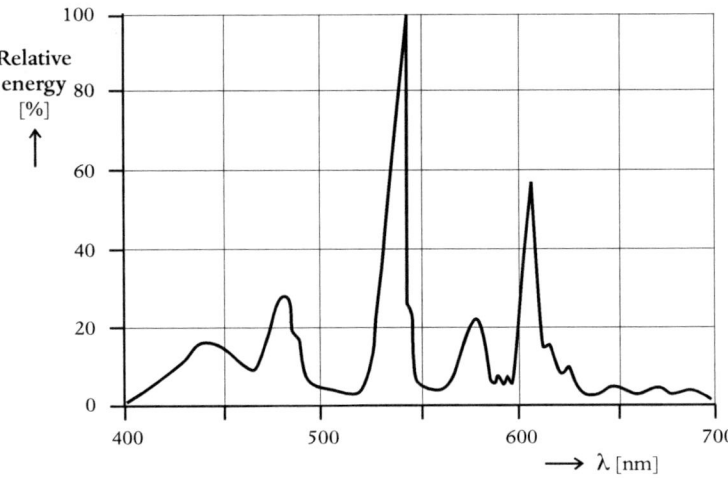

Figure 2.4.10 The emission spectrum of a mixed trichromatic phosphor for cool-coloured lamps (After Stanford, 2006, LPTB 65).

For fluorescent lamps it is found that the better the colour the lower the efficiency of the energy transformation, or the lower the efficacy of the lamps. In Table 2.4.1 some data are given for the traditional phosphors that were used in lamps that were developed before 1960.

Colour	Lamp type	Efficacy (lm/W)
level 1, very good	de luxe	<65
level 2, good	universal white	69
level 3, poor	standard warm white	83

Table 2.4.1 *Colour and lamp efficacy, old lamp types (After Ris, 1992, Table 3.4, p. 70).*

More modern lamps perform much better. The three-band lamps of the 1980s, with a colour 'very good' reach an efficacy up to about 96 lm/W (Hentschel, ed., 2002, Fig. 5.8, p. 139).

2.4.3 Types of gas-discharge lamps

(a) *Four families of lamps*

In the preceding parts of this section it is explained that there are four main families of gas-discharge lamps:
1. Low-pressure mercury lamps;
2. Low-pressure sodium lamps;
3. High-pressure mercury lamps;
4. High-pressure sodium lamps.

We will discuss these four lamp families in some detail in the following parts of this section.

First we will give some general data about the life of a number of common lamp types. In Figure 2.4.11, the lamp failures and the lamp depreciation is given for low-pressure sodium lamps, for high-pressure sodium lamps and for high-pressure mercury lamps. The data are not recent. They stem from the early 1990s, because newer data have not been made available by the industry.

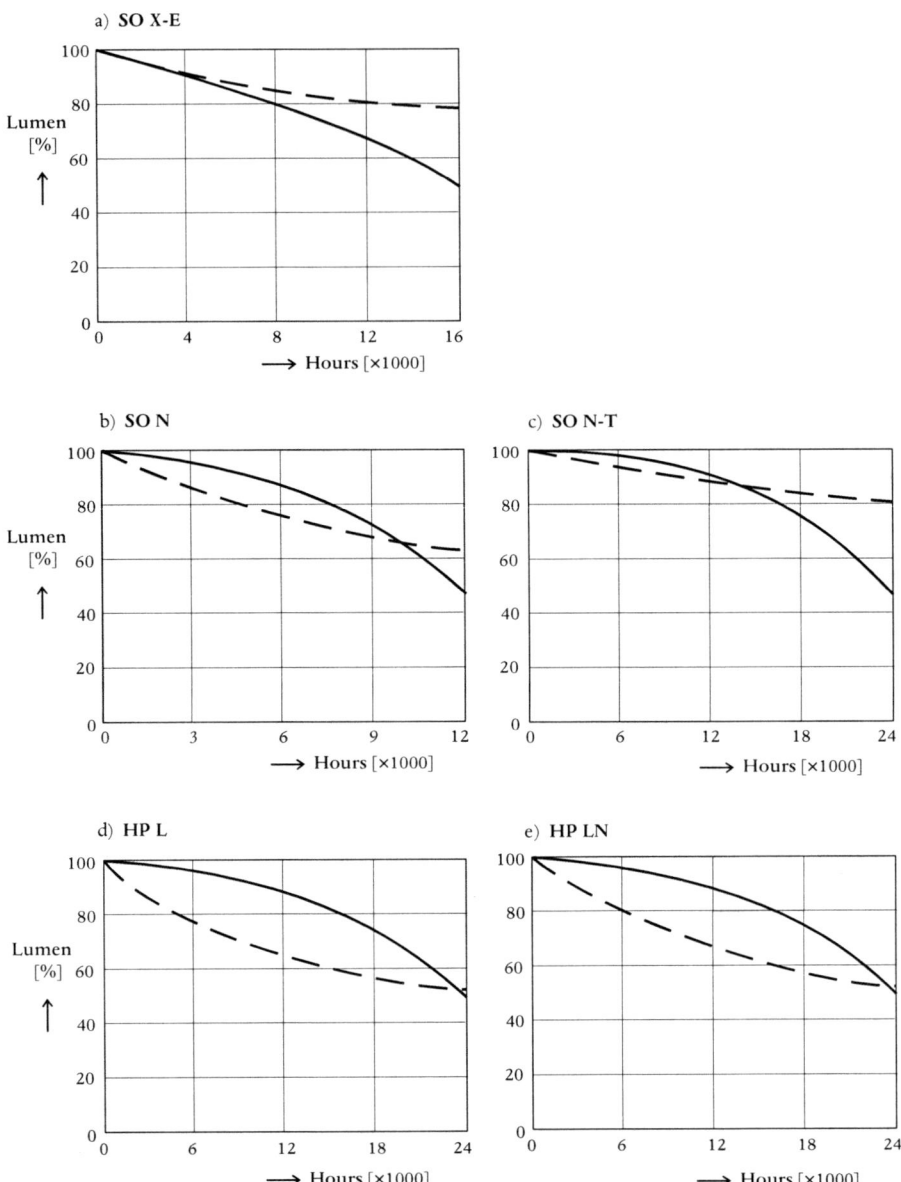

Figure 2.4.11 Lamp failures and the lamp depreciation. a: low-pressure sodium lamps; b and c: high-pressure sodium lamps; d and e: high-pressure mercury lamps. Based on data from Anon, 1993a (After Schreuder, 1998, Fig. 13.3.2).

Physical aspects of light production

(b) Low-pressure mercury lamps

Far out the most populous of the four families is that of the low-pressure mercury lamps. In a preceding part of this section it is explained that a low-pressure gas discharge has two important resonance spectral lines. Both are located in the ultra-violet part of the spectrum; their wavelengths are 254 and 185 nm respectively. The first is useless, but the second, laying in the short-wave part of the UV, is quite hazardous for causing sunburn and even skin cancer (Anon., 1986; Duchêne at al., eds., 1991; Schoon & Schreuder, 1993; Lusche, 2001; Van den Beld, 2003; NSVV, 2003, 2003a).

As is explained in sec 2.4.2, fluorescence is needed to get visible light from the UV radiation. It might be added that the glass envelope is also needed to shield the user from the hazardous UV radiation. The gas pressure is low, around 0,8 Pascal or about $7,8 \cdot 10^{-4}$ atmosphere (Meyer & Nienhuis, 1988, p. 71). The diameter of the envelope is restricted by the nature of the gas discharge and by temperature requirements. In the past, the diameter was 38 or 26 mm, and in newer types 16 mm (Van Ooyen, 2005, sec. 1-5, p. 18). Compact fluorescent lamps have an even narrower envelope of usually 12,5 mm (Van Ooyen, 2005, sec. 1.5.3.2). To get a decent amount of lumens out of the lamp, the lamp must be pretty long. A great number of lamp lengths have been standardized over the years, the traditional lengths being 2 and 4 feet (0,6 and 1,2 m; Anon., 1993, p. 16). No wonder these lamps are universally known as tubes, more specific fluorescent tubes.

The operation of fluorescent lamps is characterised by the fact that they have a rather long life. After 8000 hours the light output is still about 70 to 90% of the initial value (Van Ooyen, 2005, sec. 1-5, p. 25). The useful life depends on many aspects, like e.g. the switching frequency. The light output depends heavily on the ambient temperature; see Figure 2.4.12.

Also the ignition of the lamps depends on the ambient temperature. There are many ways to connect the lamps with their ballasts to the main electric circuit. The choice depends on economic factors, but also on the need for cold ignition. A number of the many variations is given in Van Ooyen (2005, sec. 1-5, p. 26-34).

Fluorescent lamps come in many shapes, but as regards the way they function and the way they are constructed they are all alike. The many shapes are:
- straight – the traditional tubes;
- circular;
- U-shape – often called PL lamps;
- folded up – the compact fluorescent lamps.

The last group is usually built into an envelope and integrated with the ballast, so they can be used directly as a replacement of traditional incandescent lamps. They are heavily promoted by environmental groups, because they save a lot of energy, but it is often overlooked that they are an extra burden on the waste disposal. Not just the mercury but other waste as well.

Fluorescent lamps come in many sizes. Usually this is indicated by the wattage that is dissipated by the lamps. The lamp wattages run from 3W to 150W; the lamp efficacy from around 50 lm/W to over 100 lm/W.

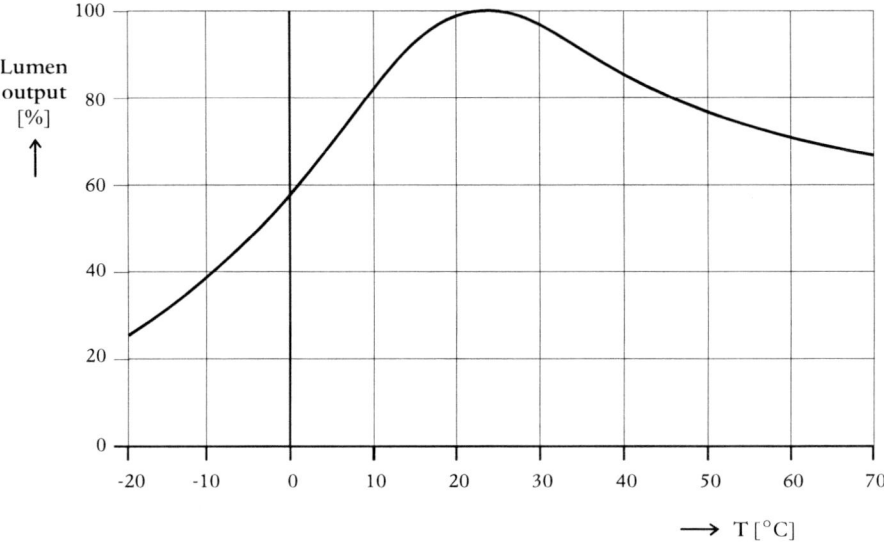

Figure 2.4.12 *The relation between light output and ambient temperature (After Van Ooyen, 2005, Fig. 1.5-20).*

(c) *Low-pressure sodium lamps*

The construction of low-pressure sodium lamps is simpler than that of fluorescent lamps. The main difference is that most energy is emitted in one line only, a line that lies close to the peak of the spectral sensitivity curve of the eye – the V_λ curve that is described in another chapter of this book. So, fluorescence is not needed.

Actually there are two resonance lines, one at 588 nm and one at 598 nm (Hentschel, ed., 2002, p. 156), or, more accurately, 589,59 and 588,99 nm respectively (Heinz, 2006a, p. 53). They are so close that in practice one reckons with one line, meaning that the light is monochromatic. This immediately puts up very severe restrictions on the fields of application of low-pressure sodium lamps. They only can be used under conditions where colour recognition is not required. In practice this means only for major traffic routes outside built-up areas. And even then, in many countries one rejects the use of the lamps because of their poor overall impression.

There is a further point. Although the lamp efficacy is the highest under standard production lamps – up to and even over 200 lm/W – their large size makes tight light control difficult, so that the efficiency of the overall lighting installation is usually lower than when high-pressure sodium lamps are used. This is explained in sec. 10.2.2. And finally, the economic life of low-pressure sodium lamps is considerably lower than that of modern high-pressure sodium lamps. So in spite of the fact that low-pressure sodium lamps have some advantages over other lamp types, their use world-wide is diminishing. As is explained in sec. 10.3.3f, the single sodium line can be filtered out. This makes low-pressure sodium lighting a favorite near astronomical observatories.

(d) High-pressure gas-discharge lamps

Although there are three main groups of high-pressure gas-discharge lamps, their functioning and their construction is very similar. They all consist of a gas-discharge tube in an outer envelope. As is explained earlier, the discharge material is either sodium or mercury. Sometimes additional discharge material with different characteristics is added. Because of the high gas pressure, the discharge tube is small. It is often called the burner. In operation, the burner gets very hot, so it must be made of a heat resistant material like e.g. quartz or a ceramic material. For ignition and operation, special gear is required.

In cold conditions, before the lamp is ignited, most discharge material is condensed. It takes some time before enough material for a proper functioning of the lamp is evaporated. The warming-up or run-up of high-pressure gas-discharge lamps takes several minutes. In Figure 2.4.13 the run-up time of a typical standard type high-pressure sodium lamp is depicted.

Because the burner gets very hot, an outer envelope is needed for protection. For thermal insulation it is not needed. In most lamp types, the envelope is clear or frosted, but sometimes the inside is covered with phosphors in order to improve the colour characteristics of the lamps.

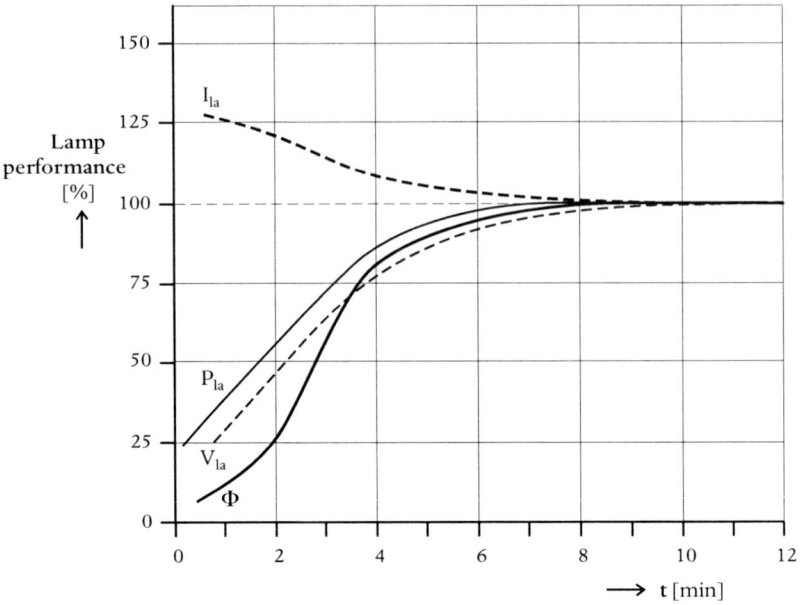

Figure 2.4.13 *The run-up time of a typical standard type high-pressure sodium lamp. I_{la} is the lamp current, V_{la} the lamp voltage. P_{la} the lamp power, and Φ the luminous flux (After Anon., 1993, Fig. 1.83).*

(e) *High-pressure mercury lamps*

The discharge of high-pressure mercury emits, in addition to the visible light, also UV radiation. By applying fluorescent materials on the inside of the envelope, the invisible UV radiation is converted into visible light. The additional visible light adds new colours to the spectrum; thus, the colour and the colour rendering properties are improved.

(f) *Metal-halide lamps*

Metal-halide discharge lamps are high-pressure mercury lamps with a clear bulb. In the discharge tube are added, in addition to the mercury, also different halide compounds of rare earth metals (Meyer & Nienhuis, 1988). The spectra emitted from the added rare earth metal vapours improve the colour, colour rendering and the efficacy of the original high-pressure mercury lamps. Metal-halide discharge lamps are used for lighting of large sized sports stadiums, squares, etc., where whitish colour and good colour rendering is necessary.

Physical aspects of light production

(g) *High-pressure sodium lamps*

As regards construction and operation, there is little difference between high-pressure mercury lamps and high-pressure sodium lamps, apart of course of the metal that is used in the gas discharge. The main difference is in the colour properties. In Figure 2.4.14 the relation between the colour characteristics and the vapour pressure in the lamp is depicted. The data run from a typical low-pressure sodium lamp to a typical high-pressure sodium lamp. The relation between colour temperature and colour rendition is explained in sec. 9.3.

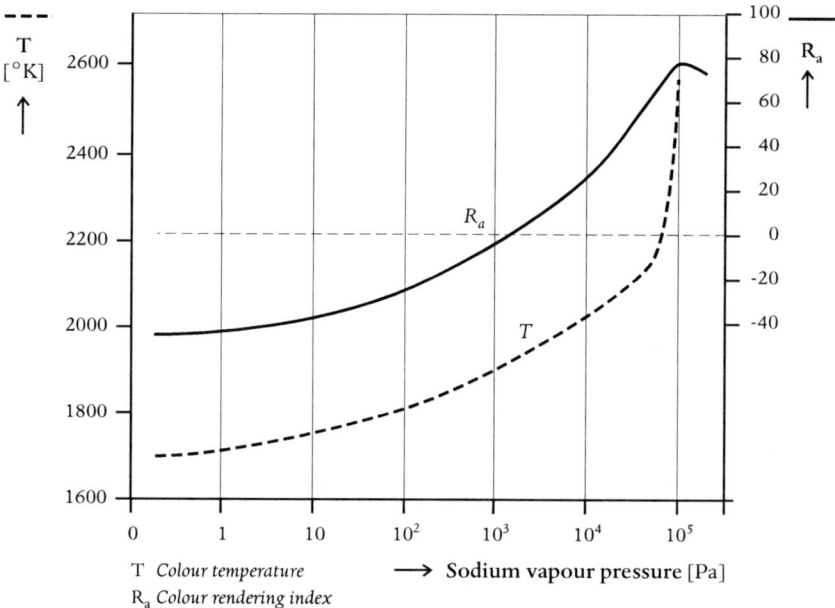

Figure 2.4.14 *The relation between the colour characteristics and the vapour pressure in a sodium lamp. T is the colour temperature in K; R_a is the colour rendering index (After Anon., 1993, Fig. 1.82).*

2.5 Semiconductor light

2.5.1 The physics of semiconductors

(a) *Intrinsic and extrinsic semiconductors*

Crystalline materials can be classified according to their resistivity. Electric conductors have a resistivity of up to 10^{-10} ohm · cm, whereas electric insulators have a resistivity of up to 10^{22} ohm · cm (Kittel, 1986, p. 159). Between those, there is a wide range of crystals that are usually called semiconductors, having typically a resistivity between 10^{-2} and 10^9 ohm · cm. Traditional semiconductors are germanium, silicon, and gallium arsenide. As a note on terminology: compound materials with a chemical formula AB, where A is a trivalent element and B is a pentavalent elements are written as III-V (three-five) components (Kittel, 1986, p. 193; Illingworth, ed., 1991, p. 427-429).

When there are no impurities, the material is called an intrinsic semiconductor (Illingworth, ed., 1991, p. 427). Semiconductors are more than just poor electric conductors or poor electric insulators. They show a very special character, which makes them the most important material of modern time. We will briefly explain why, and why in some cases these materials may act as light sources, and in other cases as amplifiers, rectifiers, or as photodetectors. Some aspects will be discussed in other parts of this book. In the explanations we will make use of the thorough, be it rather dated, treatment of Kittel (1986) and of the surveys of Illingworth, ed. (1991), Heinz & Wachtmann (2001, 2002), Heinz (2006), Schreuder (2004, 2005), and Stath (2006).

As is explained in preceding parts of this chapter, where gas discharges and fluorescence have been discussed, atoms can, in a very schematic way, be described as a system with a nucleus with a positive electric charge, surrounded by a number of electrons with negative electric charge that occupy a number of shells. Models of this kind are commonly known as the atom model of Bohr (Illingworth, ed., 1991, p. 41). Because the number of electrons equals the number of protons, in an equilibrium state, atoms are electrically neutral (Illingworth, ed., 1991, p. 24-25). Each shell contains a number of orbits, where, in an equilibrium state, the electrons may stay. Gas discharges and fluorescence, that are discussed in earlier parts of this chapter, depend on the way that electrons may jump from one shell to another.

Earlier it is explained that metals are a special type of material. They are characterized by the fact that their electron number is such that almost all fit into a restricted number of shells; almost, but not all. One or more electrons are 'left over'; they have to move into a shell with a higher energy level. When such atoms are grouped into a pattern , like e.g. a lattice or a crystal, the following happens. Those extra electrons can move freely, almost as a fluid or a gas, amongst the lattice of ions that represent the rest of the atoms when an external electric field is applied. This is called the free Fermi gas (Kittel, 1986, chapter 6). A more detailed consideration shows that the electrons in crystals are arranged in energy bands, separated by gaps for which no wavelike electron orbits exist (Kittel, 1986, p. 159). If one band is completely full, and the next higher – as regards energy levels – is completely empty, there are no free electrons. This means that there is not a possibility for an electric current to flow when an outside electric field is applied; the material is an insulator for electricity. Usually, the lower band is called the valence

Physical aspects of light production

band and the higher the conductivity band (Illingworth, ed., 1991, p. 152-154; Heinz, 2006a, Fig. 6.9, p. 75). When one band is partly filled – say between 10 and 90 percent – with electrons, a free Fermi gas can develop; such materials are electric conductors. However, when one band is almost completely filled or almost completely empty, the material becomes – at room temperature – a semiconductor (Kittel, 1986, p. 159).

At a given temperature a specific number of electrons will be thermally excited into the conductivity band. They leave behind an equal number of vacant states in the valence band. When an external electric field is applied, this will cause conduction both in the conduction band and in the valence band. The electrons in the valence band move to occupy the adjacent vacancy. The net effect is that the vacancy moves through the material as if it were a positive charge. The vacancies are known as holes. They are treated as carriers of a positive charge (Illingworth, ed., 1991, p. 427).

The characteristics of the semiconductors at different temperatures depends on the width of the 'forbidden' band between the valence band and the conductivity band. At absolute zero temperature (0 K), the semiconductor acts as an insulator. At a certain temperature the thermal motion of the electrons may allow for some of them to jump the gap, causing some 'sort of' conductivity, as is explained above. When the temperature rises, more electrons may cross the gap, until at the end the material is almost an electric conductor (Kittel, 1986, p. 184).

The picture changes completely when impurities, in very small quantities, are added to the semiconductor material. Usually, this is called doping (Illingworth, ed., 1991, p. 124). The material is called an extrinsic semiconductor (Illingworth, ed., 1991, p. 427).

Some elements may fit, because of their size, reasonably well in the lattice, but they may have a larger or a smaller number of electrons. This would mean that there is a surplus of electrons or, contrary to this, a lack of electrons. In semiconductor language, there is a surplus of 'holes', that is, holes, where an electron would fit in. In the first case, one speaks of an 'n-type' semiconductor, in the second case of a 'p-type' semiconductor (Breuer, 1994, p. 317). Where the p-type semiconductor and the n-type semiconductor meet, one speaks of a p-n junction. The operation of semiconductor diodes and of transistors is based on these phenomena.

(b) Semiconductor diodes

A diode acts basically as a rectifier. Electric current can pass only in one direction. It has only two electrodes – hence the name diode. Almost all modern diodes are semiconductor diodes (Illingworth, ed., 1991, p. 118). In a semiconductor diode, two complementary effects may be notes. The first is a photo-electric effect. When the diode is hit by electromagnetic radiation, electron-hole combinations are formed that may transport energy, but not electric charge, as they are electrically neutral (Kittel, 1986, p. 296-297). This allows to use them as photo-elements, either as a switching device or as a measuring device (Hentschel, ed., 2002, Fig. 4.14, p. 84). When many individual diodes are brought together in one device, and when they are connected in the appropriate way, so that packages of charge can move in a controlled way, they form a Charge-Coupled Device or CCD (Illingworth, ed., 1991, p. 57). Modern photography, both amateur and professional, is hardly possible without CCDs (Narisada & Schreuder, 2004, sec. 14.5.1). Almost every mobile telephone has a built-in CCD photo and video camera.

Another extremely important application of this effect is to use semiconductor diodes as a source of electric energy. By combining diodes in a large array, a considerable electric power (wattage) can be generated when the device is subjected to direct sunlight. This is of course a solar panel (Green, 1996, Schreuder, 1998a, Würfel, 1995, Zilles et al., 2000).

The second is the opposite effect: the diode will emit light, when a voltage is applied over the barrier layer in such a way that a current flows only in the forward direction (Durgin, 1996). By doping with materials that have a valence that differs from that of the carrier material, usually germanium or silicon, excess electron-hole pairs will be generated. When these excess electron-hole pairs recombine in such a way that an electron in the conduction band recombines with a hole in the valence band, a photon is emitted. This is the principle of the LED that is discussed further on.

The process is often called photoluminescence (Stath, 2006, p. 2), As mentioned earlier, at zero kelvin, all electrons are in the valence band. When hit by high-energy radiation, some of them will be transferred into the conduction band, leaving behind holes in the valence band. After a very short time the electrons and the holes will recombine again, emitting a light quantum with the same energy. See Figure 2.5.1.

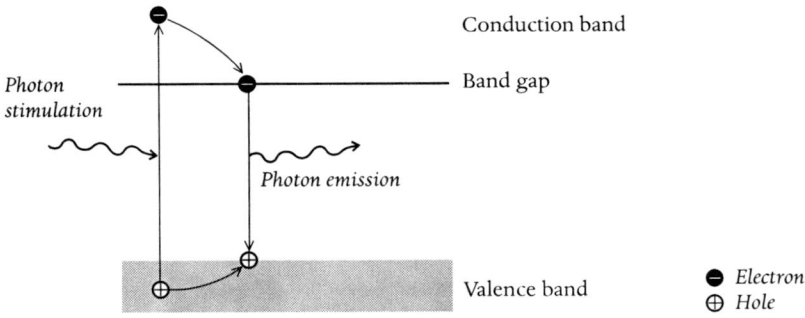

Figure 2.5.1 Photoluminescence in a semiconductor (After Stath, 2006, Fig. 3, p. 2).

Physical aspects of light production

(c) *Semiconductor light-emitting diodes*

The process that is depicted on Figure 2.5.1 must be extended to make a semiconductor light source or electro-luminescence. The device must be transformed by appropriate doping into a p-n junction where the charge carriers must be excited by the incident light. In an n-doped semiconductor there are, in energy terms, donors just below the conduction band. In a p-doped semiconductor there are acceptors, just above the valence band. See Figure 2.5.2.

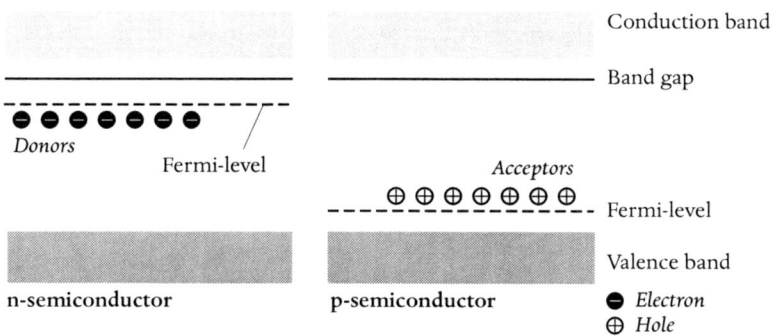

Figure 2.5.2 The bands in an n- and in a p-semiconductor (After Stath, 2006, Fig. 4, p. 2).

When the two are brought together as in a p-n junction, the surplus electrons will travel from the n-side towards the p-side. The holes will travel from the p-side towards the n-side. Two barrier layers will form, one at each side of the p-n junction. At the n-side there is a surplus of holes. Consequently, the n-side has a positive electric charge. At the p-side there is a surplus of electrons. Consequently, the p-side has a negative electric charge.

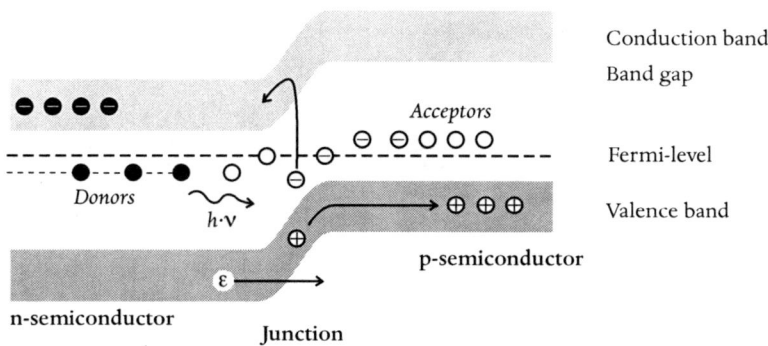

Figure 2.5.2a A p-n-junction (After Hentschel, ed., 2004, Fig. 4.14, p. 84).

In a state of thermal equilibrium, the Fermi-level of both materials must be equal. This would imply a vertical shift of the two parts of Figure 2.5.2. This is depicted in Figure 2.5.2a. This leads to a shift in the energy bands at the p-n junction. This shift acts as a threshold of the electric potential. This threshold must be conquered before the electrons and the holes have a chance to recombine. As a rule, without an external voltage, this threshold is too high, so nothing much will happen. Things change dramatically, however, when an external voltage is applied. The threshold of the electric potential is reduced, so that the electrons and the holes will recombine. When these excess electron-hole pairs recombine in such a way that an electron in the conduction band recombines with a hole in the valence band, a photon is emitted. A light source is created. This type of light source is called a Light Emitting Diode or LED. The underlying processes are described in more detail in Heinz & Wachtmann (2001, 2002), Heinz (2006a), and Stath (2006).

2.5.2 Anorganic LEDs

(a) *The construction of anorganic LEDs*

There are several types of LED. Traditionally, LEDs consist of anorganic materials. They are called anorganic LEDs. These LEDs will be discussed in this section. In the sections to follow, other types of LED are discussed, such as organic LEDs or OLEDs.

Almost all anorganic LEDs consist of a substrate of III-V compound semiconductor materials, following the designation that is given in sec. 2.5.1a (Stath, 2006, p. 1). On the substrate several extremely thin layers are deposited. The process is called epitaxy. Thin layers are made to 'grow' on the substrate in such a way that their crystal lattice structure is identical to that of the substrate (Illingworth, ed., 1991, p. 156). The function of these layers has been explained in the preceding parts of this section. In Figure 2.5.3 a simple sketch is given that shows the structure of an anorganic LED.

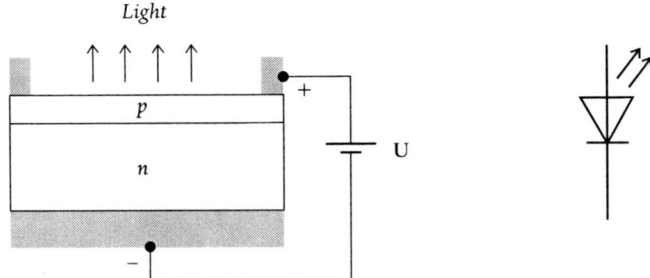

Figure 2.5.3 *A sketch of the structure of an anorganic LED (After Van Ooyen, 2005, Fig. 1.4-94).*

The substrate is attached to a heat sink, usually a layer of copper.

Physical aspects of light production 71

(b) The colour of anorganic LEDs

The colour that is emitted by an anorganic LED depends primarily on the semiconductor material of the substrate, and on the doping materials. Most common LEDs are based on Gallium Phosphide (GaP) and Gallium Arsenide Phosphide (GaAsP). Depending on the doping materials, their colours may reach from green to deep red. Gallium Nitride (GaN) may give purple and blue light (Van Ooyen 2005, table 1.5-95, p. 123; Anon., 2006a, p. 2).

Coloured LEDs are mainly used for signalling purposes. In the field of general lighting applications, into which LEDs are beginning to penetrate, white light is required. There are several ways to make white, or near-white, light with LEDs. The three most commonly in use are:

1. The use of separate LEDs of different colours. Usually, red and green and blue LEDs are used. They are clustered closely together in packages of one colour each. When viewed from a certain distance of some tens of centimeters, the array looks white. The colour impression can be changed during operation by adjusting the output of the LEDs of the individual colours. This is widely in use in decorative lighting when changing colour effects are desired. An example is depicted in Figure 2.5.4. It shows the peaks of the three LEDs in the blue, the yellow-green and the red. The band width of each colour is about 24-27 nm.
2. The use a blue-emitting GaN-based LEDs and add an appropriate phosphor that converts part of the blue light into light of longer wavelengths (Van Ooyen 2005, p. 123; Anon., 2006a, p. 3,4). An example is depicted in Figure 2.5.5. It shows the peak in the blue at about 465 nm from the direct light of the GaN-based LED and the wider band extending from around 500 to 700 nm from the phosphor.
3. The use of a near-UV-emitting LED and add a number of different phosphors. By changing the amount of the different phosphors, the colour impression can be changed from warm-white to cool-white (Anon., 2006a, p. 4)

(c) The performance of anorganic LEDs

The light intensity is proportional to the number of excess electron-hole pairs. The useful light output depends on the quality of the crystal, and particularly of its surface, whereas the colour of the light will depend on the material used (Illingworth, ed., 1991, p. 266). Fluorescent materials may be included in the device to alter the emitted colour.

The efficiency of LEDs may be illustrated by comparing the energy flow of a signal for traffic control, equipped with LEDs or with incandescent lamps. An example is depicted in Figure 2.5.6.

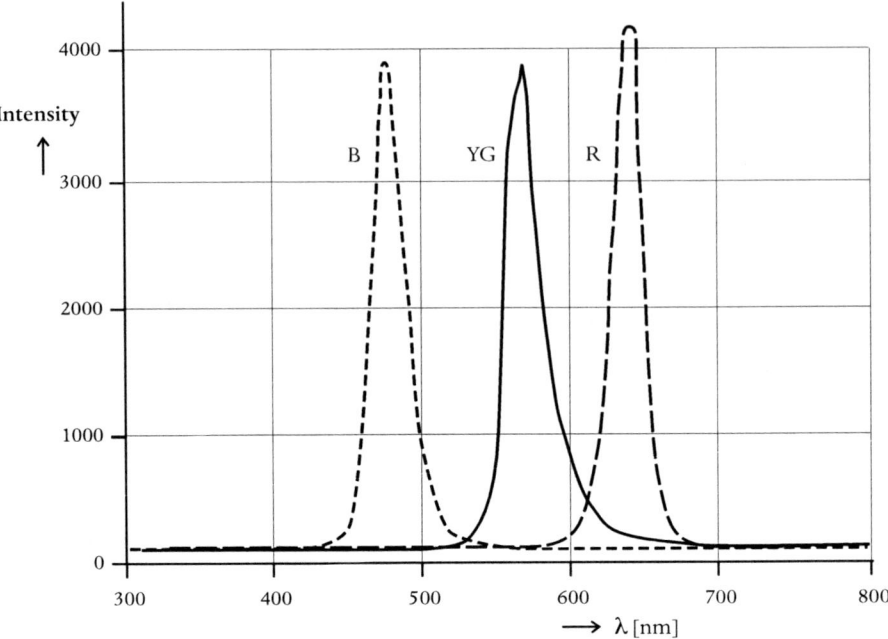

Figure 2.5.4 Spectrum of a white LED composed on three colours (Based on data from Anon., 2006a, p. 4).

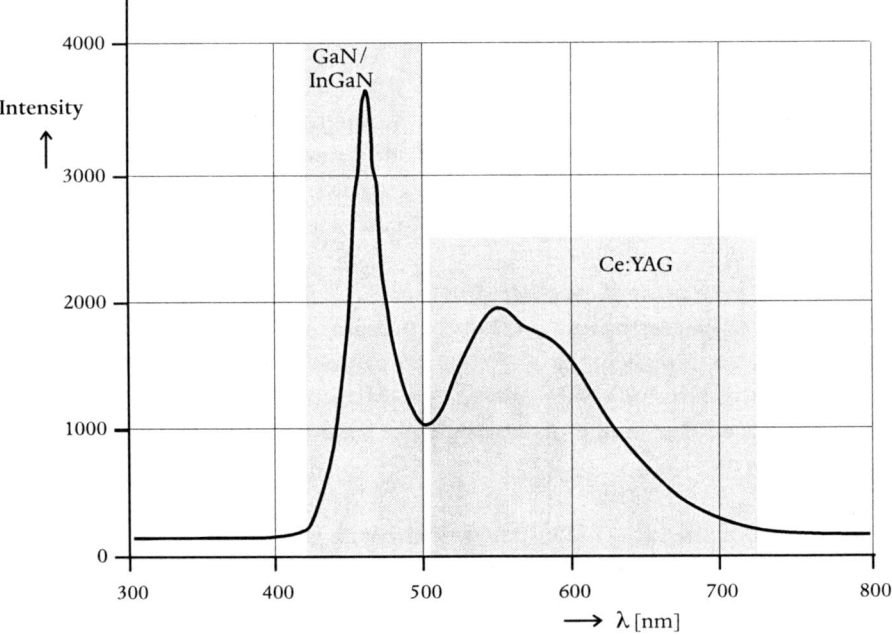

Figure 2.5.5 Spectrum of a 'white' LED (Based on data from Anon., 2006a, p. 4).

Physical aspects of light production

The technological development of LEDs has been very fast. The luminous efficiency or efficacy increased dramatically over the years as a result a.o. of the introduction of new materials. See Table 2.5.1.

Year	Efficacy (lm/W)	Chemical formula
1970	0,4	GaP
1978	2	GaAsP-N
1983	3,5	GaAlAs
1987	9	GaAlAs
1995	13	InGaAlP
2001	45	InGaAlP
2006	40	InGaN

Table 2.5.1 The efficacy of anorganic LEDs (Based on data from Stath, 2006, Fig. 1).

a) **Energy flows in LED signal [W]**

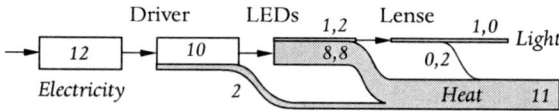

b) **Energy flows in incandescent lamp signal [W]**

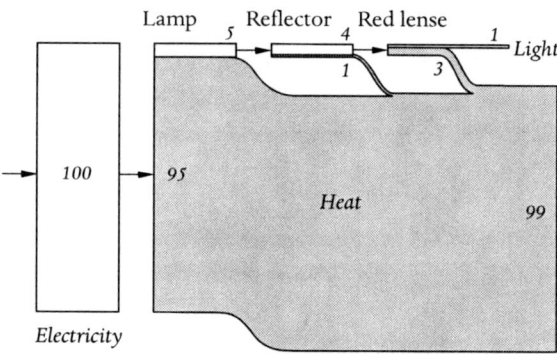

Figure 2.5.6 The energy flow in road traffic signals. (a) LED light sources (b) incandescent lamps (After Zandvliet & Van Geldermalsen, 2002, p. 4. Based on Narisada & Schreuder, 2004, Fig. 11.1.5).

The situation of about mid-2006 is indicated in Heinz (2006, 2006a), Stath (2006), and Anon. (2006a). It is to be expected, however, that these surveys will be outdated in a few years' time and that they will be only of historic interest.

(d) Use of anorganic LEDs

Until recently, LEDs were used mainly for signalling purposes. The high performance allow LEDs to be used for general lighting as well (Visser, 2000; Stath, 2006). The main advantages as compared to traditional incandescent and fluorescent lamps are (Schreuder, 2001a):

1. New developments make that the luminous efficacy is similar to that of fluorescent tubes and superior to halogen incandescent lamps. It should be noted in passing that the usual definition of luminous efficacy cannot be applied directly to LEDs. Some adaptation is needed (CIE, 1997). This point is explained in chapter 5, where photometry is discussed.
2. Their practical life is very much longer. Instead of a few thousand hours, a life span of 100 000 or 200 000 hours – 11,4 years or 23 years – is commonly quoted (Durgin, 1996, Visser, 2000). However, there is a light depreciation to be accounted for. According to the manufacturers, the relative light output is over 90% for the first 10 000 hours. It is expected to drop to about 70% at 100 000 hours (Anon., 1996; Schreuder, 1997, 1997a,b, 1999).
3. Low energy consumption. A LED array of 7 W might replace a 70 to 150 W incandescent lamp in traffic signals (Durgin, 1996; Haazebroek, 2000; Schreuder, 2006).
4. Small dimensions. LEDs measure only a few millimeters in stead of 5-15 cm for traditional lamps (Haazebroek, 2000). This makes them particularly suitable to the application of low-pollution road marking systems (Jongenotter et al., 2000). See also Bylund (2002).

Disadvantages are that the colour rendition of LEDs is sometimes inferior to that of incandescent lamps. Also, the temperature has a considerable influence on the light output and on the lamp life. These disadvantages are, however, slight when compared to the advantages.

As an example, we quote some data about a luminaire that houses 24 white high intensity LEDs. The intensity is 20 cd, the life 100 000 hours (Patlite, 2002).

Since then, the efficacy has been increased even further, as is explained in the preceding part of this section. It is suggested that 100 lm/W is attainable with white light LEDs (Stath, 2006, Figure 1, p. 1). Most manufacturers, however, are reluctant to give precise information about the performance of their products. This is a wise policy, as the performance of LEDs, contrary to most lamps, cannot be given in isolation. Their performance depends heavily on the electronic circuits that are used, on the total lighting installation, and on many external, environmental factors. Furthermore, it should be noted that LEDs usually include an optical system. All published data refer to such

Physical aspects of light production

systems, and are inclusive the optics. The LED data should not be compared to those of 'naked' lamps, but to complete lighting installations. This makes all comparisons rather difficult and even rather arbitrary, as there is no 'standard' installation for lamps.

In effect, LEDs are a half-product. The p-n junction is the actual light source. However, it is hardly ever used directly as such. LEDs that are commercially available are almost always provided with an optical system – often of dubious optical quality. Additionally, fluorescent materials are often added, particularly in 'white' LEDs. Then, in many cases, the LEDs are not provided as individual units, but are assembled in groups, clusters etc. More important, a LED requires a source of electric energy, like e.g a battery of a solar panel. Furthermore, some electronic circuitry is needed for the LED to work properly. The better the required performance of the LED, the more sophistication is needed for the circuitry. Therefore, it is almost impossible to give product information about the LED as such – apart from the fact that most manufacturers are very secretive about the characteristics of their products. And finally, the technical and commercial development of LEDs is going very fast indeed. Any technical specification or price indication given at a certain moment, will be obsolete in a very short time.

The photometry of LEDs is more complicated than that or 'ordinary' lamps (CIE, 1997). First, LEDs usually include an optical system. 'Naked' LEDs are not used. As mentioned before, all published data refer to such systems, and are inclusive the optics. The LED data should not be compared to those of 'naked' lamps, but to complete lighting installations. This makes all comparisons rather difficult and even rather arbitrary, as there is no 'standard' installation for lamps. As an example, Begemann (2005) assumed that 85% of the light emitted by a LED installation may be 'useful' light, whereas some 30% to 50% of the light of an ordinary lamp is useful. The rest are losses in the external optical system etc. For a light source efficacy of 50 lm/W for LEDs and of 120 lm/W for ordinary lamps – in both cases, high-efficiency sources -, the system efficiency would be 42,5% for the LEDs and 36 to 60% for the ordinary lamps. As a rule of thumb, one may assume that the system efficiency of installations with LEDs and with ordinary lamps are directly comparable.

2.5.3   Organic LEDs

In the preceding part of this section, anorganic LEDs have been discussed. They are called so because their base is an anorganic semiconducting substrate. Since about 1990, a new member is added to the family of semiconductor, or solid-state, light sources: the Organic Light Emitting Diode, or OLED. Their basis is not a rigid single crystal of semiconductor material, but a semiconducting polymer. A major advantage is that OLEDs can be made flexible, so they can be made in sheets of varying size and shape. As some of them, when not in operation, can be made almost completely transparent, their field of application in illuminating engineering is very wide (Visser, 2005; Heinz, 2006).

Polymers belong to what is commonly called 'plastic', and as such are studied in the branch of organic chemistry, hence their name. This does not imply that the material itself comes directly from living organisms; polymers are usually a by-product of the oil industry. The polymer in use is often a PPV (Poly p-Phenylene Vinylene Polymer; Visser, 2005). An alternative name of an OLED is a Light Emitting Polymer or LEP. We will use the more commonly used name of OLED.

The semiconducting polymer is placed between two layers. At the front a layer of transparent Indium-Tin-Oxide and a glass cover, at the rear a metal. An OLED consists of a great number of pixels that each may emit red or green or blue light: the RGB-pixels. A more detailed picture is given in Figure 2.5.7.

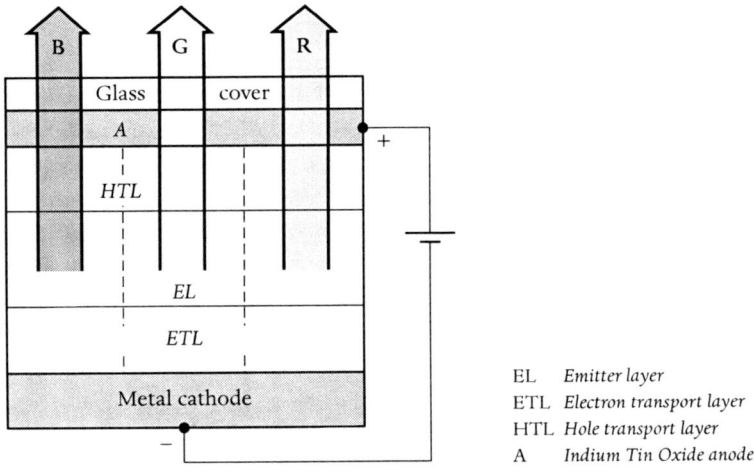

Figure 2.5.7 Schematic structure of one OLED RGB pixel (After Heinz, 2006, Fig. 1).

When an external voltage is applied, electrons will be ejected from the cathode into the electron transport layer, and holes from the anode into the hole transport layer, in analogy to the processes that have been explained earlier in this section.

The electrons and the holes get into a state of excitation. When they fall back to the ground level, a light quantum is emitted, analogous to the electro-luminescent processes that take place in an anorganic LED, such as is explained in the preceding part of this

Physical aspects of light production

section. The energy of the light quantum, and thus the colour of the emitted light, depends on the energy difference between the excited and the ground levels. It can be influenced by the selection of the material in the emitter layer. When semiconducting polymers are used, the product is called a p-OLED. Also, small molecules can be used in the polymers. Those products are called sm-OLEDs. Of those, the sm-OLEDs seem to be the most promising. In some modern methods of OLED-production, fluorescent materials are included. The efficiency of the light generation will be higher, because fluorescence is added to the electro-luminescence processes. Details about these different aspects are given in Heinz (2006, p. 3-5; 2006a, chapter 6).

For general lighting applications the light must be white or at least near-white. In principle both p-OLEDs and sm-OLEDs can be used. As is shown in Figure 2.5.8, the emitter layer must be quite thin, not more than about 100 nm. This makes it very hard to produce units that are larger than a few mm across.

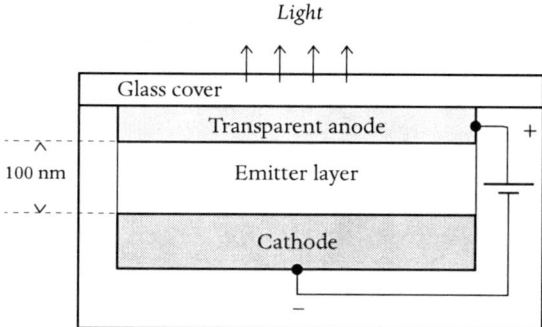

Figure 2.5.8 Schematic structure of an OLED for general lighting purposes (After Heinz, 2006, Fig. 9).

The emission spectrum of OLEDs shows wider bands than that of a LED, as is depicted in Figure 2.5.9.

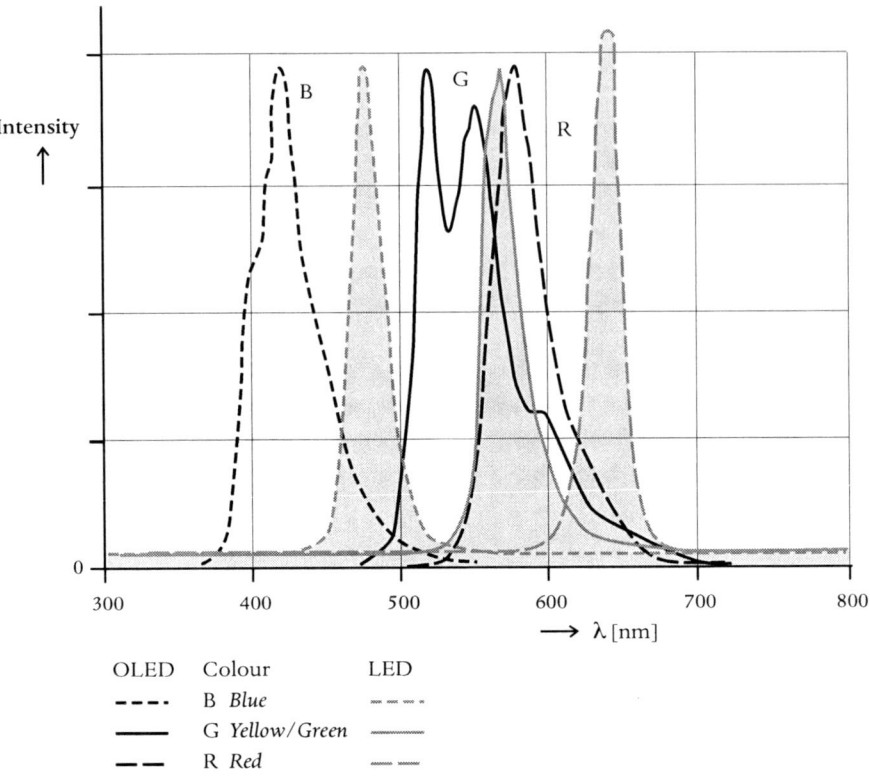

Figure 2.5.9 The emission spectra of OLEDs and LEDs (After Heinz, 2006, Fig. 8).

2.6 Conclusions

Light can be understood as the aspect of radiant energy which an observer perceives through visual sensation, but also as a physical phenomenon. For this, four different models may be used: Light as a collection of light rays; light is an electromagnetic wave; light as a stream of photons; light as fluid of power. Each model has its own use and its own applications. Light rays are straight and infinite narrow, and do not show any smaller detail. They are the basis of geometric optics. Electromagnetic waves and streams of photons appear in duplicity of light. Waves are essential to describe diffraction phenomena. The corpuscular theories of light are essential to describe photo-electric phenomena.

As regards the generation of light, three principles are useful in light applications:
1. Incandence – the basis for incandescent lamps;

Physical aspects of light production

2. Recombination of electrons and ions in a plasma – the basis for gas-discharge lamps;
3. Recombination of electrons and holes in a semiconductor – the basis for semiconductor lamps like LEDs.

Incandescent lamps are based on the laws of black-body radiation. At all temperatures, any body will emit electro-magnetic radiation, the characteristics of which depend exclusively on the temperature of the body. First the Law of Planck. That contains all other laws of radiation, a.o. Wien's displacement law that describes the wavelength at which the maximum energy is emitted by a black body at a temperature and Stefan-Boltzmann's Law that proves directly that incandescent lamps never can be 'efficient' light sources. Their visible radiation is usually around 5% of the consumed wattage. To improve the efficacy of the incandescent lamps without reducing the life of the lamps, the halogen incandescent lamp was introduced. Incandescent lamps have some advantages. They show a continuous spectrum and have an excellent colour rendition. They are very versatile. They can be connected directly to the electricity grid. They are very cheap.

There are four main families of gas-discharge lamps:
1. Low-pressure mercury lamps;
2. Low-pressure sodium lamps;
3. High-pressure mercury lamps;
4. High-pressure sodium lamps.

All have their own area of application. For interior lighting, low-pressure mercury lamps or fluorescent tubes are the most common. Road lighting is the area of sodium lamps, particularly the high-pressure sodium lamps. High-pressure mercury lamps, particularly metal halide lamps are widely used in general outdoor lighting.

More recently semiconductor light, particularly LEDs and OLEDs have become very important, mainly in signalling. It is to be expected that they will be also important in general lighting in the near future, their main advantages over other lamp types being the very long life and the low energy consumption.

References

Akutsu, H.; Watarai, Y.; Saito, N. & Mizuno, H. (1984). A new high-pressure sodium lamp with high colour acceptability. Journal of IS (1984) July, p. 341-349.

Anon. (1979). 16. Generalkonferenz für Mass und Gewicht. Paris 1979 (16th General Conference on Measures and Weights; cit.: Hentschel, 1994, p. 37).

Anon. (1986). U.V.-straling; Blootstelling van de mens aan ultraviolette straling (UV Radiation; Exposing human beings to ultraviolet radiation). 's Gravenhage, Gezondsheidsraad, 1986.

Anon. (1993). Lighting manual. Fifth edition. LIDAC. Eindhoven, Philips, 1993.
Anon. (1993a). The comprehensive lighting catalogue. Edition 3. Borehamwood, Herts., Thorn Lighting Limited, 1993.
Anon. (1996). Sending New Signals. CMO 08, Hewlett Packard, 1996.
Anon. (1997). Philips lichtcatalogus 1997/1998 (Philips lighting catalogue 1997/1998). Eindhoven, Philips Lighting, 1997.
Anon. (2001). Luxjunior. 21 – 23 September 2001, Dörnfeld/Ilmenau. Proceedings. Ilmenau, University, 2001.
Anon. (2002). Right Light 5. 5th conference on energy-efficient lighting, 29-31 May 2002 in Nice, France. Proceedings. Nice, 2002.
Anon. (2003). Lamp catalogue, Iwasaki Denki, Tokyo, 2003.
Anon. (2005). Handboek verlichtingstechniek (Lighting engineering handbook). Loose-leaf edition, 2005 issue. Deventer, Ten Hagen Stam; Den Haag, SDU, 2005.
Anon. (2006). Fluorescent lamps. Wikipedia, internet, 20 December 2006.
Anon. (2006a). Light-emitting diode. Wikipedia, internet, 24 December 2006.
Anon. (2006b). Licht 2006, 17. Lichttechnische Gemeinschafttagung; Bern, September 2006.
Anon. (2007). Lampen; Altijd een oplossing (Lamps, Always a solution). Eindhoven, Van Gansewinkel, product information. Internet, 4 January 2007.
Baer, R. (1990). Beleuchtungstechnik; Grundlagen (Essentials of illuminating engineering). Berlin, VEB Verlag Technik, 1990.
Baer, R., ed. (2006). Beleuchtungstechnik; Grundlagen. 3., vollständig überarteite Auflage (Essentials of illuminating engineering, 3rd., completely new edition). Berlin, Huss-Media, GmbH, 2006.
Barrows, W.E. (1938). Light, photometry and illuminating engineering. New York, McGraw-Hill Book Company, Inc., 1938.
Begemann, T. (2005). Gaat er een lichtje branden? (Is a light beginning to burn?). In: LED 2005, conference, 8 December 2005, Maarssen, the Netherlands, 2005.
Blüh, O. & Elder, J.D. (1955). Principles and applications of physics. Edinburg, Oliver & Boyd, 1955.
Breuer, H. (1994). DTV-Atlas zur Physik. Zwei Bänder, 4.Auflage (DTV atlas on physics, two volumes, 4th edition). München, DTV Verlag, 1994.
Bylund (2002). Energy efficient LED lighting solutions for reduced cost and improved road safety. In: Anon., 2002, p. 15-19.
CIE (1997). Measurement of LED's. Publication No. 127. Vienna, CIE, 1997.
Clugston, M.J., ed. (1998). The new Penguin dictionary of science. London, Penguin books, 1998.
Daintith, J. & Nelson, R.D. (1989). The Penguin Dictionary of mathematics. London. Penguin Books, 1989.
De Groot, J.J. & Van Vliet, J.A.J.M. (1986). The high-pressure sodium lamp. Deventer, Kluwer, 1986.
Duchêne A.S.; Lakey, J.R.A. & Repacholi, M.H., eds. (1991). IRPA-guidelines on protection against non-ionizing radiation. New York, Pergamon Press, 1991.
Durant, W. (1962). The story of philosophy; The lives and opinions of the greater philosophers. Third paperback printing. New York, Simon and Schuster, 1962.
Durgin, G, (1996). Precision lensing. Traffic Technology International '96. New York, Dialight, 1996.
Elenbaas, W., ed. (1959). Fluorescent lamps and lighting. Eindhoven, Philips Technical Library, 1959.
Elenbaas, W., ed. (1965). High pressure mercury vapour lamps and their applications. Eindhoven, Philips Technical Library, 1965.
Einstein, A. & Infeld, L. (1960). The evolution of physics. New edition. New York, Simon & Schuster, Inc., 1960.
Feynman, R. (1990). The character of physical law. Sixteenth printing. Cambridge, Mass. The M.I.T. Press, 1990.

Feynman. R.P.; Leighton, R.B. & Sands, M. (1977). The Feynman lectures on physics. Three volumes. 1963; 6th printing 1977. Reading (Mass.). Addison-Wesley Publishing Company. 1977.

Gladhill, D. (1981). Gas lighting. Shire albums 65. Aylesbury, Shire Publication Ltd., 1981.

Green, M. A. (1986). Solar cells; Operating principles, technology and system applications. University of New South Wales, 1986.

Gribbin, J. (1984). In search of Schrödinger's cat. London, Bantam, 1984.

Haazebroek, N. (2000). Ontwikkelingen op het gebied van LED lantaarns (Developments regarding LED lanterns). In: NSVV, 2000, p. 70-81.

Heinz, R. (2006). Lichterzeugung mit organischen Werkstoffen; OLEDs für Displays und Allgemeinbeleuchtung (Light generation with organic materials: OLEDs for displays and general lighting). In: Anon., 2006b.

Heinz, R. (2006a). Grundlagen der Lichterzeugung, 2. Auflage (Principles of light generation, 2nd edition). Rüthem, Highlight Verlag, 2006.

Heinz, R. & Wachtmann, K. (2001). Innovative Lichtquellen durch LED-Technologie (Innovative light sources by means of LED-technology). In: Anon., 2001, p. 199-207.

Heinz, R. & Wachtmann, K. (2002). LED-Leuchtmittel: moderne Halbleiterstrahlungsquellen im Visier. (LED light sources: looking at modern semiconductor light sources). In: Welk, ed., 2002, p. 34-40.

Hentschel, H.-J. ed. (2002). Licht und Beleuchtung; Grundlagen und Anwendungen der Lichttechnik; 5. neu bearbeitete und extended Auflage (Light and illumination; Theory and applications of lighting engineering; 5th new and extended edition). Heidelberg, Hüthig, 2002.

Hütte (1919). Des Ingenieures Taschenbuch (The manual for engineers). Berlin, Wilhelm Ernst und Sohn, 1919.

Illingworth, V., ed. (1991). The Penguin Dictionary of Physics (second edition). London, Penguin Books, 1991.

Jongenotter, E.; Buijn, H.R.; Rutte, P.J. & Schreuder, D.A. (2000). Nieuwe richting voor wegverlichting (New directions in road lighting). Verkeerskunde, 51 (2000) no 1, January, p. 32-36.

Joos, G. (1947). Theoretical physics (First edition in German 1932). London, Blackie & Sons Limited, 1947.

Keitz, H.A.E. (1967). Lichtmessungen und Lichtberechnungen. 2e. Auflage (The measurement and calculation of light. Second edition). Eindhoven, Philips Technische Bibliotheek, 1967.

Kittel, C. (1986) Introduction to solid state physics. Sixth edition. New York. John Wiley & Sons, 1986.

Kuchling, H. (1995). Taschenbuch der Physik, 15. Auflage (Survey of physics, 15th edition). Leipzig-Köln, Fachbuchverlag, 1995.

Lafferty, P. & Rowe, J, eds. (1994). Dictionary of science. London, Brockhampton Press, 1994.

Longhurst, R.S. (1964). Geometrical and physical optics (fifth impression). London, Longmans, 1964.

Lusche, D. (2001). Licht und Gesundheit (Light and health). VITT, IN 1/2001.

Meyer, Chr. & Nienhuis, H. (1988). Discharge lamps. Philips Technical Library. Deventer, Kluwer, 1988.

Mills, E. (1999). Fuel-based light: Large CO_2 source. IAEEL Newsletter 8 (1999) no 2, p. 2-9.

Minnaert, M. (1942). De natuurkunde van 't vrije veld, derde druk (The physics of the open air, third edition). Zutphen, Thieme, 1942.

Moon, P. (1961). The scientific basis of illuminating engineering (revised edition). New York, Dover Publications, Inc., 1961.

Narisada, K. & Schreuder, D.A. (2004). Light pollution handbook. Dordrecht, Springer, 2004.

Nichia Company (2000). White LED data (year estimated).

NSVV (2000). Het Nationale Lichtcongres 2000 (The National Light Conference 2000). Arnhem, NSVV, 2000.

NSVV (2003). Licht en gezondheid voor werkenden; Aanbeveling (Light and health for workers; Recommendation). Arnhem, NSVV, 2003.

NSVV (2003a). Het Nationale Lichtcongres. Ede, 12 november 2003; Syllabus (The National Light Conference. Ede, 12 November 2003; Proceedings). Arnhem, NSVV, 2003.

Oranje, P.J. (1942). Gasontladingslampen (Gas-discharge lamps). Amsterdam, Meulenhoff, 1942.

Patlite (2002). Sasaki Information sheet. Catalogue no. 157. Osaka, 2002 (year estimated).

Prins, J.A. (1945). Grondbeginselen van de hedendaagse natuurkunde, vierde druk (Fundaments of modern physics, 4th edition). Groningen, J.B. Wolters, 1945.

Ris, H.R. (1992). Beleuchtungstechnik für Praktiker (Practical illuminating engineering). Berlin, Offenbach, VDE-Verlag GmbH, 1992.

Schoon, C.C. & Schreuder, D.A. (1993). HID car headlights and road safety; A state-of-the-art report on high-pressure gas-discharge lamps with an examination of the application of UV radiation and polarised light. R-93-70. Leidschendam, SWOV, 1993.

Schreuder, D.A. (1997). The functional characteristics of road and tunnel lighting. Paper presented to the Israel National Committee on Illumination on Tuesday, 25 March 1997 at the Association of Engineers and Architects in Tel Aviv. Leidschendam, Duco Schreuder Consultancies, 1997.

Schreuder, D.A. (1997a). Visibility aspects of matrix signals. Note for Technion, Haifa, Israel. Leidschendam, Duco Schreuder Consultancies, 1997.

Schreuder, D.A. (1997b). A comparison between fiber optics (FO) and Light Emitting Diodes (LED) for variable message signs. Note for discussion at Netivei Ayalon on Monday, 24 March 1997 in Tel Aviv, Israel. Leidschendam, Duco Schreuder Consultancies, 1997.

Schreuder, D.A. (1998). Road lighting for safety. London, Thomas Telford, 1998. (Translation of: Schreuder, D.A., Openbare verlichting voor verkeer en veiligheid. Deventer, Kluwer Techniek, 1996).

Schreuder, D.A. (1998a). Functie en markt van autonome photo-voltaische openbare verlichting. Studie verricht voor Ecofys (Function and market for autonomous photo-voltaic public lighting. A study made for Ecofys). Leidschendam, Duco Schreuder Consultancies, 1998.

Schreuder, D.A. (1999). De invloed van vervuiling op de lichtsterkte van verkeerslantaarns; Een overzicht van de literatuur (The influence of dirt on the light intensity of traffic lights; A survey of the literature). Leidschendam, Duco Schreuder Consultancies, 1999.

Schreuder, D.A. (2001). Energy efficient domestic lighting for developing countries. Paper prepared for presentation at The "International Conference on Lighting Efficiency: Higher performance at Lower Costs" to be held on 19-21 January, 2001 in Dhaka, Bangladesh, organised by the Illumination Society of Bangladesh. Leidschendam, Duco Schreuder Consultancies, 2001.

Schreuder, D.A. (2001a). Principles of Cityscape Lighting applied to Europe and Asia. Paper presented at International Lightscape Conference ICiL 2001, 13 – 14 November 2001, Shanghai, P.R. China. Leidschendam, Duco Schreuder Consultancies, 2001.

Schreuder, D.A. (2004). Verlichting thuis voor de allerarmsten (Home lighting for the very poor). NSVV Nationaal Lichtcongres 11 november 2004. Arnhem, NSVV, 2004. Leidschendam, Duco Schreuder Consultancies, 2004.

Schreuder, D.A. (2005). Domestic lighting for developing countries. Prepared for publication in: UNESCO – A world of science, Paris, France. Leidschendam, Duco Schreuder Consultancies, 2005.

Schreuder, D.A. (2006). De zichtbaarheid van verkeerslichten uitgerust met Light Emitting Diodes; Interimrapport, 1 december 2006 (The visibility of traffic lights with LEDs; Interim Report, 1 December 2006). Leidschendam, Duco Schreuder Consultancies, 2006 (Not published).

Stath, N, (2006). Anorganische LEDs; Innovationen bei Halbleiter-Lichtquellen (Anorganic LEDs; Innovations in semiconductor light sources). In: Anon., 2006b.

Stanford (2006). Specification of mixed trichromatic phosphors for fluorescent lamps: LPTB 28; LPTB 50. Internet, 20 December 2006.

Sterken, C. & Manfroid, J. (1992). Astronomical photometry. Dordrecht, Kluwer, 1992.

Van den Beld, G. (2003). Aanbevelingen en aandachtspunten voor gezonde verlichting (Recommendations and points of interest of healthy lighting). In NSVV, 2003a, p. 66-72.

Van Heel, A.C.S. (1950). Inleiding in de optica; derde druk (Introduction into optics; third edition). Den Haag, Martinus Nijhoff, 1950.

Van Ooyen, M.H.F. (2005). Lichtbronnen en hulpapparatuur, herziene uitgave (Light sources and supporting gear, revised edition). Section 1.5. In: Anon., 2005.

Vermeulen, J. (2000). Personal communication.

Visser, R. (2000). LED's. p. 20 -29 in: Het Nationale Lichtcongres 2000. Arnhem, NSVV, 2000.

Visser, R. (2005). LEP's of OLED's (LEPs or OLEDs). Sec. 1.5.6.5. In: Anon., 2005.

Von Weizsäcker, C. F. (2006). The structure of Physics (Edited, revised and enlarged by Görnitz, T., and Lyre, H.). Dordrecht, Springer, 2006.

Vos, J.J.; Walraven, J. & Van Meeteren, A. (1976). Light profiles of the foveal image of a point source. Vision Research, 16 (1976) 215-219.

Wachter, A. & Hoeber, H. (2006). Compendium of theoretical physics (Translated from the German edition). New York, Springer Science+Business Media, Inc., 2006.

Welk, R., ed. (2002). Lichtlösung mit Leuchtdioden (Lighting solutions with light diodes). Licht Special 3. München, Richard Pflaum Verlag, 2002.

Worthing (1926). Physical Review, 28 (1926) 174.

Würfel, P. (1995). Physik der Solarzellen (Physics of solar cells). Heidelberg, Spektrum Akademischer Verlag, 1995.

Zandvliet, P. & Van Geldermalsen (2002). Seeing the light; LED^2 signal lamps, the new standard? Rijkswaterstaat, Information sheet, Case study (Year estimated).

Zilles, R.; Lorenzo, E. & Serpa, P. (2000). From candles to photovoltaics: A four-year experience at Iguape-Cananeia, Brazil. Progress in Photovoltaics: Research and Applications. 8 (2000) 17 August, issue 4, p. 421-434.

3 Radiometry and photometry

Photometry is a branch of the wider field of radiometry. One might call radiometry the measurement of electromagnetic radiation independent of the detector that is used in the measurements, whereas photometry refers to the measurement of electromagnetic radiation when using the human visual system as a detector.

Photometry is essential, both in visual science and in lighting engineering. To deal with matters of lighting and vision in a scientific way, the first thing is to quantify light. For this purpose, the quantities of and the units of the light are strictly and precisely defined and described in the International Lighting Vocabulary that is published by CIE. This vocabulary is based on the ISO-Standards of Weights and Measures.

When light strikes the eye, a sensation of light is provoked. The intensity of the light stimulus is called the photometric quantity, and the sensation produced by the stimulus is called brightness. The energy of electric radiation can be expressed in watts. However, only the wavelength range between about 400 nm and 800 nm can produce a sensation of light and colour.

Fundamentally, photometry is nothing else but photon counting. Each photon is counted, and weighted according to the photopic spectral luminous efficiency curve. In this chapter the different photometric quantities and units are described, and so is their measurement. We begin with the luminous flux, followed by the luminous intensity, the illuminance, and the luminance. The mathematical aspects of the luminance concept are discussed in the next chapter. In chapter 5, the measurement of light is discussed. This chapter is based to a considerable degree on Narisada & Schreuder (2004, chapter 14) and Schreuder (1998, chapter 5).

3.1 Radiometry

3.1.1 Principles of radiometry

(a) *The difference and the similarity between radiometry and photometry*

Photometry is essential, both in visual science and in lighting engineering. Photometry makes it possible to express lighting and vision phenomena in a quantitative way, to define them, to compare research results with those of other researchers or of other laboratories, and to quantify the performance of lighting equipment.

Photometry is a branch of the wider field of radiometry. One might call radiometry the measurement of electromagnetic radiation independent of the detector that is used in the measurements, whereas photometry refers to the measurement of electromagnetic radiation when using the human visual system as a detector. This means that at the one hand there is hardly any fundamental difference between radiometry and photometry, but that at the other hand the response, more in particular the spectral response, of the human visual system is an essential part of photometry. This implies that radiometry needs to be regarded as a branch of experimental natural science. The relevant methods are those of experimental physics. Photometry, however, may be regarded as a branch of applied psychology. The relevant methods are those of experimental phychophysics. The bottom line is that the accuracy that can be reached in radiometric measurement is determined by the limits of instruments, whereas the accuracy of photometric measurement is limited by the way the performance of the human visual system is determined. Radiometry usually is far more precise than photometry.

(b) *Radiant power*

Radiometry is the measurement of radiant power. As is explained sec. 2.1, light is a form of radiant power, so all that will be said here about radiometry will also apply to the measurement of light, or photometry. The relation between radiometry and photometry is the subject of this section.

Essentially, the measurement of radiant power is a conversion of energy, in this case from radiant energy into another kind of energy (Moon, 1961, p. 15). This statement may seem rather obvious, but there are a number of points to be made aware of. First, radiometry is an experimental effort. This implies that experimental errors cannot be avoided, which implies in its turn that radiometry never can be absolutely accurate. Photometry, relying on psycho-physical measurements, is even far less accurate. Secondly, as defined in this way, radiometry applies only to electromagnetic energy as far as it is considered to be a matter of waves in the electromagnetic field. As is explained in sec. 2.1, there are three more – different – models for light. These apply as well to other forms of the propagation of electromagnetic energy. It remains to be seen, in how far the definition of radiometry as given here, applies to those cases.

(c) *The basic formula of radiometry*

Radiometry is defined as the measurement of radiant power, being a conversion of energy. This implies that there are two actors involved: the sender and the receiver. This leads immediately to the following relation:

Radiometery and photometery

$$E = \int_0^{+\infty} w_\lambda \cdot J_\lambda \cdot d\lambda \qquad [3.1\text{-}1]$$

in which
- E: the irradiance, or power density;
- J_λ: spectral distribution function of the incident radiation;
- w_λ: the weighing function of the receiver, or the spectral sensitivity function of the detector;
- λ: the wavelength of the incident radiation.

Moon & Spencer (1981, p. 212) called this relation, aptly 'the fundamental equation'. We will come back to this relation in chapter 4, when discussing the mathematics of luminance. The difference between the traditional radiometry and the traditional photometry is that in traditional radiometry, J_λ equals unity for all wavelengths, whereas in traditional photometry, w_λ is the weighing function of the human visual system when operating in the photopic mode, or in daylight vision. In that case, the weighing function is conventionally called the v_λ-curve. This curve is discussed in considerable detail in sec. 7.1.1. For convenience, in Figure 3.1.1, a picture of this curve is given in this section.

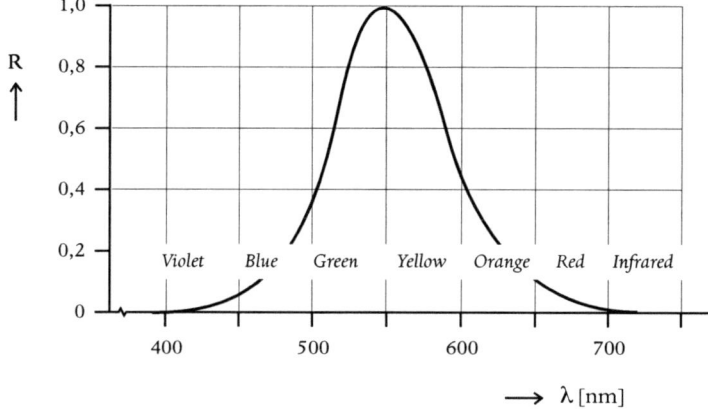

Figure 3.1.1 The v_λ-curve: the relative sensitivity curve for human photopic vision (After Narisada & Schreuder, 2004, Fig. 8.2.1).

We added the word 'traditional', because in modern science and in modern technology, the sources have all sorts of different spectral emission distribution functions, and the human observer, adapted to the photopic mode, is only one of the many detectors that are used. As an example, in modern astronomy, the human observer is obsolete in photometry. Until recently, most photometry was performed by using photographic emulsions, but since a decade or two, almost all astronomical measurements, photometry included, is done with CCDs (Narisada & Schreuder, 2004, ch. 14). Astronomical photometry is discussed in detail in Sterken & Manfroid, 1991, and Budding, 1992, where further information is given about the spectral sensitivity function of modern detectors. Examples of other detectors are given in Sterken & Manfroid (1992, sec. 1.16, Figure 1.17). Still other detectors, old and new, are described a.o. in Baer, 1990; Helbig, 1972; Ris, 1992; Unger et al., 1988.

(d) Terminology

Because there is no fundamental distinction between radiometry and photometry, they are described by the same physical concepts. However, the terminology is different. One reason is the fact that radiometry and photometry did, historically speaking develop along rather different pathways. Another reason is that radiometry and photometry sometimes serve different purposes. As an example, radiometrists are concerned in particular with the propagation of energy, whereas photometrists are concerned by the fact that most common light sources are neither point sources nor uniform emitters of light. Also, in most lighting applications, the concept of luminance is essential.

For convenience, we will list here the usual units, quantities and terms for radiometry and for photometry. See Table 3.1.1.

Radiometry prefix 'radiant'	Definition	Photometry prefix 'luminous'
flux F (W)	F	flux F (lm)
intensity I (W · sr^{-1})	$dF/d\omega$	intensity I (cd)
irradiance E (W · m^{-2})	dF/da	illuminance E (lx = lm/m^2)
radiance L (W · sr^{-1} · m^{-2})	$d^2F/d\omega\,da \cdot \cos\theta$	luminance L (cd/m^2)

Table 3.1.1 *Terminology of radiometry and photometry (After Sterken & Manfroid, 1992, Table 1.1).*

The terminology of both radiometry and photometry is based on a number of standards and regulations. Several have been mentioned in different parts of this book. Here, we will list again some of the most relevant ones: Anon., 1932, 2001; CIE, 1987, 2005; CGPM, 1979; ISO, 1992.

3.1.2 The solid angle

As is explained in sec. 3.2.3, the candela is defined in terms of luminance. This gives rise to a fundamental problem, because the mathematical expressions involved are valid only for a point source, whereas luminance only makes sense for a source larger than a point – a source with a measurable surface area. "Mathematically, a solid angle

Radiometery and photometery

must have a point at its apex; the definition of luminous intensity therefore applies strictly only to a point source" (Moon, 1961, p. 556; Anon., 1932).

In the definition, the concept solid angle is introduced. The solid angle is defined as the surface (in m²), cut out of a sphere with a radius of 1 m. Such a sphere is called an unit sphere. The unit of the solid angle is the steradian. For a different sphere radius, the same solid angle cuts out a different area. The area is proportional to the square of the sphere radius. This is depicted in Figure 3.1.2. In this figure, a solid angle with a unity value, is depicted as it cuts out a surface area of r² (m²) out of a sphere with a radius of r (m).

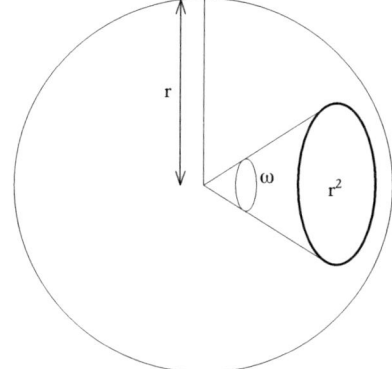

Figure 3.1.2 The solid angle (After Schreuder, 1998, Fig. 5.2.3).

Using this description of the solid angle, the relation in which the luminous intensity I is defined,

$$I = \frac{\delta \Phi}{\delta \omega} \quad\quad [3.1\text{-}2]$$

in which Φ means the luminous flux emitted within the solid angle ω, can be depicted as in Figure 3.2.1. The luminous intensity is described in section 3.2.3.

There is some controversy about the dimension of the steradian. Looking at the equation in which the steradian is defined, it would seem that it is a dimensionless quantity: the 'square meter' appears in the nominator as well as in the denominator of [3.1-1]. Some maintain, however, that 'crosswise' is different from 'lengthwise'. This would imply that the dimension of the solid angle is [meter square crosswise / meter square lengthwise] (Hentschel, ed., 2002, sec. 2.1.1, p. 17; Gall, 2004, p. 18-19). In practice, it makes no difference; it has been, however, often one of the finer points in lighting engineering examinations.

3.2 Basic photometric concepts

3.2.1 The SI-units

(a) *Seven basic units*

In 1960, at the 11th General Conference of Measures and Weight, the Système International d'Unités (the International System of Unities or SI) was adopted (Anon., 1960). It is based on seven basic units, called the SI-units:
1. The metre (m) as the unity for length;
2. The kilogram (kg) as the unity of mass;
3. The second (s) as the unity of time;
4. The ampère (A) as the unity of electric current;
5. The kelvin (K) as the unity for thermodynamic temperature;
6. The mol (mol) as unity of amount of matter;
7. The candela (cd) as the unity of luminous intensity.

From these basic units, all sorts of derived unities have been defined. The definitions of the basic units are given – in German – in Anon. (2001). The definitions themselves are, however, sometimes a little strange – to say the least (Narisada & Schreuder, 2004). We will give a few examples:
- The metre – a word that, in many languages also means a measuring device – is defined in terms of time (the length that light travels in vacuum in 1/299 792 458 seconds);
- The kilogram used the prefix kilo – which would mean 1000 basic units, not one;
- The second is, as the word says, a unit of the second level. The hour would be more appropriate;
- The candela is defined in terms of luminance. This gives rise to a fundamental problem, because the mathematical expressions, such as those that are discussed in sec. 3.2.3 are valid only for a point source, whereas luminance only makes sense for a source larger than a point – a source with a measurable surface area. As quoted earler: "Mathematically, a solid angle must have a point at its apex; the definition of luminous intensity therefore applies strictly only to a point source" (Moon, 1961, p. 556; Anon., 1932). Furthermore, the whole idea of defining the photometric units and quantities while using the V_λ-curve is reduced to the introduction of the constant 1/683. As is explained in sec. 7.2.2, this constant follows from the integration of the V_λ-curve (Hentschel, ed., 2002, p. 33). It corresponds to the ratio between the 'watt' and the 'light watt' or 'luminous watt' (Schreuder, 1998, sec. 5.2).

Radiometery and photometery

(b) The photometric units

Fundamentally, photometry is nothing else but photon counting. Each photon is counted, and weighted according to a weighing function. This weighing function is, of course, the spectral luminous efficiency curve that was described earlier. As we deal with 'official', or photopic photometry, the spectral luminous efficiency curve that we need is the one for the human visual system, viz. the V_λ-curve.

In sec. 2.1.3c, when discussing the dual nature of light, photons have been described in detail. In brief: "A photon is a quantum of electromagnetic radiation. It has an energy of h · ν, where h is the Planck-constant and ν the frequency of the radiation. Planck constant is a universal constant, having the value of $6{,}626\,076 \cdot 10^{-34}$ J·s" (Illingworth, ed., 1991, p. 350, 354). A quantum represents energy. "Energy is the quantity that is the measure of the capacity of a body or a system for doing 'work'. When a body does work, its energy decreases by an equal amount" (Illingworth, ed., 1991, p. 151). The unit of energy is joule (J). A joule equals a watt · second (Ws; Kuchling, 1995, p. 106). Thus, n photons represent the energy of n · h · ν. What interests lighting engineers is usually not the number of photons, but rather the number of photons that are emitted – or that pass – per second. So, the logical basic unit of photometry is the energy flux. When we discuss 'light', it is the energy flux weighted according to the spectral luminous efficiency curve for the human visual system. This is, for obvious reasons, called the luminous flux. So, in spite of the fact that SI considers the candela as the basic unit for photometry, we will begin our description with the luminous flux, and the related concepts of transmission, reflection, and absorption. From that, the luminous intensity follows in a natural way. This leads to the illuminance, and from there on, the different derivatives are described.

(c) Units of the ISO-photometry

The standard includes also a brief description of the units and symbols that are adopted by CIE and CIPM. These units and symbols are presently part of the ISO-standardisation. As these are the official units and symbols, we will summarize them here briefly. They are used in this book, but not throughout. In some cases, for reasons of convenience, other – usually older – units and symbols are used.

The SI photometric base unit is the candela (cd). It is defined by the 1979 CGPM-conference as: "the luminous intensity, in a given direction, of a source that emits monochromatic radiation of frequency $540 \cdot 10^{12}$ hertz and that has a radiant intensity in that direction of (1/683) watt per steradian" (CGPM, 1979).

Contrary to photometry, the fundamental physical quantity used in optical radiometry is the radiant flux or radiant power, Φ_e, measured in watts. The corresponding photometric quantity is, as is explained sec. 3.2.2, the luminous flux (Φ_v) measured in lumen (CIE, 2005, p. 1). As is explained in sec. 3.2.2a, the flux is, from theoretical point of view, a better unit to be used as the basis for photometry, just as it is for radiometry. But also is more convenient for practical lighting engineering purposes. See Narisada & Schreuder (2004, sec. 14.1.2,b), and Schreuder (1998, sec. 5.2.1).

3.2.2   The luminous flux

(a) Definition

The luminous flux is the quantity or the amount of the light. The unit is lumen (lm). Its dimension is, as is explained in the preceding part of this section, that of the watt. It is indicated by the symbol Φ. Since the luminous flux under normal lighting conditions cannot accumulate like e.g. water, and proceeds a velocity of about 300 000 km/h and instantaneously transformed into heat, the quantity of the light, therefore, is not the quantity of the accumulated amount of the light but a rate of the stream flow of the light per second. The quantity of the accumulated amount of the light is the exposure. The exposure is described as "the product of the illuminance or irradiance and the time for which the material is illuminated or irradiated" (Illingworth, ed., 1991, p. 63). The exposure is an important quantity to take into account when discussing photographic emulsions (Narisada & Schreuder, 2004, sec. sec. 14.5.1b) or CCDs (Narisada & Schreuder, 2004, sec. 14.5.2).

As has been mentioned in an earlier part of this section, the integration of the V_λ-curve gives the photometric equivalent of radiation (Hentschel, ed., 2002, p. 33). The numerical value is 1/683. This indicates that it is theoretically not possible to construct any light source that produces more than 683 lumens for each watt; in other words, the maximum value of the luminous efficacy of any light source is 683 lm/W (Schreuder, 1998, sec. 5.2, p. 46). The concept of the luminous efficacy is explained in sec. 2.2.2. If one wants to express the maximum luminous efficacy in terms of the, non-standardized, scotopic photometry, the number would be 1699 lm'/W (Hentschel, ed., 2002, p. 33). This does not imply that lamps would be more efficient in scotopic vision, only that the scotopic lumen (lm') has a different value from the photopic lumen. This difference, of course, results from the differences between the V_λ-curve and the V_λ'-curve. More details are given in sec. 7.2.3, where scotopic vision is discussed.

(b) Reflectance, transmittance

When a quantity of the luminous flux flows into any surface, a part of it is reflected, a part is transmitted, and the rest is absorbed. For an opaque surface, the transmission is, of course, zero. The reflectance is the ratio of the reflected luminous flux to the incident luminous flux under the given conditions. The transmittance is

Radiometery and photometery

the ratio of the transmitted luminous flux through the optical medium to the incident luminous flux in the given conditions. The absorbance is the ratio of the luminous flux that is absorbed by the optical medium to the incident luminous flux in the given conditions. Usually, the reflectance is indicated by R, the transmittance by T and the absorbance by A. It follows directly from the definitions that:

$$R + T + A = 1 \qquad [3.2\text{-}1]$$

The optical phenomena of reflection and transmission of the light are not simple. They are influenced by the configuration of in-flowing luminous flux and the optical media through which the light passes. The physical phenomena and the mathematical description of reflection and transmission are discussed in considerable detail in chapter 4 when describing the mathematics of the luminance.

3.2.3 The luminous intensity

Almost all light sources, both natural and man-made, emit light at different rates in different directions. Even the Sun, which approaches an ideal emitter of light, shines differently in different directions, for one reason because it is flattened, and not spherical. For all common man-made light sources, this is very much the case, because they all require a lead into the actual source to provide its energy – gas, wax, or electricity. Furthermore, they have to be supported. So there is an urgent need to be able to make a clear distinction for the action of the source into different directions. For this, the concept of luminous intensity is introduced.

The luminous intensity of a light source signifies, therefore, the intensity of the light that is emitted from the source itself. As is described in a preceding part of this section the ISO uses the unit of luminous intensity (the candela) as the basic unit for all photometry. We have pointed out the mathematical, physical, and practical difficulties that arise when doing so.

In practical photometry, a rather different definition is used. This definition follows directly from the description we have given earlier in this section. Essentially, it means the luminous flux in a certain direction. Thus, as has been explained in a preceding part of this section, the luminous intensity I is defined as:

$$I = \frac{\delta \Phi}{\delta \omega} \qquad [3.1\text{-}2]$$

in which Φ means the luminous flux emitted within the solid angle ω. The relation is depicted in Figure 3.2.1

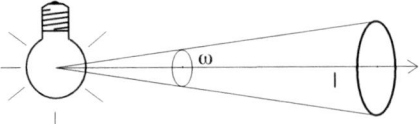

Figure 3.2.1 The luminous flux emitted within a solid angle (After Schreuder, 1998, Fig. 5.2.2).

In this relation the concept solid angle is introduced. The concept is explained in sec. 3.1.2.

Based on the relation [3.1-2], the luminous intensity is defined as the luminous flux in a certain direction. But as the concept 'directions' essentially is a line only, the limit transition from the 'difference-quotient' into a 'differential-quotient' is required. The formal definition of the luminous intensity is given in the relation [3.2-2].

$$I = \lim_{\delta\omega \to 0} \frac{\delta\Phi}{\delta\omega} = \frac{d\Phi}{d\omega} \qquad [3.2\text{-}2]$$

The concept of limit is explained in Bronstein et al. (1997, sec. 2.1.4).

3.2.4　The illuminance

(a) *Definition*

When light strikes a surface steadily, an amount of luminous flux is incident on the surface. If the surface is evenly lit, the illuminance is expressed as an areal density of the luminous flux incident the surface. The unit of the illuminance is lux. Sometimes, 'lux' is abbreviated in 'lx'. The illuminance of an area A (in m^2) is:

$$E = \Phi / A \qquad [3.2\text{-}3]$$

The dimension of the illuminance is therefore lm/m^2.

It is, of course, possible to define E for a mathematical plane through which the light shines. No physical phenomenon can be detected, and the term 'illuminance' can be somewhat misleading. It might seem better to call it the density of the luminous flux (Narisada & Schreuder, 2004, sec. 14.1.7; Schreuder, 1998, p. 46). It is interesting to mention, that it has been proposed to base a general theory of illuminating engineering on the concept of flux density (Moon & Spencer, 1981, based on Gershun, 1939). The proposals led to nothing, mainly because Moon and Spencer wanted to introduce, at the

Radiometery and photometery

same time, a completely different, new terminology (Moon, 1961, preface to the Dover edition, p. v-vii). However, from theoretical point of view, their attempts deserve more attention than they have ever received. The light field concept is explained in sec. 4.1, when discussing the mathematics of luminance.

(b) *Horizontal, vertical and semicylindrical illuminance*

For obvious reasons, if the surface is located horizontally, the illuminance is called the horizontal illuminance (E_h). If it is located vertically, the illuminance on the surface is called the vertical illuminance (E_v). At other places in this book it is explained that most standards and recommendations in applied illuminating engineering related to the quality or the quantity of light are based on the horizontal illuminance.

The illuminance on the flat surface is not suitable to express the brightness of spherical or cylindrical object, such as the face of pedestrians. In such cases, the semicylindrical illuminance is used. The semicylindrical illuminance (E_{sc}) is defined as the average value of illuminances on the surface of a cylinder. As one can see only one side of a cylinder at the time, the definition is limited to half a cylinder. The semicylindrical illuminance can be described as follows:

$$E_{sc} = \frac{I}{\pi h^2} \cdot \sin \alpha \cdot \cos^3 \alpha \cdot (1 + \cos \beta) \qquad [3.2\text{-}4]$$

The relation [3.2-4] is illustrated in Figure 3.2.2. In this figure, I signifies the luminous intensity. The figure indicated the meaning of α, β and h.

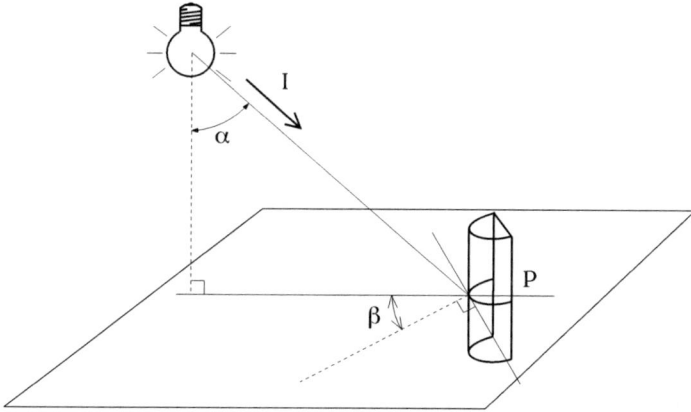

Figure 3.2.2 *The semicylindrical illuminance (After Schreuder, 1998, Fig. 5.2.1).*

(c) The average illuminance and the non-uniformity

For the characterization of many aspects of outdoor lighting installations, the average illuminance is often a crucial parameter. Although the description of the average as such is obvious, we will spend a few words on it.

When the illuminance distributions over any area, horizontal or vertical, or the incident luminous flux is not uniform, the average illuminance is often an important characteristic. It can be assessed as follows:

$$E_{ave} = \Phi_{tot} / A \qquad [3.2\text{-}5]$$

with:
    Φ_{tot}: the total luminous flux (in lumens) incident in the area;
    A: the surface area on which the luminous flux is falling (in m^2).

In most practical cases, the distribution of the illuminance is not uniform at all. This implies that the average illuminance usually is not sufficient to characterize the lighting installation. Therefore, almost all indoor and outdoor standards or recommendations on lighting installations give, apart from requirements about the average illuminance, also requirements about the degree of non-uniformity that is acceptable. These requirements are given in different fashion, like e.g. as the minimum divided by the average (E_{min}/E_{ave}), or as the minimum divided by the maximum (E_{min}/E_{max}). In this way, the degrees of non-uniformity are always numbers smaller than one, which is convenient for calculations, and which avoids difficulties when the minimum is zero. Very similar definitions of the average and the non-uniformity are used, when using the luminance as a design quality criterion.

A practical point must be added here. When measuring illuminances, e.g. on a road or a table in an office, the procedure is that one places the illuminance meter – the luxmeter – on the ground, takes all necessary precautions to avoid disturbances, and makes the measurement. Thus, the illuminance in one point on the road is measured. In this respects, there are no uncertainties other than the usual measuring errors. However, if the average road illuminance must be assessed, the point measurements must be repeated in different locations. It is essential to be very clear about the measuring grid that is going to be used. If not, severe errors may be the result, particularly if the illuminance is unevenly distributed. Obviously, such repeated point measurements are needed as well in order to assess the non-uniformity. It must be mentioned that the term 'point' is used here in a non-mathematical way.

Radiometery and photometery

As an example, when point-luminance measurements are made in road lighting, the measuring area must lay at a considerable distance in front of the observer. Usually, one takes a distance between some 60m and 160m (De Boer, 1967; Schreuder, 1967; 1998). According to the definition of the road-surface luminance, the measurements are made from a point 1,5m over the road surface. The angles between the line that connects the observer and different points in the measuring area and the vertical through these points are given in Table 3.2.1. When we assume that a modern point-luminance meter is used with a circular measuring field with a diameter of 2 minutes of arc, the actual area where the measurement is made, is not punctiform at all. As is depicted in Figure 3.2.3 and Table 3.2.1, the area can be up to nearly 0,1m wide and more than 5m long. See also Narisada & Schreuder (2004, sec. 14.5.3e), and Schreuder (1998, sec. 5.4.5).

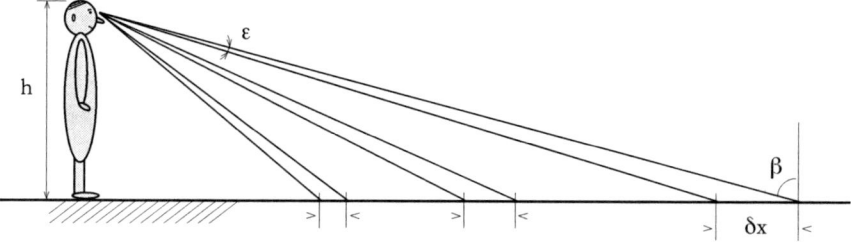

Figure 3.2.3 The measuring area increases with increasing distance (After Schreuder, 1998, Fig. 5.4.11).

	Distance to measuring point (m)		
	60	100	160
height (m)	1,5	1,5	1,5
aperture (degrees)	0,033	0,033	0,033
β (degrees)	88,568	89,141	89,463
area length (m)	2	3,3	5,3
area width (m)	0,03	0,05	0,08

Table 3.2.1 Dimensions of the measuring area

(d) *The inverse square law*

From the definition of the luminous intensity, that is given in an earlier part of this section, it can be concluded that the illuminance decreases with the square of the distance between the light source and the receiving plane. This relation is called, for obvious reasons, the inverse square law:

$$E = k \cdot (I/ r^2) \qquad [3.2\text{-}6]$$

with:
- E: the illuminance;
- I: the luminous intensity of the source;
- r: the distance;
- k: a constant, depending on the units that are used).

This formulation is from Schreuder (1998, p. 52), based on data from Breuer (1994, p. 175). This is depicted in Figure 3.2.4.

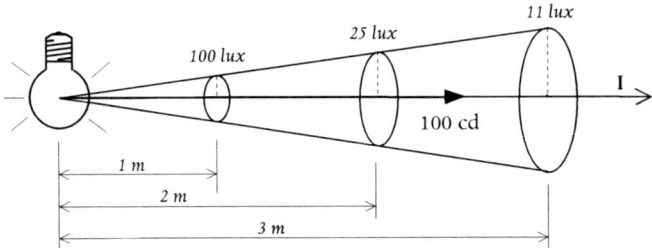

Figure 3.2.4 The inverse square law (After Schreuder, 1998, Fig. 5.3.1).

(e) *The distance law for large sources*

As can be seen directly from the definition of the luminous intensity, relation [3.2-6] is valid only for point sources. When the source is small in relation to the distance, relation [3.2-6] can still be used a an approximation. For large sources, like rows of fluorescent tubes in tunnels or in office interiors, this is not possible. Of course, one can always regard a long source as a chain of many, short sources, do the exercise for each small part separately, and add up the results. This is done in most software programmes that are used in the design of the lighting of tunnels and offices. In pre-pc times, mathematical methods were needed. Elegant descriptions are given in Bean & Simons (1968, chapters 2, 8), Moon (1961, chapters 8, 9, 10), Zijl (1951), and De Boer & Vermeulen (1967).

The discrepancy between the actual measured value, and the value that would be found if the source were a true point source, is given in the relation [3.2-7].

$$\frac{I}{I'} = \frac{r^2 + R^2}{r^2} \qquad [3.2\text{-}7]$$

with:
- I: the actual luminous intensity as measured;
- I': the approximated luminous intensity, assuming that the inverse square law is valid.
- r: the distance between the light source and the measuring point;
- R: the radius of the light source, assumed to be circular.

Radiometery and photometery

The derivation of relation [3.2-7] is given in Helbig (1972, p. 47-48). The relation is depicted in Figure 3.2.5.

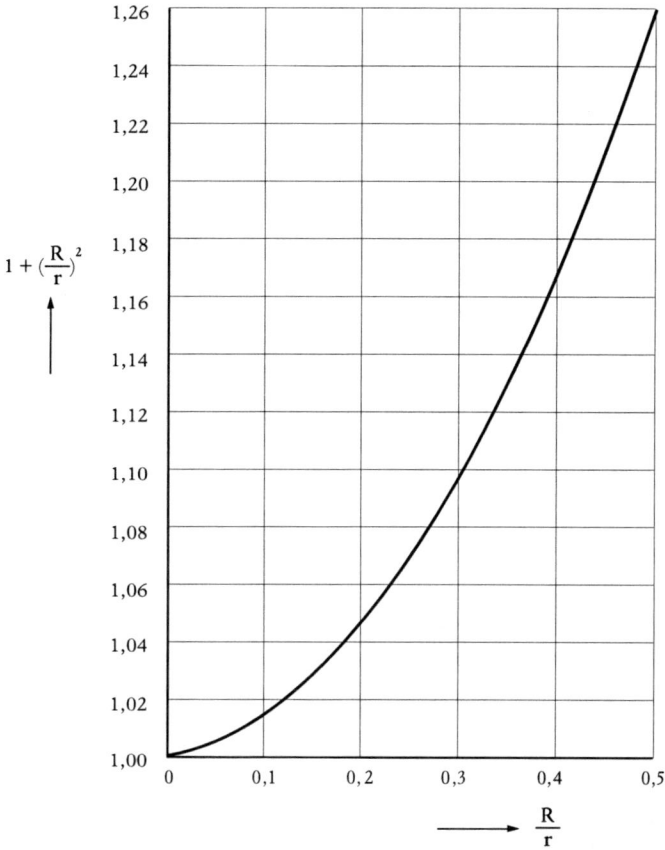

Figure 3.2.5 *Deviations from the inverse square law for non-point sources (After Schreuder, 1998, Fig. 5.3.2. Based on Helbig, 1972, p. 48). The meaning of the letters correspond to those used in relation [3.2-7].*

For easy reference, a few data are given in Table 3.2.2.

Size (R/r)	Deviation (I/I')
1/2	1,250
1/4	1,062
1/6	1,028
1/8	1,016
1/10	1,010

Table 3.2.2. *Deviations from the inverse square law for non-point sources (After Schreuder, 1998, Table 5.3.1. Based on Helbig, 1972, p. 48). The meaning of the letters correspond to those used in relation [3.2-7].*

Table 3.2.2 shows that the deviations soon become 'acceptable', particularly if we take into account that R is the radius of the light source. Usually, that corresponds with half the luminaire length. If we consider that a deviation of 1% is still acceptable – a rather stiff requirement – the distance needs to be only 5 times the length of the luminaire. Sometimes, this distance is called the photometric threshold distance (Schreuder, 1998, p. 53). See also Keitz (1967, p. 187); Walsh (1958).

(f) The distance law for bundled light

The inverse square law, as well as the approximations that have been discussed in earlier parts of this section, are valid only for diffuse ('Lambertian') light sources (Helbig, 1972, sec. 5.2). For bundled light sources, like e.g. search lights, hand-held torches, and vehicle headlamps, the inverse square law is only applicable if the distance is measured, not from the actual light source, but from its optical image. For this, one has to take into account that bundled light sources are made by placing the actual light source – e.g. the filament – in or near the focal point of a hollow mirror or a convex lens. If the mirror is a parabola and the light source is placed in the exact focal point, the beam is parallel. In optical terms, the virtual source is at a distance of minus infinity from the measuring point. As regards the inverse square law, it would mean that any finite change in the distance between the lamp and the measuring point would have no effect as 'an infinite distance plus a finite stretch is still infinite'. In practical terms, it means that the illuminance is equal at any distance from the lamp. This is depicted in Figure 3.2.6. Of course, this is precisely the reason why bundled lights, particularly if they have a parallel beam, are so useful in may applications. One example is the high beam of vehicle headlamps.

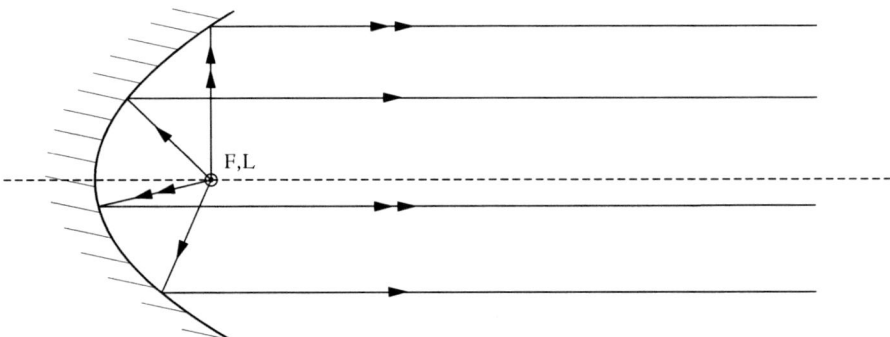

Figure 3.2.6 The beam of a bundled light with a nearly parallel bundle. F is the focus of the parabola, L the light source (After Schreuder, 1998, Fig. 5.3.3).

Radiometery and photometery

A completely different situation is met, when the light source is placed further away than the focal point from the mirror. Here, the optical image is real, and it is located at a certain, often a large, distance before the lamp. The beam is a converging one. This is depicted in Figure 3.2.7.

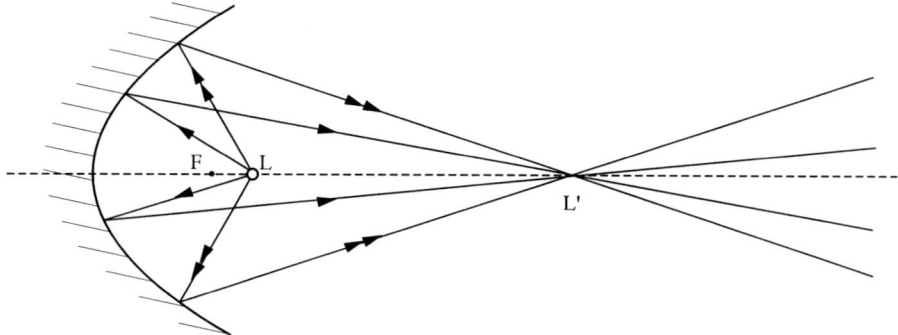

Figure 3.2.7 *The beam of a bundled light with a converging bundle. F is the focus of the parabola, L the light source, L' the optical image of the light source (After Schreuder, 1998, Fig. 5.3.4).*

When assessing the influence of the distance on the illuminance, the assessments must begin at the real optical image. This often implies that, when the distance increases, the illuminance is reduced much more steeply than the inverse square of the distance between the lamp and the measuring point.

(g) *The cosine law*

If the surface on which the light falls is tilted, the area on which the luminous flux is falling increases and, consequently, the illuminance decreases. The illuminance decreases in proportion to the cosine of the incident angle. This is called the cosine law. The law follows directly from the definition of the cosine. It can be described as follows:

$$E = \frac{I}{d^2} \cos \gamma \qquad [3.2\text{-}8]$$

in which:
   E: the illuminance;
   I: the luminous intensity of the light source;
   d: the distance;
   γ: the angle with the normal.

The relation [3.2-8] is illustrated in Figure 3.2.8.

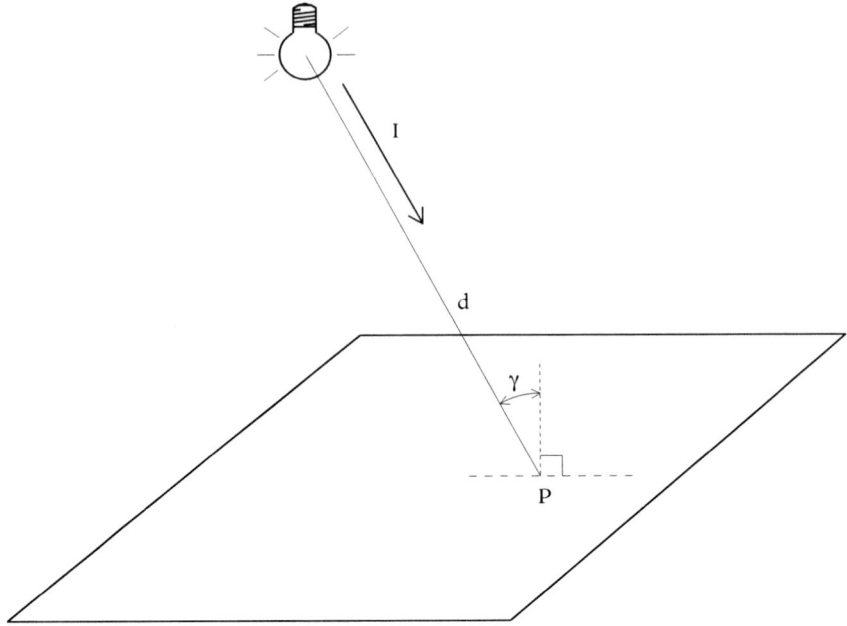

Figure 3.2.8 *The cosine law. The letters are explained in the text (After Schreuder, 1998, Fig. 5.3.5).*

(h) *The cosine to the third law*

A combination of the inverse square law and the cosine law produces the third important law in photometry: the cosine to the third law. It described the illuminance at different points in a plane. It can be described as follows:

$$E = \frac{I}{h^2} \cos^3 \gamma \qquad [3.2\text{-}9]$$

in which:
    E: the illuminance;
    I: the luminous intensity of the light source;
    h: the mounting height of the luminaries (analogous to the distance in [3.2-8]);
    γ: the angle with the normal.

The relation [3.2-9] is illustrated in Figure 3.2.9.

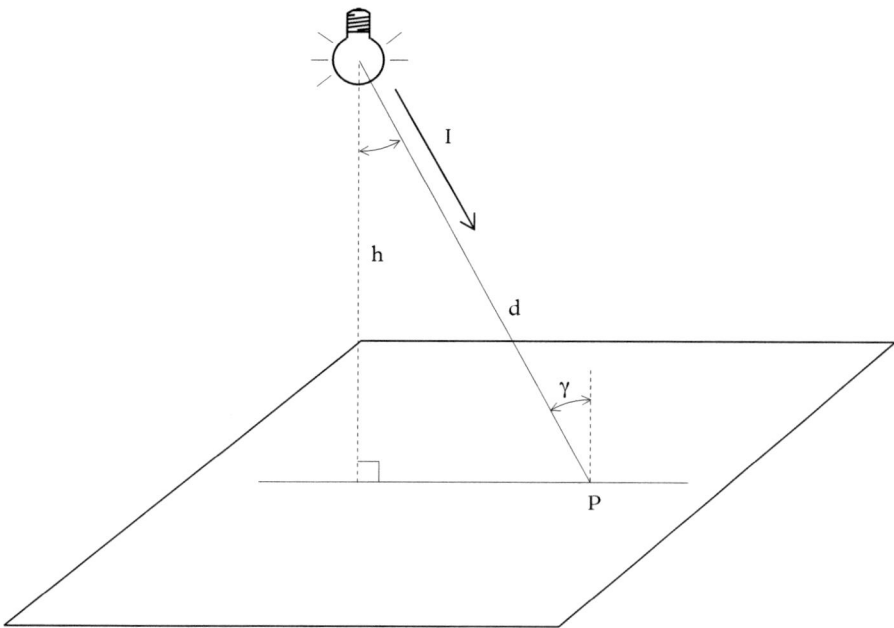

Figure 3.2.9 *The cosine to the third law. The letters are explained in the text (After Schreuder, 1998, Fig. 5.3.6).*

One may think about the cosine to the third law as follows: one cosine comes from the cosine law, and two cosines come from the inverse square law. The law has many applications in outdoor lighting, particularly if the area under consideration is flat and horizontal, and the luminaries to be used are identical.

3.2.5   The luminance

(a) *General definition*

Light only has a perceivable effect when the visual system is activated in one way or another. In other, simpler words, light is only visible once it falls on the eye. These phenomena are describes in a more philosophical way in Wright (1967).

Light is seen as the brightness of the observed object. In order to arrive at a better definition, as well as a concept that is easier to measure, luminance, indicated by L, was introduced. Luminance can be said to be the objective, measurable measure of brightness. In illuminating engineering, the concept of brightness is used in two distinct ways. The first is the colloquial equivalent of the luminance. The second is the subjective experience of the light impression. The subjective experience of the light impression is discussed in sec. 8.1.3.

In chapter 4 the mathematical and physical fundaments of the luminance concept are discussed in more detail. This section focusses on the practical aspects of photometry as regards luminance.

The definition of luminance that is given by ISO for the photometric units, is not applicable to measuring or calculating luminances in practice. As is mentioned in an earlier part of this section, the only function of the ISO-unit is to establish the numerical value of the unit of luminous intensity, the candela. For other purposes a different definition of the luminance is needed. We will follow here, by means of some quotations, the theoretical treatise given in Baer ed. (2006, sec. 1.2.2.1, p. 24-26). Details are given, as is mentioned earlier, in chapter 4.

The general definition of the luminance, as based on the CIE Standard Spectral Sensitivity Curve is:

$$L_v = K_m \int_{360nm}^{830nm} L_{e,\lambda_a} \cdot V_{\lambda,a} \cdot d\lambda \qquad [3.2\text{-}10]$$

In a similar way, the 'scotopic' luminance could be defined as:

$$L'_v = K'_m \int_{360nm}^{830nm} L'_{e,\lambda_a} \cdot V'_{\lambda,a} \cdot d\lambda \qquad [3.2\text{-}11]$$

in which:
L_v and L'_v: the photopic and the scotopic luminance;
L_{e,λ_a} and L'_{e,λ_a}: the spectral radiance for photopic and scotopic vision;
$V_{\lambda,a}$ and $V'_{\lambda,a}$: the CIE Standard Spectral Sensitivity Curves for photopic and scotopic vision;
K_m: 683 lm/W;
K'_m: 1699 lm/W.

These expressions are corrected after Schreuder (1998, p. 49-50) and are based on Hentschel (1994, p. 29). It should be noted, however, that the 'scotopic' luminance is not an ISO-unit. In sec. 7.2.4 it is pointed out that CIE not only promotes the use of the scotopic luminance but even a set of mesopic luminances (CIE, 1989, 2005a).

It only makes sense to speak of the luminance of an object, if the object itself emits light. Otherwise, it is just dark and invisible, and the brightness – and the luminance – will be zero. In a literary sense, this statement would mean that a non-object, like the open night sky, could have no luminance. In sec. 4.3 it is explained in which way the luminance of the sky will be defined.

Radiometery and photometery

(b) *The luminance of light-reflecting objects*

As mentioned earlier, a surface that does not emit light itself is invisible when no light falls on it, and the illuminance is zero. However, it is also invisible when all incident light is absorbed, so that there is no light left over to be reflected – and thus, to be observed. Only if, at least part of, the light is reflected, the object can be perceived. When more light is reflected, the object is brighter. This can be the result of two factors.

1. When more light falls on the object – when the illuminance increases – the brightness increases proportionally.
2. When, the illuminance being unchanged, the reflection is higher, also the brightness increases. The brightness will increase proportionally to the reflection.

These two factors lead to the simple equation:

$$L = R \cdot E \qquad [3.2\text{-}12]$$

in which R is the luminance factor.

It should be noted that the common ISO-unit for luminance refers, as is mentioned earlier, to a surface. It is expressed in the unit candela per square meter, which is strange, because the candela is defined for a point source only. The definition given earlier in this section in relation [3.2-10], does not circumvent this problem. In relation [3.2-10], the luminous flux is considered, that is emitted by the small element dA_1 of the luminous plane A_1. The element dA_1 is small, but not zero; relation [3.2-10] is not applicable to a point source. It should be mentioned that, actually, two planes are included in the definition: the plane A_1, from where the light is emitted – or seems to be emitted, hence the word 'virtual' in the definition – and the plane A_2, on which the light is falling. In chapter 4 the significance of these two planes will be explained.

(c) *The luminance of light emitting objects*

The luminance is defined as the part of the luminous flux $d\Phi$, that is emitted from the virtually luminous plane ($dA_1 \cdot \cos \gamma_1$), in a prescribed direction within a solid angle element $d\omega_1$ falls onto a plane. The unit is cd/m².

$$L = \frac{d^2\Phi}{dA_1 \cdot \cos \gamma_1 \cdot d\omega_1}$$

$$= \frac{I_\gamma}{A_1 \cdot \cos \gamma_1} \qquad [3.2\text{-}13]$$

The relation [3.2-13] is illustrated in chapter 4 where the mathematics of the luminance-concept is discussed.

This definition can be used to assess the luminance of an object, that emits light itself. It is explained in an earlier part of this section that, according to ISO, the luminance is derived from the luminous intensity. This is based on a light emitting surface, that is subdivided into small parts δA, that are so small, that the luminous intensity according to its actual definition may be used. As is explained earlier, this definition is valid for point sources only. The luminance of that part can be described as the ratio between the luminous intensity and the surface:

$$L = \frac{\delta I}{\delta A} \qquad [3.2\text{-}14]$$

Taking the limit for δ A towards zero, the definition of the luminance is:

$$L = \frac{dI}{dA} \qquad [3.2\text{-}15]$$

Now it is clear why the luminance is expressed in candela per square meter (cd/m^2).

3.3   Conclusions

Photometry is essential, both in visual science and in lighting engineering. Photometry makes it possible to express lighting and vision phenomena in a quantitative way. Photometry is a branch of radiometry. Radiometry is a branch of experimental natural science. The relevant methods are those of experimental physics. Photometry, is a branch of applied psychology. The relevant methods are those of experimental phychophysics. The accuracy of radiometric measurement is determined by the limits of instruments, whereas the accuracy of photometric measurement is limited by the way the performance of the human visual system is determined. Radiometry usually is far more precise than photometry.

Radiometry is the measurement of radiant power. Light is a form of radiant power. Essentially, the measurement of radiant power is a conversion of energy, in this case from radiant energy into another kind of energy. Because radiometry is an experimental effort, experimental errors cannot be avoided. There are two actors involved: the sender and the receiver.

In photometry, the concept of the solid angle is essential in many definitions. The other essential aspect of photometry as compared to radiometry is the sensitivity of the human visual system for optical radiation of different wavelengths. The corresponding function is called the 'photopic spectral luminous efficiency function', and is conventionally described by the V_λ-curve.

Fundamentally, photometry is nothing else but photon counting. Each photon is counted, and weighted according to the photopic spectral luminous efficiency curve. In this chapter, the different photometric quantities and units are described, and so is their measurement. We begin with the luminous flux, followed by the luminous intensity, the illuminance, and the luminance. The mathematical aspects of the luminance concept and the measurement of light are discussed in the chapters 9 and 10 respectively.

References

Anon. (1932). Illuminating engineering nomenclature and photometric standards. IES, 1932.
Anon. (1960). General Conference on Measures and Weights. Paris, 1960.
Anon. (2001). Die SI-Basiseinheiten; Definition, Entwicklung; Realiserung. Nachdruck 2001 (The SI-basic units; Definition, development, realization. Reprinted 2001). Braunschweig, Physikalisch-Technische Bundesanstalt, 2001.
Baer, R. (1990). Beleuchtungstechnik; Grundlagen (Fundaments of illuminating engineering). Berlin, VEB Verlag Technik, 1990.
Baer, R., ed. (2006). Beleuchtungstechnik; Grundlagen. 3., vollständig überarteite Auflage (Essentials of illuminating engineering, 3rd., completely new edition). Berlin, Huss-Media, GmbH, 2006.
Bean, A.R. & Simons, R.H. (1968). Lighting fittings performance and design. Oxford. Pergamon Press, 1968.
Breuer, H. (1994). DTV-Atlas zur Physik. Zwei Bänder, 4.Auflage (DTV atlas on physics, two volumes, 4th edition). München, DTV Verlag, 1994.
Bronstein, I.N.; Semendjajew, K.A.; Musiol, G. & Mühlig, H. (1997). Taschenbuch der Mathematik. 3. Auflage (Manual of mathematics. 3rd edition). Frankfurt am Main, Verlag Harri Deutsch, 1997.
Budding, E. (1993). An introduction to astronomical photometry. Cambridge University Press, 1993.
CIE (1967). Proceedings of the CIE Session 1967 in Washington. (Vol. A, B). Publication No. 14. Paris, CIE, 1967.
CIE (1987). International Lighting Vocabulary. 4th Edition. Publication No. 17-4. Paris, CIE, 1987.
CIE (1989). Mesopic photometry: History, special problems and practical solutions. Publication No. 81. Paris, CIE, 1989.
CIE (2005). Photometry – The CIE system of physical photometry; International Standard CIE S 010/ E:2004; ISO 23539:2005(E). Vienna, CIE, 2005.
CIE (2005a). Vision and lighting in mesopic conditions. Proceedings of the CIE Symposium '05. Leon, Spain, 21 May 2005. CIE X028. Vienna, CIE, 2005.
CGPM (1979). Comptes rendues des scéances de la 16e Conférence Générale des Poids et Mesures (CGPM). Paris, Bureau International des Poids et Mesures, 1979.
De Boer, J.B. (1967). Visual perception in road traffic and the field of vision of the motorist. Chapter 2. In: De Boer, ed., 1967.
De Boer, J.B., ed. (1967). Public lighting. Eindhoven, Centrex, 1967.
De Boer, J.B. & Vermeulen, J. (1967). Simple luminance calculations based on road surface classification. In: CIE, 1967.

Gall, D. (2004). Grundlagen der Lichttechnik; Kompendium (A compendium on the basics of illuminating engineering). München, Richard Pflaum Verlag GmbH & Co KG, 2004.

Gershun, A. (1939). The light field (original title Svetovoe pole, Moscow, 1936). Translated by Moon & Timoshenko. Journal of Mathematics and Physics, 18 (1939) No 2, May, p. 51-151.

Helbig, E. (1972). Grundlagen der Lichtmesstechnik (Fundamentals of photometry). Leipzig, Geest & Portig, 1972.

Hentschel, H.-J. (1994). Licht und Beleuchtung; Theorie und Praxis der Lichttechnik; 4. Auflage (Light and illumination; Theory and practice of lighting engineering; 4th edition). Heidelberg, Hüthig, 1994.

Hentschel, H.-J. ed. (2002). Licht und Beleuchtung; Grundlagen und Anwendungen der Lichttechnik; 5., neu bearbeitete und erweiterte Auflage (Light and illumination; Theory and applications of lighting engineering; 5th new and extended edition). Heidelberg, Hüthig, 2002.

Illingworth, V., ed. (1991). The Penguin Dictionary of Physics (second edition). London, Penguin Books, 1991.

ISO (1992). ISO Standard ISO 31-0:1992 "Quantities and units – Part 0: General principles".

Keitz, H.A.E. (1967). Lichtmessungen und Lichtberechnungen. 2e. Auflage (The measurement and calculation of light. Second edition). Eindhoven, Philips Technische Bibliotheek, 1967.

Kuchling, H. (1995). Taschenbuch der Physik (Manual for physics). 15. Auflage. Leipzig-Köln, Fachbuchverlag, 1995.

Moon, P. (1961). The scientific basis of illuminating engineering (revised edition). New York, Dover Publications, Inc., 1961.

Moon, P. & Spencer, D.E. (1981). The photic field. Cambridge, Massachusetts, The MIT Press, 1981.

Narisada, K. & Schreuder, D.A. (2004). Light pollution handbook. Dordrecht, Springer, 2004.

Ris, H.R. (1992). Beleuchtungstechnik für Praktiker (Practical lighting engineering). Berlin, Offenbach, VDE-Verlag GmbH, 1992.

Schreuder, D.A. (1967). Theoretical basis of road-lighting design. Chapter 3. In: De Boer, ed., 1967.

Schreuder, D.A. (1998). Road lighting for safety. London, Thomas Telford, 1998. (Translation of: Schreuder, D.A., Openbare verlichting voor verkeer en veiligheid. Deventer, Kluwer Techniek, 1996).

Sterken, C. & Manfroid, J. (1992). Astronomical photometry. Dordrecht, Kluwer, 1992.

Unger, S.W.; Brinks, E.; Laing, R.A.; Tritton, K.P. & Gray, P.M. (1988). Observers' Guide. Version 2.0. November 1988. La Palma, Isaac Newton Group, 1988.

Van Bommel, W.J.M. & De Boer, J.B. (1980). Road lighting. Deventer, Kluwer, 1980.

Walsh, J.W.T. (1958). Photometry (3rd edition). London, Constable, 1958. Reprinted. New York, Dover, 1965.

Wright, W.D. (1967). The rays are not coloured. London, Adam Hilger, 1967.

Zijl, H. (1951). Manual for the illuminating engineer on large size perfect diffusors. Eindhoven, Philips Industries, 1951.

4 The mathematics of luminance

As is explained in other chapters of this book, one may consider light as a stream of particles. From this idea, it is possible to describe many aspects of light in terms of a field. A field is a region under the influence of some physical agency. Fields are not a form of matter. A field can be pictorally represented by a set of curves, often referred to as field lines. The density of these lines at any given point represents the strength of the field. Field lines are just trajectories.

The field concept is also used in a much wider conation, like e.g the morphic field, which organizes its characteristic structure and pattern activity. The light field is not an electromagnetic field. It may be regarded is a morphic field, a photic field, or a field of light rays.

The light field has some similarity to a fluid. Some aspects of hydrodynamics are discussed, although light is not a Bernoulli-fluid. Still, the continuity principle holds. Diffraction may be regarded as friction in the stream.

The main subject of this chapter is the discussion of the luminance of real and virtual objects. Based on the general definition of luminance, light tubes are discussed. This leads to the concepts of geometric flux and the throughput of light. Finally, the luminance of reflecting surfaces is discussed. A definition and the derivation of the luminance factor is given.

4.1 The field concept

4.1.1 Field theory

(a) *Light fluid and light vectors*

As is explained in sec. 2.1, light can be considered as a stream of photons moving from one place to another. Each photon can be represented by a vector, where the scalar (the 'size' of the vector) corresponds to the energy of the photon, and the direction of the vector corresponds – obviously – to the direction of the light. Therefore, the flow of photons can be described in two ways, viz. as a vector field, and also in terms of fluid dynamics. We will discuss these two aspects in this section.

(b) Fields

A field is a region under the influence of some physical agency. Physical fields interrelate and interconnect matter and energy within their realm of influence. Fields are not a form of matter. In current physics, several kinds of fundamental fields are recognized, e.g. the gravitational and electromagnetic fields and the fields of quantum physics (Sheldrake, 2005). A field can be pictorally represented by a set of curves, often referred to as flux lines, force lines, or, particularly in electromagnetic fields, field lines. The density of these lines at any given point represents the strength of the field, and their direction represents the direction conventionally associated with the agency (Illingworth, ed., 1991, p. 171).

According to Einstein, field lines are just trajectories, and nothing more. "All the lines of the field indicate only how a test body would behave if brought into the vicinity of the body for which the field is constructed" (Einstein & Infeld, 1960, p. 126-127).

(c) Forces and potentials

For the description of the field concept, we will take a gravitational field as an example. We will begin with the experimental result that is laid down as Newton's Law of Gravitation

$$f = G \frac{M \cdot m}{r^2} \qquad [4.1\text{-}1]$$

in which
>   M and m: the two masses that attract each other;
>   r: their distance;
>   G: a constant of proportionality, called the universal gravitational constant (Blüh & Elder, 1955, p. 181).

In the region surrounding a mass, other masses experience a force. This region around the mass where such forces can be observed is called a field, in this case a gravitational field (Blüh & Elder, 1955, p. 184).

The force per unit mass is defined as the field intensity or the field strength and is designated by f. The force F on a mass m at a point in the field is

$$F = m \cdot f \qquad [4.1\text{-}2]$$

A field of force can conveniently be described in terms of the work W that is necessary to move a unit mass in the field against the force exerted on it by the field. If a mass m is moved away from a larger mass M through a distance $(r_1 - r_2)$ work must be done against the gravitational attraction of the masses. The potential of the field around a mass M at a distance r is defined as the work that must be done to move a unit mass from that point towards infinity (Blüh & Elder, 1955, p. 186). It is designated as U:

$$U = \frac{G \cdot M}{r} \qquad [4.1\text{-}3]$$

In the example some characteristics of the field concept are illustrated for a gravitational field according to the theories of Newton. In a later century, Maxwell did prove that electromagnetic fields show very similar characteristics.

(d) *Morphic fields*

The field concept is also used in a much wider conation, like e.g the morphic field. This is a field within and around a morphic unit or which organizes its characteristic structure and pattern activity. A morphic unit is a unit of form or organization at all levels of complexity, like e.g. atoms, cells, patters of behaviour, social groups, planetary systems, etc. A morphic unit is also called a holon. Morphic fields underlie the form and behaviour of holons at all levels of complexity (Sheldrake, 2005). The concept of morphic fields seems to be less familiar to physicists than the concept of physical fields.

4.1.2   The light field

(a) *The photic field*

In the 1930s, Parry Moon and Domina Spencer tried to put photometry mathematically speaking on a more solid base. They did base their work on previous studies, more in particular those on light fields (Gershun 1938). The result of this work is published in 1981 (Moon & Spencer, 1981). 'Fields' were an essential part of these efforts.

"Field theory is associated with such names as Euler, Laplace and others. It is one of the most powerful methods of mathematical physics. Several investigators have used it in photometrics, but most lighting experts are completely ignorant of it" (Moon & Spencer, 1981, p. 1).

It is only fair to say that this ignorance is due, at least in part, to the fact that Moon and Spencer did introduce a completely new terminology in their studies. This is always a sure way to make a complete break with the past. The efforts of Moon, in the revision of his standard work on illuminating engineering, did not have much effect to heal the rupture (Moon, 1961, p. v-viii, preface to the Dover edition). One should add that this 'ignorance' is not so obvious in other scientific and engineering disciplines that also occupy themselves with electromagnetic radiation (Sterken & Manfroid, 1992).

As will be explained further on, the light field is not an electromagnetic field. It may be regarded is a morphic field, although that term had not yet been introduced then. Moon and Spencer (1981) use the term photic field. Probably the terminology did contribute to the confusion and to the 'ignorance' because these terms were being frowned upon by some traditional scientists.

(b) Fields of light rays

In the following section we will discuss several aspects of the field theory, because they may clarify a number of questions that seem to baffle most lighting engineers. As regards the terminology, we will adhere as far as possible to the more common CIE terminology. We will use the term light field, as introduced by Gershun (1939), in stead of the term photic field that was introduced by Moon and Spencer.

One must realise that the light field is quite distinct from the electromagnetic field that is described by Maxwell's equations. The light field belongs to the geometric optics, and the electromagnetic field to the physical optics. The difference between these two is described in some detail in sec. 2.1, where the nature of light is discussed. Geometric optics are discussed in sec. 10.1. As mentioned earlier, the light field is a morphic field, whereas the electromagnetic field is a physical field.

Although, as is indicated earlier, field lines are, according to Einstein, just trajectories, and nothing more, the electromagnetic field is something real. "The electric field is produced by a changing magnetic field, quite independently, whether or not there is a wire to test its existence. The magnetic field is produced by a changing electric field, quite independently, whether or not there is a magnetic pole to test its existence" (Einstein & Infeld, 1960, p. 145).

In the light field, the field lines are the light rays that are described in sec. 10.1.1b as the elements of geometric optics.

(c) The speed of light

The term 'field' is often used rather loosely. We indicated earlier that a field is a region under the influence of some physical agency. In current physics, several kinds of fundamental fields are recognized. We mention here, more in particular, the electromagnetic field. As is explained in sec. 2.1.1, one of the ways to represent 'light' is to consider it as vibrations in the electromagnetic field. Vibrations propagate themselves with a speed that is specific for the field. These vibrations are commonly called 'light'. This is the link between the light field and the electromagnetic field. Hence we will discuss here one of the essential aspects of the electromagnetic field, viz. the speed of light.

The mathematics of luminance 113

In the electromagnetic field, the speed of propagation of disturbances is the speed of light. There is something peculiar about the speed of light. In physical terms, it is about 300 000 km/h. However, according to Einstein, it behaves like being infinite (Einstein, 1956; Einstein & Infeld, 1960). Infinity is usually described as "that which goes beyond any fixed bound" (Blackburn, 1996, p. 193). That is to say that the speed stays the same whether the source moves or is at rest. The speed stays 300 000 km/h, even if something is added or subtracted.

The fact that light speed behaves like infinity can easily be seen from traditional Newtonian dynamics. According to the well-known equation of Newton, a body will move when a force is applied:

$$K = m \cdot a \qquad [4.1\text{-}4]$$

with
    K: the force;
    m: the mass of the body;
    a; the acceleration.

When the force is applied only during a limited time interval t, [4.1-4] becomes:

$$K \cdot t = m \cdot a \cdot t \qquad [4.1\text{-}4a]$$

   $K \cdot t$ is the momentum; $a \cdot t$ is the speed v.

In an approximation, on might write:

$$a \cdot t = d^2x/dt^2 \cdot dt = dx/dt = v \qquad [4.1\text{-}4b]$$

Actually, this ought to be written differently. So, [4.1-4b] becomes

$$K \cdot t = m \cdot v \qquad [4.1\text{-}5]$$

Because photons have no mass, which means m = 0, v becomes infinity. There is no sound explanation why 'infinity' in reality means '300 000 km/h'.

There is another weird aspect of this. According to the Heisenberg uncertainty principle, the product of the uncertainties of the momentum and of the location is about constant:

$$\Delta p \cdot \Delta x \text{ equals about } h/2\pi \qquad [4.1\text{-}6]$$

(After Illingworth, ed., 1991, p. 503). Because in [9.1-4a] K · t = 0/0, it can, mathematically speaking, have any value. So, according [4.1-6], x can have any value as well, which means of course that the position of the photon in its path (the x-direction) cannot be determined. This is sometimes expressed as meaning that the photon is as long as its path from source to detector!

(d) *The basic formula of photometry*

In sec. 3.1.1c, the basic formula of radiometry is introduced. It is described as:

$$E = \int_0^{+\infty} w_\lambda \cdot J_\lambda \cdot d\lambda \qquad [4.1\text{-}7; 3.1\text{-}1]$$

in which
E: the irradiance, or power density;
J_λ spectral distribution function of the incident radiation;
w_λ: the weighing function of the receiver, or the spectral sensitivity function of the detector;
λ: the wavelength of the incident radiation.

As has been mentioned, Moon & Spencer (1981, p. 212) called this relation 'the fundamental equation'. Here we will look closer into this relation, using the description of Gall (2004, sec. 6.1, p. 37-39). We will begin this discussion by the introduction of the concept of the perfect diffuser. The perfect diffuser is also called the Lambertian radiator (Gall, 2004, sec. 6.3, p. 40-41). A Lambertian radiator is a radiator that shows the same luminance in all directions:

$$L(\gamma_1) = L = \text{constant} \qquad [4.1\text{-}8]$$

It is depicted in Figure 4.1.1.

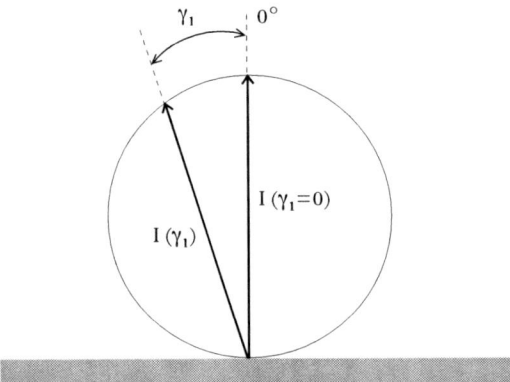

Figure 4.1.1 A perfect diffuser or Lambertian radiator (After Gall, 2004, Fig. A19a, p. 40).

The mathematics of luminance

A perfect diffuser is defined as an object – or rather the surface of an object – that is characterised by the fact that the luminance is the same in all directions. This implies that the luminous intensity of an area element dA_K of this surface differs in all directions. In a direction that makes an angle γ to the normal on the surface, the apparent or projected size of the surface element dA_p is reduced, as it seems to be foreshortened. The significance of the terms in the following discussion is explained in Figures 4.1.1, 4.1.2 and 4.1.3.

As can be seen from Figure 4.1.2, its apparent size $dA_p = dA_K \cdot \cos\gamma$.

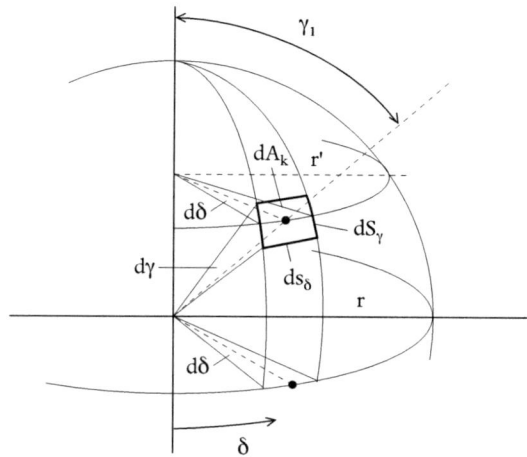

Figure 4.1.2 The projection of a solid angle (After Gall, 2004, Fig. A9, p. 20).

In the integration, one must take into account that the spherical element $\delta\omega$ gets 'narrower' when one nears the top of the half-sphere. This is expressed as follows:

$$\Omega_p = \int_\delta \int_\gamma \sin\gamma \cdot \cos\gamma \cdot d\gamma \cdot d\delta \qquad [4.1\text{-}9]$$

Next, we need to reconsider the luminance concept. We will follow here the description of the luminance used by Gall (2004, sec. 5.6, p. 35). The luminance concept is depicted in Figure 4.1.3.

From this figure, it follows that

$$L_v = \frac{d^2\Phi_v}{dA_{p_1} \cdot d\Omega_1} = \frac{dI_v(\gamma_1)}{dA_{p_1}} \qquad [4.1\text{-}10]$$

in which:

L_v: the luminance (the subscript v means visual effect);
A_{p_1}: the projected surface element (p means projected);
Φ_v, γ_1, and Ω_1 follow from Figure 4.1.3.

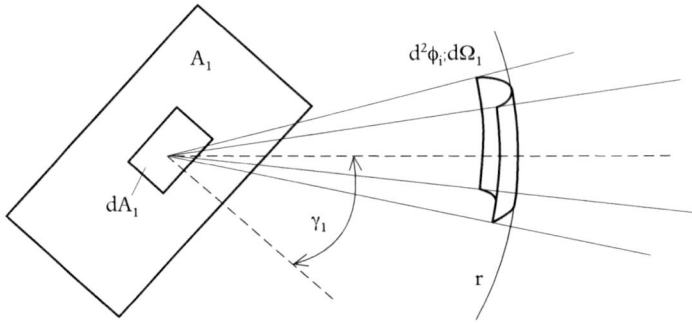

Figure 4.1.3 The definition of the luminance (After Gall, 2004, Fig. A17, sec, 5.6, p. 35).

From this the basic formula of photometry can be described. This description is based on the explanations given by Gall (2004, sec. 6.1, p. 37-39). We consider the exchange of light between two surfaces, where the index 1 stands for the light-emitting surface, and the index 2 for the illuminated surface. From the definition of the luminance given in [4.1-10], the following expression for the luminous flux can be derived. From this the basic formula of photometry can be described. This description is based on the explanations given by Gall (2004, sec. 6.1, p. 37-39). We consider the exchange of light between two surfaces, where the index 1 stands for the light-emitting surface, and the index 2 for the illuminated surface.

In [4.1-10] it is stated that

$$L_v = \frac{d^2\Phi_v}{dA_{p_1} \cdot d\Omega_1} = \frac{dI_v(\gamma_1)}{dA_{p_1}} \quad\quad [4.1\text{-}10]$$

From this, it can be derived that

$$d^2\Phi_1 = L_1(\gamma_1) \cdot dA_1 \cdot \cos\gamma_1 \cdot d\Omega_1 \quad\quad [4.1\text{-}10a]$$

In this is

$$d\Omega_1 = \frac{dA_2 \cdot \cos\gamma_2 \cdot \Omega_0}{r^2} \quad\quad [4.1\text{-}11a]$$

$$d\Omega_1 \cdot \cos\gamma_1 = d\Omega_{p_1} \quad\quad [4.1\text{-}12a]$$

The mathematics of luminance

and also

$$d\Omega_2 = \frac{dA_1 \cdot \cos \gamma_1 \cdot \Omega_0}{r^2} \qquad [4.1\text{-}11b]$$

$$d\Omega_2 \cdot \cos \gamma_2 = d\Omega_{P_2} \qquad [4.1\text{-}12b]$$

As is explained in sec. 4.3.2c, the double integral is the geometric flux G:

$$G = \int_{A_1} \int_{\Omega_1} dA_1 \cdot \cos \gamma_1 \cdot d\Omega_1 \qquad [4.1\text{-}13]$$

When we consider the fraction $d^2\Phi_1$ of the luminous flux from dA_1 that hits dA_2, we have as indicated earlier:

$$d^2\Phi_1 = L_1(\gamma_1) \cdot dA_1 \cdot \cos \gamma_1 \cdot d\Omega_1 \qquad [4.1\text{-}10a]$$

When we compare this with the fraction $d^2\Phi_2$ of the luminous flux from dA_2 that hits dA_2, we have:

$$d^2\Phi_2 = L_2(\gamma_2) \cdot dA_2 \cdot \cos \gamma_2 \cdot d\Omega_2 \qquad [4.1\text{-}10c]$$

From [4.1-10a] and [4.1-10c] it follows that the exchange of the luminous flux between the surfaces A_1 and A_2 is proportional to the lumimances L_1 and L_2. Thus:

$$\frac{d^2\Phi_1}{d^2\Phi_2} = \frac{L_1(\gamma_1)}{L_2(\gamma_2)} \qquad [4.1\text{-}14]$$

In Figure 4.1.4, the terminology for the basic formula of photometry is depicted.

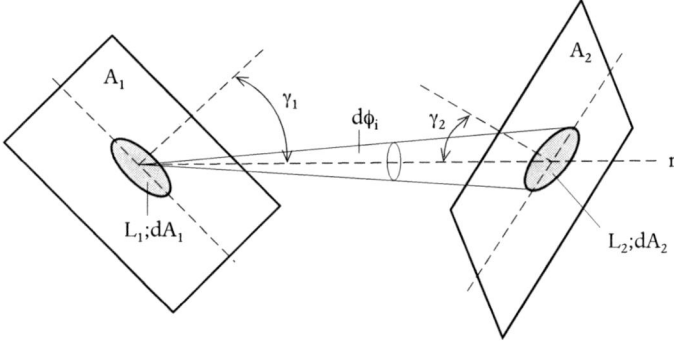

Figure 4.1.4 Terminology for the basic formula of photometry (After Gall, 2004, Fig. 18).

The basic formula of photometry can be described in different ways. The most common way is

$$d^2\Phi_1 = L(\gamma_1) \cdot dA_{p_2} \cdot d\Omega_2 \qquad [4.1\text{-}15]$$

4.2 Some aspects of hydrodynamics

4.2.1 The continuity principle

The continuity principle is one of the fundamental concepts of physics (Wachter & Hoeber, 2006, sec. 2.1.1, p. 116). For fluids, it is succinctly described as follows: "For continuous motion, the increase in mass of fluid in any time interval δt within a closed surface in the fluid is equal to the difference of the mass flow in and the mass flow out through the surface" (Illingworth, ed., 1991, p. 85). For incompressible fluids, this results from the Bernoulli law of hydrodynamics, that will be explained in a further part of this section (Gerlach, ed., 1964, p. 232).

Here, it is stated for fluids. As is explained in Narisada & Schreuder, 2004, sec. 14.1.7a, it can be applied equally well to any other form of moving particles or waves. As regards the electric charges in electromagnetism, it can be written as:

$$\text{div } i + \frac{\delta \rho}{\delta t} = 0 \qquad [4.2\text{-}1]$$

in which:
  i is the current density;
  ρ is the charge density;
  t is the time.
  (Gerlach, ed., 1964, p. 94).

It can be shown that this relation follows from the well-known Maxwell equations. The basic concept of the Maxwell equations is explained in Illingworth, ed., 1991, p. 291-292 and in Gerlach, ed., 1964, p. 97-99. The complete derivation of the Maxwell equations is given by Joos in the chapter on quasi-stationary fields (Joos, 1947, chapter XVI, more in particular p. 299, 314-315), and also in Wachter & Hoeber (2006, sec. 2.2).

The continuity principle means for the Maxwell equations that the current density i is invariant in time i.e. the equations do not change when t is replaced by -t, which means of course a reverse of the flow direction. More generally it can be stated that the equations do not change when t is replaced by t', where t is the time for a stationary object and t' the time for a moving object (Feynman et al., 1977, p. 15-2).

4.2.2 Bernoulli-fluids

As is explained earlier, a field can be pictorally represented by a set of curves, usually referred to as field lines, lines of flux. or lines of force. The density of these lines at any given point represents the strength of the field, and their direction represents the direction conventionally associated with the agency (Illingworth, ed., 1991, p. 171). The speed of the propagation of the disturbances is no concern; the field is a static collection of force lines. This is the way the field concept can be applied in geometric optics. As is explained in sec. 10.1, geometric optics are the basis of almost all optical design, lighting design included. With this in mind, force lines can conveniently be considered as light rays. The point that light rays have no physical meaning can be disregarded at the scale of traditional optical and lighting design.

There are a number of aspects in lighting applications where the 'model' of the electromagnetic field, consisting of light rays, is not adequate. The model concept is explained in sec. 1.5.1. In different sections of this book, a number of cases are mentioned. For these, it seems to be more to the point to look for the other way to describe light: as a stream of particles moving from one place to another. As is explained in sec. 2.1.3c, these particles are called photons. The flow of photons can be described in terms of fluid dynamics. However, there is a problem. Traditional fluid dynamics discuss the speed of incompressible fluids. Its basic theorem is the theorem of Bernoulli:

$$p + \frac{\rho}{2} \cdot v^2 = \text{constant} = p_0 \qquad [4.2\text{-}2]$$

with:
   p in Pascal
   ρ in kg/m^3
   v in m/s

This representation is for incompressible fluids and horizontal flow without friction (After Kuchling, 1995, p. 159, eq. 10.8). In a further part of this section, we will come back to the implications of this assumption, when discussing diffraction. In other words, the elementary particles of a Bernoulli-fluid are 'regular' particles like atoms of molecules. Their main characteristic is that they have mass. Furthermore, the mass is constant for speeds much lower than the speed of light. According to the well-known relations of the theories of the special relativity, the mass m of a moving particle is determined by its rest mass m_0 and its speed (Illingworth, ed., 1991, p. 405).

$$m = \frac{m_0}{\sqrt{(1 - v^2/c^2)}} \qquad [4.2\text{-}3]$$

It must be noted that $m = m_0$ for low speeds when $v \ll c$. The 'light fluid', however, is different. The elementary particles of light are photons that have a rest mass of zero, and that have a constant speed equal to the speed of light. If one would insert these values in equation [4.2-3], the result would be $m = 0/0$, which, mathematically speaking, has no meaning. One of the implications is that the assumption of incompressibility must be abandoned.

A complete, mathematically rigid, derivation of the formulae of hydrodynamics would lead is too far away from the subject matter of this book. A full mathematical treatment can be found e.g. in Joos (1947, chapter IX), where the fundamental equation of hydrodynamics is given:

$$\mathbf{G} = \frac{1}{\rho} \cdot \text{grad } p \qquad [4.2\text{-}4]$$

in which
  G: the body force acting on an element of the fluid;
  ρ: the density of the fluid;
  p the normal pressure. The pressure is normal, e.g. perpendicular to the surface, because in a fluid the shear can be neglected (After Joos, 1947, p. 161; p. 180, eq. 1a).

4.2.3 The equation of continuity

Equally important as the fundamental equation of hydrodynamics is another equation, called the equation of continuity (Joos, 1947, p. 184). The continuity principle is discussed earlier in this section.

The equation of continuity indicates that in an equilibrium condition, the mass of fluid passing outward through the surface of a volume element must be equal to the decrease of the amount of fluid within the volume element – barring sources or sinks. When the relevant surface integral is converted into a volume integral using Gauss's theorem. The Gauss's theorem links surface integrals to volume integrals. Its derivation is given in Joos (1947, chapter I, sec. 8, p. 23-25). Since the equation must hold for every element of volume, we have:

$$\text{div } \rho \, \mathbf{v} = -\frac{\partial \rho}{\partial t} \qquad [4.2\text{-}5]$$

The mathematics of luminance

in which

  ρ: the density of the fluid;
  v: velocity;
  t: time.

After Joos (1947, p. 185, eq. 12). From these two equations the Bernoulli-theorem, that holds for incompressible fluids, can be derived (Joos, 1947, p. 185-187). Because the 'light fluid' cannot be considered as being incompressible, we will not follow this. However, we will come back to the equation of continuity in another part of this section.

4.2.4 Friction and diffraction

(a) The width of a light ray

In an earlier part of this section, we discussed the Bernoulli-fluid. One of its characteristics is that the fluid flows without friction. The same condition is assumed for the 'light fluid', be it that the concept of friction has to be adapted. When beams are narrow, the simplified concept of photons that hurl through empty space without any mutual interaction, falls short. The ensuing phenomena can be described adequately by using the alternative way to describe light, i.e. by electromagnetic waves. The wave model allows us to apply the well-known phenomena of diffraction. One of the effects of diffraction is that pencils of light – or light beams – only can be considered as propagating themselves along straight lines if they are much wider than the wavelength. According to Prins (1945, p. 51), the width of a beam must be at least 10 times the wavelength, before the loss of energy at the sides of the beam can be neglected. This is depicted in Figure 4.2.1.

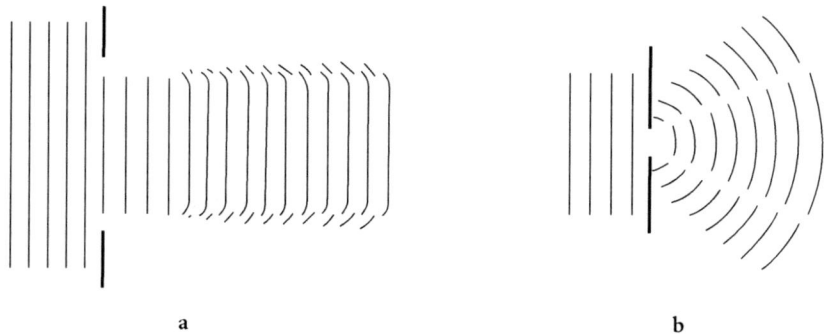

Figure 4.2.1 *Diffraction limits the minimum width of a light beam (After Prins, 1945, Fig. 34).*

(b) Diffraction

In the study of the visual acuity of the human visual system, point sources form an important class of test objects. The colloquial definition of a point source is simply a source that is so small that no extension can be seen. Stars are point sources. However, this definition cannot be used in a more precise consideration of the visual detection of point sources. Due to the diffraction – and to other optical aberrations as well – a point source will never be seen as a mathematical 'point'. We will give a brief explanation here. Details are given in Narisada & Schreuder, 2004, sec. 4.1.4c., where the visual acuity for point sources is discussed.

From theoretical optics it is well-known that, when light hits a screen after passing through a small aperture, a ring-shaped pattern will be seen. This pattern is called the diffraction pattern. For an arrangement with circular symmetry, the illuminance at a point P on the screen is given by the relation:

$$E = E_0 \cdot \left| \frac{J_1(2m)}{m} \right|^2 \qquad [4.2\text{-}6]$$

with

$$m = \frac{\pi \delta}{2 \lambda} \cdot \frac{r_0}{z_0} \qquad [4.2\text{-}6a]$$

in which
  δ: diameter of light source;
  r_0: displacement of P in the screen from the axis;
  λ: wavelength of the light;
  z_0: the distance between the source and the screen;
  $J_1(2m)$: the Bessel function of the first order;
  E: illuminance in P;
  E_0: illuminance for $r_0 = 0$.

For the calculation of the pattern it is assumed that a large number (n) of oscillators of equal strength are spread evenly over the aperture (Feynman et al., 1977, Vol. I. p. 30-1). Their phase (ϕ) differs for different positions on the screen. This, of course, is the cause for the diffraction pattern to arise. The resulting intensity on the screen shows a pronounced maximum at phase zero – straight ahead. For increasing phase, the intensity on the screen passes through a series of equally spaced minima with maxima in between. The intensity in the minima is zero; the intensity of the maxima decreases rapidly with increasing phase. The intensity at the first maximum is less than 5% of that of the maximum at phase zero (0,0047; Feynman et al., 1977, Vol. I. p. 30-2). See Figure 4.2.2. Measurements of the retinal distribution are described by Vos et al. (1976). Some of these considerations have also been discussed in sec. 2.1.3a.

The mathematics of luminance

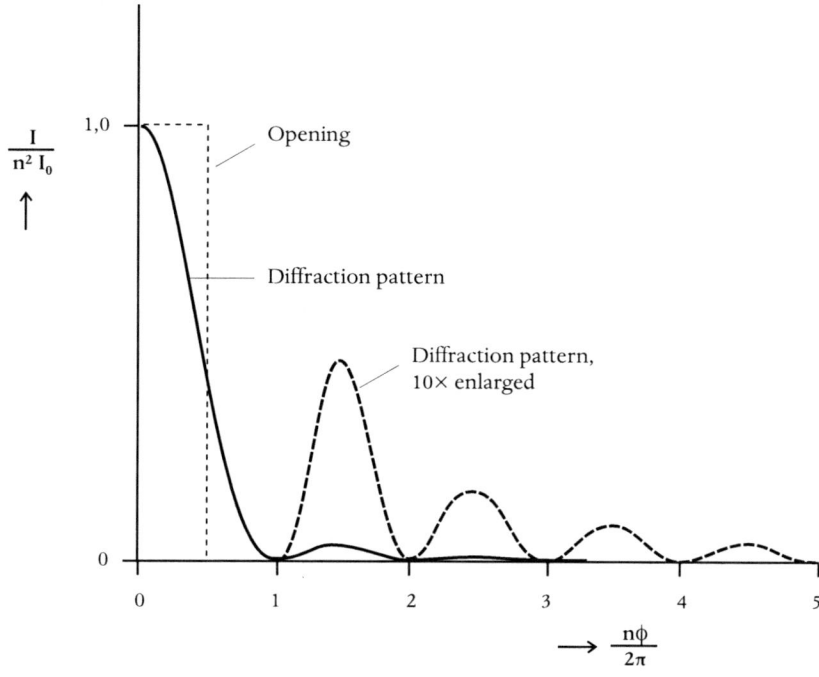

Figure 4.2.2 *The calculated refraction pattern (After Narisada & Schreuder, 2004, Fig. 9.1.12. Based on Feynman et al., 1977, Vol. I. Fig. 30-2).*

(c) The minimum separable

For the practice of determining visual acuity an important test condition is called the 'minimum separable'. It is the minimum distance between two points of light. i.e. two neighboring stars (Moon, 1961, table XVIII; see Narisada & Schreuder, 2004, Table 9.1.4). Crudely speaking, the apparent size of a point source is said as being equal to the width of the first spread-out maximum, that is the place where the first minimum occurs, or, more precisely, the diameter of the first diffraction ring. The minimum distance is determined by the refraction in the eye in such a way that the two objects seem to be situated at consecutive maxima, that are, as indicated earlier, equidistant in angular measure.

Using the pattern as given in Figure 4.2.2, the minimum separable can be calculated. See Figure 4.2.3.

The detection of point sources is discussed in detail in Narisada & Schreuder (2004, sec. 9.1.7), where several examples are worked out.

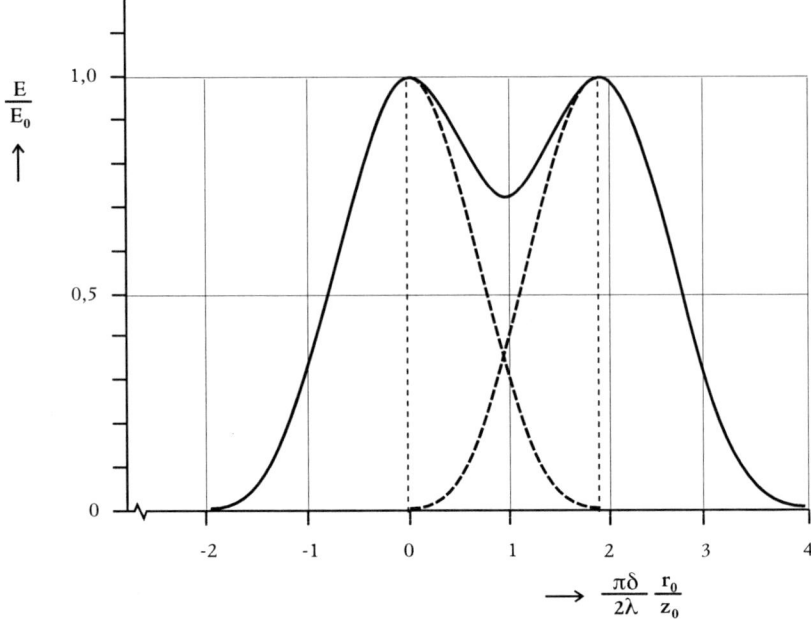

Figure 4.2.3 *The calculated retinal illumination caused by two light sources at minimum distance (After Narisada & Schreuder, 2004, Fig. 9.1.12. Based on Moon, 1961, Fig. 12.17).*

4.3 The luminance of real and virtual objects

4.3.1 The need for a proper definition of luminance

The subject matter of this section covers one of the most fundamental aspects of photometry, viz. the precise definition of luminance. The subject is usually not treated properly in most lighting engineering textbook – if at all. It must be stressed, however, that the proper definition of the luminance is essential in most engineering applications. As a matter of fact, many lighting engineers disregard these problems in their daily design work. This disregard may occasionally have disastrous results.

The discussions of the subject matter in this section is based on a number of sources, that all deal with the same aspects of electromagnetic radiation, be it in different ways. Because the subject is from a mathematical point of view more complex than most other parts of this book, we have tried to summarize these sources in such way that the fundamental aspects are made clear.

The sources we have mentioned include a.o. Gall (2004), Helbig (1972), Hentschel (1994, sec. 2.2, p. 21-24), Hentschel, ed. (2002, sec. 2.2, p. 24-28), Moon & Spencer (1981, sec. 1.06), Narisada & Schreuder (2004, sec. 14.1.7 a,b,c), Reeb (1962, sec. 2.314), Sterken & Manfroid (1992, sec. 1.2.1), and Zijl (1951).

4.3.2 The general definition of luminance

(a) The direction aspects of the luminance

As is explained in sec. 3.2.5, the luminance is usually expressed in terms of the luminous intensity per square unit area, viz. cd/m^2. This definition makes it difficult to apply it to a mathematical surface, which may, or may not, correspond to a physical surface, such as the night sky (Narisada & Schreuder, 2004, sec. 14.1.7a). It is not straightforward to indicate what a square meter of open sky actually does mean. In order to clarify this, we need to consider some of the principles of photometry. Of course, the principles of photometry are the same as those of radiometry, as far as the spectral luminous efficiency curve – the v_λ-curve – is taken into account.

The basic notion is that, as indicated earlier, photometry is essentially just nothing but 'counting photons'. The spectral luminous efficiency curve provides the weighing factors for photons of different 'colour' – or wavelength, or energy. These weighing factors are explained in sec. 7.1.1, where visual perception is discussed.

The luminance describes the geometrical distribution of luminous flux with respect to position and direction. It equals the flux per unit of projected area, and per unit of solid angle (Born & Wolf, 1964, as referred to by Sterken & Manfroid, 1992, p. 8). It must be stressed as essential, that the area is always the apparent area of the emitting surface as seen by the observer – or by the measuring device.

The geometry is depicted in Figure 4.3.1.

(b) Light tubes

The next step is the consideration of the direction of the flow. In sec. 4.2.1, we have explained the continuity principle. In an equilibrium condition, the mass of fluid passing outward through the surface of a volume element must be equal to the decrease of the amount of fluid within the volume element. In simple words, what goes in will go out. It is customary, however, to explain the continuity principle as the fact that the energy or mass flow at the two ends of a tube are equally large.

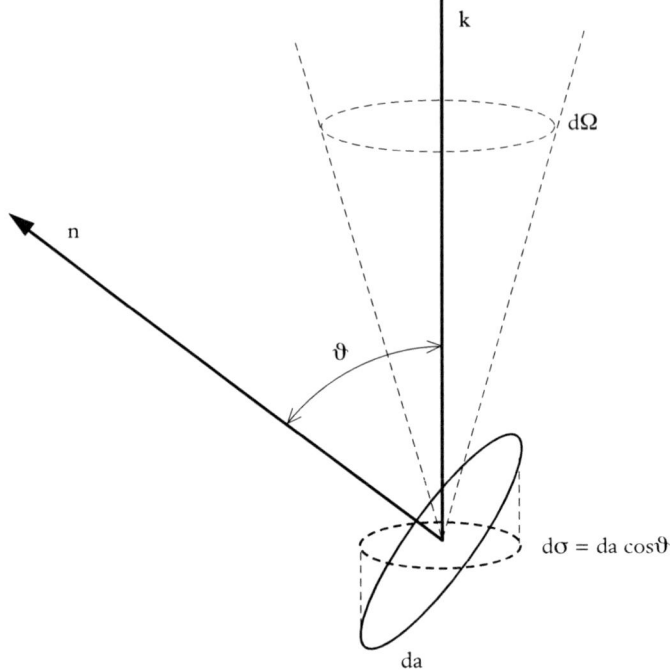

Figure 4.3.1 *The luminance in the direction k is the ratio between the luminous intensity in that direction, and the projected area (After Narisada & Schreuder, 2004, Fig. 14.1.13. Based on Sterken & Manfroid, 1992, Fig. 1.4).*

In most cases, the calculations are made in such a way that it might seem that at both ends, energy or mass goes in. This is, of course, not consistent with the continuity principle, nor with reality. As a matter of fact, it is nothing but the result of a convention about which direction is considered positive or negative. In this convention, it must be kept in mind that hydrodynamics is a special case of the theories of elasticity, be it with special boundary conditions of the elasticity coefficients. In the theory of elasticity, stresses normal to a closed surface are reckoned positive when in the outward direction, inward pressures need to be given a negative sign (Joos, 1947, p. 180, footnote).

Earlier in this section when explaining the continuity principle in terms of the Maxwell equations, it was indicated that the current density i is invariant in time i.e. the equations do not change when t is replaced by -t (Feynman et al., 1977, p. 15-2). This would mean that the flow is 'inward' but with a negative sign, which means of course 'outward'. It seems to be a complicated way to express things, but it agrees with the conventions of theoretical and mathematical physics.

The mathematics of luminance

(c) The physics of light tubes

Moon and Spencer did derive the characteristics of the light tubes directly from Maxwell's equations (Moon & Spencer, 1981, p. 3-12). We will follow here this approach, but focus for practical reasons on the equally strict, but more accessible, approach of Reeb (1962, sec. 2.314, p. 16-18).

As is explained in sec. 4.1.1b, a field is a region under the influence of some physical agency. A physical field is a portion of space in which a given physical phenomenon occurs (Illingworth, ed., 1991, p. 171; Moon & Spencer, 1981). This can be described by field lines, or by a field map. The field map consists of tubes of flux. As will be explained later, the definition of a light tube includes the fact that the radiant power (in watts) through a tube is invariant.

We will begin with the definition of the light tube by making use of Figure 4.3.2. The significance of A, dA, ε, and dq follow directly from that figure.

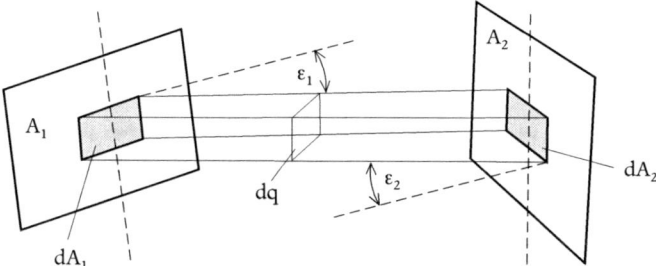

Figure 4.3.2 *The definition of the light tube (After Reeb, 1962, Fig. 11, p. 17).*

(d) *The definition of the geometric flux*

The definition of the geometric flux is based on the following geometrical considerations, that are in part based on Reeb (1962, sec. 2.314) and on Narisada & Schreuder (2004, sec. 14.1.7). Geometric considerations means that the central issue is the part of the electromagnetic field that is considered in the exchange of radiation between two surfaces. In most cases, a certain quantity of radiation flux – or energy – can be allotted to a volume element.

Usually, the fact that the limiting surfaces of the tubes are inclined with respect to the axis of the tube is taken into account (Reeb, 1962, p. 16-17). However, this involves only the cosine of the angle between the normal to the surface and the axis, as is depicted in Figure 4.3.1 and 4.3.2. As the inclination is not of any fundamental relevance, we will look only to the projected areas on a plane perpendicular to the axis. In this way, the awkward cosinus disappears. In fact they do not disappear, but as we consider only the normal to the surface, all cosines have the value of 1, so they can be disregarded.

First, we will consider the exchange of radiant flux between two surface elements A_1 and A_2. If a surface element of A_1 is projected in a parallel way onto A_2, one gets a prismatic, elementary part of the radiation field. This elementary volume has been depicted in Figure 4.3.2. It is called an elementary light tube or a light pencil.

The cross-section of the pencil is

$$dq = dA_1 \cdot \cos \varepsilon_1 = dA_2 \cdot \cos \varepsilon_2 \qquad [4.3\text{-}1]$$

In this relation, dq is very small but not infinitesimally small, as is the case for all differentials in physics. As is explained in the preceding part of this section, when considering only the area perpendicular to the line connecting dA_1 and dA_2, the cosinus of both ε_1 and ε_2 equal unity, so they can be disregarded. We will write da_1 for dA_1 when $\cos \varepsilon_1 = 1$, and da_2 for dA_2 when $\cos \varepsilon_2 = 1$. The distance r between any part of da_1 to any part of da_2 is the same. This implies that it is possible to allot a quantity of radiation to the elementary light tube. This radiation flows from a_1 to a_2. According to what was explained in an earlier part of this section, when the invariance of time and of direction was discussed, the same quantity of radiation flows from a_2 to a_1. The total exchange of radiation between a_1 to a_2 is therefore equal to the sum of all elementary light pencils, that are considered as being independent of each other. Effects of diffraction are disregarded.

The size of a single elementary pencil is proportional to da_1 as well as to the solid angle $d\omega_1$ with its apex in da_1 and which includes da_2. This solid angle measures:

$$d\omega_1 = \frac{da_2}{r_2} \qquad [4.3\text{-}2a]$$

In a similar way, the solid angle $d\omega_2$ with its apex in da_2 and which includes da_1, measures:

$$d\omega_2 = \frac{da_1}{r_2} \qquad [4.3\text{-}2b]$$

These solid angles are depicted in Figure 4.3.3.

Integration over all relevant elementary light pencils yields the total volume relevant fo the radiation exchange:

The mathematics of luminance

$$G = \int\int_{A_1\,\omega_1} dA_1 \cdot d\omega_1 \qquad [4.3\text{-}3]$$

Combining [4.3-3] with [4.3-2a] and [4.3-2b] the geometric flux can be defined as:

$$G = \int\int_{A\,\omega} dA \cdot d\omega \qquad [4.3\text{-}4]$$

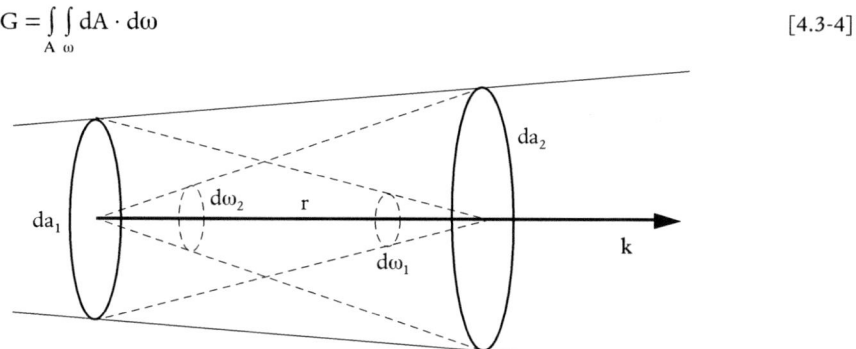

Figure 4.3.3 Description of the solid angles in [4.3-2a] and in [4.3-2b] (After Narisada & Schreuder, 2004, Fig. 14.1.14. Based on Sterken & Manfroid, 1992, Fig. 1.5).

(e) *The throughput of light*

In sec. 4.2.2 it is explained that, in traditional fluid dynamics, the basic theorem is that of Bernoulli. The Bernoulli-theorem holds for fluids that are incompressible and that flow without friction. The theorem relates the speed to the cross-section of the canal. Also it is explained that light, although it shows some characteristics that usually are linked to fluids, is not a Bernoulli-fluid: the speed is constant. So the assumption of incompressibility must be abandoned. However, in another aspect the continuity principle is still valid. In a light tube, light is added or subtracted only if there are sources or sinks. If this is not the case, the light that comes into the tube at the one end will get out at the other end.

This fact allows us to define the throughput of light. From Figure 4.3.3 it follows, that the radiant flux is invariant along the light beam. In Figure 4.3.3, the luminance is, according to the usual definition, the luminous intensity **k** divided by the surface dσ that is projected on a plane perpendicular to **k**. The luminance L can be written as:

$$L = \lim_{d\sigma \to 0} \frac{dI}{d\sigma} \qquad [4.3\text{-}4a]$$

or as:

$$L = \lim_{d\sigma \to 0,\, d\Omega \to 0} \frac{d_2 F}{d\Omega \cdot d\sigma} \qquad [4.3\text{-}4b]$$

The relations [4.3-4a] and [4.3-4b] are based on Sterken & Manfroid, 1992, p. 9. See also Narisada & Schreuder, 2004, sec. 14.1.7c.

As has been adstructed in Figure 4.3.3, the luminous flux is invariant along the light beam. As is indicated earlier, being a light pencil, where no radiation is entered nor lost through the outer mantle surface, the fluxes through both end are equal. The flux through surface S_1 is:

$$dF_1 = L_1 \cdot d\sigma_1 \cdot d\Omega_1 \qquad [4.3\text{-}5a]$$

The flux through surface S_2 is:

$$dF_2 = L_2 \cdot d\sigma_2 \cdot d\Omega_2 \qquad [4.3\text{-}5b]$$

From Figure 4.3.3, it follows that

$$d\sigma_1 \cdot d\Omega_1 = d\sigma_2 \cdot d\Omega_2 \qquad [4.3\text{-}6]$$

From [4.3-5a], [4.3-5b] and [4.3-6], it follows that $F_1 = F_2$. Along a light pencil, the decrease in divergence $d\Omega$ is exactly compensated by the increase of the cross-section area $d\sigma$ (Narisada & Schreuder, 2004, p. 780; Sterken & Manfroid, 1992, p. 9).

From this, the throughput can be defined. The throughput (sometimes written as 'thruput') is, of course, the luminous flux passing through the light tube. It is, as in indicated earlier, invariant. It can be written in different ways. We will conclude this section by presenting the description given by Narisada & Schreuder, 2004, sec. 14.1.7c, based on Sterken & Manfroid, 1992, sec. 1.2.1. The description is adstructed in Figure 4.3.4.

Figure 4.3.4 describes the luminous flux carried from a surface with an area da_1 at S to a receiving surface of area da_2 at R. Again taking only the scalar value of F, it follows from Figure 4.3.2, that:

$$dF = L \cdot d\sigma_1 \cdot d\Omega_2$$
$$= L \cdot da_1 \cdot \cos\theta_1 \cdot da_2 \cdot \frac{\cos\theta_2}{r^2}$$
$$= L \cdot d\sigma_2 \cdot d\Omega_1$$

The mathematics of luminance

The quantity

$$d\sigma_1 \cdot d\Omega_2 = d\sigma_2 \cdot d\Omega_1 = da_1 \cdot \cos\theta_1 \cdot da_2 \cdot \frac{\cos\theta_2}{r^2}$$

is called the throughput. It is purely a geometric notion. What it actually means is that along a pencil of light, the divergence is in fact constant, and for two opening angles, the flux per unit area goes down as much as the angles increase, hence the conservation of throughput (Schwarz, 2003). It should be noted that, for reasons of clarity, the solid angle $d\Omega_2$ is not included in Figure 4.3.4. See for this Figure 4.3.3. This discussion is adapted from Sterken & Manfroid (1992, p. 7-9). The same considerations are described in Hentschel, ed. (2002, sec. 2.2, p. 24-28) and Reeb (1962, sec. 2.314).

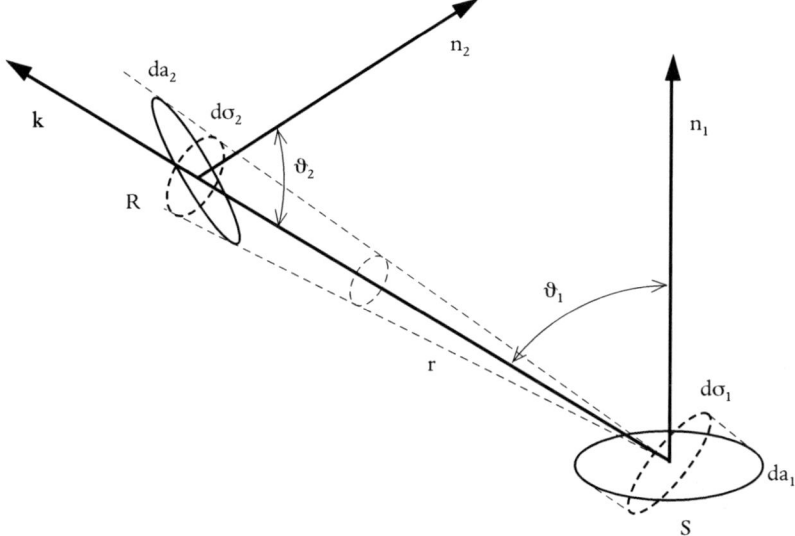

Figure 4.3.4 *The throughput of a light beam is conserved (After Narisada & Schreuder, 2004, Fig. 14.1.15. Based on Sterken & Manfroid, 1992, Fig. 1.6).*

4.4 The luminance of reflecting surfaces

4.4.1 The use of surface luminances

One of the most common parts in lighting design and its engineering aspects is the determination of the luminance of a specific surface when the illuminance is known. In road lighting this is the central issue of the 'luminance engineering' approach, where it is assumed that the road surface luminance is the basic quantity that determines the quality of the lighting installation (De Boer, 1967; De Boer, ed., 1967; Schreuder, 1964, 1967; Van Bommel & De Boer, 1980). This assumption is the basis of almost all standards and recommendations of national and international organisations for road and tunnel lighting. We just may mention a few international documents: CEN (2000; 2002, 2003); CIE (1965, 1992, 1995); Also we will add a few documents from the Netherlands: NSVV (1963, 1991, 2002, 2003); Schreuder (1964a). The different aspects of road lighting design are discussed in chapter 11.

In many indoor applications, the same is true. In the design of most utility rooms, such as offices, the most important design aspect is the assessment of the luminance of different parts of the field of view. Most important are the luminance of the visual task, and that of its direct surroundings. These two determine the contrast between the task and the background. The luminance of the walls and of the ground surface follow in importance. These two determine to a large degree the adaptation level, and thus the lower limit of the visual performance. In all cases, it regards the luminance of physical surfaces. These subjects are treated in detail in the 'classical' textbooks of Baer (1990), Baer, ed. (2006), Boyce (1981), De Boer & Fischer (1981), Hentschel (1994), and Hentschel, ed. (2002). See also CIE (1975, 1982), and Schreuder (2006).

Of course, other luminances, that are not directly related to physical surfaces, are important as well in interior lighting installations. These luminances include that of light sources, luminaires, and windows. In more recent times, the luminance of computer display screens is of particular interest.

4.4.2 Definition of reflection

In sec. 3.2.2c it is explained that, when light hits a surface, a part of it is reflected, a part is transmitted, and the rest is absorbed. For an opaque surface, the transmission is, of course, zero. The reflectance is the ratio of the reflected luminous flux to the incident luminous flux under the given conditions. The transmittance is the ratio of the transmitted luminous flux through the optical medium to the incident luminous flux in the given conditions. The absorbance is the ratio of the luminous flux that is absorbed by the optical medium to the incident luminous flux in the given conditions. Usually, the reflectance is indicated by R, the transmittance by T, and the absorbance by A. The definition of these factors makes use of the luminous flux involved. The following definition are used:

$$T = \frac{\phi_T}{\phi_{[total]}}$$

The mathematics of luminance

$$R = \frac{\phi_R}{\phi_{[total]}}$$

$$A = \frac{\phi_A}{\phi_{[total]}}$$

in which:
- ϕ_T: the transmitted flux;
- ϕ_R: the reflected flux;
- ϕ_A: the absorbed flux;
- $\phi_{[total]}$: the incident flux.

It should be noted that all three factors are dimensionless (Baer, ed., 2006, sec. 1.2.2.2, p. 27).

Because light is not 'destroyed' nor 'created', the total amount is constant. From this, it follows directly that:

$$R + T + A = 1 \qquad [3.2\text{-}1]$$

More precisely put, this in fact describes the first law of thermodynamics. This law is also called the law of the conservation of energy. The law runs: "In a closed system, the sum of all forms of energy is constant" (Kuchling, 1995, p. 287). In a general form it can be written as:

$$Q = \delta U + A \qquad [4.4\text{-}1]$$

with:
- Q: the absorbed energy (i.e. heat);
- δU: the increase of the internal energy;
- A: the external energy ('work')

After Schreuder, 1998, eq. 5.3.5; Breuer 1994, p. 109).

For obvious reasons, an object only can be observed if at least some of the incident light is reflected. To be more precise, the reflected light is observed.

As is explained in sec. 3.2.5, where the practical issues of luminance are discussed, brightness is used as a first indication of luminance. It is also indicated that, when more light is reflected, the object is brighter. This can be the result of two factors.
1. When more light falls on the object – when the illuminance increases – the brightness increases proportionally.
2. When, the illuminance being unchanged, the reflection is higher, also the brightness increases. The brightness will increase proportionally to the reflection.

4.4.3 The luminance factor

(a) The reflection factor and the luminance factor
In view of the proportionality that is explained in the preceding part of this section one might be tempted to write L = R · E. Because the reflection factor R is dimensionless, it would mean that L and E would need to have the same dimension. However, it is explained in the preceding parts of this chapter that this is not the case. So, one must adapt the simple relation:

L = q · E   [4.4-2]

Because q determines the luminance for a given illuminance, it is usually called the luminance factor. In further parts of this section we will discuss the luminance factor in more detail.

(b) A description of the luminance factor
The reflection factor, just as the transmission factor and the absorption factor that have been introduced earlier in this section, 'overall' factors. They are related to the light or radiant flux in total, independent of the directions that are involved. As is mentioned earlier, they follow from the First Law of thermodynamics, the law of energy conservation.

In sec. 2.1.3c it is explained that electromagnetic radiation can, in many cases, be considered as a stream of particles, the photons. Photons move at a fixed speed – by definition the speed of light – in a certain direction. Although they have a rest-mass of zero, they convey energy. It is therefore possible – even necessary – to consider them as a vector. As is well-known, a vector is characterised by a scalar – just a number – and a direction. In the case of photons, the vector represents work with the unit watt. As is explained in sec. 3.1.1, when radiation hits a surface, the effect is called the irradiance. Its dimension is watt/square meter. In case of visible light, by definition the photopic photometry is used, which involves the spectral sensitivity curve or v_λ-curve of the human visual system. The effect is called the illuminance. Its dimension is lumen/square meter.

In the description of the luminance factor that is given here, the illuminance over the relevant surface element is considered. As is explained in sec. 3.1.4, when discussing the illuminance, it is indicated that the illuminance decreases when the source is further away from the surface, or when the light hits the surface obliquely – the well-known inverse-square and cosine laws. When we deal with the illuminance on the plane itself, these influences have no meaning and can be disregarded.

The mathematics of luminance

(c) The luminance factor of a perfect diffuser

When the object is a perfect diffuser, all light energy is reflected. There is no transmission and no absorption. A perfect diffuser is defined as an object – or rather the surface of an object – that is characterised by the fact that the luminance is the same in all directions. Sometimes the term 'perfect diffuser' is also used for surfaces that have a constant luminance in all directions, but that absorb some light. In order to avoid misunderstanding, we will call such surfaces quasi-perfect diffusers with a reflection factor R where $R < 1$.

As is explained in sec. 4.1.2d, the constant luminance implies that the luminous intensity of an area element dA of this surface differs in all directions. In a direction that makes an angle γ to the normal on the surface, the apparent size of the surface element dA is reduced, as it seems to be foreshortened. See Figure 4.1.1. As can be seen from that Figure 4.1.1, its apparent size is $dA \cdot \cos \gamma$.

A more strict mathematical treatise of the case where the luminance factor depends on the direction, is discussed in Baer, ed. (2006, sec. 1.2.2.1; 1.2.2.2), Gall (2004, sec. 7.2.2.2.1, p. 63-64), Hentschel, ed. (2002, sec. 2.1.1, p. 17-22), Reeb (1962, sec. 2.314; 2.321; 2.431), and Stevens (1969, p. 34-37). In the following, we will make use of the description given by Reeb.

(d) The definition of the luminance factor

In discussing the luminance factor, we have to deal with the total luminous flux of the incident and of the reflected light. It should be noted that the derivation of the reflection formula is very similar to the derivation of the basic formula of photometry that is given in sec. 4.1.2c. The reason is that in both cases the essence is the exchange of luminous flux between two surfaces.

All reflected light must be integrated over the sphere around the point on the surface under consideration. The definition of a perfect diffuser implies that no light passes the boundary plane. It is sufficient, therefore, to consider only the half-sphere 'over' the surface.

The quantities that are used in the derivation of the luminance factor are depicted in Figure 4.4.1.

As is explained earlier, for a perfect diffuser the emission does not depend on the direction. In order to avoid misunderstanding, in sec. 4.4.3c, such surfaces have been called quasi-perfect diffusers with a reflection factor R where $R < 1$.

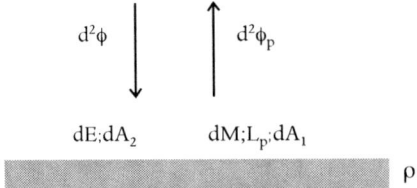

Figure 4.4.1 The derivation of the luminance factor (After Gall, 2004, Fig A26, p. 63).

In sec. 4.3.2b it is explained that the radiation density is defined as the physical quantity that characterises the homogeneous radiation in an elementary light tube. However, it is not only a characteristic of the radiation of a surface but also the invariant aspect of the radiation itself.

In sec. 4.5.2d, when discussing the geometric flux, this is defined as:

$$G = \int^A \int^\omega dA \cdot d\omega \quad\quad [4.3\text{-}4]$$

From this the radiation flux can be defined. This is indicated by $d\Phi$, and it represents the power that flows from the surface element da_1 towards the surface element da_2. It depends not only on the geometric flux that is defined earlier, but also on the radiation intensity in the light tube that is called L (Reeb, 1962, sec. 2.41, p. 23-24).

Integration over all relevant elementary light pencils yields the total radiation flux

$$\Phi = \int^{A_1} \int^{\omega_1} L \cdot dA \cdot d\omega_1 \quad\quad [4.4\text{-}3a]$$

From [4.4-3a] it can be deduced that:

$$L = \frac{d^2 \Phi}{dA \cdot d\omega} \quad\quad [4.4\text{-}3b]$$

For a Lambertian emitter L has the same value in all directions. So [4.4-3a] can be written as

$$\Phi = L \cdot \int^A \int^\omega dA \cdot d\omega \quad\quad [4.4\text{-}3c]$$

For a large distance between the surfaces A_1 and A_2 the solid angle ω_1 of the surface A_2 as seen from A_1 is the same for all surface elements dA_1, [4.4-3c] can be written as

$$\Phi = L \cdot \omega_1 \int^A dA \cdot d\omega \quad\quad [4.4\text{-}3d]$$

The mathematics of luminance

As mentioned earlier, Φ is called the flux. When we focus on visual perception, it is called the luminous flux. Per unit of surface area, it is called the specific luminous flux, sometimes written as M (Reeb, 1962, sec. 2.431, p. 25-26). From [4.4-3d] follows that;

$$M = \frac{d\Phi}{dA}$$

or:

$$M = \int^{\omega} L \, d\omega \qquad [4.4\text{-}4]$$

From the description of the solid angle it follows that for a half-sphere,

$$\int^{\omega} d\omega = \pi$$

So for a Lambertian emitter:

$$M = \pi \cdot L \qquad [4.4\text{-}5]$$

In common photometric terms, the specific luminous flux is nothing but the illuminance, and the radiation intensity is nothing but the luminance.

In this way the common reflection formula for a perfect diffuser is derived; for a quasi-perfect diffuser with a reflectance of ρ, one finds the even more familiar relation:

$$L = \frac{\rho}{\pi} \cdot E \qquad [4.4\text{-}5a]$$

4.4.4 The luminance factor of practical materials

Many practical materials, like e.g. road surfaces, do not show such circular symmetry as regards their reflection properties. They require the vectorial approach. As is indicated earlier, the luminance of a non-emitting surface is determined by two quantities: the amount of incident light and the reflection characteristic of the surface. Both quantities have a vectorial character, showing both magnitude and direction (without being, mathematically speaking, 'true' vectors, as vector summation usually is not valid). The direction of each vector can be described with two angles; this means four directional parameters in total, plus two scalars for the length of the vectors. The product of two vectors, is of course, a tensor. See also Schreuder (1967, sec. 3.3.1; 1998, sec. 10.4). The theory of vectors and of vector analysis is discussed in great detail in Feynman et al. (1977, sec. I-11 and sec. II-12). The theory of tensor and tensor calculus is also discussed in great detail in Feynman et al. (1977, sec. II-33), and Wachter & Hoeber (2006, Appendix A1).

If the material is isotropic, it is sufficient to introduce only one angle to define the direction. In this treatise, γ is used, the angle to the normal.

The light reflection of road surfaces is discussed in detail in sec. 10.4.1, where a number of examples are given.

4.5 Conclusions

Physical fields interrelate and interconnect matter and energy within their realm of influence. Field lines are just trajectories, and nothing more. The light field is quite distinct from the electromagnetic field that is described by Maxwell's equations. The light field belongs to the geometric optics, and the electromagnetic field to the physical optics.

The continuity principle is one of the fundamental concepts of physics. Traditional fluid dynamics discuss the speed of incompressible fluids, based on theorem of Bernoulli. The light fluid is different. The elementary particles of light are photons, that have a rest mass of zero, and that have a constant speed equal to the speed of light. Both Bernoulli-fluids and light fluids have no friction. Real fluids, however, have friction. In light fluids, friction shows itself as diffraction.

The luminance describes the geometrical distribution of luminous flux with respect to position and direction. It equals the flux per unit of projected area, and per unit of solid angle. From this, the geometric flux and the related concept of throughput are derived.

When light hits the surface of an object, the way it is propagated is described by the transmission factor, the reflection factor, and the absorption factor. According to the law of the conservation of energy, the sum total of the three is always 1. Finally, the luminance factor of perfect diffusers and of that of practical materials is described.

References

Baer, R. (1990). Beleuchtungstechnik; Grundlagen (Fundaments of illuminating engineering). Berlin, VEB Verlag Technik, 1990.

Baer, R., ed. (2006). Beleuchtungstechnik; Grundlagen. 3., vollständig überarteite Auflage (Essentials of illuminating engineering, 3rd., completely new edition). Berlin, Huss-Media, GmbH, 2006.

Blackburn, S. (1996). The Oxford dictionary of philosophy. Oxford, Oxford University Press, 1996.

Blüh, O. & Elder, J.D. (1955). Principles and applications of physics. Edinburg, Oliver & Boyd, 1955.
Born, M. & Wolf, E. (1964). Principles of optics. Oxford, Pergamon Press, 1964.
Boyce, P.C. (1981). Human factors in lighting. Appl. Sci. Publ., 1981.
Breuer, H. (1994). DTV-Atlas zur Physik. Zwei Bänder, 4.Auflage (DTV atlas on physics, two volumes, 4th edition). München, DTV Verlag, 1994.
CEN (2000). Technical report on lighting of road traffic tunnels. CEN/TC169/WG6. 12 th Draft, October 2000. Brussels, Central Sectretariat CEN, 2000.
CEN (2002). Road lighting. European Standard. EN 13201-1..4. Brussels, Central Sectretariat CEN, 2002 (year estimated).
CEN (2003). Lighting applications – Tunnel lighting. CEN Report CR 14380. Brussels, CEN, 2003,
CIE (1965). International recommendations for the lighting of public thoroughfares. Publication No. 12. Paris, CIE, 1965.
CIE (1975). Guide on interior lighting. Publication No. 29. Paris, CIE, 1975.
CIE (1982). Guide on interior lighting. Publ. No. 29-2. Paris, CIE, 1982.
CIE (1992). Guide for the lighting of urban areas. Publication No. 92. Paris, CIE, 1992.
CIE (1995). Recommendations for the lighting of roads for motor and pedestrian traffic. Technical Report. Publication No. 115-1995. Vienna, CIE, 1995.
De Boer, J.B. (1967). Visual perception in road traffic and the field of vision of the motorist. Chapter 2. In: De Boer, ed., 1967.
De Boer, J.B. & Fischer, D. (1981). Interior lighting (second revised edition). Deventer, Kluwer, 1981.
De Boer, J.B., ed. (1967). Public lighting. Eindhoven, Centrex, 1967.
Einstein, A. (1956). Über die spezielle und allgemeine Relativitätstheorie (On the special and general theory of relativity). Braunschweig, Vieweg & Sohn, 1956.
Einstein, A. & Infeld, L. (1960). The evolution of physics. New edition. New York, Simon & Schuster, Inc., 1960.
Feynman, R.P.; Leighton, R.B. & Sands, M. (1977). The Feynman lectures on physics. Three volumes. 1963; 6th printing 1977. Reading (Mass.). Addison-Wesley Publishing Company. 1977.
Gall, D. (2004). Grundlagen der Lichttechtnik; Kompendium (Basics of illuminating engineering; A compendium). München, Pflaum Verlag, 2004.
Gerlach, W., ed. (1964). Physik (Physics). Das Fischer Lexikon FL11. Frankurt am Main. Fischer Bücherei GmbH, 1964.
Gershun, A. (1939). The light field (original title Svetovoe pole, Moscow, 1936) Translated by Moon & Timoshenko. Journal of Mathematics and Physics, 18 (1939) No 2, May, p. 51-151.
Helbig, E. (1972). Grundlagen der Lichtmesstechnik (Fundaments of photometry). Leipzig, Geest & Portig, 1972.
Hentschel, H.-J. (1994). Licht und Beleuchtung; Theorie und Praxis der Lichttechnik; 4. Auflage (Light and illumination; Theory and practice of lighting engineering; 4th edition). Heidelberg, Hüthig, 1994.
Hentschel, H.-J. ed. (2002). Licht und Beleuchtung; Grundlagen und Anwendungen der Lichttechnik; 5., neu bearbeitete und erweiterte Auflage (Light and illumination; Theory and applications of lighting engineering; 5th new and extended edition). Heidelberg, Hüthig, 2002.
Illingworth, V., ed. (1991). The Penguin Dictionary of Physics (second edition). London, Penguin Books, 1991.
Joos, G. (1947). Theoretical physics (reprinted). London, Blackie & Son Limited, 1947. Translated from the First German Edition of 1932.
Keitz, H.A.E. (1967). Lichtmessungen und Lichtberechnungen. 2e. Auflage (Measuring and calculating light, 2nd edition). Eindhoven, Philips Technische Bibliotheek, 1967.
Kuchling, H. (1995). Taschenbuch der Physik (Manual for physics). 15. Auflage. Leipzig-Köln, Fachbuchverlag, 1995.

Moon, P. (1961). The scientific basis of illuminating engineering (revised edition). New York, Dover Publications, Inc., 1961.
Moon, P. & Spencer, D.E. (1981). The photic field. Cambridge, Massachusetts, The MIT Press, 1981.
Narisada, K. & Schreuder, D.A. (2004). Light pollution handbook. Dordrecht, Springer, 2004.
NSVV (1963). Aanbevelingen voor tunnelverlichting (Recommendations for tunnel lighting). Eletrotechniek. 41 (1963), 23; 46.
NSVV (1991). Aanbevelingen voor de verlichting van lange tunnels voor het gemotoriseerde verkeer (Recommendations for the lighting of long vehicular tunnels). Arnhem, NSVV, 1991.
NSVV (2002). Richtlijnen voor openbare verlichting; Deel 1: Prestatie-eisen. Nederlandse Praktijkrichtlijn 13201-1 (Guidelines for public lighting; Part 1: Performance requirements. Practical Guidelines for the Netherlands 13201-1). Arnhem, NSVV, 2002.
NSVV (2003). Aanbevelingen voor tunnelverlichting (Recommendations for tunnel lighting). Arhnem, NSVV, 2003.
Prins, J.A. (1945). Grondbeginselen van de hedendaagse natuurkunde. Vierde druk (Fundaments of modern physics. 4th edition), Groningen, J.B. Wolters, 1945.
Reeb, O. (1962). Grundlagen der Photometrie (Fundaments of photometry). Karlsruhe, Verlag G. Braun, 1962.
Schreuder, D.A. (1964). De luminantietechniek in de straatverlichting (The luminance-technology in road lighting). De Ingenieur, 76 (1964) E89-E99.
Schreuder, D.A. (1964a). The lighting of vehicular traffic tunnels. Eindhoven, Centrex, 1964.
Schreuder, D.A. (1967). Theoretical basis of road-lighting design. Chapter 3. In: De Boer, ed., 1967.
Schreuder, D.A. (1998). Road lighting for safety. London, Thomas Telford, 1998. (Translation of: Schreuder, D.A., Openbare verlichting voor verkeer en veiligheid. Deventer, Kluwer Techniek, 1996).
Schreuder, D.A. (2006). Gerontopsychologische overwegingen bij het ontwerpen van binnenverlichting (Gerontopsychologal considerations in interior lighting design). NSVV Nationaal Lichtcongres 2006, 23 november 2006. Ede, NSVV, 2006.
Schwarz, H.E. (2003). Personal communication.
Sheldrake, R. (2005). Glossary. Internet, www.sheldrake.org., 2005.
Sterken, C. & Manfroid, J. (1992). Astronomical photometry. Dordrecht, Kluwer, 1992.
Stevens, W.R. (1969). Building physics: Lighting – seeing in the artificial environment. Oxford, Pergamon Press, 1969.
Van Bommel, W.J.M. & De Boer, J.B. (1980). Road lighting. Deventer, Kluwer, 1980.
Von Weizsäcker, C. F. (2006). The structure of Physics (Edited, revised and enlarged by Görnitz, T. & Lyre, H.). Dordrecht, Springer, 2006.
Vos, J.J.; Walraven, J. & Van Meeteren, A. (1976). Light profiles of the foveal image of a point source. Vision Research. 16 (1976) 215-219.
Wachter, A. & Hoeber, H. (2006). Compendium of theoretical physics (Translated from the German edition). New York, Springer Science+Business Media, Inc., 2006.
Zijl, H. (1951). Manual for the illuminating engineer on large size perfect diffusors. Eindhoven, Philips Industries, 1951.

5 Practical photometry

In Chapter 3, the fundamentals of radiometry and photometry, the relation between the two are also discussed. This chapter deals with the practical implications of photometry. After a brief history of photometry, the measurement of units in different scales are explained. The definition of measurement, the general aspects of it, the accuracy when measuring in different scales, and the concept of calibration is explained.

Details about visual photometry are given, explaining its essential role in standardization, and in the determination of the V_λ-curve. It is explained that modern traditional objective photometry is instrumental photometry, using detectors like barrier-layer cells, gas-filled and vacuum photocells, and photomultipliers. Finally, it is explained that modern objective photometry is based on the use of Charge-Coupled Devices, or CCDs. The chapter ends with some remarks about CCD data extraction and data processing.

5.1 General aspects of photometry

5.1.1 Five stages in the history of photometry

Photometry is, as the word says, the measurement of light. What light is, and the relation between photometry and radiometry is discussed in Chapter 3. A description of the notion of measurement is given in a further part of this section.

The history of photometry in astronomy goes back several thousands of years. The Greek philosopher Hipparchus classified, in about 120 B.C., the stars that can be seen by the naked eye in six classes, called magnitudes. This system was adopted by Ptolemy, taken from the Almagest, and later supported by many other sources (Schaefer, 1993). Other branches of photometry are more recent. The first important treatises on photometry happened to be published in the same year (Lambert, 1760; Bouguer, 1760). A short survey of the history of photometry is given by Reeb (1962, p. 3). According to Reeb, major steps in photometry have been made by Liebenthal (1907) and Walsh (1926). See also Moon & Spencer, 1981, p. 1.

Because subjective and objective photometry serve different purposes, not only in the past but also at present, it is interesting to consider what happened over the years with the precision of photometric measurements. The precision – or the lack of precision – that can be reached in visual photometry is depicted in Figure 5.1.1.

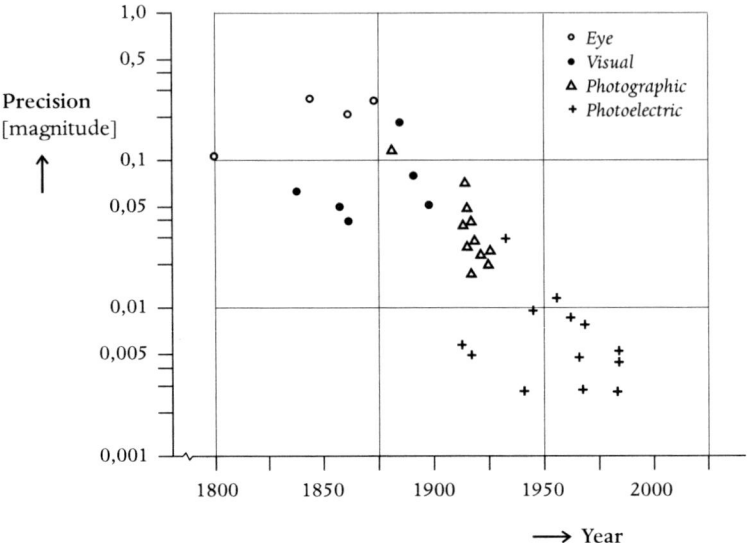

Figure 5.1.1 *Evolution of the precision of photometric measurements (After Narisada & Schreuder, 2004, Figure 14.2.1. Based on data from Sterken & Manfroid, 1992, Figure 1.15, and Young, 1984).*

Notes to Figure 5.1.1:
- 'Eye' means visual estimated aided by telescopes only.
- 'Visual' refers to all methods where the eye, as a detector, is assisted by other means like e.g. attenuation wedges, comparison lamps, etc.

Over the centuries several methods of photometry have been used:
1. Counting stars as used by the astronomers in ancient Greece;
2. Qualitative photometry as used by painters from the Renaissance downward, who made estimates of brightness;
3. Quantitative visual photometry as introduced by the 'Fathers' of illuminating engineering: Lambert, Bouguer, and others;
4. Objective quantitative photometry using photocells, like e.g. selenium barrier cells, phototubes, photomultipliers etc. This is the standard method that is used from the first half of the 20th century. It is still used in most cases. The methods quoted under (3) and (4) always relate to separate light sources, direct or indirect ones;

Practical photometry 143

5. Objective quantitative photometry using imaging techniques like CCD cameras. Here in essence all lighting parameters within the field of view can be assessed simultaneously.

5.1.2 Definition of measurement

(a) A general description of measurement

Stevens gives probably the most general, and also a rather precise description of the concept of measurement: "Measurement involves the process of linking the formal model called the number system to some discriminable aspect of objects or events" (Stevens, 1951, p. 22). Still more precise but less general is the following description: "Measurement is the assignment to aspects of objects or events of elements drawn from the formal system to which the postulates of algebra apply" (Stevens, 1951, p. 22-23).

Some of these postulates relate to the well-known communicative, the associative, and distributive rules (Stevens, 1951, Table 4, p. 15). Algebra is the branch of mathematics that deals with the general properties of numbers, and the generalizations arising therefrom (Daintith & Nelson, 1989, p.13). It seems that the general concept of number is usually not well defined. A nice description is: "Something or somebody distinguished by a numerical symbol" (Smith & O'Loughlin, eds., 1948, p. 727). The next step would be to define a 'numerical symbol'. Because this might lead to an unending series of definitions, we will stop here. For practical purposes, it seems to be sufficient to state that measuring is to quantify an object or event.

There is more to it. In sec. 1.5.3 we have introduced four different types scales. Each of them will require a different way to describe the concept of measurement. Details are given in Stevens (1951, p. 23-30).

From the definitions that have been given earlier, it follows that nominal scales are non-quantitative scales, whereas the other three, viz. ordinal scales, interval scales, and metric scales are quantitative scales. Therefore, they need to be treated separately.

(b) Measuring in nominal scales

As a nominal scale does not include any quantitative measure, the concept of measurement is restricted to the ability to discern clearly and univocally between individual elements of the sample. As an example, when labelling composers, Albeniz may be (a); Bach (b); Chopin (c) etc. It does not make any sense at all to maintain that 'a' is better than 'b', etc. Clouds or sand heaps cannot be classified in this way in a nominal scale, because they are not always separate them individually. Often it is not possible to tell where the one ends and the other begins. The description of measurement as given by Stevens does not apply to nominal scales as the number system is not involved.

(c) Measuring in ordinal scales

An ordinal scale is a scale where items are scaled according to their magnitude, irrespective of how much their quantitative difference is. All one can say is that a higher class represent larger items, but it is not possible to say how much larger. As regards the measurements, ordinal scales must be regarded as quantitative scales; the quantification is, however, in fact restricted to one out of three statements:'a is larger than b'; 'a equals b', or 'a is smaller than b'. In visual photometry, many measurements relate to equality. As is explained in a further part of this section, the test criterion is the fact that the brightness of a test field equals that of a comparison field. The description of measurement as given by Stevens does not apply in full to ordinal scales. The number system is involved in so far that the equality or non-equality of two numbers is assessed. The numbers themselves, however, play no role. Only a limited number of the postulates of algebra apply.

It is of course possible to attribute elements of interval or metric scales to the individual elements of an ordinal scale. Interval and metric scales also obey the rule of 'greater-equal-smaller' of the ordinal scale. It is necessary, however, to stress the point that this process is an attribution only. The attributed elements of the interval or metric scale may be subject to the arithmic procedures that are allowed for such scales; the data, however, stay an ordinal scale. In sec. 1.5.3, the pitfalls that must be avoided are explained when using the discomfort glare scale as an example. Details are given in Stevens (1951, p. 26-27); Adrian & Schreuder (1971); De Boer & Schreuder (1967). The discomfort glare is discussed in sec. 8.4.

(d) Measuring in quantitative scales

As has already been explained earlier, there are two kinds of quantitative scales: interval scales and metric scales. An interval scale is an ordinal scale where the differences between the levels are equal. A metric scale is an interval scale with a real 'zero' point. In both cases, the description of measurement as given by Stevens does apply in full, because numbers are linked to some discriminable aspect of objects. In sec. 1.5.3, the temperature scales of Celsius and Kelvin are used as an example of an interval scale and a metric scale respectively. The numbers are the 'degrees Celsius' or the 'kelvin'; the discriminable aspect is the length of a mercury column or the electric resistance.

Practical photometry

(e) The accuracy when measuring in different scales

In should be noted that the accuracy of the measurements is of equal importance in all cases where quantitative scales are used. It is a misconception to believe that the measurements do not need to be accurate when one deals 'only' with an ordinal scale. As an example, history books indicate that there has been one 'draw' in the famous Oxford-Cambridge Boat Race. In the 1860s, the end result was determined by looking at the boats when they approached the finish. The umpire, who was, according to the story, somewhat for the worse by drink, could not see the difference, so a draw was called. This unfortunate occurrence did prompt the introduction of the measurement of the actual rowing time. In some recent races an accuracy in the milliseconds range has been called for. Questions of measuring accuracy are explained in sec, 5.3.3e.

5.1.3 The relation between radiometry and photometry reconsidered

Photometry is a branch of the wider field of radiometry. One might call radiometry the measurement of electromagnetic radiation independent of the detector that is used in the measurements, whereas photometry refers to the measurement of electromagnetic radiation when using the human visual system as a detector. This means that on the one hand there is hardly any fundamental difference between radiometry and photometry, but that on the other hand the response, more in particular the spectral response, of the human visual system is an essential part of photometry. This implies that radiometry needs to be regarded as a branch of experimental natural science. The relevant methods are those of experimental physics. Photometry, however, may be regarded as a branch of applied psychology. The relevant methods are those of experimental phychophysics. The bottom line is that the accuracy that can be reached in radiometric measurement is determined by the limits of instruments, whereas the accuracy of photometric measurement is limited by the way the performance of the human visual system is determined.

As is explained in sec. 7.2, it is customary to express the performance of the human visual system by the spectral response. This means that one determines the reaction of the visual system for light of different wavelengths. The conventional way to depict this relation is by means of the V_λ-curve, that is introduced in sec. 7.2.2. It is explained in sec. 7.2.4, when discussing mesopic vision, that this reaction depends on the overall brightness of the scene – more precisely stated, on the adaptation level. One might define a separate V_λ-curve for each value of the adaptation level. For photometry, where it is essential to be able to compare the results of measurements made by different methods, different observers, and different test facilities, only one V_λ-curve is allowed. All measures, units and quantities that are used in photometry, and that are defined by

ISO and CIE, are based of this one curve. This curve is, as is explained sec. 7.2.2b, called the curve of day vision or the spectral sensitivity curve for photopic vision.

One may be tempted to add the word 'traditional', because in modern science and in modern technology, the sources have all sorts of different spectral emission distribution functions, and, more important, the human observer, adapted to the photopic mode, is only one of the many detectors that are used. As an example, in modern astronomy, the human observer is obsolete in photometry. Until recently, most photometry was performed by using photographic emulsions, but since a decade or two, almost all astronomical measurements, photometry included, is done with CCDs (Narisada & Schreuder, 2004, Chapter 14). Astronomical photometry is discussed in detail in Sterken & Manfroid (1991) and Budding (1992), where further information is given about the spectral sensitivity function of modern detectors.

5.1.4 Calibration

(a) Gauging and calibration

In essence, photometry is, as mentioned earlier, nothing but counting photons. In some cases, this is exactly what is done, e.g. when using photomultipliers for the measurement of extremely low light levels. The use of photomultiplier tubes is explained in sec. 5.3.2e. Because a photon is a quantum of energy, counting photons is to measure energy. When a stream of photons is measured, as is usually the case, one measures energy per unit of time, or 'work'. This is a common way to assess the total energy that is emitted by a light source. The device to do so is a bolometer, a device that is mentioned in sec. 5.2.2a. A bolometer is the essential measuring device in radiometry. Photometry, however, takes the spectral response of the detector into account. As is mentioned already several times, the detector is the human visual system and the spectral response is represented by the V_λ-curve. This implies that absolute photometry cannot be done directly; all photometry is a comparison to a standard. Over the years, defining a standard that could be used as a real intentional standard did prove to be the most demanding challenge for theoretical and practical photometry.

(b) Standards

As is explained in sec. 3.2.1, the ISO system has introduced the candela as the standard unit on which all photometric units and quantities are based. Until recently, it was the candle itself that was used as the standard.

For several decades, the most common type of 'candle' was the Hefner lamp, introduced in 1884 (Keitz, 1967, p. 26). It is described in some detail by Barrows (1938, p. 30-32). It is a carefully designed lamp that burns Amyl Acetate ($C_7H_{14}O_2$). The lamp is depicted in Figure 5.1.2.

Practical photometry

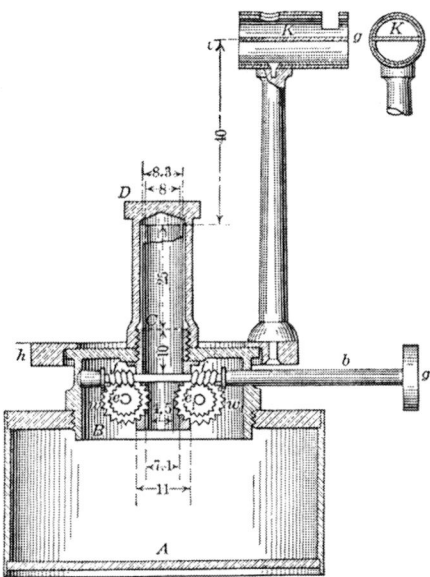

Figure 5.1.2 The Hefner lamp (After Barrows, 1938, Figure 2.1).

Because the lamp is defined in objective dimensions of lengths etc., it can be reproduced at will to any desired accuracy. So is the fuel. The only important variables that cannot be fixed are the humidity of the air and the barometric pressure. According to Barrows this could lead to discrepancies up to 8,5% (Barrows, 1938, p. 31-32). This means that the Hefner lamp is not suitable for accurate photometry.

As an alternative to the Hefner lamps, some countries used a set of carefully selected and preserved incandescent lamps as a primary standard. In 1909, the international candle was based on this (Keitz, 1967, p. 26). It was called 'international', although it was used only in a small number of countries (Reeb, 1962, p. 81). It was found that such devices are very useful as a secondary standard, but not as a primary standard (Barrows, 1938, p. 36; CIE, 2002). Some of the reasons are:
- there existed only one set, so everyone had to travel to Washington DC to calibrate substandards;
- incandescent lamps deteriorate over time, particularly if they are used. And of course they must be used in order to be able to calibrate substandards.

In hind sight it seems to be rather obvious to use a radiometric standard as the 'real' standard for photometry, the more so because the V_λ-curve needs to be defined precisely in any case. The first steps to do so were made in 1941 (Reeb, 1962, p. 61). The thermal radiator or black-body radiator was used. The reason is that the black-body radiator is defined in absolute terms by the laws of Planck, Raleigh, Stephan-Boltzmann, and

Wien, as is explained in sec. 2.3.1a, where incandescence is discussed. Based on these considerations, the candela was defined as "1/60 of the luminous intensity that is emitted by 1cm² a black body at the temperature of solidifying platinum, perpendicular to the surface" (Keitz, 1967, p. 26).

There was a problem, however. In real life there are no perfectly black bodies. As is explained in sec. 2.3.1b, all real life bodies are 'grey bodies'. The only physical device that is, at least to fair approximation, a black body is a hollow room. Such a hollow room has been used for the definition of the candela. A device based on this idea is depicted in sec. 2.3.1b, Figure 2.3.3.

There were more problems, both of a fundamental and a practical nature. In sec. 3.2.3 it is pointed out that the description provides a definition of the luminance, but not of the luminous intensity: "The luminance that is found has the value of 60 stilb" (Reeb, 1962, p. 60). It proved that the device was prone to all sorts of measuring errors: "The method can be used only in very well equipped laboratories" (Reeb, 1962, p. 60-61).

Further considerations did lead to a further definition of the 'candle' or 'candela'. It goes without saying that this multitude of successive standards does not help to clarify the situation for manufacturers, designers, or researchers. It is not clear what at a certain moment in time is the exact and correct position, nor is it possible to compare recent studies with earlier ones. Finally, no one can truly say that the current definition is the final one. As a matter of fact, at the moment the assessment and validity of the V_λ-curve is under scrutiny. Nonetheless, in 1979 the candela, with a new definition, has been introduced as one of the seven basic units of the official units and symbols that form the ISO International standardization. The 1979 CGPM-definition of the candela is: "the luminous intensity, in a given direction, of a source that emits monochromatic radiation of frequency $540 \cdot 10^{12}$ hertz and that has a radiant intensity in that direction of (1/683) watt per steradian" (CGPM, 1979).

5.2 Traditional subjective photometry

5.2.1 Brightness estimation

One might consider the classification of cosmic objects as the beginning of photometry. As is mentioned earlier, the Greek philosopher Hipparchus classified, in about 120 B.C., the stars that can be seen by the naked eye in six classes, called magnitudes. The brightness of the stars were estimated during twilight, when, after sunset, first the brightest stars became visible and later the fainter stars. Therefore, the

Practical photometry

magnitudes represent equal steps in visual perception. As is explained in sec. 7.3.2 and 7.4.3, when discussing the Weber-fraction and Fechner's Law, equal steps in experience usually correspond with equal steps in the logarithm of the stimulus. Thus, it is often stated that 'the senses work logarithmically'. Not a very precise expression, but basically true.

It is generally accepted that the logarithmical intervals in stellar magnitudes was noted for the first time by Pogson (Sterken & Manfroid, 1992, p. 24). Pogson also selected the value of 2,512 as the photometric ratio between consecutive magnitudes. This value has been adopted by the astronomical world. As $2,512^5 = 100$, this agreement means that a star of zero magnitude is exactly 100 times brighter as a star of 5th magnitude. Details of stellar magnitudes and the way they are used in today's astronomy are given in Herrmann (1993, p. 31); Narisada & Schreuder (2004, sec. 14.2.2), and Weigert & Wendker (1989).

The magnitude scale has no 'zero point', because the logarithm of zero is minus infinity. It is essentially an interval scale, as is explained in sec. 1.5.3. Because an interval scale is awkward to work with, a zero point has been added by defining the brightness of the pole star Polaris as being 2,12 Magnitude (Stumpff, ed., 1957, p. 106).

One might think that measurements by means of star counting is a primitive way of doing, that has been made obsolete already a long time ago. This, however, is not true at all. In contrary, star counting is about the only way to make large scale surveys of sky glow and other light pollution effects.

In amateur astronomy, it is a generally accepted rule that under favorable conditions, the limiting magnitude for stars is about 6,0 magnitude (Schaefer, 2003). See also Narisada & Schreuder (2004, sec. 9.1.7g).

Since a number of years, a major programme is underway, called "Globe at night" (Anon., 2006). The principle is that observers were asked to indicate which are the weakest stars in the constellation of Orion that could just be seen. In the 2006 survey, 46 000 observations were gathered from 18 000 observers in 96 countries on every continent – except from Antarctica, because Orion is invisible there (Anon., 2006).

5.2.2 Visual photometry

(a) *The sensitivity of the eye*

As is explained earlier, in radiometry, the total power (in watts) of the radiation is measured directly, e.g. by means of a bolometer (Budding, 1993; Hentschel, ed., 2002, sec. 4.3.1). In photometry, the response of the human visual system must be taken into account.

The human visual system is very sensitive to radiation. Photons that hit the photoreceptor will be absorbed; part of an absorbed photon will give rise to specific photochemical reactions that are discussed in more detail in sec. 6.3.5. Under optimal conditions, it is sufficient that within the time span of several milliseconds, three to four quanta hit one particular rod in order to cause a sensation of light (Weale, 1968, p. 27). For the cones, a much larger number of quanta is required.

The visual system is not equally sensitive to light of all wavelengths. Furthermore, the sensitivity depends on whether the rods or the cones, or possibly both, are active at the same time. Moreover, there are three types of cones, each with their own sensitivity. In this respect, it must be kept in mind that the retina, that is the light-sensitive area of the eye, contains a several different 'families' of light detectors. In the more common day-time vision, the cones are active, whereas in night-time vision, it is mainly the rods that are active. Details on the characteristics of the cones and the rods, as well as on cone vision and rod vision, and on the related colour vision, are discussed in Chapter 9.

In spite of these complicating factors, it is possible to determine the sensitivity of the visual system as a whole for different wavelengths. The physiological and psycho-physical reasons for this are explained in Chapter 6, where the physiology of vision is discussed.

The response to light of different wavelengths can be measured in different ways. A major problem in standardising the eye sensitivity is, that each method gives different results. The methods are discussed in a further part of this section.

As mentioned earlier the proliferation of lamps, mostly gas lamps at that time, that took place since the 1880s and 1890s, required an international, time independent, reproducible standard against which the performance of lighting equipment could be measured and certified. In a preceding part of this section, several standards have been discussed.

It has also been pointed out that a standard light source is not enough. As has been stressed at several places already, the response of a detection system for radiation depends on the spectral emission of the radiation source and on the spectral sensitivity of the detector. The first is being taken care of by an appropriate radiation standard; the second must be well-defined as well. For this, the standard observer is introduced. This is not a sinecure, as is pointed out in sec. 7.1.1a when psychophysical measuring methods are discussed. There always is a large inter-subject spread in psychophysical study – no two persons are identical.

There was another problem. Around the beginning of the 20th century the current laboratory equipment did not allow precise measurement of blue light. This fact turned out to have severe consequences for the day-to-day photometry of light sources: even in photopic photometry, blue emitting high-pressure mercury lamps were systematically underrated as regards their photometric performance, like e.g. the lamp efficacy (Gall & Wuttke, 1985; Narisada & Schreuder, 2004, sec. 8.2.1 and 8.2.2). Lamp efficacy is explained in another section of this book.

(b) *The photopic V_λ-curve*

In spite of all this, CIE defined in 1924 a 'standard observer'. One did start with 'common' vision, where only cones were involved. Cone-vision is called photopic vision, but also day-time vision. The standard observer for cone-vision is described in the "CIE Standard spectral luminous efficiency function V(lambda) for photopic vision" (CIE, 1924, p. 67; p. 232). See also Baer, ed. (2006, sec. 1.2.1, p. 18). In 1951, CIE defined also a 'standard observer for scotopic vision' (CIE, 1951, Vol 1, sec 4; Vol 3, p. 37).

The basis of this 'standard observer for photopic vision' was that the observer adapted fully to high levels of ambient luminances. Another proviso is that the standard observer is defined doing the observations within a two-degree field. Later it was found that the size of the measuring field has important consequences. This is explained in sec. 6.3.5d, where the spatial distribution of rods and cones over the retina is discussed. In the definition of the standard observer, the average of 200 observers was used (Barrows, 1938, p. 2). Moon (1961, p. 49-50) gives more information about the way this standard observer was defined.

In the 1924 definition of the standard observer, a standard spectral luminous efficiency function is introduced. This is commonly described as the V_λ-curve that has been mentioned already several times.

In mathematics, a function is defined in the following way. We will take two variables and call them x and y. When at each given value of x one and only one value of y can be given, one calls y a function of x. This is written as:

$$y = f(x) \qquad [5.2\text{-}1]$$

For obvious reasons, y is called the independent variable and x the dependent variable (Bronstein et al., 1997, sec. 2.1.1.1, p. 43). In more general terms one may call y a function of x if there is a well defined recipe or prescription that determines the value or values of y for different values of x (De Bruin, 1949, chapter II). About the only requirement for the recipe is that is must be well defined. Because the CIE-definition given earlier is not well-defined, it is questionable whether it is mathematically speaking justified to call V_λ a 'function'. In this book, however, we will disregard this mathematical refinement and adhere to the custom to do so.

The 'Standard Visibility Curve' is usually presented in two ways:
1. As a curve, depicting the relative sensitivity versus the wavelength. The curve, called the 'V_λ-curve', this is shown in Figure 5.2.1.
2. As a table, giving the same relation. A shortened version is given in Table 5.2.1. A more complete table is given by Judd (1951, table 5, p. 819). The data are depicted in Figure 5.2.2.

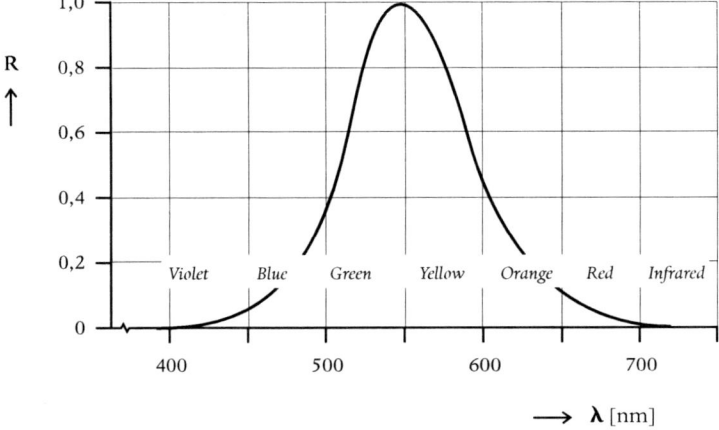

Figure 5.2.1 The V_λ-curve (After Narisada & Schreuder, 2004, Fig. 8.2.1).

CIE has produced a certified version of the table that gives the value of the sensitivity up to intervals of 1 nm (CIE, 1990). In that report, the 1988 modified two-degree spectral luminous efficiency function is presented. This modification is a supplement to, but not a replacement of, the 1924 V_λ-curve (CIE, 1924). It has been stressed that the modification refers only to wavelengths below 460 nm.

Practical photometry 153

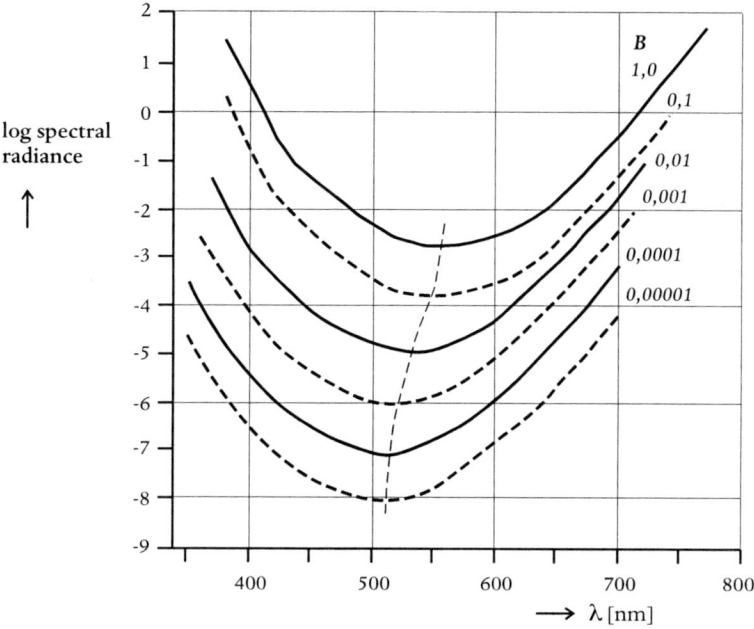

Figure 5.2.2 Contours of equal brightness corresponding to constant values of the adaptation level. Contours at 10 dB intervals (After Judd, 1951, figure 2. Based on data from Weaver, 1949).

Wavelength (nm)	Relative sensitivity
400	0,0004
450	0,038
500	0,323
550	0,995
600	0,631
650	0,107
700	0,0041
750	0,00012

Table 5.2.1 The relative sensitivity of the average normal eye in steps of 50 nm (After Narisada & Schreuder, 2004, table 8.2.1. Based on data from Barrows, 1938, Table 1.1).

Note: as per definition, the sensitivity is set at unity for a wavelength of 556 nm.

It should be stressed that the presentations in graphical form (Figure 5.2.1) as well as that in table-form (Table 5.2.1) represent exactly the same data. One is therefore completely free to use the one or the other. Furthermore, it cannot be stressed strongly enough that the Standard Visibility Curve is the central issue that links the photometric units to the other physical units. It is therefore no surprise that one is very reluctant to include any corrections in the visibility curve, such as would be needed to correct the underrating of deep blue colours. Such a correction would have its repercussions, not only in the lighting industry, but in the whole structure of physical measurements and nomenclature. And finally, it must be stressed that all photometry is based on the Standard Visibility Curve; it is therefore defined exclusively for photopic vision, for the standard observer, and for the two-degree field of observation. Any change in any of these assumptions would change the whole structure.

(c) *The determination of the V_λ-curve*

The spectral luminous efficiency curves have been derived by means of psychophysical experiments where the stimuli of the light and the sensation of brightness are compared. Traditionally, this is called visual photometry. As a matter of fact, the determination of the V_λ-curve is, after the star counting, probably the only area where visual photometry is still used. Even more, it is probably the only area where visual photometry is essential. It cannot be replaced by objective photometry because the detectors must somehow be calibrated to the V_λ-curve.

As mentioned further on, the determination of the V_λ-curve is essentially heterochromatic photometry (Reeb, 1962, p. 114). This means that two areas in the field of view are presented to an observer. One area represents the 'known' brightness, the other the brightness to be assessed. What 'known' means here is not self-evident. We will come back to that point later on. The area with the 'known' brightness is fixed, the brightness of the other area can be adjusted by the observer. The two areas differ in colour. It is the task of the observer to adjust the brightness of the variable area in such a way that the two areas seem to have the same brightness. This step introduces the subjective aspect in visual photometry, which consequently is called subjective photometry. In this way the ratio between the brightnesses of areas that differ in colour can be assessed. The way the actual measurements are made is explained in a further part of this section where flicker photometry is discussed. Flicker phenomena, or the reaction of the visual system to periodic light stimuli is discussed in sec. 8.1.2b. It should be stressed here that flicker phenomena differ from the phenomena related to the speed of observation, although there are similarities between them.

It is tempting to think that a graph that depicts these ratios in relation to the wavelength would represent the V_λ-curve, at least in a relative scale. This, however, is not completely correct. The trouble is the adjustment of the brightness. As one is interested in a quantitative V_λ-curve, the adjustment must also be quantifiable. Usually the adjustment

Practical photometry

is done by inserting a neutral filter of known attenuation into the light path in one of the two areas. As the two areas differ in colour, both the neutrality and the degree of attenuation of the filters can only be measured by means of heterochromatic photometry, which can only be done if the V_λ-curve is known – a circular argument! On these grounds it is mentioned that it is questionable whether it is mathematically speaking justified to call V_λ a mathematical function.

Of course in practice the matters are less serious than they might seem to be. Over the years and decades, the accuracy has been gradually improved by means of an iterative process. Some misgivings are, however, still there. Compared to the other problems that have been mentioned by CIE, they may be disregarded – at least they are disregarded in this book.

(d) *Heterochromatic and isochromatic photometry*

The problems that have been discussed in the preceding part of this section are linked to heterochromatic photometry. In isochromatic photometry, where the two parts of the field of view have te same colour, the attenuation can be assessed simply by applying the inverse square law that is explained in sec. 3.2.4d. The inverse square law states that the illuminance on a plane decreases with the square of the distance between the light source and the plane. This law is the basis for the photometer bench, the most common apparatus in any photometric laboratory. It is depicted in Figure 5.2.3. The photometer bench is the basic set-up for the split-field photometry that is discussed in the next part of this section.

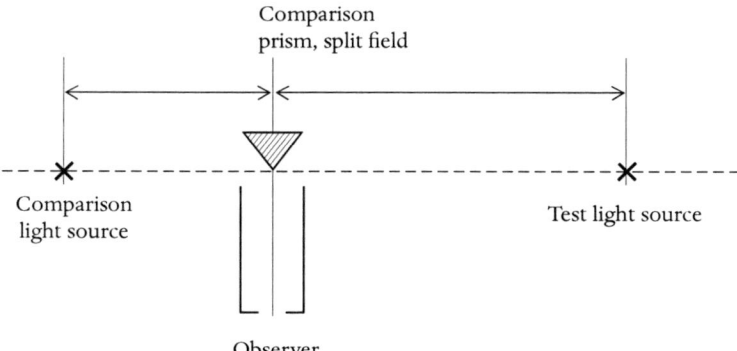

Figure 5.2.3 *The principle of the photometer bench (After Keitz, 1967, figure 137, p. 247).*

Heterochromatic photometry has the additional, special problem that the two parts of the field not only differ in brightness but also in colour, just as the term would suggest.

It is known from experience that it is not possible, even not for a trained observer, to estimate the degree on inequality of two parts of the field of view if the colour differs markedly. This is also the case if the two parts join together. The reason is, of course, that the borderline between the two parts does not disappear by equal brightness: the colour difference remains. The only feasible solution is to use flicker photometry that is explained in one of the further parts of this section.

(e) *The contrast method of photometry*

The principle of all visual photometric systems is the same. The visual field is divided in two parts. One part is illuminated by the source that must be measured. The other part is illuminated by a standard light source, the lighting characteristics of which are exactly known. This method is known as split-field photometry.

There are two main principles of split-field photometry. In the first, and by far the oldest, the two parts of the field of view, the test field and the comparison field, are presented simultaneously to the observer. A number of old, but still interesting, constructions is described in detail by Keitz (1967, sec XIV.2). See also Barrows (1938, chapters 4 and 5), de Boer (1951), Zwikker (1933, chapter II), Hütte (1919), and Narisada & Schreuder (2004, sec. 14.5.1).

The criterion of equality of the two parts of the field of view is the subjective disappearance of the border between the two parts of the field. Clearly it is essential for this criterion that the two parts join precisely. The subjective disappearance of the border can be described by the fact that the contrast between the two parts is below the threshold of the contrast sensitivity. The contrast sensitivity is discussed in sec. 7.4.3. Consequently, one may call this type of photometry, where the two parts of the field of view are observed simultaneously the contrast method of visual photometry. As mentioned earlier, this method is particularly suitable for isochromatic photometry. As such it has been over many decades the standard method for almost all photometric work in research, in the lighting industry, and practical illuminating engineering. However, since cheap, fast, sensitive, and reliable photocells of all sorts became available toward the end of the 20th century, visual photometry ended up in the museum, and it is not to be found any more in the laboratory.

For some, that was the end of the romantic days of illuminating engineering. There is more to it, however. It was also the end of what one might call the traditional 'hands-on' photometry. When researchers and designers become to rely exclusively on computer-type machinery, there is a real risk that the atavistic 'gut-feeling' disappears. This may sound far-fetched, but gut-feeling is essential as a first-line warning that something is wrong. One might compare it to the advance warning in medical diagnosis of looking

Practical photometry

at the tongue or feeling the pulse. As we have mentioned several times already, it is one of the aims of this book to help readers to avoid pitfalls. In this context, some practical experience during schooling with visual photometry may help to put things into perspective. It is mentioned in the Introduction that education is not a subject matter of this book. So let these few remarks be sufficient.

(f) Flicker effects

As mentioned earlier, the way the ratio between the brightnesses of areas that differ in colour can be assessed, is an important aspect of subjective photometry. A major tool is flicker photometry. Before we can explain the way flicker photometry is used, the reactions of the visual system to periodic light stimuli, the flicker phenomena, have to be discussed.

A periodic stimulus can be observed only if the stimulus is intense enough to be observed when the observation time is unlimited:

$$t \cdot (i - i_u) = c \qquad [5.2\text{-}1]$$

with
  t: observation time;
  i: stimulus intensity;
  i_u: the stimulus intensity for unlimited observation time;
  c: a physiologically determined constant.

[5.2-1] is called the Hoorwegs' Rule.

Also, a periodic stimulus can be observed only if the product of stimulus intensity i and observation time t reaches a specific minimum value. If the luminance is L_s:

$$t \cdot (L_s - L_u) = 0{,}21 \cdot L_u \qquad [5.2\text{-}2]$$

[5.2-2] is called Blondel-Rey's Law. For small objects one may replace the luminance L of the object by the illuminance E caused by the object at the pupil of the eye. This is Ricco-Piper's Law that is discussed in sec. 7.4.3d.

For stimuli of a duration where $t < 0{,}05$ second, [5.2-2] becomes

$$L_s \cdot t = 0{,}21 \cdot L_u \qquad [5.2\text{-}3]$$

[5.2-3] is called Bunsen-Roscoe's Law.

It has been found that the detection of the periodic stimulus depends on its wave-form (Jantzen, 1960; Schreuder, 1964, Schmidt-Clausen, 1968).

When the frequency of the light pulses is above a certain value, the pulses cannot be discerned separately any more. This pulse frequency is called the critical flicker-fusion frequency or CFF. The CFF depends on many variables. but primarily on the adaptation luminance L_a:

$$CFF = a \cdot \log L_a + b \qquad [5.2\text{-}4]$$

with a and b constants that depend on the conditions of observation, particularly whether the observations are made in the photopic or the scotopic region. [5.2-4] is called Talbot's Law. It is based on the studies of Ferry and Porter from the early 1900s. In Figure 5.2.4, the relation between the CFF and the adaptation luminance is depicted. WF is a factor that depends on the wave form. WF = 0,637 for rectangular pulses. WF = 0,500 for a sinusoidal wave and WF = 0,15 for a common fluorescent tube (Hentschel, ed., 2002, sec. 3.4, p. 61).

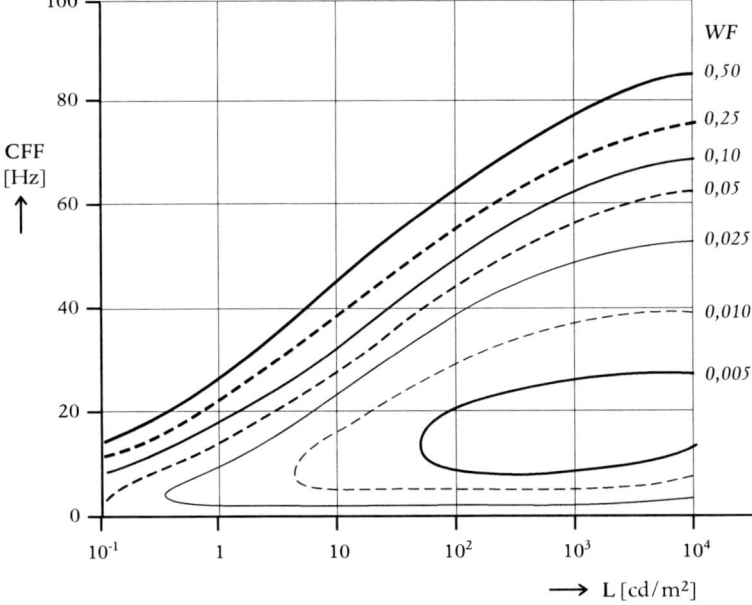

Figure 5.2.4 The relation between CFF and adaptation luminance, with the wave form WF as a parameter (After Hentschel, ed., 2002, fig. 3.14, p. 64. Based on Kelly (1961-1962)).

Practical photometry

The CFF as a function of retinal illuminance for a two-degree test field and a surround of different size is depicted in Figure 5.2.5.

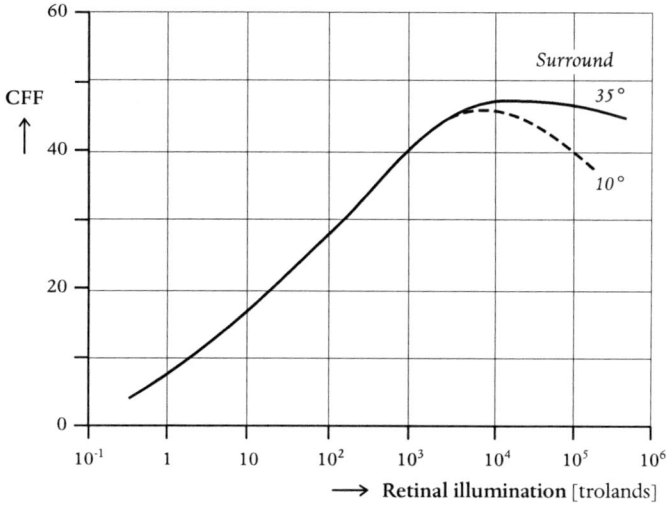

Figure 5.2.5 *CFF as a function of retinal illuminance (After Bartley, 1951, fig. 44 left, p. 960-961. Based on data from Hecht, 1938).*

There is a clear difference in the behaviour of cones and rods when subjected to intermittent light. Furthermore, rods react differently to light of different wavelengths. This is depicted in Figure 5.2.6.

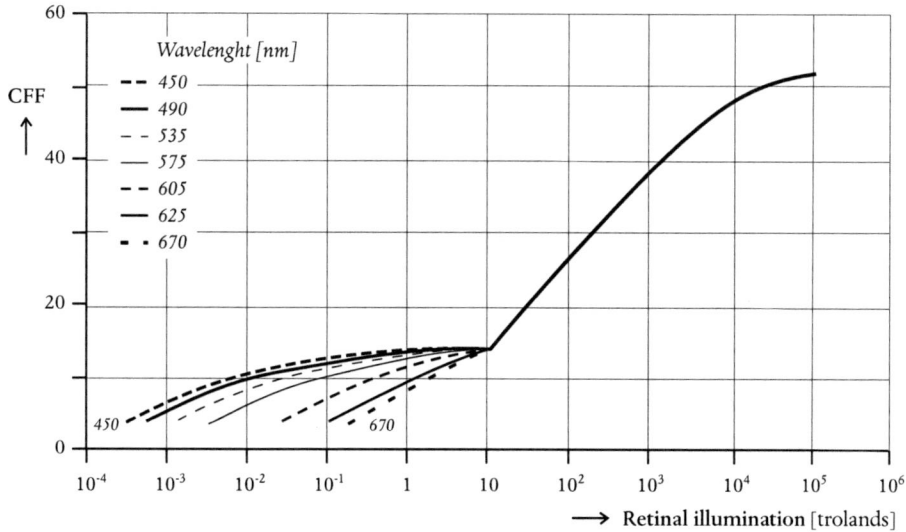

Figure 5.2.6 *Values of CFF an relation to retinal illuminance for light of different wavelengths (After Bartley, 1951, fig. 45 right. p. 961. Based on data from Hecht & Shlaer, 1936).*

Some of the underlying anatomical and neuronal aspects are given in Ruch, 1951, p. 150.

(g) Flicker photometry

As mentioned above, the determination of the V_λ-curve is essentially heterochromatic photometry (Reeb, 1962, p. 114). This means that two areas in the field of view are presented to an observer. One area represents the 'known' brightness, the other the brightness to be assessed. In heterochromatic photometry, the two areas differ in colour. It is the task of the observer to adjust the brightness of the variable area in such a way that the two areas seem to have the same brightness. In this part of this section, the way the actual measurements are made are explained.

It is mentioned above that it is known from experience that it is not possible, even not for a trained observer, to estimate the degree on inequality of two parts of the field of view if the colour differs markedly, even if the two parts join together. The only feasible solution is to use flicker photometry.

Flicker photometry can give the answer only as a result of a curious characteristic of the human visual system. Assume we have two separate fields of view that differ both in brightness and in colour, and that can be presented to an observer the one after the other. When the two fields are presented at a certain, low, frequency, the observer can see them separately. When the frequency is made higher, the two fields seem to blend in a certain fashion. A distinct experience of flicker will arise. As mentioned earlier, when the frequency is made still higher, the two fields blend completely into one; the flicker experience disappears. The corresponding frequency is usually called the critical flicker-fusion frequency or CFF.

Now the curious characteristic of the human visual system that was alluded to above is the fact that with increasing flicker frequency, first the experience of flickering colours disappear, and only later the experience of flickering brightness (Reeb, 1962, p. 124; Keitz, 1967, p. 217). The basis for flicker photometry stems from the early 20th century (Ives, 1912; Nutting, 1914; So, 1920). It should be stressed that without this curious phenomenon, heterochromatic photometry would have been hardly possible at all.

Over the decades, many ingenious constructions have been made to realise a simple and reliable flicker photometer. As these many proposals are mainly of historic interest, we will only refer to a number of surveys (Barrows, 1938; Bouma, 1946; Helbig, 1972; Hentschel, ed., 2002; Hütte, 1919; Keitz, 1967; Reeb, 1962). We will mention only two examples: the Bunsen 'grease spot' photometer, an example of simplicity. A grease spot

Practical photometry

on a piece of paper makes it translucent. By lighting it from two sides, and by looking at the two sides alternatively, a simple flicker photometer can be made. See Figure 5.2.7.

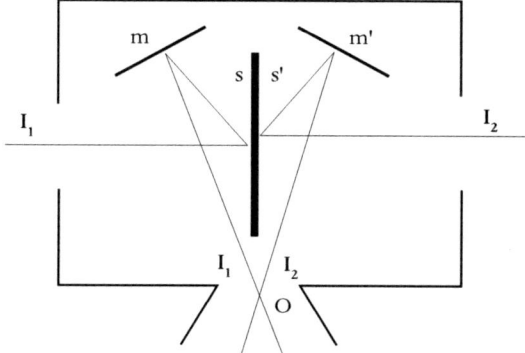

Figure 5.2.7 The Bunsen 'grease spot' flicker photometer (After Barrows, 1938, fig. 5-7, p. 119).

The other is Bechstein meter, more suitable for accurate measurements. See Figure 5.2.8.

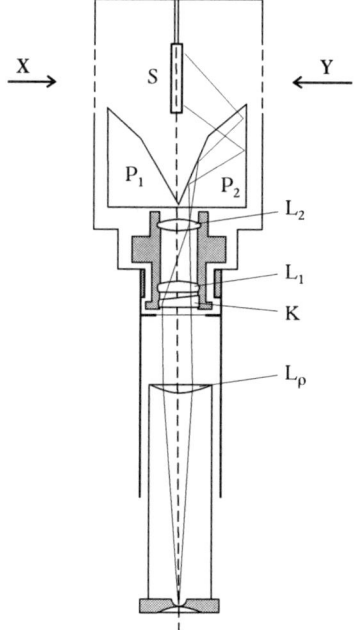

Figure 5.2.8 The Bechstein flicker photometer (After Reeb, 1962, fig. 50a, p. 125).

5.3 Traditional objective photometry

5.3.1 Instrumental photometry

(a) *Counting photons*

As mentioned earlier, measuring radiation is nothing but 'counting photons' (Cinzano, 1997, sec. 3.1.1; p. 107). Therefore, the basic unit of radiometry should be the number of photons. As we discuss 'measurements' here, this number must be specified as the number of photons that hit the receiver. The receiver is usually called the detector. In the next section, different types of detector are discussed in some more detail.

In sec. 2.1.3c, photons are briefly mentioned as the elements of electromagnetic radiation. Photons have an energy of $h \cdot \nu$, where h is the Planck-constant and ν the frequency of the radiation (Illingworth, ed., 1991, p. 350). The energy of n photons equals $n \cdot h \cdot \nu$. In communication engineering, the photon flow is supposed to contain some information, more than just its presence. The flow is called the signal; the signal it contains is the message. To make it clear: when there is no message, communication engineers lose interest in sending anything at all. In illuminating engineering, things are somewhat different. Illuminating engineers are interested in sending light, not as a carrier of some message, but as light itself – as a photon or energy flux. What interests lighting engineers is usually not the number of photons, but rather the number of photons that are emitted – or that pass – per second. The dimension is energy per unit of time, or 'power'. So, the logical basic unit of radiometry is the energy flux.

(b) *Sensors; Control and decision-making systems*

As regards the terminology, there is some difference between sensors and detectors. Although the two are often considered as synonyms, a sensor is the functional element, usually in a control loop. A detector is the hardware equivalent of a sensor.

Before we can explain the role of sensors and detectors, we need to discuss briefly control systems and decision-making systems. Details can be found in Narisada & Schreuder (2004, sec. 10.4.1 and 10.4.2 b,c). A sensor is the functional element, usually in a control loop. Control systems are characterised by a closed loop or by feedback (Anand & Zmood, 1995; Schwippert, 2006). This means that the output of the system that is subjected to control, influences the input – hence the term closed loop, contrary to systems that are erroneously called open-loop systems. Here the output has no influence on the input; there is no loop at all.

Basically a control loop consists of the following elements:
- a sensor that observes the output;

- data processing;
- control, the comparison to the goal;
- feedback;
- display, where the results of the process are made visible on a screen or in another way.

In steering or 'governing' a State, a car, a chemical plant, or the lighting of an office, the first thing to do is to define the goal of the activity. It is best to explain the related aspects in terms of steering a boat. For a boating trip, the destination is the goal of the exercise. From the goal, the direction in which one must travel to arrive at the goal must be defined. In control systems, this is usually called the 'Sollwert'. Then, after setting off, during the whole trip, the course must be checked at regular intervals. This checking is done by comparing the values of the direction at the specific time (the so-called 'Istwert') to the Sollwert. If there is a discrepancy greater then a predetermined margin, the course must be corrected. This is the feed-back loop.

Decision making models are a special case of control systems. As described briefly in Schreuder (1998, p. 138) the most simple model of a decision making process is the simple S-R model, the notorious Stimulus-Response model of the behavioristic theories of old. It can describe the behaviour of many automatic 'robotic' machines, including the reflex of many living organisms. However, when decisions play a role, these have to be an integral part of the model. Schreuder (1973) introduced for this the S – D – R model, where D signifies the decision. In order to be able to describe more complex decisions, two more elements have to be added. The first is the memory, where the desired outcome of the process is compared to the experience gained in earlier, similar situations. The second is the feed-back loop, which allows to compare the outcome of the decision process to the objectives of the process.

A classic collection of this material is given in the proceedings of a series of NASA-sponsored conferences (Anon., 1969). A contribution of major importance was that of Krendel & McRuer (1969). See also Barwell (1973), Broadbent (1958), Krendel & McRuer (1968), OECD (1972), and Schreuder (1973).

In the foregoing, the wording is that of steering a sailing boat; the idea can be applied to all control tasks. The whole process is described in some detail in Schreuder (1985). Comparing the Istwert to the Sollwert at regular intervals and correcting the course accordingly is pro-active control; just waiting until the trip is concluded and, in case the goal is missed, to embark on a new trip to reach the goal this time, is reactive control. It is obvious that pro-active control is to be preferred over reactive control (Narisada & Schreuder, 2004, sec. 10.4.2c, p. 407-408). Reactive processes are hardly 'decisions'. The system simply reacts to happenings in the outside world. In terms of the descriptions given earlier in this section, they are open loop systems: the outcome has no influence on the input. They are called 'reactive decisions'; they often have the characteristics of conditioned reflexes. In psychology they are often called stress reducing, or tension reducing, motives or actions. Often they have a heuristic character. Pro-active decisions describe the selection of a certain strategy. In terms of the descriptions given earlier in this section, they are closed loop systems: the outcome has a definitive influence on the input. As is explained earlier, they look like the decisions made in steering a vessel: a course is set, and deviations from the course are corrected well before the risk of collisions or of grounding is acute. Therefore, they are called pro-active decisions. As they depend mainly on cognition rather than on experience, they can be improved primarily by education and schooling.

(c) *The S/N ratio*

Data transmission is a form of communication engineering. The aim of communication engineering is, to send a signal from the sender to the receiver in such a way that the receiver can use the message. This is a description in very general terms; in traditional photometry, the sender is the photocell; the receiver is the measuring device like e.g. a galvanometer.

In almost all cases, the communication is hindered by noise from the background. In communication engineering, the word 'noise' is used only if it is composed of the same stuff as the message. Because according to this description the signal and the noise consist of the same stuff, basically 'work' (in wattage), the relation between the two can be expressed by a dimensionless number, the signal-to-noise ratio or 'S/N ratio'.

There are many sources of noise. Most of them can be avoided or at least reduced by an appropriate way to construct and operate the measuring devices. One sort of noise, however, cannot be avoided: the photon noise. Photon noise, or shot noise is the ultimate source of noise that cannot be overcome, because it is an essential aspect of the 'graininess' of the Universe. These fluctuations can also be classified as electromagnetic noise. In fact, these fluctuations did give rise to the whole idea of 'noise' in the field of communication engineering (Bok, 1948). Photon noise is a major source of disturbance in astronomy, just because in that area most other sources of error and fluctuations have been mastered. One might safely say that astronomy is the field of the most precise photometry. Astronomical photometry is discussed in considerable detail in Narisada & Schreuder (2004, sec. 14.2.3). See also Budding (1993); Crawford (1997); Sterken & Manfroid (1992).

5.3.2   Detectors

(a) Photocells

As is mentioned earlier, photocells have the same properties as the human visual system, in so far that they measure power (in watts) and not energy, like photographic emulsions (Reeb, 1962, p. 166-167). This means that photocells require a separate storage for the information.

Photocells convert the incoming light flux into electric current, using some sort of photo-electric principle. Therefore, they are often called photo-electric cells. There are three different processes that can be involved in the photo-electric effect. All three rest on the absorption of photons by a crystal so that the free electrons in the crystal become more energetic (Hentschel, ed., 2002, sec. 4.3.2; Narisada & Schreuder, 2004, sec. 14.5.1).

1. The free electrons may leave the crystal. This is often called the external photo-effect;
2. The electric conductivity of the crystal is increased. This is often called the internal photo-effect;
3. The free electrons may cross over a 'forbidden' zone. This is often called the barrier-layer photo-effect.

The concept of the forbidden zone in crystals is explained in sec. 2.5, where the structure and characteristics of semiconductors is discussed. In photometry, several types of photo-electric cells are in use. The most important ones will be described in further parts of this section.

(b) Barrier-layer photo-effect

Barrier-layer photocells are, as in indicated in an earlier part of this section, on the fact that, when free electrons in the crystal are hit by incoming photons, they become more energetic, so that they may cross over a 'forbidden' zone.

A barrier-layer photocell is, in fact, a semiconductor. Their use preceded the formulation of the physical theory of semiconductors for many decades. The physical characteristics semiconductors is explained in sec. 2.5. See also Narisada & Schreuder (2004, sec. 14.5) and Hentschel ed. (2002, p. 84-92). Details about the physical aspects of the barrier-layer photo-effect are given in Simon & Suhrmann (1958) and Feynman et al., (1977).

The photocurrent is proportional to the illuminance on the cell. It is interesting to note, that there is no need of an external source of electric energy. In the past, most barrier cells were made of selenium (Schreuder, 1967, Keitz, 1967). The selenium is 'doped'

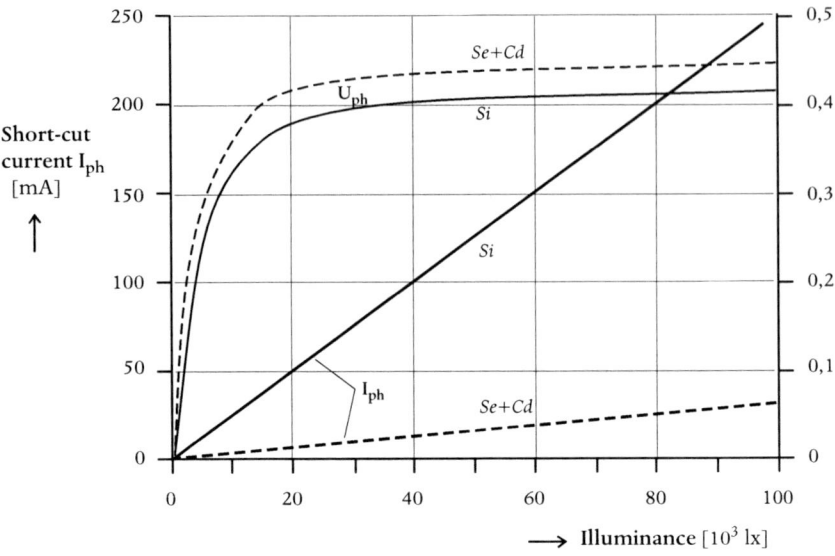

Figure 5.3.1 The photocurrent and the photovoltage versus the illuminance on the cell (After Narisada & Schreuder, 2004, figure 14.5.2).

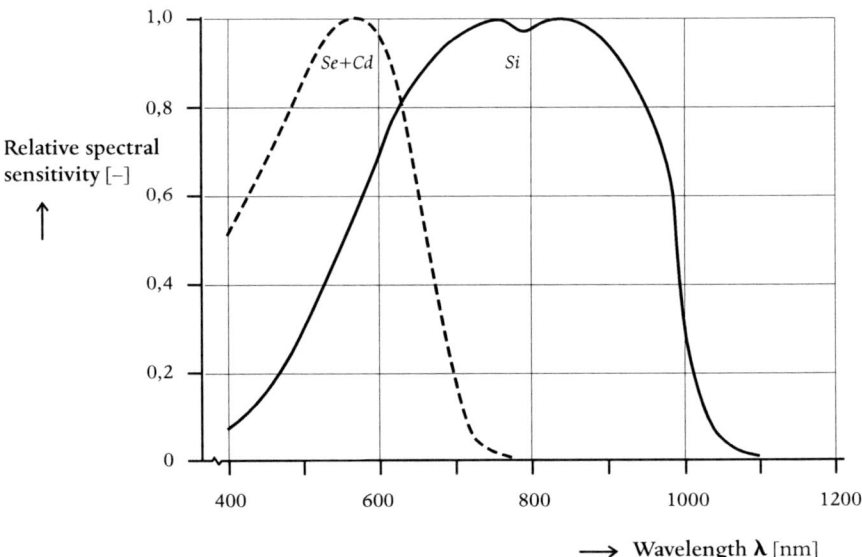

Figure 5.3.2 The relative spectral sensitivity of Se+Cd-cells and Si-cells (After Narisada & Schreuder, 2004, figure 14.5.3).

Practical photometry

with cadmium. The cells are usually designated as Se+Cd-cells (Hentschel, 1994, p. 80). Most modern cells are made of silicium. The photocurrent of Si-cells is much higher than that of Se-cells. This is depicted in Figure 5.3.1, where the photocurrent is given for short-cut condition, as well as the photovoltage for an open cell. The photocurrent depends heavily on the external resistance of the circuit.

The spectral sensitivity of Se+Cd-cells is quite different from the Si-cells. This is depicted in Figure 5.3.2.

(c) *Photocells for internal photo-effects*

Some crystals change their electric resistance, when they are hit by light, and absorb photons. The resistance is different for different light levels. This effect is, as is explained earlier in this section, an internal photo-effect. The effect is described in a summary form in Hentschel ed. (2002, p. 81-83. For details about the theory, see again Feynman et al. (1977). For cells that are used in practice, the resistance in the dark may be as large as 100 MΩ, and at 1000 lux about 100 Ω (Schwippert, 2006, sec. 4.5.1, p. 225). The size of the cell is not indicated. Photoresistors are not accurate enough for measuring purposes, because their change in resistance depends on many other factors, apart from the illuminance. Photoresistors are, however, very well suited for switching purposes and for several types of control tasks.

Photodiodes are similar to the barrier cells, that are discussed in an earlier part of this section. In fact, they are rectifiers. The current can flow in one direction, but it is blocked in the other direction. The p-n junction is responsible for that. This blocking can, in part, be lifted if the p-n junction is hit by light; the energy of the photons may knock some electrons out of the p-semiconductor, through the junction, into the n-semiconductor. When a voltage is applied, the photocurrent is proportional to the illuminance. Contrary to barrier cells, photodiodes require an external source of electric energy. At present, photodiodes are the most common photocells, either as single units or as assemblies in CCDs. CCDs will be discussed in a separate part of this section.

Phototransistors are, like all transistors, amplifiers. Many are made of gallium arsenide GaAs. Because of their non-linear character, they are very well suitable for switching purposes, but less for measurements. Their main application is as detectors in CD and DVD players.

(d) *Photocells for the external photo-effect*

When a surface is hit by photons, their energy may knock electrons out of the material. As is explained earlier in this section, this is often called the external photo-effect. The photo-electric effect is described in some detail in Sterken & Manfroid (1992, sec. 4.1, p. 65-67). See also Hentschel (1994) and Hentschel, ed. (2002). The theory goes back, as is well known, on Einstein. As a matter of fact, it was the formulation of the correct theoretical explanation of this effect, that gained him his Nobel prize for physics. When a light quantum with energy $h \cdot v$ strikes the surface, an electron with a smaller energy $h \cdot v_0$ can be emitted. For reasons of conservation of energy, $h \cdot v_0$ never can be larger than $h \cdot v$. In general:

$$h \cdot v = e_0 \cdot U_m + h \cdot v_0 \qquad [5.3\text{-}1]$$

with:
    h: the Planck-constant;
    e_0: the electrical charge of an (of any) electron;
    U_m: the voltage related to the exit electron. (After Hentschel, 1994, p. 73, equation 4.9)

Because the kinetic energy of the electron is $(m \cdot v) / 2 = c_0 \cdot U_m$; and because $v \cdot c_0 = \lambda$, with c_0 the universal speed of light, we may write:

$$U_a = (h \cdot v_0) / c_0 \qquad [5.3\text{-}2]$$

with:
    U_a: the exit potential of the electron. (After Hentschel, 1994, p. 74, eq. 4.10).

This effect is used in photocell tubes, where the light is directed towards a cathode, from which the electrons are knocked out. An anode is added. As a result of the voltage difference between the cathode and the anode, a photocurrent will flow, when light strikes the cathode. It was the main contribution of Einstein that only the number of knocked-out electrons, but not their frequency – or wavelength – depends on the illuminance. Thus, the photocurrent depends on the illuminance of the cathode. Also, the photocurrent depends on the voltage difference U_b between the cathode and the anode. In a vacuum photocell tube, the photocurrent reaches a maximum, a saturation value, but in a gas-filled photocell tube, ionization and recombination allow to a steady, non-linear increase of the photocurrent for an increasing voltage difference. These characteristics make photocell tubes more suitable for switching purposes than for measuring purposes (Schreuder, 1998, sec. 5.4; Narisada & Schreuder, 2004, sec. 14.5.1f).

(e) Photomultipliers

    In an earlier part of this section, the external photo-effect is described briefly, as well as its application in photocell tubes. An important class of detectors is based, at least in part, on the same effect: the photomultipliers. Photomultipliers are described

Practical photometry 169

in summary in Hentschel ed. (2002, p. 79), and in detail in Sterken & Manfroid (1992, sec. 4.3, p. 67-68).

The first part of a photomultiplier resembles a vacuum photocell tube, such as is discussed in an earlier part of this section. When photons of sufficient energy hit a cathode, electrons are knocked out. In a vacuum photocell tube, these electrons are accelerated by a voltage difference, and arrive at the anode, where they may give rise to a photocurrent in an external circuit. In a photomultiplier, the electrons that are knocked out of the cathode, are called 'primary electrons', for reasons that will be made clear hereafter. The primary electrons are accelerated by a voltage difference as well, but they do not arrive at the anode, but at an intermediate surface, called the dynode. Here, secondary emission takes place, which means that, again, electrons are knocked out of the surface, but this time by the primary electrons, and not by the incident photons. These electrons are, for obvious reasons, called 'secondary electrons'. By sending the secondary electrons to another dynode, this effect may be repeated. Of course, a photomultiplier multiplies light only if the number of secondary electrons is larger than the number of primary electrons. The secondary emission process can be described by the secondary emission coefficient δ. This is the ratio between the secondary electrons and the incident electrons (Sterken & Manfroid, 1992, p. 71). On theoretical grounds, one may expect hat the maximum value per dynode is between 200 and 1000. It can be approximated by:

$$\delta = A \cdot E^\alpha \quad\quad [5.3\text{-}3]$$

with:
- A: constant;
- E: interstage voltage;
- α: a coefficient, determined by dynode material and geometry. Usually α is about 0,7 (After Sterken & Manfroid, 1992, eq. 4.3, p. 72).

It seems, however, that this theoretical maximum is not easy to reach. For practice, a factor of 4 per dynode is given by Sterken & Manfroid (1992, p. 66). This does not seem very impressive, but if 10 dynodes are used, the total amplification – or gain – of the device is approximatively 4^{10}, which corresponds to about 10^6. Similar data are given by Budding (1993). It seems that per stage a factor of 3 is used. For a 12-stage photomultiplier, the number of anode electrons per released cathode electron, the gain, is 3^{12}. This equals to about the same number, viz.: 10^6 (Budding, 1993, p. 123).

In the end, the electrons arrive at the anode, where they may give rise to a photocurrent in an external circuit, just as in the case of a photocell tube.

Photomultipliers are particularly suited to measure very small amounts of light at very high frequencies. The inertia of the device is limited only by the time the electrons need to cross from the cathode to the anode. This time is in the order of 10^{-9} seconds or even less (Hentschel, ed., 2002, p. 80).

5.3.3 Measuring photometric quantities

(a) Basic considerations

In sec. 3.2.1, the different quantities are defined that, together, constitute the set of photometric units and quantities. In the ISO photometric system the luminous intensity is defined as the primary quantity, its unit being the candela. The candela is one of the seven primary units of the SI-system. In sec. 3.2.1 it is explained that, for practical use, it is more convenient to define the description of the SI-system in the following order:
- the luminous flux;
- the luminous intensity;
- the illuminance;
- the luminance;
- the luminance factor.

Photometry is a specialized profession, that requires skill, good equipment, and a very large dose of experience. There are many things that can go wrong; there are many pitfalls. We will refer here to a number of outstanding standard works that deal with the different issues with great authority and in great detail: Baer, ed. (2006); Barrows (1938); Gall (2004); Helbig (1972); Hentschel, ed. (2002); Keitz (1967); Moon (1961); Narisada & Schreuder (2004); Reeb (1962); Ris (1992); Schreuder (1967, 1998); Sterken & Manfroid (1992); Van Bommel & De Boer (1980); Walsh (1926).

It has been stated earlier, that photometry is, in essence, just 'counting photons'. Counting photons, of course, is measuring the flow of energy, or the energy per unit of time, or 'power' – expressed in watts. As has been mentioned already several times, we deal here only with photons out of the range of visual experience. These are quanta of electromagnetic radiation, weighted according to the V_λ-curve. In other words, counting such photons corresponds to the measurement of the illuminance. This is reflected in the 'toolbox', that is used by practical photometrists: its main tool is the device to measure illuminances – for short, the luxmeter.

(b) Luxmeters

Not only is the luxmeter the primary tool of the photometrist, the illuminance is also the main criterion to describe the characteristics of a lighting installation. When

Practical photometry

one knows the lighting level – or rather, the illuminance level – in a lighting installation, one can derive the other criteria of quality, at least by approximation.

More important, luxmeters are very simple instruments, that are easy to construct and easy to maintain. Apart from the top-grade instruments, they usually have a modest price. In fact, the practical standard in an engineering photometric laboratory usually is a calibrated luxmeter, that is used to check all other equipment.

Luxmeters exist in all ranges of precision, size, handling and, of course, price. Simple pocket-size meters use a Se+Cd-cell, and no external electric power supply is needed. Their range is from about 100 to about 100 000 lux, that is, they are suited for higher class interior lighting, and for daytime use outdoors. Their overall level of error is at least 10%. They range in price from some 20 euro to a few hundred euro.

More advanced types of luxmeters use silicium photodiodes, that require an external power source. They range from intermediate equipment, which can measure down to maybe 0,1 lux, with a precision of a few percent, and a price of one or two thousand euros, to the top levels, that measure from the microlux levels up to the megalux levels, with a precision better that 0,1%. They may cost up to 100 000 euro. For top-level instruments, great care about temperature, humidity, and precision of the mains voltage as well as positioning of the cell is essential. They require, apart from a affluent budget, expert operators to reach the top quality results. Precise photometry is not a anybody's job. There are many manufacturers of such equipment in the world.

Luminous intensities of light sources are usually measured with standard luxmeters. The only thing to do is to measure the distance between the meter and the source very precisely. The high precision is needed, because, according to the 'inverse square law', that is discussed in sec. 3.2.4d, a slight error in the distance will result in a large error in the measured value of the luminous intensity. The 'distance law', that is discussed in sec. 3.2.4, requires that the distance between the meter and the source must be at least 5 times, and preferably 10 times or more, the dimensions of the source.

(c) Luminance meters

The measurement of the luminance requires a somewhat different arrangement of the equipment. Essentially, luminance meters are just plain luxmeters, where the angular range of the light that is taken into account, is limited and precisely restricted. This area represents the measuring field: the luminance is measured for all elements that show up within the measuring field. The only thing to do is to calibrate the meter in luminance values in stead of in illuminance values. The calibration as such is simple enough; the reason one may do this is, however, not simple. Some of the fundamental problems are explained in detail in Chapter 4, where the mathematics of the luminance are discussed. Most traditional luminance meters have silicium photodiodes, or in special cases, photomultipliers as a detector. As regards the optical system, usually luminance meters have a lens and diaphragm – 'stop' – system that limits the measuring field. This may be somewhere between 1 and 5 degrees in diameter. If the measuring field is much smaller, one may speak of a 'spot meter'. The measuring field may go dow to some 5, or even 2, minutes of arc. Because the narrow measuring field restricts the number of photons that may reach the detector, usually quite complicated electronic gear is needed. Medium priced luminance meters, costing several thousand euro, may go down to about 0,1 cd/m^2, which is sufficient for most general lighting installations both indoors and outdoors. However, in road lighting, the requirements are usually much higher. As an example, in residential streets the darkest spots on the road may have an illuminance of about 0,1 lux, which corresponds to about 0,005 cd/m^2 (CEN, 2002; CIE, 1992, 1995; NSVV, 1990, 2002). Meters that can measure such low values accurately, must have a lower limit that is about a factor 10 better, that is a luminance of about 0,0005 cd/m^2. Particularly for spot meters, the number of photons that may reach the detector is very small, so that the requirements for the equipment is very high. Such meters are difficult to use, they need careful handling and frequent calibration, and are quite costly. Prices range to 100 000 euro or considerably more.

In recent times, CCD cameras are used in photometry. As is explained in sec. 5.4.1, CCDs did open a new field in photometry to such an extent that is seems to be justified to speak of 'modern photometry'.

(d) Measuring the luminous flux, and the light distribution
The luminous flux can measured in exactly the same way as the luminous intensity, more in particular the luminous intensity distribution of a light source, like e.g. the light distribution of a luminaire. At a certain distance, the intensity is measured. Actually, as is explained earlier, it is a measurement of the illuminance at that particular location. This is repeated for so many directions as is relevant for the type of light source. If the light distribution is needed, the values of the intensity in different directions are plotted. If the luminous flux is needed, all values are integrated. The integration process can, of course, also be made by integrating directly all light that is coming from the source. This is done by placing the source in a integrating photometric sphere, often called the Ulbricht sphere. If the inside of the sphere is covered by a material that reflects all incident light, by the action of multiple reflections, the luminance of the interior will be completely uniform over the complete surface, independent of the light distribution of the source. Measuring this luminance yields directly the total luminous flux of the source. Details may be found in several of the standard works that have been mentioned

Practical photometry 173

earlier like e.g Barrows (1938); Hentschel, ed. (2002); Keitz (1967); Reeb (1962), and Ris (1992).

(e) Accuracy

When dealing with measuring systems, one of the criteria is the minimum required accuracy of the measuring equipment. International bodies have set up standards, regulations and recommendations to do so. See e.g. CEN (2002) and CIE (1987). They exist for almost all possible types of measurements. Usually, classes are defined, and for each individual measuring apparatus it is possible to indicate to what class it belongs. In Germany, the current classification in given in DIN 5032 part 7 (DIN, 1985, 1985a; Hentschel, ed., 2002, sec. 4.4.1). Many European countries follow the German standards. Standardization is important for practical applications, because in many cases, e.g. by granting permission to set up outdoor lighting installations, the limiting values are set down to which the installations must agree. Such limits have little value if there is no indication about the class of accuracy to which the measurements must be made. This is particularly important if there is disagreement between the parties involved, and if court action is considered. Court actions may, of course, lead to convictions and damage claims. A further implication is, that the quality of the testing is important. Testing of instruments must be performed by test houses that are 'certified' to do so. There are special ISO-Standards that describe in detail the underlying processes and considerations.

Measurements never can be absolutely correct. That is, of course, usually not necessary. In Narisada & Schreuder (2004, Table 14.5.1) some qualitative suggestions have been given as to how accurate the measurements, and thus the equipment to be used, must be for some specific fields of application.

However, there is a sort of 'ground rule'. Many call it the Law of Constant Misery. Almost without any exception, accurate measurements are more costly than casual measurements: costly in money for the equipment, costly in measuring time and costly in expertise of the operators. So it is worth while to insure that the measurements are not more accurate than necessary. In the following, some suggestions given for the degree that is required for outdoor lighting installations. As can easily be seen, road lighting served as an example for this:
1. Errors in input data for calculations: less than 5%;
2. Differences between measurements and calculations: less than 2%;
3. Errors in measurements (instrument errors)
   – horizontal illuminance: less than 3%;
   – vertical illuminance: less than 5%;
4. Luminance factors: less than dan 5%;

5. Luminances
   - road surface luminance: less than 10%;
   - point luminances: less than 15%;
   - veiling luminances: less than 25%;
6. Photometric measurements:
   - illuminances: cosine correction: better than 99%; colour correction: better than 95%;
   - temperature correction: better than 99%;
   - luminances: colour correction: better than 95%; temperature correction: better than 99%.

These values are quoted from Schreuder (1998; sec. 10.7). They are based on draft documents that were made for discussion in several working groups of CIE and CEN (Schreuder, 1995).

In practice, the accuracy of field illuminance measurements with is usually not better than 10% (Van Bommel & De Boer, 1980, p. 212). Field luminance measurements with spotmeters are usually not better than 20% (Van Bommel & De Boer, 1980, p. 215, referring to Walthert, 1977).

As an example, we will give two classifications of photometric measuring equipment, one for illuminance measurements and another for luminance measurements. The tables are quoted from Ris (1992). They are based on the German standards for photometry (DIN, 1985; 1985a). See Table 5.3.1 and Table 5.3.2. In both cases, four classes are introduced that refer to the field of application of the measuring equipment:
1. Class L, the highest, for laboratory use;
2. Class A, high, also for laboratory use;
3. Class B, average, for practical applications;
4. Class C, low, where no specification of the use is given.

With modern equipment, it is possible to agree with these requirements. This will be illustrated by two examples that are related to luxmeters.

Practical photometry

Criterion	Unit	Error limits for equipment class			
		L	A	B	C
V_λ-correction	%	1,5	3	6	9
UV and IR sensitivity	%	0,2	1	2	4
cosine correction	%	1	1,5	3	6
linearity error	%	0,2	1	2	5
display error	%	0,2	3	4,5	7,5
fatigue	%	0,1	0,5	1	2
temperature coefficient	%/K	0,1	0,2	1	2
change in range	%	0,1	0,5	1	2
total error	%	3	5	10	20
lower frequency limit	Hz	40	40	40	40
upper frequency limit	Hz	105	104	104	103

Table 5.3.1 Accuracy requirements for illuminate measurements (After Ris, 1992, table 10.2, p. 335).

Criterion	Unit	Error limits for equipment class			
		L	A	B	C
V_λ-correction	%	2	3	6	9
UV and IR sensitivity	%	0,2	1	2	4
cosine correction	%	2	3	6	9
linearity error	%	0,2	1	2	5
display error	%	0,2	3	4,5	7,5
fatigue	%	0,1	0,5	1	2
temperature coefficient	%/K	0,1	0,2	1	2
change in range	%	0,1	0,5	1	2
focus error	%	9,4	1	1	1
total error	%	5	7,5	10	20
lower frequency limit	Hz	40	40	40	40
upper frequency limit	Hz	105	104	104	103

Table 5.3.2 Accuracy requirements for luminance measurements (After Ris, 1992, table 10.2, p. 335).

(f) *Examples of cosine and colour corrections for lux-meters*

As is explained in sec. 3.2.4g, the illuminance on a surface decreases with the cosine of the angle between the direction of light incidence and the normal to the surface. This is called the cosine law. Because photocells usually are somewhat glossy, the measured value will be less than the value that corresponds to the cosine law. Consequently, a some sort of cosine correction is required. In Figure 5.3.3 one possible solution is depicted.

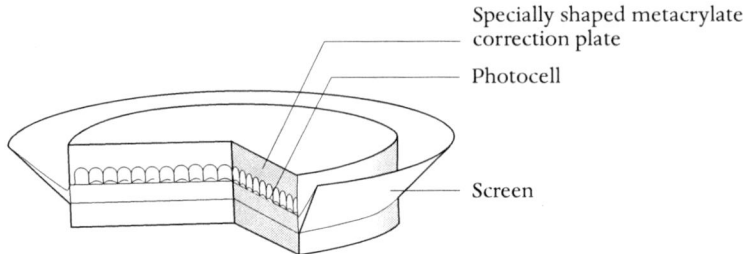

Figure 5.3.3 An example of cosine-corrected barrier-cell. (After Schreuder, 1967, Fig. 8.6, p. 440)

In Figure 5.3.4, the degree of correction is given of the device that is depicted in Figure 5.3.3.

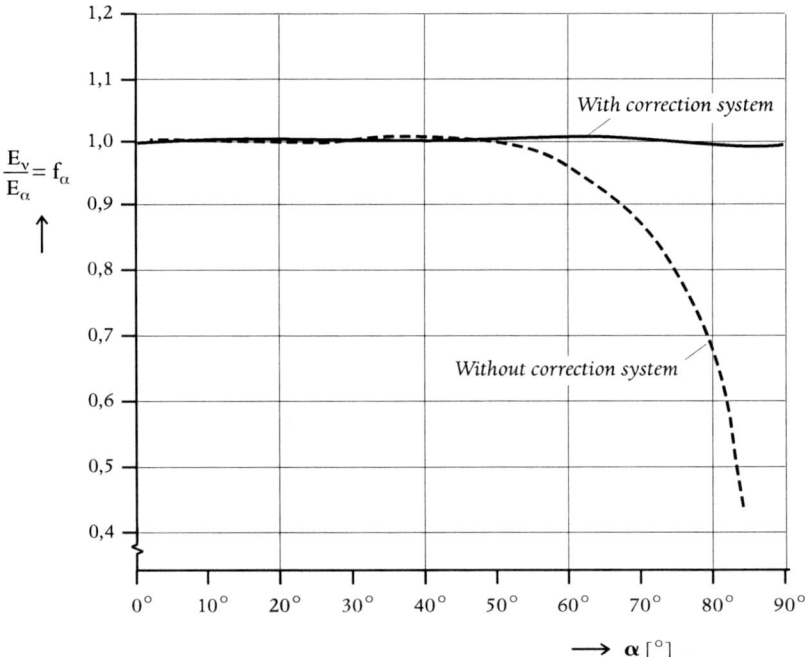

Figure 5.3.4 The degree of correction of the device depicted in Figure 5.3.3. (After Schreuder, 1967. Fig. 8.8, p. 441).

Practical photometry

In Figure 5.3.5 an example is given of the colour correction of a photocell for a lux-meter.

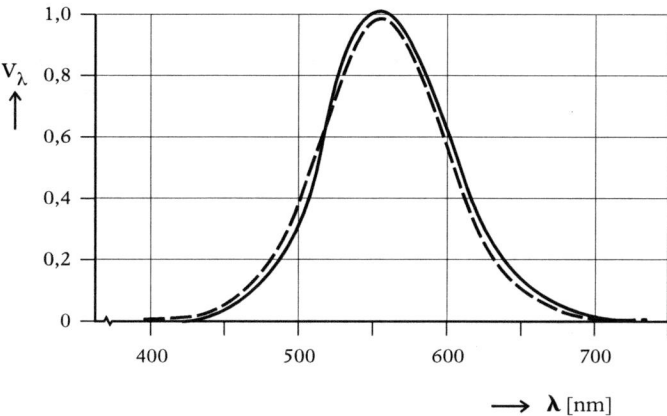

Figure 5.3.5 *The colour correction of a photocell (After Ris, 1992, Fig. 10.3, p. 331).*

5.4 Modern objective photometry

5.4.1 CCDs

(a) *CCDs for taking pictures*

Charge-Coupled Devices, or CCDs, are the most recent light detectors that are used on a large scale. They combine the capability of two-dimensional imaging of photographic plates with the linearity and the sensitivity of silicium photodiodes. The linearity is, as is mentioned earlier, within a few percent over at least 10 decades. CCDs are at the heart of all video cameras; they are used in many modern amateur still photo cameras. For that purpose, they have several additional advantages. The picture can be seen and judged immediately after being taken; development and other chemical processes are not needed. If needed, it can be send, by cellular telephone, immediately to any place in the world, an advantage for professional press photographers. This means also that the same 'material' can be used over and over again, which is often considered as a way of saving money. Also, the picture, once taken, can be transferred to a computer, where a large variety of processes can be applied to change the original picture. This capability is used on a large scale by professional photographers, but it remains to be seen in how far amateurs actually do this. In the past, the limited resolution could be a disadvantage. Modern, top line cameras boast over 10 million 'pixels' on their CCD. Modern cellular telephones often have some 3 or 4 million pixels. In a further part of this section we will explain what 'pixel' means. A top class 'chemical' camera, with 24·36 mm film and a resolution of 110 lines per mm, will be equivalent to 10 454 400 pixels. The gap has been closed already by several professional CCD-cameras.

One further advantage of CCDs for photometry is their sensibility over other media like e.g. photographic plates. This pays off especially in astronomical observations.

In summary it is stated that a CCD will 'use' about half of the impinging photons, as compared to one in hundred for the human visual system (Anon., 2007). This is a summary statement where no details are given. In view of the photon yield of about one in three that has been found by Weale (1949) for rod vision, one may assume that the statement given here refers to cone vision. The photon yield, or quantum efficiency QE is discussed in Narisada & Schreuder (2004, sec. 14.5.2c). Already in 1994 is was found that photographic plates detect only about 0,1% of the light entering a telescope, and that, overall, CCDs are 30 times more sensitive than photographic plates (Lafferty & Rowe, eds., 1994, p. 112). It is only for large-scale survey purposes as are used in astronomical, that the storing capacity of a photographic plate is still superior (Adams et al., 1980; Howell, 2001). 'Schmidt-plates' are equivalent to about one Giga-pixel. "Giga-pixel CCDs are still far away" (Sterken & Manfroid, 1992, p. 212).

(b) The properties of CCDs

A CCD consists of an array of tiny photodiodes made of silicium (Sterken & Manfroid, 1992, p. 192). Pixel size varies from 0,3 micron to 0,7 micron (Schwippert, 2006, p. 263). The array is usually, but not always, two-dimensional. One-dimensional, or line scan CCDs have been used in measuring the luminance of road surfaces in road lighting (Schreuder, 1996; 1998, p. 71; Schreuder & Van de Velde, 1995). The individual photodiodes are called picture elements or pixels. Present-day CCDs generally come in sizes from 512 by 512 pixels to arrays as large as 4096 by 4096 pixels, totalling 262 144 pixels and 16 777 216 pixels respectively (Howell, 2001, p. 2; Schwippert, 2006, p. 263).

In CCDs, the pixels may be regarded as capacitors, that are charged when exposed to light. The charge packets, one for each pixel, are caused to flow from one capacitor to another. This is why the array is called a Charge-Coupled Device. The charge packets pass through readout electronics, that detect and measure each charge in a serial fashion (Howell, 2001, Chapter 2). They have many advantages in weight, power consumption, noise characteristics, linearity, spectral response, and others (Howell, 2001, p. 7). Apart from these, the other main advantages of CCDs over other means to measure and store light are: the large dynamic range of about 10^{10} and the high quantum efficiency (Sterken & Manfroid, 1992, p. 191). A major disadvantage is bleeding. Bleeding occurs when a pixel is overexposed and overflows into neighboring pixels (Howell, 2001, p. 17,

Practical photometry

footnote). It can be compared to irradiation in optical systems and to glare in visual perception.

(c) The performance of CCDs

CCDs have a number of essential characteristics, that are too specialized to discuss here. These characteristics are discussed in detail in Budding (1993, sec. 5.2.6, p. 129-132); Sterken & Manfroid (1992, Chapter 13); Howell (2001, Chapters 2 and 3), and Narisada & Schreuder (2004, sec. 14.5.2).

The performance of a CCD luminance measuring system depends primarily on the luminance range one is interested in. For areas where the overall luminance levels are high, such as the lighting of offices or tunnels, the measurements are straight-forward. They have been used on a large scale for almost two decades now. See e.g. Frank & Damasky (1990); Huijben (2002); Narisada & Schreuder (2004, sec. 14.5.2), and Serres (1990).

In modern equipment, low luminances can be measured by using long integration times and by adding up the result of neighboring pixels. This is called binning (Howell, 2001, sec. 3.4, p. 36). The consequences are clear: long exposure times allow only the measurement of static scenes, and binning reduces the resolution. The effects can be dramatic, however. See Table 5.4.1. For further details, see Gall et al. (2002) and Schmidt & Krüger (2000).

Time	L_{max}			L_{min} (including binning)		
		binning	1x1	3x3	7x7	18x18
0.1 ms	28.000	1400				
5 s	0,558		0,028	$9 \cdot 10^{-3}$	$4 \cdot 10^{-3}$	$1,6 \cdot 10^{-3}$

Table 5.4.1 CCD measurements at low luminances for different integration times. All values in cd/m^2 (After Fischbach, 2001, Table 4, and Narisada & Schreuder, 2004, Table 14.5.4).

The main sources of 'noise' in a CCD image are:
1. The noise of the signal;
2. The noise of the background;
3. The detector 'readout noise'.

Sources (1) and (2) are statistical noise sources, that follow, like as all shot-noise phenomena, Poisson distribution. The signal-to-noise ratio (S/N) is described in sec. 5.3.1c. See also Narisada & Schreuder (2004, sec. 14.2.3e, equations [14.2.13] and [14.2.14]), and Tritton (1997).

(d) CCD data extraction and data processing

CCD imaging starts with the extraction of the information from the CCD. The information regarding the position on the CCD is already in digital form by means of pixel column and row. In consumer colour cameras, the colour information is contained in the three types of pixels for the three basic colours. Basically, the 'amount' of light per pixel is in analogous form. In the electronic hardware, this amount is digitized, e.g. in 255 levels. The comparison of the three types of colour yields the colour information. This is essentially enough to present pictures. Cameras usually are delivered with software packages that, together with one of the many computer software packages, allow further manipulation of the image. Sometimes, some of this manipulation is called 'cosmetics'. They are needed to get a beautiful, full colour picture that is worth showing. CCD imaging is discussed in detail in Howell (2001, Chapter 4, p. 47-74) and Unger et al. (1988, sec. 4.2, p. 77-85).

The following discussion about data extraction is bases in part on Narisada & Schreuder (2004, sec. 14.5.2e), and also on Howell (2001) and Vermaelen (2003). Data extraction includes several steps. Usually, the first steps are to get rid of all disturbances that are not coming from the heavenly bodies. This includes faulty pixels or faulty pixel lines, bias, thermal, UV, and cosmic ray noise etc. There are standard ways to take these out (Howell, 2001). The next step is to extract unwanted and disturbing information, like disturbing spectral lines, compensating for dark current, etc. The second step compensates for 'bias' like e.g. compensating for offset, uncorrected dark current, etc. The next steps are to reduce image noise, e.g. by averaging a number of subsequent images, and to homogenize the field of exposure by comparing it to a homogene plane of reference. These compensations and corrections are more complicated and include complex mathematical procedures, like e.g. the deconvolution of Fourier and Bessel function. Such corrections are related to errors in the telescope optical system, like spherical aberrations and astigmatism, as well as bringing all exposures to the same scale in order to be able to compare them.

Recently, CCD imaging is used on an increasing scale for light measurements, particularly for luminance measurements in the field, e.g in lighting application, both indoors and outdoors. For this, some additional steps are needed. The first is, that locations in the field must be allotted to locations on the CCD. Further, the levels that are delivered for each pixel, must be calibrated in luminance values. This can be done in a variety of ways, the most common being a comparison with a luminance standard. When this is done, the luminance of any point in the field, that now corresponds to a specific pixel, can be quantified. Lines of equal luminance can be included by making all points of which the luminance equals that of the 'line' are made pure white. If needed, false colours

Practical photometry 181

can be added for luminance ranges. Special software is developed that allows arithmetic manipulations of the data, e.g. assessing the maximum, the minimum or the average of the luminances in the field, and that allow different ways to present the data. Usually, this sort of software is developed by the manufacturer of the measuring equipment, and is equipment-specific. In most cases, the process stops here. Lighting engineers, designers, planners and architects are in most cases satisfied with the type of data mentioned earlier, like the maximum, the minimum or the average of the luminances in the field, the glare, and the false-colour images. Then comes the most difficult step: pattern recognition. Pattern recognition is discussed in Narisada & Schreuder (2004, sec. 10.3.1i). See also Schreuder (2008).

In astronomical photometry, data extraction and data processing are much more complicated undertakings, because astronomers want to have much more, and much more detailed, information than the overall information which satisfies lighting designers (Narisada & Schreuder, 2004, sec. 14.5.2e).

5.4.2   CCDs in photometry

Usually, CCDs are applied in the measurement of luminance. In this type of application, a digital camera is used; basically, any type of commercially available camera could be used. For reasons of calibration, as well as for the application at low luminances, special cameras have been developed. Appropriate, dedicated software allows a number of manipulations that are essential to assess the luminance and the luminance distribution of lighted surroundings. To name a few: the luminance of individual points can be measured, similar to the measurement with a 'spotmeter'. The same can be done, of course, for a great many points in the same surroundings. This allows one to asses the average luminance in any area of the field, as well as the non-uniformity of the luminance distribution. Also, it is possible to select single lines in the picture, as is done in 'line scan photometry'. Finally, lines of equal luminance can be assessed , and added to the picture.

5.5   Conclusions

Photometry is essential, both in visual science and in lighting engineering. Photometry makes it possible to express lighting and vision phenomena in a quantitative way. Photometry is a branch of radiometry. Radiometry is a branch of experimental natural science.

The fundamentals of radiometry and photometry, as well as the relation between the two are discussed in Chapter 3. This chapter deals with the practical implications of photometry.

It is concluded that the accuracy of photometric measurements is lower than the accuracy that is common in other areas of measurement, like e.g radiometry that is discussed in Chapter 3. The reason is that radiometry is a branch of experimental natural science. The relevant methods are those of experimental physics, whereas photometry is a branch of applied psychology. The relevant methods are those of experimental phychophysics. The accuracy of radiometric measurement is determined by the limits of instruments, notably by the signal-to-noise ratio, whereas the accuracy of photometric measurement is limited by the way the performance of the human visual system is determined, notably by the assessment of the spectral sensitivity of the human visual system.

An essential aspect of all quantitative measurements, like e.g the measurement of phenomena that are expressed in ordinal or other quantitative scales is calibration.

Gauging and calibration require standards. Physical standards are adequately defined in international agreements to a precision that always can be made better than the precision requirements of the actual measurements. Physiological standards are defined in terms of a hypothetical 'standard observer', and therefore are always fraught with errors. More important, the standard changes over time, not because of alterations in the definitions, but because of improvements in measuring accuracy.

References

Adams, M.; Christian, C.; Mould, J.; Stryker, L. & Tody, D. (1980). Stellar magnitudes from digital pictures. Kitt Peak National Observatory publications, 1980.

Adrian, W. & Schreuder, D.A. (1971). A modification of the method for the appraisal of glare in street lighting. In: CIE, 1972.

Anand, D.K. & Zmood, R.B. (1995). Introduction to control systems. 3rd edition. Oxford, Butterworth-Heinemann Ltd., 1995.

Anon. (1969). Fourth annual NASA-University Conference on Manual control, University of Michigan, Ann Arbor, March 21-23, 1968. NASA SP-192. Washington, DC., NASA, 1969.

Anon. (1997). Control of light pollution – measurements, standards and practice. Conference organized by Commission 50 of the International Astronomical Union and Technical Committee TC 4-21 of la Commission Internationale de l'Eclairage CIE at The Hague, Netherlands, on August 20, 1994. The observatory, 117 (1997) 10-36.

Anon. (2006). The next star hunt is on. IDA Newsletter no 67. November 2006.

Anon. (2007). CCD; Sterrenkunde in Nederland (CCD; Astronomy in the Netherlands). Internet Encyclopedie. 1 February 2007.

Baer, R. (1990). Beleuchtungstechnik; Grundlagen (Essentials of illuminating engineering). Berlin, VEB Verlag Technik, 1990.

Baer, R., ed. (2006). Beleuchtungstechnik; Grundlagen. 3., vollständig überarbeitete Auflage (Essentials of illuminating engineering, 3rd., completely new edition). Berlin, Huss-Media, GmbH, 2006.
Barrows, W.E. (1938). Light, photometry and illuminating engineering. New York, McGraw-Hill Book Company, Inc., 1938.
Bartley, H. (1951). The psychology of vision, Chapter 24, p. 921-984. In: Stevens, S.S., ed., 1951.
Barwell, F.T. (1973). Automation in control and transport. Oxford, Pergamon Press, 1973.
Bok, S.T. (1948). Cybernetica (Cybernetics). Utrecht, Spectrum Aula, 1948.
Bouguer, P. (1760). Traité d'optique sur la gradation de la lumière (On the optics of the gradations of light). Paris, 1760.
Bouma, P.J. (1946). Kleuren en kleurenindrukken (Colours and colour impressions). Amsterdam, Meulenhoff, 1946.
Broadbent, D.E. (1958). Perception and Communication. London, Pergamon Press, 1958.
Bronstein, I.N.; Semendjajew, K.A.; Musiol, G. & Mühlig, H. (1997). Taschenbuch der Mathematik. 3. Auflage (Manual of mathematics. 3rd edition. Translated from the Russian original). Frankfurt am Main, Verlag Harri Deutsch, 1997.
Budding, E. (1993). An introduction to astronomical photometry. Cambridge University Press, 1993.
CEN (2002). Road lighting. European Standard. EN 13201-1..4. Brussels, Central Sectretariat CEN, 2002 (year estimated).
Cinzano, P. (1997). Inquinamento luminoso e protezione del cielo notturno (Light pollution and the protection of the night sky). Venezia, Institutio Veneto di Scienze, Lettere ed Arti. Memorie, Classe di Scienze Fisiche, Matematiche e Naturali, Vol. XXXVIII, 1997.
CGPM (1979). Comptes rendues des scéances de la 16e Conférence Générale des Poids et Mesures CGPM (Minutes of the meetings of the 16th General Conference on Measures and Weights CGPM). Paris, Bureau International des Poids et Mesures, 1979.
CIE (1924). Proceedings of the Commission Internationale de l'Eclairage, Geneva, 1924.
CIE (1951). CIE Proceedings 1951. Paris, CIE, 1951.
CIE (1972). Compte rendue, 17e Session. Barcelona, September 1971. Publication No. 21a. Paris, CIE, 1972.
CIE (1987). Methods of characterizing illuminance meters and luminance meters: Performance, characteristics and specifications. Publication No. 69. Vienna, CIE, 1987.
CIE (1990). CIE 1988 2° spectral luminous efficiency function for photopic vision. Publication No. 86-1990. Vienna, CIE, 1990.
CIE (1992). Guide for the lighting of urban areas. Publication No. 92. Paris, CIE, 1992.
CIE (1995). Recommendations for the lighting of roads for motor and pedestrian traffic. Technical Report. Publication No. 115-1995. Vienna, CIE, 1995.
CIE (2002). The use of tungsten filament lamps as secondary standard sources. Publication No. 149. Vienna, CIE, 2002.
Crawford, D.L. (1997). Terminology and units in lighting and astronomy. In: Anon, 1997.
Daintith, J. & Nelson, R.D. (1989). The Penguin Dictionary of mathematics. London. Penguin Books, 1989.
De Boer, J.B. (1951). Fundamental experiments of visibility and admissible glare in road lighting. Stockholm, CIE, 1951.
De Boer, J.B. & Schreuder, D.A. (1967). Glare as a criterion for quality in street lighting. Trans. Illum. Engn. Soc. (London). 32 (1967) 117-128.
De Boer, J.B., ed. (1967). Public lighting. Eindhoven, Centrex, 1967.
De Bruin, N.G. (1949). Beknopt leerboek der differentiaal- en integraalrekening (Short textbook on differential and integral calculus). Amsterdam, N.V. Noord-Hollandsche Uitgevers Maatschappij, 1949.

DIN (1985). Photometer; Begriffe, Eigenschaften und deren Kennzeichnung. DIN 5032 Lichtmessung. Teil 6 (Photometer; definitions, characteristics and their indication. DIN, 5032 Photometry. Part 6). Berlin, DIN, 1985.

DIN (1985a). Klasseneinteilung von Beleuchtungsstärke- und Leuchtdichtemessgeräte. DIN 5032 Lichtmessung. Teil 7 (Classification of illuminance and luminance measuring equipment. DIN, 5032 Photometry. Part 7). Berlin, DIN, 1985.

Feynman, R.P.; Leighton, R.B. & Sands, M. (1977). The Feynman lectures on physics. Three volumes. 1963; 6th printing 1977. Reading (Mass.), Addison-Wesley Publishing Company, 1977.

Fischbach, I. (2001). Bewertung von Sichtverhältnissen im nächtlichen Strassenverkehr mit Leuchdichteanalysatoren (Assessment of the visibility conditions in nighttime road traffic using luminance analyzing equipment). Ilmenau, TechnoTeam Bildverarbeitung GmbH, 2001 (Year estimated).

Frank, H. & Damasky, J. (1990). Entwickling eines mobilen Meßsystems zur Untersuchung der lichttechnischen Eigenschaften des Straßenraumes bei Dunkelheit (Development of a mobile measuring system to investigate the lighttechnical characteristics of the street scene in darkness). In: NSVV, 1990a.

Gall, D. (2004). Grundlagen der Lichttechnik; Kompendium (A compendium on the basics of illuminating engineering). München, Richard Pflaum Verlag GmbH & Co KG, 2004.

Gall, D.; Krüger, U.; Schmidt, F. & Wolf, S. (2002). Moderne Möglichkeiten zur Messung und Bewertung von Beleuchtungsparametern. Herbstkonferenz 2002 der GfA e.V. (Modern possibilities to measure and assess lighting parameters. Fall Conference, 2002. GfA, e.V.). Ilmenau, Technical University, 2002.

Gall, D. & Wuttke, V. (1985). Ist das Licht der Natrium-hochdrucklampe 'dunkler' als das der weissen Lichtquellen? (Is the light of high-pressure sodium lamps more 'dark' than that of white light sources?). Elektro-Praktiker 39 (1985) 11, p. 386-388.

Hecht, S. (1938). The nature of the visual process. Bull. N.Y. Acad. Med. 14 (1938) 21-45.

Hecht, S. & Shlaer, S. (1936). Intermittent stimulation by light. V: The relation between intensity and critical flicker frequency for different parts of the spectrum. J. Gen. Physiol. 19 (1936) 965-977.

Helbig, E. (1972). Grundlagen der Lichtmesstechnik (Fundaments of photometry). Leipzig, Geest & Portig, 1972.

Hentschel, H.-J. (1994). Licht und Beleuchtung; Theorie und Praxis der Lichttechnik, 4. Auflage (Light and illumination; Theory and practice of lighting engineering, 4th edition). Heidelberg, Hüthig, 1994.

Hentschel, H.-J. ed. (2002). Licht und Beleuchtung; Grundlagen und Anwendungen der Lichttechnik; 5. neu bearbeitete und erweiterte Auflage (Light and illumination; Theory and applications of lighting engineering; 5th new and extended edition). Heidelberg, Hüthig, 2002.

Herrmann, J. (1993). DTV-Atlas zur Astronomie, 11. Auflage (DTV-atlas on astronomy, 11th edition). München, DTV Verlag, 1993.

Howell, S.B. (2001). Handbook of CCD astronomy, reprinted 2001. Cambridge (UK). Cambridge University Press, 2001.

Hütte (1919). Des Ingenieures Taschenbuch (The manual for engineers). Berlin, Wilhelm Ernst und Sohn, 1919.

Huijben, J.W. (2002). Lumimanties van de hemel, verlichtingssterkten en luminantiefactoren van materialen bij tunnelingangen (Sky luminances, illuminances and luminance factors of materials near tunnel entrances) Utrecht, Bouwdienst Rijkswaterstaat, 27 juni 2002.

Illingworth, V., ed. (1991). The Penguin Dictionary of Physics (second edition). London, Penguin Books, 1991.

Ives, H.E. (1912). Studies on the photometry of lights of different colours. Philos. Mag. 24 (1912) p. 149, 352, 744, 845, 853.

Jantzen, R. (1960). Flimmerwirkung der Verkehrsbeleuchtung (Flicker effects caused by road traffic lighting). Summary in Lichttechnik, 12 (1960) 211
Judd, D.B. (1951). Basic correlates of visual stimuli. Chapter 22. In: Stevens, ed., 1951.
Keitz, H.A.E. (1967). Lichtmessungen und Lichtberechnungen. 2e. Auflage (The measurement and calculation of light. Second edition). Eindhoven, Philips Technische Bibliotheek, 1967.
Kelly, D.H. (1961/1962). Visual response to time dependent stimuli. Journ. Opt. Soc. Amer. 51 (1961) 421-429; 747-754; 52 (1962) 8-95.
Krendel, E.S. & McRuer, D.T. (1968). Psychological and physiological skill development; A control engineering model. Proc. 4th Annual Conference on Manual Control, 1968.
Krendel, E.S. & McRuer, D.T. (1969). Psychological and physiological skill development; A control engineering model. In: Anon, 1969, Chapter 15.
Lafferty, P. & Rowe, J, eds. (1994). Dictionary of science. London, Brockhampton Press, 1994.
Lambert, J.H. (1760). Photometria sive de mensura et gradibus luminis, colorum et umbrae (Photometry, or the measurement and grades of light, colour and shades). Augsburg, 1769.
Liebenthal, E. (1907). Praktische Photometrie (Practical photometry). Braunschweig, 1907.
Moon, P. (1961). The scientific basis of illuminating engineering (revised edition). New York, Dover Publications, Inc., 1961
Moon, P. & Spencer, D.E. (1981). The photic field. Cambridge, Massachusetts, The MIT Press, 1981.
Narisada, K. & Schreuder, D.A. (2004). Light pollution handbook. Dordrecht, Springer, 2004.
Nutting, P.G. (1914). The visibility of radiation. Trans. Illum. Eng. Amer. 9 (1914) 633.
NSVV (1990). Aanbevelingen voor openbare verlichting (Recommendations for public lighting). Arnhem, NSVV, 1990.
NSVV (1990a). Licht90. Tagungsberichte Gemeinschaftstagung, Rotterdam, 21-23 Mai, 1990 (Licht90. Proceedings of joint meeting, Rotterdam, 21-23 May 1990). Arnhem, NSVV, 1990.
NSVV (2002). Richtlijnen voor openbare verlichting; Deel 1: Prestatie-eisen. Nederlandse Praktijkrichtlijn 13201-1 (Guidelines for public lighting; Part 1: Performance requirements. Practical Guidelines for the Netherlands 13201-1). Arnhem, NSVV, 2002.
OECD (1972). Symposium on road user perception and decision making. Rome, OECD, 1972.
Reeb, O. (1962). Grundlagen der Photometrie (Fundaments of photometry). Karlsruhe, Verlag G. Braun, 1962.
Ris, H.R. (1992). Beleuchtungstechnik für Praktiker (Practical illuminating engineering). Berlin, Offenbach, VDE-Verlag GmbH, 1992.
Rossi, G. ed. (1996). International workshop and intercomparison of luminance CCD measurement systems (draft). Liege, Belgium, 23 September 1994. CIE, 1996.
Ruch. T.C. (1951). Sensory mechanisms. Chapter 4, p. 120-153. In: Stevens. ed., 1951.
Schaefer, B.E. (1993). Vistas in Astronomy. 36 (1993) 311.
Schaefer, B.E. (2003). Personal communication.
Schmidt, F. & Krüger, U. (2000). Einsatz von Standard-CCD-Matrizen für fotometrische Messungen – Anwendung und Design von Kameras mit hoher Auflösung und Genauigkeit (Use of standard CCD arrays for photometric measurements – application and design of cameras with high resolution and accuracy). Ilmenau, TechnoTeam Bildverarbeitung GmbH, 2000 (Year estimated).
Schmidt-Clausen, H.-J. (1968). Über das Wahrnehmen verschiedenartiger Lichtimpulse bei veränderlichichen Umfeldleuchtdichten (On the observation of different light pulses by different levels of the surrounding luminance). Darmstadt, Dissertation TH, 1968.
Schreuder, D.A. (1964). The lighting of vehicular traffic tunnels. Eindhoven, Centrex, 1964.
Schreuder, D.A. (1967). Measurements. Chapter 8. In: De Boer, ed., 1967.
Schreuder, D.A. (1973). De motivatie tot voertuiggebruik (The motivation for vehicle usage). Haarlem, Internationale Faculteit, 1973. Leidschendam, Duco Schreuder Consultancies, 1973/1998.

Schreuder, D.A. (1985). Regelen, beheersen en sturen ... bijvoorbeeld in het wegverkeer (Govern, command and control ... e.g. in road traffic). R-85-27. Leidschendam, SWOV, 1985. Also in: Wegen 59 (1985) 217-220.

Schreuder, D.A. (1995). Tolerances in the measurements. Contribution to CEN/TC169/WG6. Standard for the lighting of road traffic tunnels. Draft. 21 March 1995. Leidschendam, Duco Schreuder Consultancies, 1995.

Schreuder, D.A. (1996). A CCD line-scan system for road luminance measurement. Traffic Engineering and Control, 37 (1996) 208-209.

Schreuder, D.A. (1998). Road lighting for safety. London, Thomas Telford, 1998 (Translation of "Openbare verlichting voor verkeer en veiligheid", Deventer, Kluwer Techniek, 1996).

Schreuder, D.A. (2008). Looking and seeing; A holistic approach to vision. Dordrecht, Springer, 2008 (in preparation)

Schreuder, D.A. & Van de Velde, A. (1995). A CCD line-scan measuring system for road surface luminance. In: Rossi, ed. (1995).

Schwippert, G.A. (2006). Sensoren (Sensors). Chapter 4. In: Schwippert, ed, 2006.

Schwippert, G.A., ed. (2006). Sensoren: Theorie en toepassingen (Sensors; Theory and applications). Deventer, Uitgeverij Nassau, 2006.

Serres, A.-M. (1990). Les images pour les études de visibilité de nuit (Images to investigate the night time visibility). Bull Liais. Labo. P. et Ch. 165 (1990) jan-fév. 65-72.

Simon, H. & Suhrmann, R. (1958). Der lichtelektrische Effekt und seine Anwendung, 2. Auflage (The photoelectric effect and its application, second edition). Berlin, Springer, 1958.

Smith, A.H. & O'Loughlin, J.L.N., eds. (1948). Odhams Dictionary of the English Language (reprinted). London, Odhams Press, 1948.

So, M. (1920). Visibility of radiation throughout the spectrum. Proc. Phys. Math. Soc. Japan. 2 (1920) 177.

Sterken, C. & Manfroid, J. (1992). Astronomical photometry. Dordrecht, Kluwer, 1992.

Stevens, S.S. (1951). Mathematics, measurement and psychophysics. Chapter 1. In: Stevens, ed., 1951.

Stevens, S.S., ed. (1951). Handbook of experimental psychology. New York, John Wiley and Sons, Inc, 1951.

Stumpff, K., ed. (1957). Astromomie (Astronomy). Das Fischer Lexikon 4. Frankurt am Main. Fischer Bücherei GmbH, 1957. Reprinted 1961.

Tritton, K.P. (1997). Astronomical requirements for limiting light pollution. In Anon., 1997, p. 10-13.

Unger, S.W.; Brinks, E.; Laing, R.A.; Tritton, K.P. & Gray, P.M. (1988). Observers' Guide. Version 2.0. November 1988. La Palma, Isaac Newton Group, 1988.

Vermaelen, S. (2003). Private communication.

Walsh, J.W.T. (1926). Photometry. London, Constable, 1926. Reprinted, New York, Dover, 1965.

Walthert, R. (1977). Lichtmessungen auf Strassen und in Sportanlagen. Bericht SLG-Tagung Lichtmesstechnik. (Light measurements of roads and sports lighting installations. Proceedings SLG meting on photometry). Switzerland, 1977.

Weale, R.A. (1968). From sight to light. Edinburgh and London, Oliver and Boyd, 1968.

Weaver, K.S. (1949). A provisional standard observer for low level photometry. Journ. Opt. Soc, Amer. 39 (1949) 278.

Weigert, A. & Wendker, H.J. (1989). Astronomie und Astrophysik – ein Grundkurs, 2. Auflage (Astronomy and astrophysics – a primer, 2nd edition). VCH Verlagsgesellschaft, Weinheim (D), 1989.

Young, A.T. (1984). Proc. workshop improvements to photometry. NASA Conference Publ. vol. 2350, p. 8.

Zwikker, C. (1933). Beknopte verlichtingsleer (A summary of illuminating engineering). Amsterdam, De Paltrok, 1933.

6 The human observer; physical and anatomical aspects of vision

Human beings are characterised by being conscious of themselves and of the world around them. In daily life most information is visual information. All information from the outside world that reaches into the brain gets there via the senses. Being an essential part of all sense organs, the nervous system is an essential part of the visual sense. The structure of the nervous system is the subject of this chapter.

In essence, the optical elements of the eye work together to form an optical image of the outside world on the retina. The retina converts the incoming light into electrical pulses that are propagated along the optical nerve systems. They ultimately reach the optical cortex.

Apart from the image forming processes, light in the eye is involved in several non-image forming effects of light. These are of particular interest for the biological clock, and more in general for the adjustment of the human organism to the outside world. These effects have been dealt with only very briefly here.

Some of the age-effects that are relevant for outdoor lighting are discussed at various places in this, and in the following chapters.

6.1 The ability to see

Human beings are characterised by having a consciousness of themselves and of the world around them. Brain functions are essential, although not sufficient to explain consciousness – whatever some authors do claim. All information from the outside world that reaches into the brain, gets there via the senses. Although other senses are very important in life, in this book we focus on the visual sense – the ability to see.

One may discern between the reality 'out there' and the reality 'in here'. All knowledge about the reality 'out there' comes to us via the senses. As senses are never fully exact, we have to be satisfied by an approximation. Visual perception will in all likelihood provide a quite large amount of useful information about the world 'out there' but one never can rely on it in order to come to hard conclusions about the outer world. Sensory perception is the only window we have towards the reality 'out there'. We have to be satisfied that the window distorts the view (Schreuder, 2008).

6.2 The nervous system

6.2.1 The structure of the nerve cells

(a) Neurones

The nervous system consists of cells, just like every animal tissue. The essential cell is the nerve-cell or neuron. Some neurones classify as the largest cells in the human body. They may stretch for over one metre (Gregory ed., 1987, p. 514). Far more than that in a whale or an elephant. In a giant whale, some axons may reach a length of up to 12 metres (Gregory, ed., 1987, p. 550).

Neuro-physiological research has explained much during the last couple of decades, but many unanswered questions remain. Many details can be found in the standard books of Crick (1989, 1994), Damasio (1990, 1994), Dennett (1993, 2006), Edelman (1992, 2004), or Greenfield (1997, 1999, 2000, 2000a, 2001). See also Hubel (1990, chapter 4); Hentschel (1994, sec. 1.4); Hentschel, ed. (2002); Narisada & Schreuder (2004, sec. 8.1), and Schreuder (1998, sec. 6.4).

The anatomy of the neurones is reasonably well-known. As mentioned earlier, a neurone or nerve cell is a single cell. Its special characteristic is, however, a 'spur', the axon. The axons are extremely thin, measuring usually between 10^{-3} and 10^{-4} mm but often very long, as mentioned above, up to one metre in length. Both ends of the cell, at the one side the cell body and at the other side the axon have smaller protuberances, the dendrites. The dendrites play an essential role in the transmission of the electric pulses that carry the information that must be transferred by the nerve system. Nerve cells form chains. The dendrites of the axon of one cell are in close contact with the dendrites on the cell body of the next cell. The transmission in the cell is along the axon. In Figure 6.2.1 a neuron is depicted; it must be noted that the axon is omitted from this representation.

Without memory, a neuron can exist in only two states, on and off, and cannot change between them in less than about 1 millisecond. That is not much of a unit. However, nerve cells have second messenger systems, working on slower time scales that operate on ionic membrane channels (Barlow, 2004, p. 24).

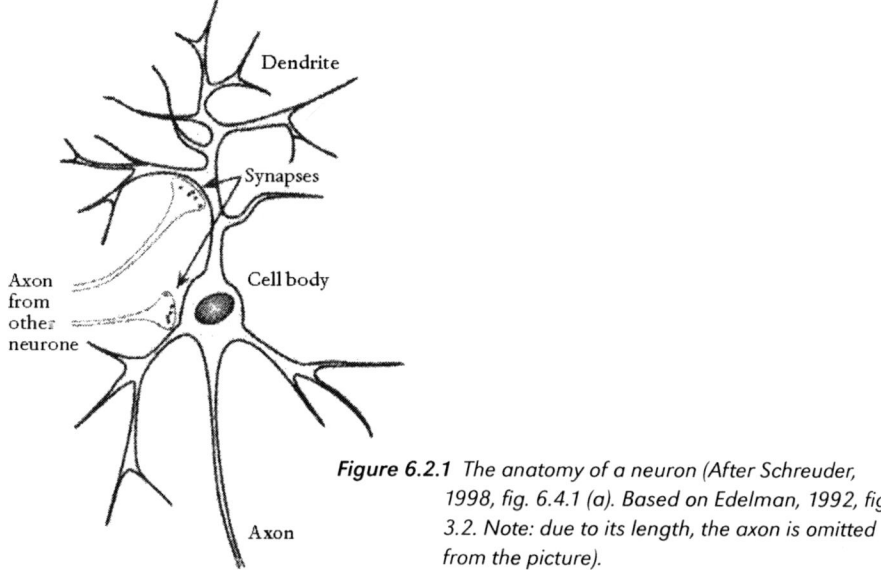

Figure 6.2.1 The anatomy of a neuron (After Schreuder, 1998, fig. 6.4.1 (a). Based on Edelman, 1992, fig 3.2. Note: due to its length, the axon is omitted from the picture).

(b) *Synapses*

Nerve cells form chains. The dendrites of the axon of one cell are in close contact with the dendrites on the cell body of the next cell, the small gap is called the synapse. At the synapse, the two cells are near, but they are not in contact. They are separated by an extremely narrow gap of about $25 \cdot 10^{-6}$ mm: the synaptic gap. In Figure 6.2.2 the synapse is summarized.

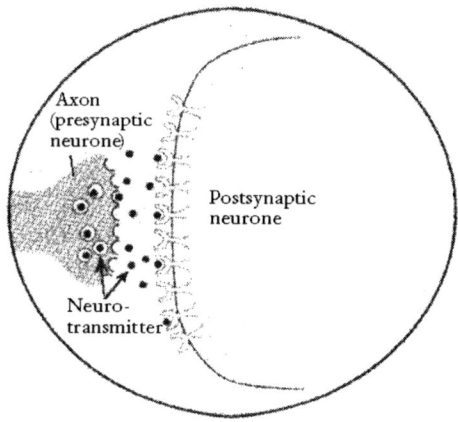

Figure 6.2.2 A schematic representation of the synapse. (After Schreuder, 1998, fig. 6.4.1 (b). Based on Edelman, 1992, fig 3.2.)

At the 'sending' side of the synapse, the energy of the electric pulse is transformed into the chemical energy of the neurotransmitter. This chemical substance seems to cross the synaptic gap. At the other end, the 'receiving' end, the chemical energy is transformed back into electric energy in the next cell. In that cell, an electric pulse is created that is transferred along the dendrites and the axon to the next cell as the spikes mentioned in Figure 6.3.7. In this way, the information can the transferred along the chain of nerve cells. The process involving the neurotransmitters is described in some detail in Gregory, ed., 2004, p. 657-673.

Neurotransmitters are highly potent substances. They are released from their storage sites in the ends of the dendrites close to the synapse. They may diffuse over a certain distance until they encounter neurotransmitter receptors with which they are designed to specifically interact. Once bound to the neurotransmitter in question, the neuroreceptor, a large protein molecule that spans the 10 nm thick membrane, shows a change in structure that shows a 'hole' or a 'passage' right through the neuroreceptor molecule, through which only a particular ion can pass, e.g. Na^+, K^+, or Cl^-. These ions are propelled by the prevailing electric fields across the membrane. Their movement generates excitatory or inhibitory synaptic potentials. As a result, the target neurone will be triggered to fire its own action potential. The afore-mentioned channel is open only for a very brief period of time, e.g. 1 microsecond. This is a very crude description of only one of the processes that may take place at the synapse. One particular fact that may be mentioned is the large number op neurotransmitters that may be involved. In Gregory, ed. (2004, table 1, p. 661), 48 different neurotransmitters are listed. There are theories that suppose to explain the large number (Gregory, ed., 1987, p. 560). Another striking fact is that amongst the fifty or so neurotransmitters, a considerable number of substances are listed that are known to be involved in psychological processes; like e.g. glutamate, histamine, adrenaline, dopamine, noradrenaline, serotonin, melatonin, and endorphin. Several of these are even used in psychotherapy (Damasio, 1990).

About photoreceptor transmitters, the following scheme seems to be well-established. Glutamate is the transmitter in the direct pathway, that is the pathway from photoreceptor to bipolar cells and from bipolar cells to ganglion cells. Gamma-aminobutyric acid (GABA) is the main transmitter in the lateral and feedback pathways that are mediated by the horizontal cells in the outer retina and by the amacrine cells in the inner retina (Wilson, 2004, p. 279). Bipolar cells, ganglion cells, and amacrine cells are explained in other parts of this section.

6.2.2   The central nervous system

The life of animals, humans included, consists of receiving information and responding to it (Gregory, ed., 1987, p. 514). A useful description is given by Küppers, 1987, p. 17:

life = material stuff + information

Animals made up of more than a few cells have a nervous system. The fact that many plants and animals cope very well without a nervous system of any kind is interesting but falls outside the scope of this book.

Broadly speaking the nervous system of vertebrate animals consist of two parts:
1. The autonomous nervous system, also called the peripheral nervous system. The autonomous nerve system consists in its turn of two parts, the sympathic system and the parasympathic system. Both are always active. The sympathic system is responsible for temperature regulation. It controls the distribution of the blood throughout the body. It contributes to the behaviour that is related to threat and aggression. The parasympathic system is the main controller of the alimentary canal (Gregory, ed., 1987, p. 524; 2004, p. 81). Many reactions could classify as unconditional and conditional reflexes, that in some cases are triggered by the non-image-forming effects of light that strikes the eye. These effects are discussed in detail in CIE (2006); Narisada & Schreuder (2004, sec. 4.5.1); NSVV (2003, 2006); Schreuder (2006, 2008), and Van den Beld (2003).
2. The central nervous system consisting of the spinal chord and the brain.

In further parts of this chapter, the different parts of the central nervous system, and in particular the optical nervous pathways as well as the brain will be discussed in more detail.

6.3   The anatomy of the human visual system

6.3.1   The overall anatomy

Considerations of the anatomy and the physiology of the visual sense begin with a broad subdivision in aspects directly related to the visual function, and to other aspects. Usually, it is assumed that the visual function is concentrated in the eye ball and the neural pathways. As a result, almost all treaties on vision are restricted to these bodily parts. However, a more close inspection shows that the picture is more complex. The eye socket, the muscular systems, the eye lids, the lachrymal fluids, and other aspects are involved in the visual process. Notably, many play a role visual tracking, in depth perception, and in the perception of movement. Small involuntary eye movements, called saccades, are essential in the visual process. In the following chapters we will come back to several of these aspects. At the other hand, the retina and the neural systems have a number of non-visual functions, including the non-image-forming functions that have been mentioned earlier. In this part of this section we will focus on the main image-forming parts of the visual system.

The human visual system, as well as that of animals, is very complicated. The traditional ideas about the anatomy have been discussed in detail in textbooks like e.g. Davies, ed., (1969) and Perkins & Hill, eds. (1977). See also Gregory (1965) and more in particular Hubel (1990). A complete but concise survey of almost all aspects of vision is given by Gregory, ed. (1987, 2004). Much more detailed is the gargantuan work of Chalupa & Werner, eds. (2004). The overview given here is based on Schreuder (1998, Chapter 6), and Narisada & Schreuder (2004, sec 8.1). For many details. we will refer to Perkins & Hill, eds. (1977), and to Chalupa & Werner, eds. (2004)

In essence, the optical elements of the eye work together to form an optical image of the outside world on the retina. The retina converts the incoming light – the incoming light quanta – into electrical pulses that are propagated along the optical nerve systems. They ultimately reach the brain, more precisely the optical cortex. The different parts will be discussed in more detail in the following parts of this chapter; first the anatomy and after that the functions. We will begin to give an overview of the optical characteristics of the different parts of the eye; see Table 6.3.1.

Characteristic	Quantity
Refractive index	
air	1,0
water (20° C)	1,333
cornea	1,376
aqueous humour	1,336
lens averaged	1,413
vitreous humour	1,336
Curvature in mm	
cornea	7,8
front of lens	10,0 – 5,33
rear of lens	6,0 – 5,33

Characteristic	Quantity
Refraction diopters	
cornea	43,08
front of lens	7,7 – 16,5
rear of lens	12,83 – 16,5
total system	59,74 – 70,54

Table 6.3.1 Geometric and optical data of the human eye in a simplified model (After Hentschel, ed. 2002, table 1.2. Based on data of Gullstrand, 1911).

More modern data are given in Augustin (2007, table 42.1). However, the considerations did not change very much over the years. The values in Table 6.3.1 are given for accommodation to infinity. In some cases the values for maximum accommodation are added. The diopter is a measure of the focal length that is in use by ophthalmologists and optometrist. It is the reciprocal of the focal length; its dimension is m^{-1} (Clugston, ed., 1998, p. 154-155). 1 diopter corresponds to a focal length of 1 m.

For comparison, the refractive index of air and water are included in Table 6.3.1. To be precise, the refractive index of vacuum is per definition equal to unity. The refraction process is explained in sec. 10.1.1b.

The human visual system consists of five major elements:
1. The optical elements of the eye;
2. The retina and the photoreceptors;
3. The nerve cells (neurons) in the eye;
4. The visual nerve tracts;
5. The brain, notably the visual cortex.

Of these, (1), (2) and (3) are within the eye-ball, whereas (4) and (5) are located within the skull. The optical elements make an optical image of the exterior world on the retina. The optical elements of the eye are:
1. The cornea, which forms the outer layer of the eye;
2. The eye lens;
3. The iris with its opening, the pupil.

In Figure 6.3.1, a broad sketch of these parts is given. The anatomy of the human eye is described in great detail in Augustin (2007, chapter 43).

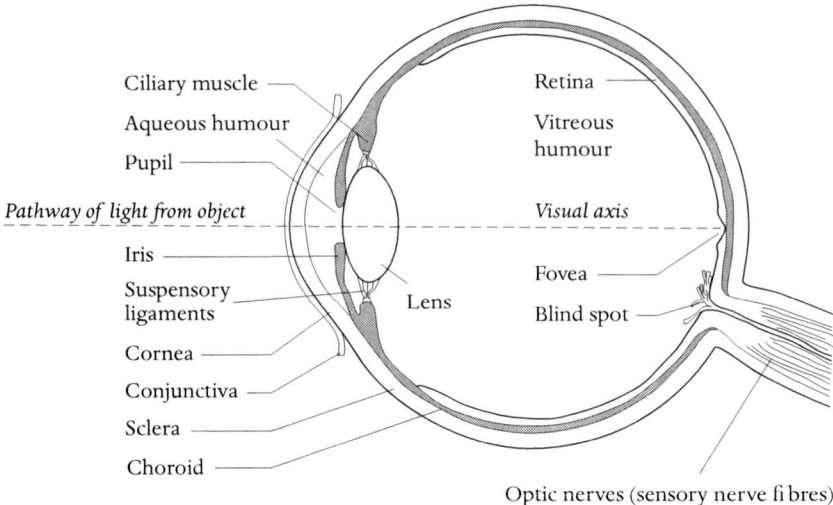

Figure 6.3.1 The elements of the human visual system (After Clugston, ed., 1998, p. 273).

The eye-ball is, as the word says, an almost spherical organ of about 25 mm diameter. In the front part of the eye the cornea, the iris and the lens are located; at its rear, the retina. The rest is filled with transparent fluids. Between the cornea and the lens the room is filled with the aqueous humour, the middle parts of the eye are filled with the vitreous humour. We will discuss the different parts very briefly here. But before that, we will mention the 'blind spot'.

As can be seen from Figure 6.3.1, the blind spot is the place where the optical nerves leave the eye-ball. Because the eye nerves lay in front of the retina, at that point there cannot be located any photoreceptors. Hence the name 'blind spot'. A part of the visual field where no observations can be made is called a scotoma. Usually, scotomata result from retinal damage or retinal malfunctions (Augustin, 2007, sec. 4.4, p. 410-412). In some way, the blind spot can be regarded as a scotoma as well. This physiological scotoma lays on a horizontal line in the field of view at 15° temporal, its diameter being 6° (Augustin, 2007, p. 1187).

One peculiarity of the blind spot is that it is not perceived as such unless special attention is paid. This effect is often called gestalt completion across it. It appears that holes in the retina resulting in visual scotoma are compensated for by a massive reorganization of cortical fields. It has been found that a deafferented cell quickly resumed firing when light fell on adjacent regions of the retina. One way this may be explained is to assume that collateral signals are fed to deafferented neurons through the horizontal long-distance connections that are mentioned in sec. 6.4.2 (Spilmann & Ehrenstein, 2004, p. 1584).

A key feature in many aspects of visual perception in the filling-in process. This process is related to the curious effects of the perception of boundaries. It is well-known from many visual illusions, often boundaries are perceived even when they do not exist in reality. A brief description of some of these effects and the related Gestalt laws are given in Narisada & Schreuder (2004, secs. 10.2.1 and 10.2.2, p. 380-384). See also Gregory (1965), Glaser (1997), Metzger (1953), and Schreuder (2008).

But also, boundaries are often not seen even if they exist. Still, it is clear that we may perceive real things. This seems to be a contradiction. When trying to solve this contradiction, it is assumed that the process begins by the brain adding, and pooling signals from cortical cells that are sensitive to opposing contrast polarities, that is positive and negative contrasts Next comes the filling-in process that is mentioned earlier (Grossberg, 2004, p. 1626, 1627). This process involves the attenuation of colour and brightness signals except at regions with considerable gradients – the original boundary. The process of filling in is also part of many of the visual illusions that are mentioned already. In effect, the same appears to happen at the physiological scotoma of the blind spot. It is concluded that: "Filling-in clarifies how our brains perceive brightnesses and colors across the retinal black spot" (Grossberg, 2004, p. 1627). To be fair, it would be better to say that it clarifies nothing at all, but only adds to the enigma of visual perception, which, as has been mentioned earlier is often be regarded as a 'almost completely unknown (Greenfield, 1997, 2000; Narisada & Schreuder, 2004, sec. 8.1.3; Schreuder, 2008).

6.3.2 The optical elements, the cornea

The optical image is made in the first instance by the cornea. Apart from being the outermost layer of the eye, it is also the main refractive element of just over 43 diopters. See Table 6.3.1. The reason that it is the cornea, and not the lens, that contributes most to the optical imaging is that the differences in the refractive indices between the cornea and its adjoining optical media – at least at one side – is much larger than the corresponding differences at the lens. See Table 6.3.1.

The anatomy of the cornea is described in great detail in Sherrard (1977). A striking feature of the cornea is its high degree of transparency. Basically, the anatomy of the cornea is rather simple. It consists of four layers. The outermost is the epithelium. It measures some 50 – 100 microns in thickness. It is some five or six cells thick. The cells divide continuously, the new ones pressing the old ones outward. Inside that the main body of the cornea is formed by the stroma. It is about 0,5 mm thick, and it consists of a vast number of very thin long fibers of collagen. The fibers are regularly arranged in layers into lamellae. To the inside of the stroma two very thin layers follow. The first is Descemet's membrane. It is 8 – 10 micron thick and it consists fibers. Finally the still thinner endothelium, a 4-5 micron thick monolayer of cells. The cornea consists of living tissues, unlike hair or nails. However, there are no blood vessels in the cornea, so the nurture of the corneal cells must be taken care of by diffusion.

The intra-ocular pressure, which is normally in the range of 16 – 20 mm mercury, presses the cornea into its characteristic dome shape. The shape is non-spherical, which may help to achieve a reasonable sharp image with only one refractive surface.

6.3.3   The optical elements, the eye lens

(a)   The anatomy of the eye lens
The role of the lens is mainly to adjust the focal length of the optical system, allowing sharp images of objects at different distances. The human eye lens consists of living cells that form fibers, laying inside a collagenous capsule. The lens grows throughout life within the capsule, never shedding cells. The newly made cells are laid down on the outside of the existing fibers, just under the capsule. The outermost part of the lens is therefore composed of the youngest cells. The lens nucleus contains the oldest fibers. One must take this fact into account when considering old-age maladies like cataract (Augustin, 2007; Schreuder, 1998, sec. 7.5.3; Mann & Pirie, 1950; Stilma, 1995). Also, the prevalence of old fibers in the nucleus of the lens is responsible for the stiffening of the lens that causes the loss of accommodation in old people. The accommodation is discussed in a further part of this section. The same holds when considering the colour of the lens, particularly in old age. The overall anatomical structure is depicted in Figure 6.3.2.

In Figure 6.3.2, the capsule and the cells that form the fibers are shown. A, B, C, and D show the different levels in the lens, and the way the cells progressively branch off and become interwoven.

The lens has no blood supply. It is largely dependent upon the aqueous humour slowly flowing over its front surface for the supply of its nutrients. Therefore the unimpeded flow is essential for the health of the eye. If this flow is restricted or interrupted, serious maladies may arize, like e.g. glaucoma, a troubling of the lens that eventually may lead to blindness (Schreuder, 1998, sec. 7.5.3; Langerhorst, 1995).

The human observer; physical and anatomical aspects of vision

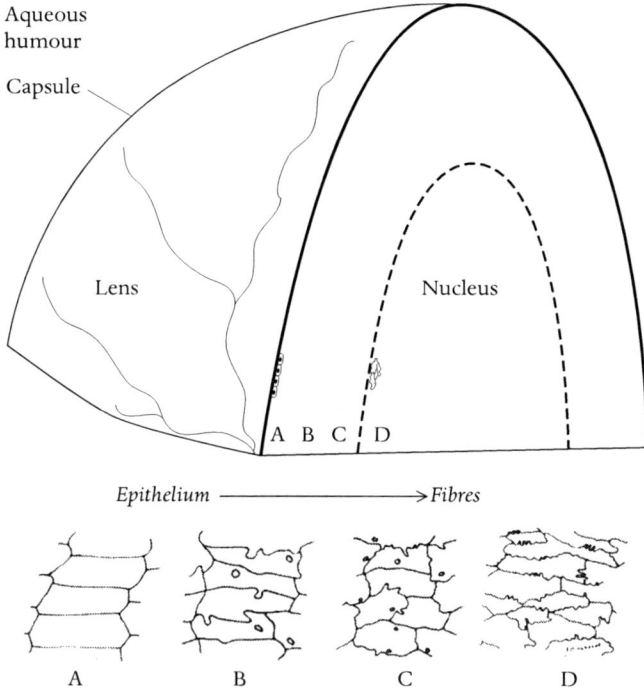

Figure 6.3.2 Diagrammic cross section of the adult human lens. (After Van Heyningen, 1977, Fig. 1, p. 36.)

(b) *Accommodation*

The focal length corresponds – depending on the state of accommodation – from about 20 to just over 30 diopters (Table 6.3.1).

The most common of visual deficiencies are usually called refraction errors. They refer to the fact that the focal lens and the depth of the eye – the length of the eye ball – are not in good agreement. The focal length of the lens in all eyes is nearly the same. It is the length of the eye ball that causes the errors. When the eye is too long, people suffer from near-sightedness or myopia; when the eye is too short, they suffer from far-sightedness or hypermetropy. See Schreuder, 1998, Figure 7.5.1.

It is explained in sec. 6.3.3 that in the eye lens the lens nucleus contains the oldest fibers. One must take this fact into account when considering old-age maladies like cataract. Also, the prevalence of old fibers in the nucleus of the lens is responsible for the stiffening of the lens that causes the loss of accommodation in old people. As the lens looses most of its flexibility when people get older, people cannot accommodate easily any more. Older people cannot focus easily on nearby objects. They suffer from presbyopia and they need reading glasses. The diminishing accommodation depth is given in Figure 6.3.3. See also Augustin, 2007, figure 42.5.

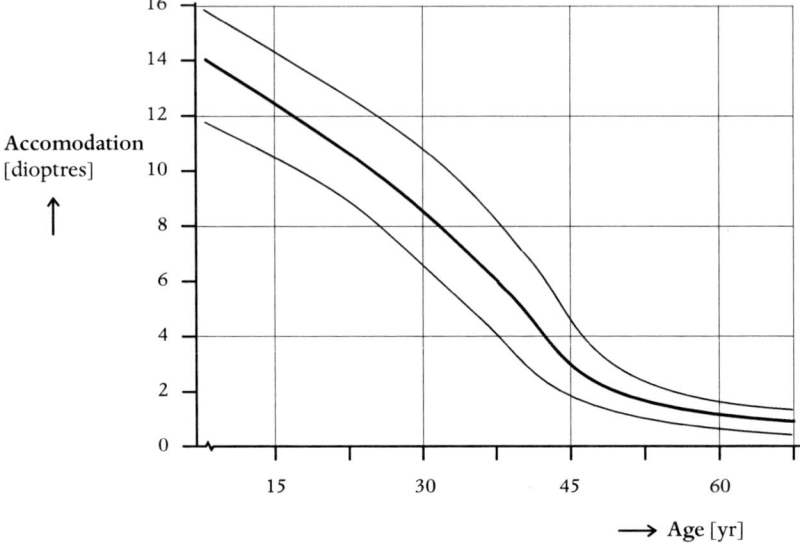

Figure 6.3.3 Depth of accommodation versus age. (After Narisada & Schreuder, 2004, fig. 8.1.2.)

Another way of expressing the loss of accommodation is to indicate the nearest point that still can be focussed by the unaided, healthy eye. See Table 6.3.2.

Age	Near-by point (cm)
10	7
20	10
30	14
40	22
50	40
60	100
70	400

Table 6.3.2 The near-by point in relation to age.

(c) *Fourier optics*

Fourier optics is the assessment of optical images based on Fourier analysis and Fourier transform. Fourier optics describe succinctly the effects of diffraction and aberrations (Williams & Hofer, 2004, p. 795). It refers to wave-form analysis. It is therefore an inextricable element of the wave form theory of light that is discussed in sec. 2.1.3a. Fourier analysis is based on the fact that it is possible to express any single-valued periodic function as a summation of sinusoidal components of frequencies that are multiples of the ground frequency of the function. Such a summation is called a Fourier series, and the analysis of a periodic function into its simple harmonic components is a Fourier analysis (Illingworth, ed., 1991, p. 181). A Fourier transform is a mathematical operation by which a function expressed in terms of one variable x may be related to a function of a different variable s. Many such pairs are useful, more in particular 'time' and 'frequency' when wave phenomena, such as light, are studied (Illingworth, ed., 1991, p. 182). A rather detailed summary of the Fourier series and Fourier analysis is given in Bronstein et al. eds. (1997, sec. 7.4, p. 382-387), and of Fourier transforms in Bronstein et al. eds. (1997, sec. 15.3, p. 674-683). Further details on Fourier-analysis are given in Wachter & Hoeber (2006, Appendix A4). A more basic discussion is given Longhurst (1964, sec. 6-5, p. 100-104). Applications in diffraction phenomena and in image formation are given in Longhurst (1964, p. 226 and p. 300-301 respectively).

The effects of diffraction can be expressed in mathematical form:

$$r_0 = 1{,}22\, \lambda/d \qquad [6.3\text{-}2]$$

with
r_0: the diameter of the fist light disk – the Airy disk that is explained further on;
λ: the wavelengths of the light;
d: the diameter of the pupil.

(d) *The point spread function or PSF*
The application of Fourier optics on the formation of the retinal image is discussed in detail in Williams & Hofer (2004). The first step is the definition of the point spread function or PSF.

The PSF is defined as the retinal light distribution that is the effect of a point source of light positioned outside the eye (Van den Berg, 1995, p. 52). The PSF is essentially a function of θ alone because the nature of the particles in the eye that cause the scatter, ensures that the stray light function is isotropic, which means that it shows a circular symmetry along the optical axis. This angle is usually described as the glare angle.

In many cases the analysis of the PSF is restricted to the way the light is distributed over the retina. This is called the optical PSF. However, the efficiency of the light that strikes the retina at an oblique angle is less than that of the light that strikes the retina perpendicular to its surface. This is called the Stiles-Crawford effect that is explained in a further part of this section.

When the Stiles-Crawford effect is taken into account, the adapted PSF is called the functional PSF. Obviously, only this functional PSF is relevant for visual performance considerations. One of the main areas of interest of the functional PSF is in the assessment of disability glare.

The PSF is the Fourier transform of the pupil function. The pupil function includes the diffraction, the achromatic aberration (Porter et al., 2001), and the Stiles-Crawford effect. Still, the PSF depends in the first instance on the pupil diameter. For a pupil diameter of about 2 mm, the diffraction dominates. This can be shown by the resulting the well-known Airy-disk pattern (Illingworth, ed., 1991, p. 115; Williams & Hofer, 2004. fig. 50.6, p. 798). For a larger diameter of the pupil, the aberrations dominate. It seems that an optimum is reached for a pupil diameter of about 3 mm (Williams & Hofer, 2004, p. 799, Campbell & Gubisch, 1966).

(e) Optical aberrations; monochromatic aberrations

Just like all other optical devices, the eye shows aberrations. The refractive errors that have been explained earlier do not classify as optical aberrations. The theory of optical abberations is given in detail in the 'classic' treaties of Van Heel (1950) and Longhurst (1964). See also Blüh & Elder (1955), and Feynman et al. (1977, Volume 1, Chapter 27). A simplified discussion is given in Kuchling (1995, Chapter 25) and Breuer (1994). The main aberrations are the monochromatic aberrations and the heterochromatic aberrations.

The monochromatic aberrations refer to these imaging errors that are not dependent upon the wavelength of the light. They result in images that are not sharp.

The errors decrease when the pupil is smaller. In the next part of this section some characteristics of the pupil system are explained.

(f) Optical aberrations; heterochromatic aberrations

The other monochromatic aberrations will not be discussed here because they are not important in the considerations about outdoor lighting. This contrary to the heterochromatic aberrations. What is mentioned earlier, refers to monochromatic light. Heterochromatic aberrations result from the fact that the refractive index of optical media depends on the wavelength of the light. The refractive index for blue light, having a short wavelength, is higher than that for red light, having a longer wavelength. Thus, the focal length of an uncorrected lens, such as the eye-lens, for blue light is shorter than that for red light. The axial chromatic aberrations of the lens cause a difference of about 2,25 diopters over the spectrum between 400 nm and 700 nm (Williams & Hofer, 2004, p. 800; Thibos, 1987; Marcos et al., 1999). The eye is myopic ('short sighted') for blue light and hypermetrope ('long sighted') for red light. See for details Augustin (2007, chapter 18); Mann & Pirie (1950); Schreuder (1998), and Stilma & Voorn, eds. (1995). Additionally, the apparent size of the image in blue light is usually smaller that in red light. This aspect is depicted in Figure 6.3.4.

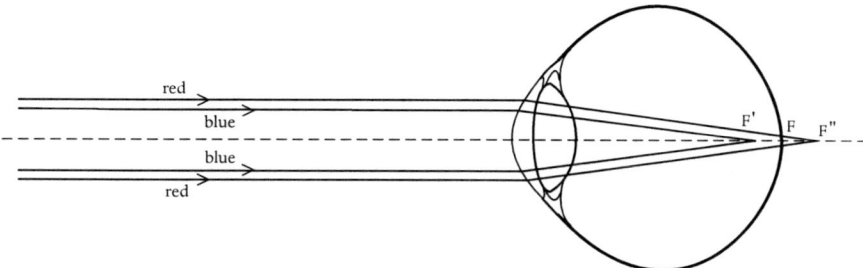

Figure 6.3.4 The effect of heterochromatic aberrations. F' is the focus for blue light, F" the focus for red light, and F the 'average' focus. (After Schreuder, 1998, figure 6.1.4.)

The heterochromatic aberrations cannot be reduced by a reduction of the pupil diameter (Schreuder, 1998, sec. 6.1.2). They can, however, be reduced or even completely eliminated when monochromatic light is used. As the word says, monochromatic light contains only one single wavelength.

As an example we will mention low-pressure sodium lamps. These lamps are essentially monochromatic. As is explained in secs. 2.4.3c and 10.3.3f, these lamps prove in many respects to be superior to other light sources. This superiority has two aspects. The lack of heterochromatic aberrations results in a better visual performance (Van Bommel & De Boer, 1980, p. 59). Also, visual comfort is often considered as being superior to that of the greenish-white light that is emitted by – uncorrected or colour-corrected – high-pressure mercury lamps (De Boer, 1951; Schreuder, 1967). For this reason, these lamps have been applied in many countries for road lighting, particularly for traffic route lighting. This is indeed a very old issue (Arndt, 1933; Schreuder, 1962; De Boer, 1967; NSVV, 1957). Another point is that the monochromatic low-pressure sodium lamps are often used in outdoor lighting near astronomical observatories, because stray light of these lamps can be filtered out, contrary to white light (Narisada & Schreuder, 2004, sec. 1.2.5e, p. 20).

High-pressure mercury lamps, either uncorrected or corrected, are gradually being phased out. In industrialized countries, they are almost completely replaced – in new installations at last – by high-pressure sodium lamps. This is done for economic reasons, mainly because of their limited lamp efficacy. Also, environmental reason play a role because mercury is one of the most dangerous heavy metals that are a direct threat to health.

As is explained in sec. 2.4.1c, both high-pressure mercury lamps and high-pressure sodium lamps have an emission spectrum with many lines and bands, so that their spectrum is almost a continuum. In the time of De Boer, they both would classify as 'white'. Nowadays, however, the designation 'white' is restricted to metal halide lamps and fluorescent tubes, that approach real white very closely. More recent investigations did show that the white light of fluorescent tubes and metal halide lamps is preferred over the 'whitish' light of high-pressure mercury or high-pressure sodium lamps (Van Tilborg, 1991).

6.3.4 The optical elements, the iris

(a) The anatomy of the iris

In humans, the iris and its opening, the pupil are the most visible parts of the eye. The iris is divided in two major regions. The first is the pupillary zone that borders to the pupil. The rest is the ciliary zone. Anatomically speaking the iris consists of a number of layers. Its main component is formed by pigmented fibrovascular tissue known as a stroma. The stroma connects a sphincter muscle, which contracts the pupil, and a set of dilator muscles which open it. The back surface is covered by an epithelial layer two cells thick. The front surface has no epithelium. The iris is usually strongly pigmented with colours ranging from brown to green, blue, grey, and hazel. Despite the wide range of colours, there is only one pigment involved, called melanin. The actual iris colour consists of the combined effects of pigmentation of the stroma and of the back epithelium, and of blood vessels within the iris stroma. More in particular in blue eyes, optical phenomena like diffraction and Rayleigh scattering play a role (Anon., 2007).

The pigmentation is genetically determined. It correlates strongly with the race. As is explained in sec. 8.3.2h, the genetically determined colour of the iris plays an important role in the effects of disability glare (Franssen & Coppens, 2007). Disability glare is discussed in sec. 8.3.

The pupil reacts to light: the pupil light reflex. This reflex is governed by two antagonistic muscles, a sphincter and a dilator, by means of parasympathetic control (Barbur, 2004, p. 641, 642). The contraction and dilatation of the pupil opening are governed by certain light sensitive ganglion cells, discussed in sec. 6.3.5e. These ganglion cells form a part of the parasympathic nervous system that governs a great number of bodily functions. Some of these functions are explained in sec. 6.3.5e. These ganglion cells have a spectral sensitivity curve that differs from the V_λ-curve of the cones as well as from the V'_λ-curve of the rods in the retina.

It should be noted that, contrary to common belief, the changes in pupil diameter do not contribute much to the adaptation. The diameter may change from about 2,8 mm to about 7 mm, or about a factor 6 in area. The relation between the pupil diameter and the ambient light level (adaptation level) is given in Figure 6.3.5.

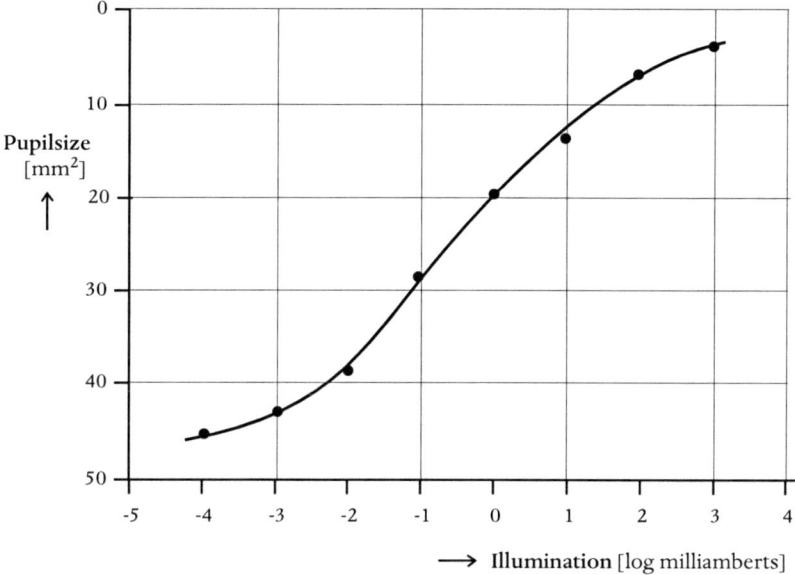

Figure 6.3.5 *The relation between the pupil area and the adaptation level. Based on data from Bartley, 1951, fig 64, p. 981 and Reeves 1920.*

(b) *The Stiles-Crawford effect*

In the preceding part of this section, it is noted that changes in pupil diameter do not contribute much to the adaptation. This effect – or rather, this lack of effect – is reinforced by the Stiles-Crawford effect. As is explained in sec. 6.3.5b,c, the photoreceptors are elongated in shape; as the words suggest, the rods more than the cones. In a sense, photoreceptors may be regarded as light tubes. Being elongated means that the light that falls more or less perpendicular to the retinal plane, that is, more or less in the same direction as the photoreceptor axis, can penetrate more easily into the receptors than light that falls obliquely on the retinal plane. Oblique light rays are less effective in provoking a sensation of light than perpendicular light rays. It can be easily seen that oblique light rays correspond to light that enters the eye via the outer area of the pupil. This is called the Stiles-Crawford effect. It should be mentioned that the light-tube model is an over-simplification. The relevant processes are much more complicated. They are discussed in detail in Burns & Lamb (2004).

The Stiles-Crawford effect takes the lower efficiency of the transmission through the lens edges into account (Helbig, 1972; Weale, 1961). See for further details Narisada & Schreuder (2004, sec. 8.1.2) and Schreuder (1998, sec. 7.2.5). The numerical influence of the Stiles-Crawford effect is given in Table 6.3.3.

Distance from centre of pupil (mm)	Relative effectiveness (%)
0	100
1	90
2	83
3	50
4	20

Table 6.3.3 The effectiveness of rays entering the pupil at different points (approximated values, averages for the different directions). (After Moon, 1961, figure 12.05, based on data of Stiles & Crawford, 1933.)

6.3.5 The optical elements, the retina

(a) The anatomy of the retina

The retina is a thin organ, located at the rear of the eye. It hugs the rear outer cover of the eye, the sclera. It is a complex organ that houses not only the light-sensitive receptors that are essential for vision, but also the nerve system that connects them to the visual nerve tract, and the blood vessels that take care of the energy and oxygen supply for the receptors to work properly. The visual nerve tract is discussed in sec. 6.4.2. Additionally, different nerve cells like e.g. the ganglion cells that are discussed in sec. 6.3.5e. Finally, support systems and other organs, not directly involved in the visual process, are located in the retina.

Anatomically speaking, the most striking characteristic of the retina is that it seems to be reversed. In Figure 6.3.6, a schematic sketch of the human retina is depicted. It should be noted that in this figure the light comes from below.

The human observer; physical and anatomical aspects of vision

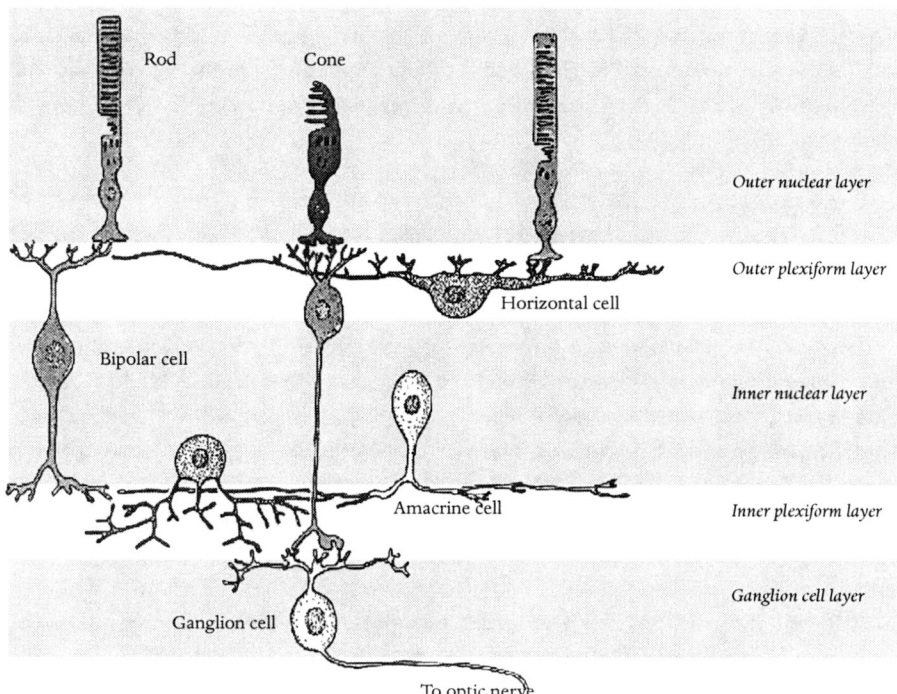

Figure 6.3.6 *A schematic cross section of the human retina. The light falls into the retina from below. (After Barnstable, 2004, figure 3.3.)*

As the main function of the retina is to convert the energy of the incoming light quanta into nerve pulses, it would seem more logical to expect that the light sensitive elements would be turned towards the light, and that the exit – the nerves – as well as the processing elements, and the supply an support systems would be at the rear. Whatever the cause – if there is a cause for this – the retina is what it is. That means that the light has to traverse a number of semi-transparent layers before it reaches the photodetectors. As is to be expected, this adds to the optical aberrations in the visual system, primarily as a considerable increase scatter of the incoming light. The light scatter in the fundus is one of the major contributions to the overall intraocular stray light, or entoptic stray light (Franssen & Coppens, 2007; Van den Berg, 1995; Vos, 1963, 1983, 1999, 2003). More details are given in sec. 8.3, where disability glare is discussed.

There is more to it, however. In a preceding part of this section it is mentioned that the pupil reaction is governed by a certain kind of ganglion cells. These cells are located about halfway down in the retina, thus before, or 'above', the photodetectors that are discussed in the next part of this section. These, or similar, ganglion cells are involved in the non-image-forming aspects of the visual system. They are essential in many biological functions, such as e.g. the timing of the biological clock. They are sensitive to light, but their spectral sensitivity curve differs markedly from the V_λ curve as well as from the V'_λ curve that image-forming aspects of vision (Schreuder, 2006, 2008; Van den Beld, 2003).

(b) The photoreceptors

The photoreceptors in the retina collect the incoming light and convert the energy into electrical pulses. Usually, the process is called visual transduction (Burns & Lamb, 2004). This process is rather complicated and it is even today not completely understood. We refer here to a number of books, where the interested reader may find the details, primarily Chalupa & Werner, eds., 2004; Greenfield, 1997, 2000, 2000a; Gregory, ed., 1987, p. 798-804, and Gregory, ed., 2004. The processes involved are quite complicated indeed. The whole chapter devoted by Hubel to these phenomena is in itself only a summary (Hubel, 1990, Chapter 3). The complete state of knowledge of the field as of the year 2004 is given in 1700-page heavy handbook of Chalupa & Werner, eds., 2004.

Basically, a light quantum that is absorbed by a photoreceptor will, as a result of the energy transfer related to this absorption, change the structure of one of the molecules of the visual pigments or photopigments. A very detailed description is given by Burns & Lamb, 2004. See also Gregory, ed., 1987, p. 153. The general term for these pigments is rhodopsin. In the past, this term was reserved for rod-operation, but nowadays it used for all visual pigments (Burns & Lamb, 2004, p. 215-216). Further details are given in Augustin (2007, figure 42.8, p. 1180).

It is a process similar to ionization: one or more electrons are 'kicked out' of the molecule. After recombination, an electrical pulse may be generated. This simplified description is based on Hentschel, ed. (2002, sec. 1.4).

The electric pulses are processed in the ganglion cells. The process is explained in sec. 6.2.1a. This process is very comprehensive; what comes out of the retina is very different indeed from the light that falls on the photoreceptors. The incident light, the stimulus, results in the creation of electric pulses, called spikes. The frequency in which spikes are fired is proportional to the intensity of the incident light. See Figure 6.3.7.

Often these processes the electric pulses are transferred to the optical nerve that transmits them from the eye towards the visual cortex, the part of he brain were in a completely unknown fashion, the electric pulses are converted into a conscious experience called 'seeing'. Although some neuro-scientists – and others – pretend otherwise, the process

The human observer; physical and anatomical aspects of vision   207

is at present almost completely unknown (Greenfield, 1997, 2000). The optical nerve system, and the brain are discussed in brief in secs. 6.4.2 and 6.4.3.

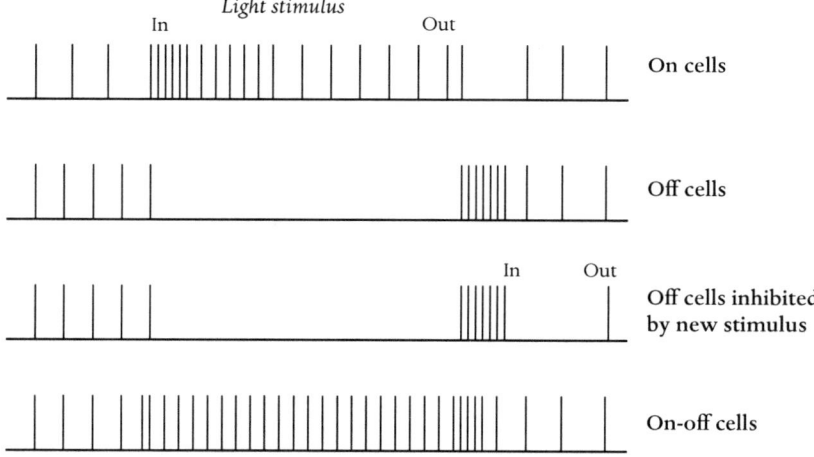

Figure 6.3.7 Examples of bursts in so-called 'on' and 'off' receptors (After Schreuder, 1998, Figure 6.15, based on data from Hentschel, 1994, fig. 1.10).

(c) Cones and rods

One might find in the normal human retina, as well as in that of many, but definitely not in all other animals, two main types of photoreceptors. They are commonly known as cones and rods. These names seem to stem from their overall shape, although one needs quite a lot of phantasy to see it! A, still simplified, sketch is given in Figure 6.3.8.

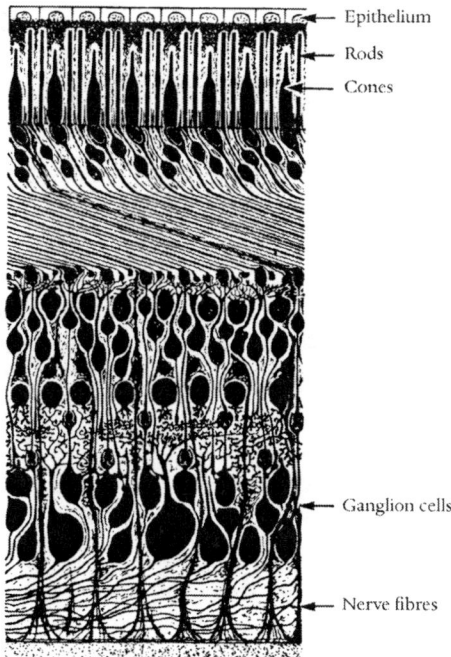

Figure 6.3.8 The human retina (After Schreuder, 1998, figure 6.1.6. Based on Anon. 1993).

When this figure is schematized even further, the shapes may be more easily recognised. See Figure 6.3.9. In an earlier part of this section, when discussing the Stiles-Crawford effect, it was suggested that photoreceptors seem to have some characteristics that make them look like light tubes.

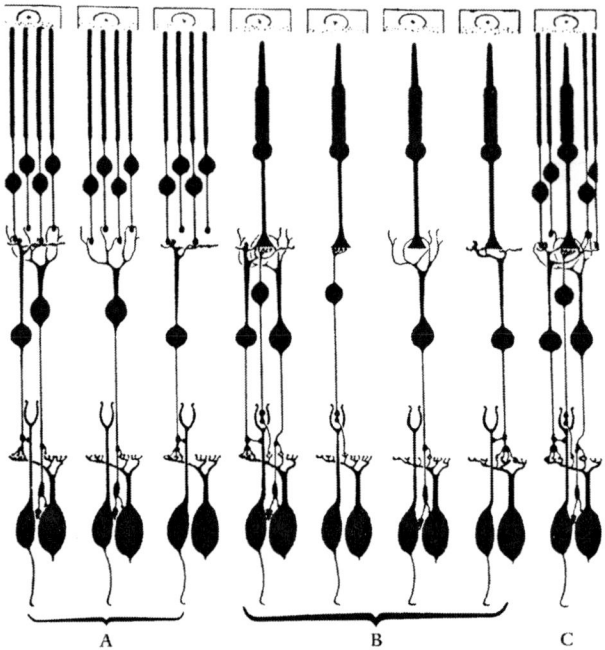

Figure 6.3.9 The human retina, showing some of the many neural links. (After Schreuder, 1998, figure 6.1.7. Based on Anon. 1993.)

It should be noted that in the Figures 6.3.8 and 6.3.9 the direction of the light that hits the retina is from the bottom to the top.

Photoreceptors are very complicated cells. See Augustin (2007, figure 42.7). Some of the internal details are depicted in Figure 6.3.10.

Although the overall anatomy of cones and rods shows some similarities, their operation shows very important differences:
1. The sensitivity of rods is much greater than that of the cones. This is explained in sec. 7.4.2, where the adaptation is discussed;
2. The rod system is not capable to discern colours, whereas the cone system, because of the interconnection of at least three different types of cones, can differentiate between different colours. This is explained in some detail in Chapter 9, where colour vision and its aberrations are discussed;

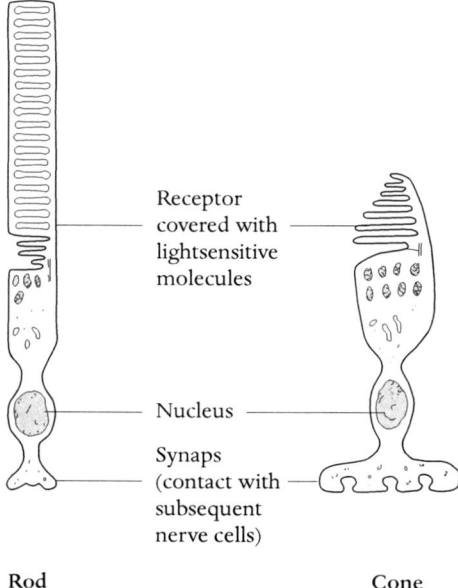

Figure 6.3.10 Some internal details of cones and rods (After Bergsma 1977, p. 44).

3. The peak of the spectral sensitivity curve of rods is shifted towards the blue end of the spectrum compared to that of the cones. This is explained in some detail in sec. 7.4.2, where mesopic vision and the Purkinje shift are discussed;
4. The spatial distribution over the retina differs very considerably. The cones are mainly concentrated in the fovea centralis, whereas the rods are absent in the fovea centralis, but are abundant in the periphery. This is explained the next part of this section. The implications are explained in sec. 7.4.4, where the spatial distribution of the visual acuity is discussed.

(d) *The spatial distribution of rods and cones*

The human retina contains a very large number of receptors, approximately $1{,}2 \times 10^8$ rods and about 6×10^6 cones. The rods are distributed over the whole retina, except in the central area. This central area is called the fovea centralis, the macula or the yellow spot. In the fovea, where the rods are absent, the cones are very densely packed. As is explained in sec. 7.4.4, where the visual acuity is discussed, this is the reason that sharp seeing can be achieved only in the central are of the field of view – an area about $2°$ in diameter. This value of $2°$ of the foveal field of view is relevant for many practical lighting engineering applications, like e.g. the lighting of streets and tunnels. Seeing very dim stimuli can be, however, done best in the outer area or periphery of the field of view. This fact has been known for millennia in applied astronomy. Dim stars disappear from view when one tries to focus on them. The spatial distribution of the rods and cones over the retina is depicted in Figure 6.3.11.

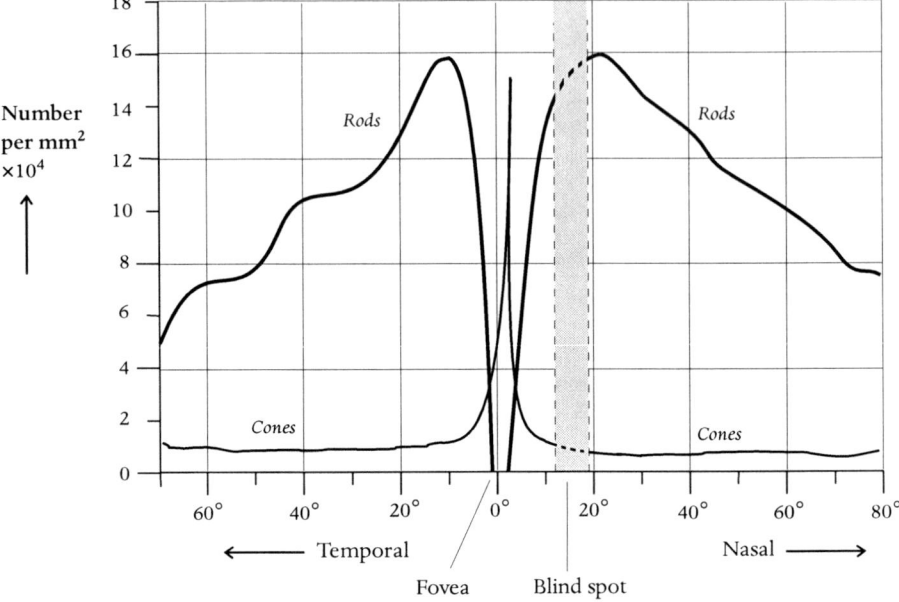

Figure 6.3.11 *The spatial distribution of photoreceptors. (After Narisada & Schreuder, 2004, Figure 8.2.4. Based on data from Osterberg, 1935, and Pirenne, 1972.)*

As mentioned earlier, this results in the fact that the visual acuity is greatest in the fovea centralis and lowest in the far periphery. This is depicted in Figure 6.3.12.

(e) *Retinal ganglion cells*

As is mentioned earlier, the retinal receptors convert light into a neural signal. This signal is processed by the retinal network and transmitted by the axons of the ganglion cells to the brain (Gregory, ed., 2004, p. 931-937). In sec. 6.2.1 it is explained what an axon is.

In the brain, most nerve cells generate an action potential. In sec. 6.2.1 it is explained what an action potential is. However, in the retina only the ganglion cells do so. The retinal receptors and the other cells in the retina exhibit graded shifts in membrane potential in response to changing light levels.

The human observer; physical and anatomical aspects of vision 211

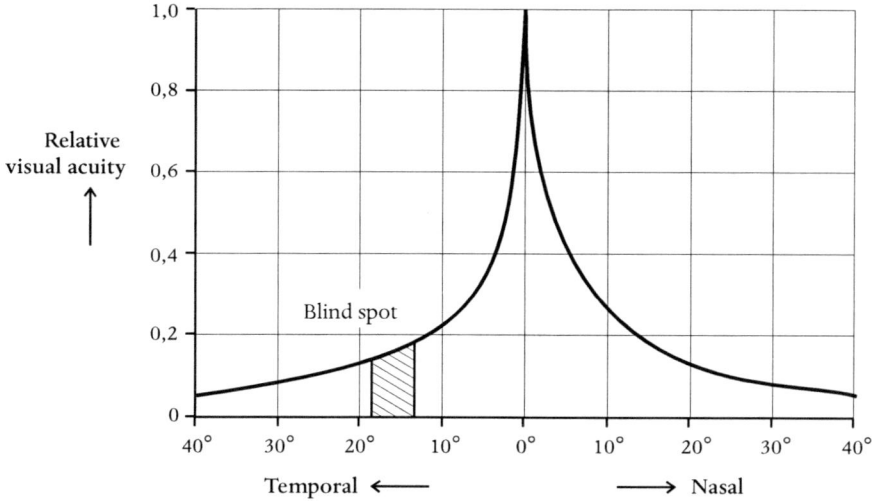

Figure 6.3.12 The relation between visual acuity and retinal location. (After Narisada & Schreuder, 2004, Figure 8.2.5. Based on data from Wertheim, 1894.)

6.4 The optical nerve tracts

6.4.1 Image forming and non-image forming effects of light

Light may fall on any part of the body of plants, animals, and humans. As regards plants, the main influence of light is related to photosynthesis. However, many plants adjust their position to the direction of the sun; sunflowers are called 'tourne-sol' in French, meaning just that.

The human skin is sensitive to radiation, mainly to infra-red and ultra-violet. However, for humans, the reactions to light that enters they eye seem to be more important. In part, these reactions may be classified as image forming and in part as non-image forming. The importance of the image-forming effects for lighting is evident. This is the case for outdoor lighting, the main subject of this book, as well as for indoor and decorative lighting. The importance of the non-image forming effects for lighting may be less evident. However, lighting, particularly indoor lighting, has many more functions than just to make objects visible. It is generally accepted that the lighting of indoor working areas influences the health and the general well-being of the people working there. We mentioned briefly the influence on the biological clock. These effects are discussed in great detail in several other publications that have been mentioned already several times. See e.g. CIE (2006); Narisada & Schreuder (2004, sec. 4.5.1); NSVV (2003, 2006); Schreuder (2006, 2008), and Van den Beld (2003).

In this book we will concentrate on the image forming effects. We will just mention the fact that image-forming effects and non-image forming effects each have their own neural pathways. The non-image forming effects are discussed in more detail in CIE (2006); Narisada & Schreuder (2004, sec. 4.5.1); NSVV (2003, 2006); Schreuder (2006, 2008), and Van den Beld (2003).

6.4.2 The visual neural pathways

(a) The organization of the retinal visual system

The retinal receptors convert the pattern of light that represents the visual image into neural signals. Considered in this way, the first part of the visual neural pathway is located in the retina. The retinal receptors make synaptic contact with bipolar cells which in turn make synaptic contact with the retinal ganglion cells. In sec. 6.2.1b the synapses and synaptic contacts are discussed in more detail. As mentioned earlier, the signal is transmitted by the axons of the retinal ganglion cells eventually to the brain. These are serial connections. At the same time, two other groups of cells, the horizontal cells and the amacrine cells make laterally directed connections that control the transfer of information through parallel connections. The neural organization of the retina is summarized in Figure 6.4.1.

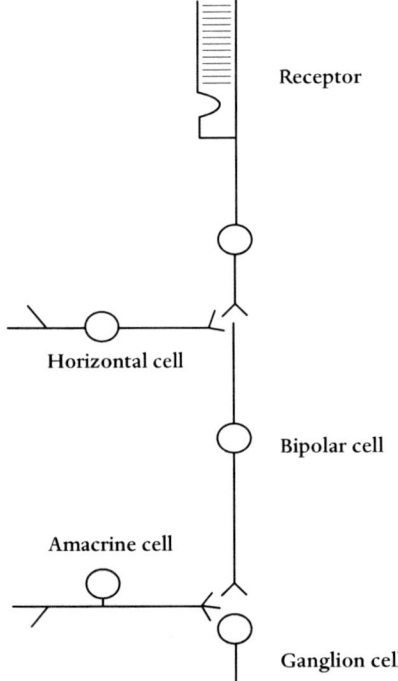

Figure 6.4.1 *Interconnections between cells in the retina. (After Gregory, ed., 2004, p. 931, Figure 1.)*

The various synaptic interactions occurring in the neural circuits of the retina determine the response of retinal ganglion cells to light stimuli. Following the signal from the bipolar cells, the retinal ganglion cells exhibit two kinds of response. In one, cells are excited by an increase in light level over the background level. In the other, cells are excited by a decrease. For obvious reasons these retinal ganglion cells are called on-cells and off-cells respectively. In addition to these 'on' and 'off' cells, further subdivision in ganglion cells are made. These subdivisions are related to colour reception or to spatial summation processes (Gregory, ed., 2004, p. 931-932).

There are two basic circuits in the retina that relay cone and rod signals to the retinal ganglion cells. Cone signals modulate 'on' and 'off' cone bipolar cells that excite 'on' and 'off' cone ganglion cells. Rod signals modulate cone terminals and relay single-photon signals via a private rod bipolar cell that excites amacrine cells. Amacrine cells inhibit 'on' and 'off' ganglion cells (Sterling, 2004, p. 236).

(b) *The optical nerve*
Neurones are usually bundled into nerve tracts. The nerve tract, relevant for vision, is the eye-nerve or nervus opticus. It starts at the retina and runs right through the skull towards the visual cortex that is located at the rear end of the brain. A very schematic representation of the visual pathways is given in Figure 6.4.2.

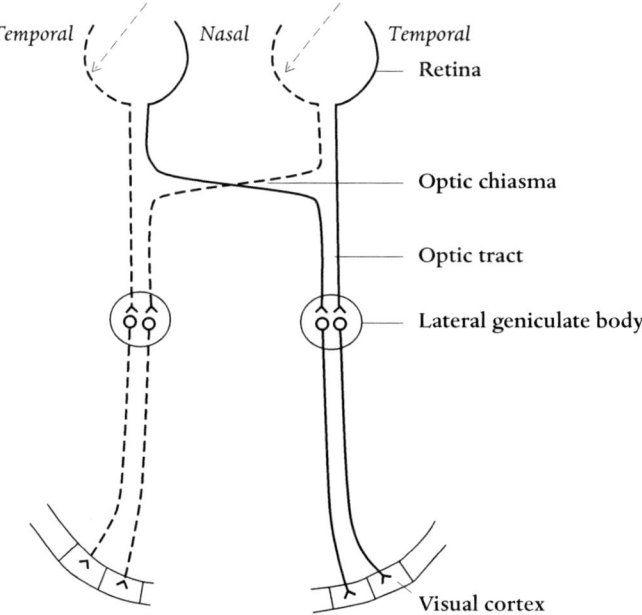

Figure 6.4.2 *Schematic diagram of the visual pathways as seen by looking down from above the head. (After Gregory, ed., 2004, p. 933, Figure 3).*

The most striking aspect in this diagram of the visual pathway is the crossing in the optical chiasma, which implies that the right side of the brain 'sees' the left side of the visual world, and vice versa (Augustin, 2007, figure 38.1 p. 1124, figure 42.6, p. 1177; Feynman et al., 1977, Vol 1, Figure 36-4).

In Figure 6.4.3, the same pathway is depicted in somewhat more detail. Here it strikes that in the optical chiasma the right hand sides of the field of vision of the two eyes are combined, and also the two left-hand sides. What happens with the signal in the brain is discussed in sec. 6.4.3. See also Narisada & Schreuder (2004, sec 10.1.4), Greenfield (1999, p. 210), and Schreuder (2008).

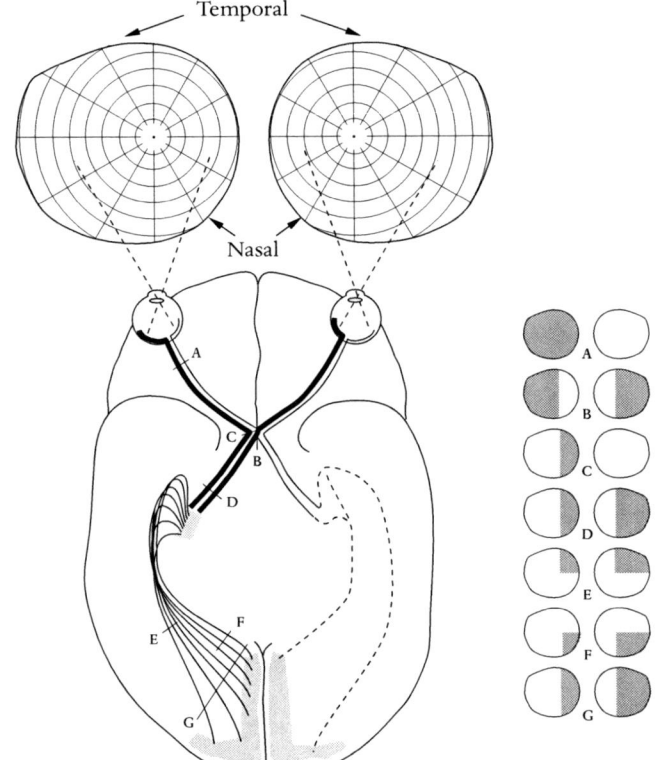

Figure 6.4.3 *The neural connections from the eye to the visual cortex. (After Narisada & Schreuder, 2004, figure 8.1.5. Based on Feynman et al., 1977, Vol 1, Figure 36-4.)*

More details about these neural connections are given in Augustin (2007, figure 38.1)

(c) *Pathways for rod vision*

In general terms, the rod terminal contains only one active zone. However, the organization of rods depends crucially on the light level. In this respect, three levels are discerned.
1. Starlight levels;
2. Moonlight levels;
3. Twilight levels.

These categories are given by Sterling (2004, p. 247-249). The luminance levels in the outside field of view that may be supposed to correspond with these categories are:
1. below 10^{-4} cd/m^2;
2. between 10^{-4} cd/m^2 and 10^{-2} cd/m^2;
3. between 10^{-2} cd/m^2 and 1 cd/m^2.

These values are taken from Narisada & Schreuder (2004, sec. 8.2.3b) and are based on the study of Hopkinson (1969). As is explained in sec. 7.2.4a, at values above about 0,1 cd/m^2, one enters the areas of high mesopic and photopic vision, where the action of the cones predominates over the actions of the rods.

At the starlight level, the rod system employs many detectors but catches only a few photons. The signal in each detector is essentially binary: usually zero and rarely one. At the moonlight level the detectors still operate binary, but with few zeros and many ones. In the twilight condition each detector catches more than one photon and thus has a coarsely graded signal. In each of these three conditions a different circuit is in operation.

In starlight, the voltage evoked by one single photon rises only modestly, by a factor of 4, above the dark noise level. This dark noise represents the neural activities in the retina that take place even in absolute darkness. In passing we may mention that darkness can be described as a visual phenomenon on the base of these activities (Heavan & Buxton, 2007; Schreuder, 2008; Van Essen, 1939).

The next neuron in the circuit, the rod bipolar cell pools the input of some 20 to 120 rods. By introducing a threshold in a non-linear synapse, the dark noise will be eliminated, but also the single-photon responses. However, the overall sensitivity is markedly improved (Sterling, 2004, fig. 17.11; Field & Rieke, 2002). This is completely in line with the earlier findings that in the dark adapted eye one would need at least around three photons to provoke an impression of light (Weale, 1968; Narisada & Schreuder, 2004, sec. 8.2.1).

In the moonlight condition, when the photon density rises slightly, a second rod pathway comes into play. Some rods have synaptic connections with the dendrites of an 'off' cone bipolar cell. These synapses amplify signals in a linear way. They transfer all photon events. When summed, the signal will exceed the dark noise level (Sterling, 2002; Field & Rieke, 2002). Although only a few rods manage to contact a cone bipolar dendrite, signals may still be pooled. It should be mentioned that these pathways have been found only in rodents, so it is not clear whether human visual performance may benefit from it (Sterling, 2004, p. 248).

When in twilight conditions the photon density reaches the photon integration time for rods, rod signals are processed by come circuits. Each rod terminal forms a junction with two cone terminals, and every cone terminal is contacted by about 40 rods. Rod signals have been observed in recordings from cones (Nelson, 1977). The electric coupling from many rods to each cone allows the graded rod signal to be filtered and relayed by the same circuits used by the graded cone signals (Sterling, 2004, p. 249).

This may be in some way the cause for several phenomena that are well-known for lighting application. It has been observed that light that hits the periphery of the retina may contribute considerably to glare effects, although the contribution to the light impression is not easy to observe. This is found in some experiments regarding tunnel entrance lighting (Schreuder, 1964) as well as in some curious phenomena that have been found in applications of sun-glasses (Alferdinck, 1997). The most curious thing is that 'Inuit-sunscreens' where almost the complete periphery is shielded but vision straight ahead is not impeded, help at all! (Schreuder, 1997; Vos, 1987).

A second effect where cones and rods seem to interact is mesopic vision that is discussed in sec. 7.2.4.

(d) Pathways for daylight levels

In most sensory systems the graded potentials of receptor cells are converted into action potentials, either directly or after the interpolation of single synapse (Sterling, 2004, p. 249). This is depicted in Figure 6.4.4.

This requires special mechanisms to filter out noise and redundant signals. All senses with the exception of the visual sense employ complex mechanical or chemical filters before transduction to the nerve tracts. The visual sense, however, employs only mild optical filtering. For example, optics in the human fovea cuts off frequencies at 60 c/s. This precisely matches the foveal sampling rate at 120 c/s. According to the Nyquist rule there must be one detector for each half cycle (Sterling, 2004, p. 250 footnote). This refers to the Nyquist limit known from sampling theory (Shannon, 1949). The highest

The human observer; physical and anatomical aspects of vision   217

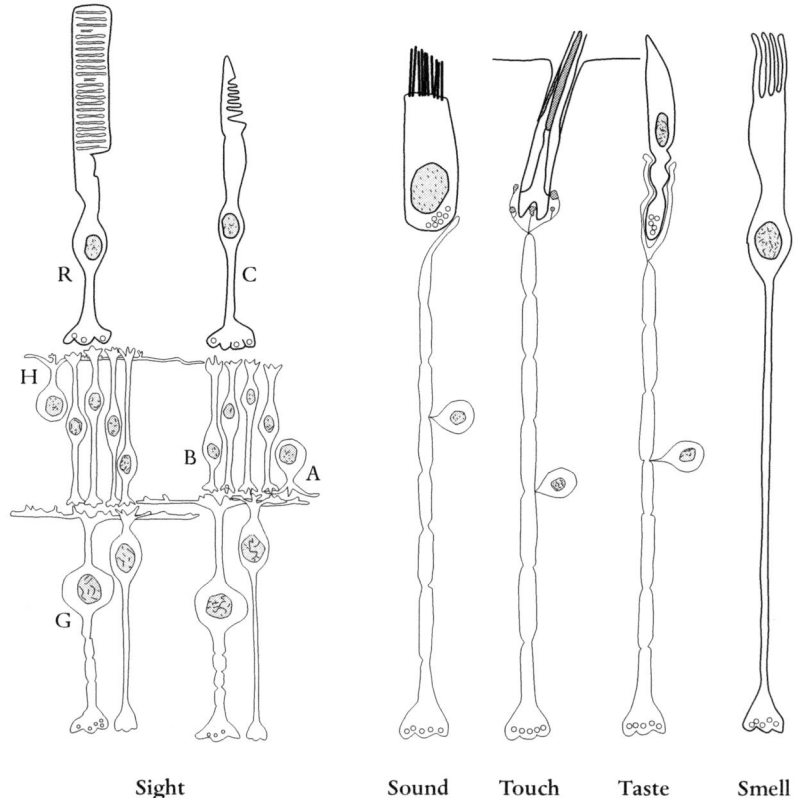

Figure 6.4.4 *Neural processing at the site of the transduction. (After Sterling, 2004, fig. 17.3, p. 237.)*

spatial frequency for a one-dimensional signal that is adequately sampled is half the sampling frequency (Williams & Hofer, 2004, p. 803). This frequency is also known as the critical flicker-fusion frequency that is discussed in sec. 8.1.2b.

Lacking vigorous preneural filtering or extensive molecular filtering by the transductor, the retina requires neural circuits to perform spatial and temporal bandpass filtering at the first synapse. The photovoltage arriving at the cone synaptic terminal spreads to neighboring cone terminals. This coupling filters the intrinsic cone noise (Sterling, 2004, p. 250). It has been found that in steady light the coupling reduces the noise more than the signal.

The next essential step in the attenuation of the signal is to eliminate the redundancies in the signal. Removing them allows the cone synapse to use its dynamic range for the essential differences between adjacent regions. This form of highpass filtering subtracts the background; it enhances the contrast. The next step, the final integration is accomplished in the horizontal cells. The spines of the horizontal cells form paired lateral elements, the so-called AMPA receptors. The horizontal cells pool their signals electrically with neighboring cells (Sterling, 2004, p. 251-252). Further details are given in Kamermans et al. (2001); De Vries et al. (2002), and Haverkamp et al. (2000).

(e) Movie tracks

A new approach into the way the visual system, that begins with photons that strike the photoreceptors in the retina and ends with a consciously experienced image of the outside world, makes use of the concept of movie tracks. Research finding suggest that the human visual system uses twelve different separate movie tracks. A summary of these findings and of their implications is given in a recent survey paper in Scientific American (Werblin & Roska, 2007).

Each of these movie tracks can be though of as a distinct abstractions of the visual world. Each track embodies a limited representation of one aspect of the scene that streams to the brain. These tracks are continuously updated by the retina. Each of these movie tracks is transmitted by its own population of fibers within the optical nerve to specific higher centres in the brain. Different features of the visual image such as motion, colour, depth, and form are processed in various regions in the cortex. Some of these centres are discussed in sec. 6.4.3a.

As is explained in sec. 6.3.5a, the outer layer of the retina contains the rods and the cones. These photoreceptors are connected to ten different kinds of bipolar cells. These project their axons into another layer that lies more towards the interior of the eye. This layer is called the inner plexiform layer. This layer is discussed in sec. 6.3.5a. Actually, this layer consists of ten distinct parallel strata. Each bipolar cell type delivers signals to just a few of these strata.

At the innermost side of the plexiform layer are twelve different types of ganglion cells. Most types project their dendrites into one distinct stratum where they receive excitary input from a limited number of bipolar cell types. Each ganglion cell type is responsible for one of the twelve movie tracks that have been mentioned earlier. The ganglion cells project the signals of each such movie track into the optical brain for further processing. The signals emitted by the bipolar cells are modulated by the amacrine cells. Some amacrine cells inhibit horizontally the communication between distant ganglion cells in the same stratum. Others inhibit vertically any signals between strata. The signals result in the end from the excitation produced by the bipolar cells and the inhibition produced by the amacrine cells. In total, there have been found 27 different types of amacrine cells. In the end, each of the twelve ganglion cell types sends continuously, as time advances, one specific movie track with specific features to the brain.

The movie tracks are only approximations. Still, it seems clear that the preprocessing of the information in the retina splits the visual world into a dozen discrete components. These components travel, intact and separately, to distinct visual brain regions. The scientific challenge to understand vision is to find out how the brain interprets these packets of information and to generate from them a magnificent, seamless view of reality (Werblin & Roska, 2007).

6.4.3 The anatomy of the brain

(a) The main structure of the brain

As is explained in sec. 6.2.2, where the anatomy of the central nervous system is discussed, it is stated that the life of animals, humans included, consists of receiving information and responding to it. In animals made up of more than a few cells, this requires a nervous system (Gregory, ed., 1987, p. 514). It is possible to make a model of a brain that evolves from simple to complicated. In fact, two models: the one describes the evolution of the brain in different species. This is called the phylogenesis. The other describes a similar evolution of the brain in different stages of the growth of an individual. This is called ontogenesis or organogenesis. Apart from these remarks, we will abstain in this book from all discussions about evolution and Darwinism (Schreuder, 2008).

As regards the sheer size of the brain as an organ, we will quote a few numbers from Goodenough (1998, p. 95). The brain of every human being contains about 10^{11} neurons. Their axons have a collective length of about half a million kilometres. As each neuron contacts on the average about 1000 other neurons, there are about 10^{14} synapses in the average human brain!

(b) Brain anatomy and brain functions

The brain has three main parts. The medulla oblongata plus the pons together form the hindbrain (Gregory, ed., 2004, p. 116-127). From the roof of the hindbrain the cerebellum developed, first becoming important in birds. Further forward – towards the beak – is the midbrain and still further forward the forebrain.

The reticular formation is a part of the hindbrain (Gregory, ed., 1987, p. 524, fig. 9). It represents a very 'old' part of the brain, essential for vital functions like breathing. It controls heart and blood pressure. It influences the biological clock, governing sleeping and waking, relaxation as well as vigilance and alertness. Vigilance and alertness are discussed in Schreuder (1998, sec. 8.1.2). See also Schreuder (2008).

The hypothalamus is concerned in many more aspects that are essential for human life, like e.g. with drives and emotions. Close-by is the thalamus. This organ relays signals from and to the brain The pituitary gland controls the production of the body's hormones (Lafferty & Rowe, eds., 1994, p. 82).

As mentioned earlier, all vertebrate animals have a medulla oblongata or brain stem. In terms of evolution, it may be regarded as the 'oldest' part of the brain; it is dominant in animal species like fish and frogs. In humans, it controls vital reflexes like respiration, heart beat, coughing, and swallowing. Vertebrate animals have a limbic system which controls emotional behaviour and motivation. In mammals the limbic system is quite pronounced. The hippocampus is important for memory and spatial orientation; the hypothalamus controls, as mentioned earlier, sleep-wake rhythms, body temperature, impulse to eat, fighting, and sexual behaviour. As may be seen from animal and human behavioral studies, the two are closely related (Straub et al., eds., 1997; Gleitman, 1995; Gleitman et al., 2004).

The limbic system is the brain part between the brain-stem and the neocortex. It is usually understood to include amongst others the hippocampus, the olfactory regions, and the hypothalamus (Gregory, ed., 1987, p. 435; 2004, p. 529). The olfactory regions are related to smell (Gregory, ed., 1987, p. 719-720; 2004, p. 848-850).

Finally the cerebrum. The cerebrum coordinates the senses, and is responsible for learning and other higher mental faculties (Lafferty & Rowe, eds., 1994, p. 111). More details about the anatomy and the functions of the cerebrum are given in the next part of this section.

(c) The cerebrum
In humans, the cerebrum is the most important part of the forebrain. The cerebrum consists of several layers, the most important being the outer layer or cortex. This outer layer consists mainly of the axons of the nerve cells. In view of its colour it is often called the 'white matter'. The inner layers contain the cell bodies. It is often called the 'grey matter'. As mentioned earlier, the outer layer forms the cortex of neocortex. It is generally accepted that it plays an important role in functions like e.g. memory, hearing, sight, speech, voluntary muscle action, thinking and reasoning (Clugston, ed., 1998, p. 81). Several of these functions are closely related to the human consciousness (Schreuder, 2008). The cerebrum consists of two halves. These two halves are anatomically more or less symmetric but in function they are quite different. Each

half of the cerebrum consists of a number of lobes (Greenfield, 2001, p. 14; Gregory, ed., 1987, p. 521, fig. 6):
1. The frontal lobes. These are located, as the name suggests, at the front end of the brain, directly behind the brow;
2. The temporal lobes, located at the side of the brain;
3. The parietal lobes, also located at the side of the brain just over the temporal lobes;
4. The occipital lobes, located at the rear of the brain. For vision, the occipital lobes are the most important.

(d) The cortex

As is mentioned earlier, the optical nerve ends in the cortex. The cortex is the outermost layer of the large brain or cerebrum (Greenfield, 1997, p. 6). It is only a few millimeters thick and it lays directly under the skull. Generally speaking, specific bodily or mental functions are linked to activities in certain areas in the cortex. However, it is not justified to consider the activities as the cause of the functions. As mentioned earlier, the brain functions related to vision are located in the rear part of the cortex, in the occipital lobes. This section is therefore called the visual cortex.

By means of a PET-scan (Positron Emission Tomography) one may have some idea what brain areas seem to be involved in specific bodily or mental activities. The method is indirect, because only the blood supply to certain brain parts is measured. A close look at human behaviour suggests that the allocation of functions to areas of the brain is far from specific. "The brain is made up of anatomically distinct regions, but these regions are not autonomous minibrains; rather, they constitute a cohesive and integrated system organised for the most part in a mysterious way" (Greenfield, 1997, p. 39). It might be better to speak of 'systems' (Damasio, 1994, p. 15). It is generally assumed that the visual cortex contains five different 'maps' of the visual field, called, without much phantasy V1, V2, V3, V4 and V5. It is not immediately clear in which way this subdivision helps to clarify the visual process.

In the following, we will concentrate on the visual sense, being the central theme of this book. This discussion is based in part on Greenfield (2001, chapter 4, p. 65-79). As mentioned earlier, the visual information gathered and preprocessed in the eye goes from the retina to the nucleus geniculatus lateralis or NGL, one at each side. So there are two NGLs. From there to the primary visual cortex where a first processing takes place. After that is finally goes to the surrounding associative cortex. The visual cortex is part of the occipital lobes, located at the rear of the brain.

As has been mentioned earlier, each eye has its own optical nerve. In the optical chiasma, a part the signals crosses, so that the signals from the left half of both eyes are combined; so are the signals from the right half of both eyes.

The visual system knows two systems that work more or less in parallel. As is mentioned in sec. 6.3.5e, the systems have to do with different kinds of ganglion cells in the retina. One might call them system A and system B. System A is related to movement. System B consists of two sub-systems, one related to colour perception, and the other with the perception of shapes.

The visual cortex is subdivided in several regions, called V1 to V5. Each of these regions is subdivided again in sub-regions related to the systems A and B respectively. To a certain extent, these parts of the visual system seem to operate independently of each other. However, at each level they inter-communicate, and there is feed-back. Some of these interconnections and the related feed-back loops are depicted in Figure 6.4.5.

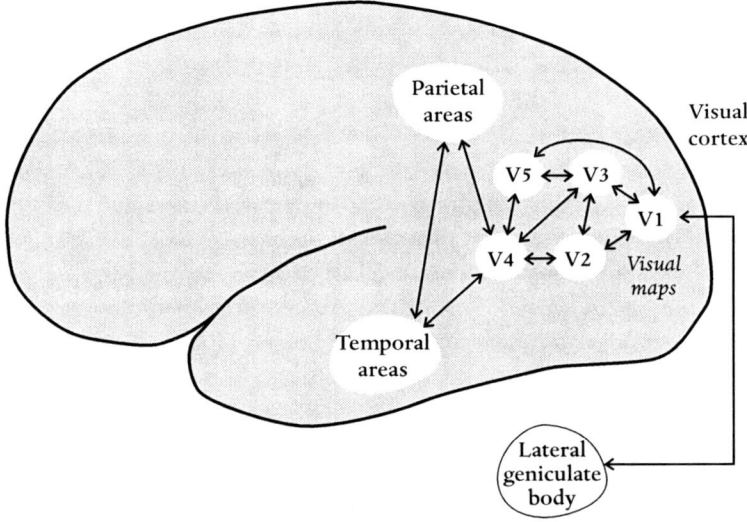

Figure 6.4.5 *Some of the neural interrelations within the visual system. (After Edelman, 1992, fig. 9-2, p. 86.)*

And then there is synesthesia, a sort of 'cross-talk' between the signals from different sense organs. Some regard this as just some form of over-active associations, like that e.g. colours have specific smells. However, there are reasons to believe that synesthesia is a true sense operation that attributes in animals to the overall world-image. There is no need to assume that this only takes place in primitive animals of past aeons. Many posh health-spas work on the same principles!

There is one more thing in visual perception that is worth-while to mention. According to Edelman, there is passive looking and active, conscious perception (Edelman 1992, 2004). The concept of qualia is incorporated in this idea. Qualia can be regarded as properties of sensory experience (Edelman, 2004; Anon., 2007a).

Many believe that vision and fantasies involve the same, or at least almost the same, brain areas, be it that the direction is opposite. In vision the signals proceed from the eye towards the cortex, whereas in fantasies the signals go from the brain in the direction of the eye (Kosslyn & Rosenberg, 2002). This would imply that the information coming from the eyes is not sufficient to form a complete, richly patterned, picture of the outside world. Our mind seems to fill the gaps left by the optical systems. Normally 'seeing' is a combination of sight and phantasy. It must be remembered that phantasy is fed to a large degree by past experiences and the related associations. The outside world is a projection that we make ourselves. And finally, there are reasons te believe that the prefrontal cortex located in the frontal lobes plays a decisive but still largely unknown role. Some more details, including the role of mirror neurons is discussed in Schreuder (2008).

6.5 Conclusions

Human beings are characterised by being conscious of themselves and of the world around them. All information from the outside world reaches into the brain. For most people most information is visual information. The nervous system is an essential part of the visual sense. The practical starting point must be the function of lighting.

The human visual system consists of five major elements:
- The optical elements of the eye with the cornea, the lens and the iris;
- The retina and the photoreceptors;
- The neurons in the eye;
- The visual nerve tracts;
- The brain, notably the visual cortex.

The optical elements of the eye work together to form an optical image of the outside world on the retina. The retina converts the incoming light into electrical pulses that are propagated along the optical nerve systems. They ultimately reach the optical cortex. Apart from the image forming processes, light in the eye is involved in several non-image forming effects of light. These are of particular interest for the biological clock, and more in general for the adjustment of the human organism to the outside world.

In conclusion it can be stated that the human visual system is well-adapted to the main task of converting optical information in the form of photons into electric pulses that in their turn are converted in the brain into a consciously experienced image of the outside world.

Two essential provisos must be made. First, by personal and societal adaptations, also blind people are perfectly capable to live a full, socially fulfilling life. The second proviso is that the human visual system can work adequately as a collector of information only in daylight or near-daylight. In the dark, additional provisions are essential. Here, outdoor lighting is essential. Only outdoor lighting allows the outside environment to be used also when natural daylight is absent.

References

Alferdinck, J.W.A.M. (1997). De toepasbaarheid van gele LED's bij informatiedragers langs de weg in relatie tot de spectrale transmissie van zonnebrillen (The applicability of yellow LED's in information carriers related to the spectral transmission of sunglasses) TNO Rapport TM-97-C021. Soesterberg, TM/TNO, 1997.

Anon. (1995). Symposium Openbare Verlichting, 22 februari 1995 (Symposium Public Lighting, 22 February 1995). Utrecht.

Anon. (2007). Iris (Anatomy). Wikipedia. Internet, 12 March 2007.

Anon. (2007a). The enigma of qualia. Wikipedia. Internet, 17 April 2007.

Arndt, W. (1933). Über das Sehen bei Natriumdampf und Glülampenlicht (Seeing by light of sodium-vapour and incandescent lamps). Das Licht, 3 (1933) 213.

Augustin, A. (2007). Augenheilkunde. 3., komplett überarbeitete und erweiterte Auflage (Ophthalmology. 3rd completely revised and extended edition). Berlin, Springer, 2007.

Barbur, J.L. (2004). Learning from the pupil: Studies of basic mechanisms and clinical applications. Chapter 39, p. 641-656. In: Chalupa & Werner, eds., 2004.

Barlow, H. (2004). The role of single-unit analysis in the past and future of neurobiology. Chapter 2. p. 14-29. In: Chalupa & Werner, eds., 2004.

Barnstable, C.J. (2004). Molecular regulation of vertebrate retinal development. Chapter 3. p. 33-45. In: Chalupa & Werner, eds., 2004.

Bartley, S.H. (1951). The psychophysiology of vision. Chapter 24. In: Stevens ed., 1951.

Bergsma, A. (1997). Het brein, ons innerlijk universum; derde druk (The brain, our inner universe; third edition). Utrecht, TELEAC, 1997.

Blüh, O. & Elder, J.D. (1955). Principles and applications of physics. Edinburgh, Oliver & Boyd, 1955.

Breuer, H. (1994). DTV-Atlas zur Physik, 2 Bände. 4. Auflage (DTV atlas for physics. 2 volumes, 4th edition). München, Deutsche Taschenbuchverlag DTV, 1994.

Bronstein, I.N.; Semendjajew, K.A.; Musiol, G. & Mühlig, H. (1997). Taschenbuch der Mathematik. 3. Auflage (Manual of mathematics. 3rd edition). Frankfurt am Main, Verlag Harri Deutsch, 1997.

Burns, M.E. & Lamb, T.D. (2004). Visual transduction by rod and cone photoreceptors. Chapter 16, p. 215-233. In: Chalupa & Werner, eds, 2004.

Cambpbell, F.W. & Gubisch, R.W. (1966). Optical quality of the human eye. J. Physiol. (London). 186 (1966) 558-578.
Chalupa, L.M. & Werner, J.S., eds. (2004). The visual neurosciences (Two volumes). Cambridge (Mass). MIT Press, 2004.
CIE (1999). Proceedings of the 24th Session of the CIE, Warsaw, 24-30 June 1999. Publication Nr. 133. Vienna, CIE, 1999.
CIE (2006). Proceedings of the 2nd Expert Symposium on Lighting and Health. Ottawa, 7-8 September 2006. Publication CIE X312:2006. Vienna, CIE, 2006.
Clugston, M.J., ed. (1998). The new Penguin dictionary of science. London, Penguin books, 1998.
Crick, F. (1989). What mad pursuit; A personal view of scientific discovery. Harmondsworth, Penguin Books, 1989.
Crick, F. (1994). The astonishing hypothesis; The scientific search for the soul. London, Simon & Schuster, 1994 (Touchstone Books, 1995).
Damasio, O. (1990). Awakenings (revised edition). Reading, Berks., Cox & Wyman Ltd., Picador, 1990.
Damasio, A. (1994). Descartes' error: Emotion, reason and the human brain. New York, Avon Books, 1994.
Davies, D.V., ed. (1969). Gray's Anatomy. Thirty-fourth edition, Second impression. London, Longmans, Green and Co, Ltd., 1969.
Davson, H. (1972). Physiology of the eye. New York, Acad. Press, 1972.
Dennett, D.C. (1993). Consciousness explained. London, Penguin Books, 1993.
Dennett, D.C. (2006). De betovering van het geloof; Religie als een natuurlijk fenomen (Translation of: Breaking the spell. Religion as a natural phenomenon; 2005). Amsterdam, Uitgeverij Contact, 2006.
De Vries, S.H.; Qi, X; Smith, R.G.; Makous, W., & Sterling, P. (2002). Electrical coupling between mammalian cones. Curr. Biol., 12 (2002) 1900-1907.
Edelman, G. (1992). Bright air, brilliant fire; On the matter of the mind. London, Penguin Books, 1992.
Edelman, G. (2004). Wider than the sky: The phenomenal gift of consciousness. Cambridge, Yale University Press 2004.
Feynman, R.P.; Leighton, R.B. & Sands, M. (1977). The Feynman lectures on physics. Three volumes. 1963; 6th printing 1977. Reading (Mass.), Addison-Wesley Publishing Company, 1977.
Field, G.D. & Rieke, F (2002). Nonlinear signal transfer from mouse rods to bipolar cells and implications for visual sensitivity. Neuron, 34 (2002) 773-785.
Franssen, L. & Coppens, J.E. (2007). Straylight in the eye; Scattered papers. Doctoral Thesis, University of Amsterdam. Amsterdam, 2007.
Glaser, W.R. (1997). Wahrnehmung (Perception). In: Straub et al., eds., 1997, Chapter III-1, p. 225-248.
Gleitman, H. (1995). Psychology. Fourth edition. New York, W.W. Norton & Company, 1995.
Gleitman, H.; Fridlund, A.J. & Reisberg, D. (2004). Psychology. Sixth edition. New York, W.W. Norton, 2004
Goodenough, U. (1998). The sacred depths of nature. New York, Oxford. Oxford University Press, 1998.
Gould, S.J. (1980). The panda's thumb. New York, W.W. Norton & Company, 1980.
Gould, S. J. (1989). Wonderful life. London Penguin Books, 1989.
Greenfield, S. (1997). The human brain; A guided tour. London, Weidefeld & Nicholson, 1997.
Greenfield, S, (1999). How might the brain generate consciousness? In: Rose, ed., 1999, Chapter 12.
Greenfield, S. (2000). The private life of the brain. London, Penguin Books, 2000.
Greenfield, S. (2000a). Brain story. London, BBC Worldwide Limited, 2000.
Greenfield, S. (2001). Brain story. De ontsluiting van onze raadselachtige binnenwereld (Brain story; opening up our mysterious inner world. Translation of Greenfield, 2000a). Baarn, Bosch & Keuning, 2001.

Gregory, R.L. (1965). Visuele waarneming; de psychologie van het zien (Visual perception; The psychology of seeing). Wereldakademie; de Haan/Meulenhoff, 1965.
Gregory, R.L., ed. (1987). The Oxford companion to the mind. Oxford, Oxford University Press, 1987.
Gregory, R.L., ed. (2004). The Oxford companion to the mind. Second edition. Oxford, Oxford University Press, 1987.
Grossberg, S. (2004). Visual boundaries and surfaces. Chapter 110, p. 1624-1639. In: Chalupa & Werner, eds., 2004.
Gullstrand, A. (1911). Einführung in die Methoden der Dioptrik des Auges des Menschen; Band IIIa (Introduction into the methods of dioptric measurements of the human eye; Volume IIIa). In: Tigerstedt, 1911.
Haverkamp, S.; Grünert. U., & Wässle. H. (2000). The cone pedicle, a complex synapse in the retina. Neuron, 27 (2000) 85-95.
Helbig, E. (1972). Grundlagen der Lichtmesstechnik (Fundaments of photometry). Leipzig, Geest & Portig, 1972.
Heavan, R. & Buxton, S. (2007). Duisternis belicht. Spiritueel ontwaken in de duisternis via meditatie (Darkness illuminated. Waking up in darkness via meditation; translation). Deventer, Ankh-Hermes, 2007.
Hentschel, H.-J. (1994). Licht und Beleuchtung; Theorie und Praxis der Lichttechnik; 4. Auflage (Light and illumination; Theory and practice of lighting engineering, 4th edition). Heidelberg, Hüthig, 1994.
Hentschel, H.-J. ed. (2002). Licht und Beleuchtung; Grundlagen und Anwendungen der Lichttechnik; 5. neu bearbeitete und erweiterte Auflage (Light and illumination; Theory and applications of lighting engineering; 5th new and extended edition). Heidelberg, Hüthig, 2002.
Hopkinson, R.G. (1969). Lighting and seeing. London, William Heinemann, 1969.
Hubel, D.H. (1990). Visuele informatie; Schakelingen in onze hersenen (Visual information; The switchboard in our brain). Wetenschappelijke Bibliotheek, deel 21. Maastricht, Natuur en Techniek, 1990 (translation of: Eye, Brain and Vision. New York, The Scientific American Library, 1988).
Illingworth, V., ed. (1991). The Penguin Dictionary of Physics (second edition). London, Penguin Books, 1991.
Kamermans. M.; Fahrenfort, I.; Schulz, K.; Janssen-Bienhold, U.; Sjoerdsma, T. & Weiler, R. (2001). Hemichannel-mediated inhibition in the outer retina. Science, 292 (2001) 1178-1180.
Kosslyn, S.M. & Rosenberg. R.S. (2002). Fundamental psychology; The brain, the person, the world. Boston (MA), Allyn & Bacon, 2002.
Krech, D.; Crutchfield, R.S. & Livson, N. (1969). Elements of psychology (second edition). New York, Alfred Knopf, 1969.
Kuchling, H. (1995). Taschenbuch der Physik, 15. Auflage (Survey of physics, 15th edition). Leipzig-Köln, Fachbuchverlag, 1995.
Küppers, B.-O. (1987). Inwieweit lassen sich die Lebenserscheinungen physikalisch-chemisch erklären? (In how far can life be explained by physics or chemistry?). Introduction to Küppers, ed., 1987, p. 9-34.
Küppers, B.-O., ed. (1987). Leben = Physik + Chemie? (Life = physics + chemistry?). München, Piper, SP 599, 1987.
Lafferty, P. & Rowe, J, eds. (1994). Dictionary of science. London, Brockhampton Press, 1994.
Langerhorst, C.L. (1995). Glaucoom (Glaucoma). Chapter 14. In: Stilma & Voorn, eds., 1995.
Longhurst, R.S. (1964). Geometrical and physical optics (fifth impression). London, Longmans, 1964.
Mann, I. & Pirie, A. (1950). The science of seeing. Harmondsworth, Penguin Books. Pelican A 157, 1950 (Revised edition).
Marcos, S.; Burns, S.A.; Moreno-Barriusop, E. & Navarro, R. (1999). A new approach to the study of ocular chromatic aberrations. Vision Res. 39(26) (1999) 4309-4323.
Metzger, W. (1953). Gesetze des Sehens (Laws of vision). Frankfurt am Main, Waldemar Kramer, 1953 (Original edition 1938).

Moon, P. (1961). The scientific basis of illuminating engineering (revised edition). New York, Dover Publications, Inc., 1961

Narisada, K. & Schreuder, D.A. (2004). Light pollution handbook. Dordrecht, Springer, 2004.

Nelson, R. (1977). Cat cones have rod input: A comparison of the response properties of cones and horizontal cell bodies in the retina of the cat. Journ. Comp. Neurol., 172 (1977) 109-136.

NSVV (1957). Aanbevelingen voor openbare verlichting (Recommendations for public lighting). Moormans Periodieke Pers, Den Haag, 1957 (year estimated).

NSVV (2003). Licht en gezondheid voor werkenden; Aanbeveling (Light and health for workers; Recommendation). Arnhem, NSVV, 2003.

NSVV (2003a). Het Nationale Lichtcongres. Ede, 12 november 2003; Syllabus (The National Light Conference. Ede, 12 November 2003; Proceedings). Arnhem, NSVV, 2003.

NSVV (2006). Licht, welzijn en de ouder wordende mens (Light, well-being, and people of advancing age). Ede, NSVV, 2006.

Osterberg, G. (1935). Topology of the layer of rods and cones in the human retina. Acta Suppl. 6. Kopenhagen, 1935.

Perkins, E.S. & Hill, D.W., eds. (1977). Scientific foundations of ophthalmology. London William Heinemann Medical Books Ltd., 1977.

Pirenne, M.H. (1972). Rods and cones. In: Davson, 1972, vol. II, p. 13.

Porter, J; Guirao, A; Cox, I. & Williams, D.R. (2001). Monochromatic aberrations of the human eye in a large population. Journ. Opt. Soc. Amer 18(8) (2001) 1793-1803.

Reeves. P. (1920). The response of the average pupil to various intensities of light. J. Opt. Soc. Amer. 4 (1920) 35-43.

Rose, S., ed. (1999). From brains to consciousness? London, Penguin Books, 1999.

Schreuder, D.A. (1962). Warum Beleuchtung mit Natriumdampflampen? (Why lighting with sodium-vapour lamps?). Elektrizitätsverwertung, 37 (1962) 7: 191-195.

Schreuder, D.A. (1991). Visibility aspects of the driving task: Foresight in driving. A theoretical note. R-91-71. Leidschendam, SWOV, 1991.

Schreuder, D.A. (1997). Zonnebrillen en verkeerslichten met LED's (Sunglasses and traffic signals with LED's). Leidschendam, Duco Schreuder Consultancies, 1997.

Schreuder, D.A. (1998). Road lighting for safety. London, Thomas Telford, 1998 (Translation of "Openbare verlichting voor verkeer en veiligheid", Deventer, Kluwer Techniek, 1996).

Schreuder, D.A. (2006). Gerontopsychologische overwegingen bij het ontwerpen van binnenverlichting (Gerontopsycholagal considerations in interior lighting design). NSVV Nationaal Lichtcongres 2006, 23 november 2006. Ede, NSVV, 2006.

Schreuder, D.A. (2008). Looking and seeing; A holistic approach to vision. Dordrecht, Springer, 2008 (in preparation).

Shannon, C.E. (1949). Communication in the presence of noise. Proc. IRE, 37 (1949), 10.

Sherrard, E.S. (1977). The cornea: structure and transparancy. Chapter 5. In: Perkins & Hill, eds., 1977.

Shevell, S.K., ed. (2003). The science of color. Second edition (first edition 1953). OSA, Optical Society of America. Amsterdam, Elsevier, 2003.

Spillmann, L. & Ehrenstein, W.H. (2004). Gestalt factors in the visual neurosciences. Chapter 106. p. 1573-1589. In: Chalupa & Werner, eds., 2004.

Sterling, P. (2004). How retinal circuits optimize the transfer of visual information. Chapter 17, p. 234-259. In: Chalupa & Werner, eds., 2004.

Stevens, S.S., ed. (1951). Handbook of experimental psychology. New York, John Wiley and Sons, Inc, 1951.

Stiles, W.S. & Crawford, B.H. (1933). Luminous efficiency of rays entering the eye pupil at different points. Proc. Roy. Soc. 112B (1933) 428.

Stilma, J.S. (1995). Cataract. Chapter 13. In: Stilma & Voorn, eds., 1995.
Stilma, J.S. & Voorn, Th. B., eds. (1995). Praktische oogheelkunde. Eerste druk, tweede oplage met correcties (Practical ophthalmology. First edition, second impression with corrections). Houten, Bohn, Stafleu, Van Loghum, 1995.
Straub, J.; Kempf, W. & Werbik, H., eds. (1997). Psychologie, Eine Einführung (Psychology, An introduction). DTV 2990. München, Deutscher Taschenbuch Verlag GmbH & Co. KG. 1997.
Thibos, L.N. (1987). Calculation of the influence of the lateral chromatic aberration on image quality across the visual field. Journ. Opt. Soc. Amer. A4 (1987) 1673-1680.
Tigerstedt (1911). Handbuch der physiologischr Methodik (Handbook of physiological methodoly). Leipzig, Hirzel, 1911.
Van den Beld, G. (2003). Aanbevelingen en aandachtspunten voor gezonde verlichting (Recommendations and points of interest of healthy lighting). In: NSVV, 2003a, p. 66-72.
Van den Berg, T.J.T.P. (1995). Analysis of intraocular straylight, especially in relation to age. Optometry and Vision Science. 72 (1995) no. 2, p. 52-59.
Van der Neut (2007). Kunnen dieren denken? (Can animals think?). Pychiologie Magazine, 26 (2007) april p. 34-40.
Van Essen, J. (1939). Etude psychophysiologique sur l'obscurité (Psychophysiological study on darkness). Archives Néerlandaises de physiologie et de phonétiqyue expérimentale (1939) p. 487-554 (Year estimated).
Van Heel, A.C.S. (1950). Inleiding in de optica; derde druk (Introduction into optics, third edition). Den Haag, Martinus Nijhoff, 1950.
Van Heyningen, R. (1977). The biochemistry of the lens: Selected topics. Chapter 6. In: Perkins & Hill, eds., 1977.
Van Tilborg, A.D.M. (1991). Evaluatie van de verlichtingsproeven in Utrecht (Evaluation of lighting experiments in Utrecht). Utrecht, Energiebedrijf, 1991 (not published; see Anon., 1995).
Vos, J.J. (1963). On mechanisms of glare. Doctoral Thesis, University Utrecht, 1963.
Vos, J.J. (1983). Verblinding bij tunnelingangen I: De invloed van strooilicht in het oog (Glare at tunnel entrances I: The influence of the straylight in the eye). IZF 1983 C-8. Soesterberg, IZF/TNO, 1983.
Vos, J.J. (1987). Het nut van een zonnebril, in het bijzonder bij buitensport (The benefit of sunglasses, in particular for outdoor sports). Visus 3 (1987) no 3, p. 4-7.
Vos, J.J. (1999). Glare today in historical perspective: Towards a new CIE glare observer and a new glare nomenclature. In: CIE, 1999, Volume 1 part 1, p. 38-42.
Vos, J.J. (2003). Reflections on glare. Lighting Res. Technol. 35 (2003) 163-176.
Vos, J.J.; Walraven, J. & Van Meeteren, A. (1976). Light profiles of the foveal image of a point source. Vision Research, 16 (1976) 215-219.
Wachter, A. & Hoeber, H. (2006). Compendium of theoretical physics (Translated from the German edition). New York, Springer Science+Business Media, Inc., 2006.
Weale, R.A. (1968). From sight to light. Edinburgh and London, Oliver and Boyd, 1968.
Werblin, F. & Roska, B. (2007). The movie in your eyes. Scientific American. 296 (2007) no. 4, p. 55-61.
Wertheim, T. (1894). Über die indirekte Sehschärfe (On the indirect visual acuity). Z. Pschychol. u. Physiol. Sinnesorgane. 7 (1894) 172-187.
Williams, D.R. & Hofer, H. (2004). Formation and acquisition of the retinal image. Chapter 50, p. 795-810. In: Chalupa & Werner, eds., 2004.
Wilson, M. (2004). Retinal synapses. Chapter 19. p. 279-303 In: Chalupa & Werner, eds., 2004.
Yogananda, P. (1950). Autobiography of a Yogi. London, Rider and Company, 1950, p. 128.

7 The human observer; visual performance aspects

The Chapters 6, 7, 8, and 9 deal with the human observer. In Chapter 6, the physical and anatomical aspects of vision are discussed, more in particular the structure of the nervous system. This chapter concentrates on the visual performance aspects, and in Chapter 8 the visual perception. Colour vision is discussed in Chapter 9.

Chapter 7 deals with the different functions of the human visual system, concentrating on the image-forming aspects of the activities of the visual sense. A central issue is the duplicity theorem, stating that the main optical elements are the cones and the rods, each having their own characteristics. The cones are mainly active in daytime, allowing for photopic vision, and the rods, being active in dim conditions, for scotopic vision. The spectral sensitivity curves for photopic, mesopic, and scotopic vision are discussed in relation to human and visual performance. The primary visual functions are discussed, viz. the adaptation, the contrast sensitivity, and the visual acuity.

7.1 The functions of the human visual system

7.1.1 The sensitivity of the eye

(a) *Standard observers*

Earlier it has been explained that the sensitivity of the human visual system is the basis for photometry, and that photometry is a branch of the wider field of radiometry. Ultimately, in both cases the power (wattage) is measured. Radiometry is the measurement of electromagnetic radiation independent of the detector, whereas photometry refers to the measurement of that part of the electromagnetic radiation for which the human visual system is sensitive. i.e. the part that may provoke an experience of light. Obviously, the human visual system is a detector.

This is a crucial point. When the response of human beings to any sort of physical stimulus is to be assessed, the human system must be used as a detector. There is no other way; any 'objective' detector must in the last instance be calibrated against the human detection system. The human observer turns up in all and every photometric unit.

In radiometry, the total power of the radiation is measured directly, e.g. by means of a bolometer (Budding, 1993; Hentschel, 1994, sec. 4.3.3.; Hentschel, ed., 2002; Keitz, 1967; Sterken & Manfroid, 1992; Walsh, 1958; Weigert & Wendker, 1989). In photometry, the response of the human visual system must be taken into account. However, it is well-known that the visual system is not equally sensitive to light of all wavelengths. Furthermore, the sensitivity depends on whether the rods or the cones, or possibly both, are active at the same time. To overcome this, the standard observer that is discussed in the next parts of this section has been introduced. This is not a sinecure, as is pointed out in sec. 3.1.1a, where psychophysical measuring methods are discussed. There always is a large inter-subject spread in psychophysical study – no two persons are identical (Boyce, 2003; Le Grand, 1956; Schober, 1960).

In the course of the 19th century, many new lamp types were developed, and added to the ancient arsenal of candles, oil lamps, and torches. The proliferation of those lamp types, in particular gas lamps and carbon-arc lamps, and later electric incandescent lamps that took place around the 1880s and 1890s, required an international, time independent, reproducible standard against which the performance of lighting equipment could be measured and certified. In the beginning, the same or at least similar light sources, operating under standardized conditions, were used. Problems arose because the performance of those lamps could not reproduced nor duplicated, it was decided in the 1920s that the eye sensitivity itself would serve as the standard. The CIE defined in 1924 a 'standard observer' (CIE, 1924, p. 67; p. 232). As is mentioned in sec. 5.2.2a, the standard did refer to daylight vision or photopic vision. In practice this means that only cones are involved in the light perception. Later, other quasi-standards were added. Daylight vision means that the observer is fully adapted to bright light levels, like those in the outdoors during the day. The key word is 'adapted'. Essentially, adaptation is a dynamic effect (Reeves, 2004). However, if the conditions do not change too rapidly, one may speak of a steady state. In agreement with that one may speak of the state of adaptation. Adaptation as a visual phenomenon is discussed in sec. 7.4.2.

It should be stressed that the standard observer for photopic vision is not one particular person. One might wonder whether there is on the whole world one individual who fulfills the description of the standard observer for photopic vision. The standard observer is in fact nothing but a table of spectral sensitivity numbers, that can be used as a basis for standardization work, as well as for measurements and calculations. Another factor

The human observer; visual performance aspects

is that the standard observer is observing in a 2° field. This approach also allows to avoid the question whether it is in principle possible to measure and quantify experiences.

As is mentioned in sec. 5.2.2, where visual photometry is discussed, there is a problem. Around the beginning of the 20th century the current laboratory equipment did not allow precise measurement of blue light.

The CIE did recognise this problem. A correction was introduced, usually called the Judd modification of the 2° function. A CIE Report published in 1990 proposed to call it the 'CIE 1988 modified 2° spectral luminous efficiency function for photopic vision' (CIE, 1990). The Judd modification has been assessed by using a 10° field in stead of the common 2° field (CIE, 1988). The consequences of the choice between the two fields are discussed in a further part of this section. The modification must be regarded as a supplement to, not a replacement of, the 1924 CIE V_λ-function. This proviso was considered necessary to avoid the need to change the V_λ-function itself. The consequences of a replacement would be very wide-ranging. It has been stressed several times that the Standard Visibility Curve is the central issue that links the photometric units to the other physical units. It is included in the definition of the candela, one of the seven fundamental quantities of ISO. It would have its repercussions, not only in the lighting industry, but in the whole structure of physical measurements and nomenclature. See also Narisada & Schreuder (2004, sec. 8.2.1 and 8.2.2). It should be mentioned that for most practical cases the discrepancies are not dramatic, because the rods are not very much involved in photopic vision. However, at wavelengths of around 500 nm, that means in the green region of the spectrum, the results of the 10° field photometry can be up to 30% higher than those of the 2° field photometry. This can be concluded from Figure 7.2.2 where the different relevant V_λ-curves are depicted.

(b) *Assessing the sensitivity of the eye*

It is well-known that the visual system is not equally sensitive to light of all wavelengths. Amongst others, the sensitivity depends on whether the rods or the cones, or possibly both, are active at the same time. Moreover, there are three types of cones, each with their own sensitivity. In spite of these complicating factors, it is possible to determine the sensitivity of the visual system as a whole for different wavelengths.

The response to light of different wavelengths can be measured in different ways. A major problem in standardizing the eye sensitivity is, that each method gives different results (Boynton, 1979, chapter 9; Boynton & Kaiser, 1978). Several methods are explained in sec. 5.2.2, where visual photometry is discussed.

The spectral luminous efficiency curve that is relevant for the CIE standard observer for photopic vision is called the V_λ-function. Further details about this curve are given in sec. 5.2.2b. As has been stressed already several times, this function valid only for the CIE standard observer, and for conditions of high ambient luminances. It is the basis for all photometric standards, and is relevant for measurements and calculations. The standard observer for cone-vision is described in the "CIE Standard spectral luminous efficiency function V(lambda) for photopic vision" (CIE, 1924, p. 67; p. 232). See also Baer, ed. (2006, p. 18).

The spectral luminous efficiency curve has been derived by means of psychophysical experiments where the stimuli of the light and the sensation of brightness are compared. As mentioned in sec. 5.2.2d, the determination of the V_λ-curve is essentially heterochromatic photometry. This means that, in a photometer, two areas are presented in the field of view to an observer. One area represents the standard condition, to which the other must be made equal. In practice the matters are less serious than they might seem to be. Over the years and decades, the accuracy has been gradually improved by means of an iterative process.

(c) Two-degree and ten-degree photometry

It has been mentioned earlier that the definition of the CIE Standard observer, and consequently the definition of the CIE V_λ-function includes the 2° field of observation. This has not been changed by the introduction of the Judd modification that is explained in an earlier part of this section. In spite of that, many measurements have been made with a 10° field. The reason was that one did hope that the larger amount of light – the higher number of photons – that passes through a large field, would help to improve the overall accuracy of the measurements, particularly at dim ambient lighting. However, as is explained in sec. 6.3.5d, where the spatial distribution of rods and cones over the retina is discussed, the cones are clustered in and near the fovea centralis. This is depicted in Figure 6.3.11. There are no rods to speak of in the fovea. Usually the diameter of the fovea is taken as about two degrees. Thus, 2° photometry would focus on cones, just what is needed if photopic vision is studied. Around the fovea, at angles more than one degree off-centre, the cone density drops of steeply, whereas the rod density stays virtually the same. This implies that 2° photometry is valid for photopic vision, but 10° photometry is not, the cones and rods having different spectral sensitivities. As is mentioned earlier the consequences are not dramatic for most practical lighting applications.

As a matter of fact, CIE advises to use the Judd modification and the related 10° field only for monochromatic point sources (CIE, 1988).

These remarks relate to the establishment of the V_λ-curve. For other applications, like e.g colorimetry, other considerations have been made. In 1964, CIE introduced a Standard Colorimetric Observer (CIE, 1986). The CIE 1964 Standard Colorimetric Observer was defined for a 10° field. Originally, it had no photometric counterpart. It was found that the 1964 Standard Colorimetric Observer function with the 10° field still could be used (CIE, 2005a).

Also, CIE defined the colour-matching functions (CIE, 2006). As a final step, these considerations and results were laid down in a CIE Standard (CIE, 2006). Further details can be found in a new draft report of CIE (CIE, 2006a).

7.2 The sensitivity of the human visual system

7.2.1 The duplicity theorem

As is explained in sec. 6.3.5b, the retina of most vertebrate animals, humans included, contains two quite distinct types of photoreceptors: viz. the cones and the rods. Although the overall anatomy of cones and rods shows some similarities, their operation shows very important differences:
1. The sensitivity to light of rods is much greater than that of the cones;
2. The rod system is not capable to discern colours, whereas the cone system, consisting of at least three different types of cones, can differentiate between different colours;
3. The peak of the spectral sensitivity curve of rods is shifted towards the blue end of the spectrum compared to that of the cones.

These differences are so great that it seems justified to speak of two distinct visual systems. This approach is called the duplicity theory. One of the first to point this out was Schultze in 1866. Nowadays the duplicity theory is generally accepted (Reeves, 2005, p. 851).

7.2.2 Photopic vision

(a) *Three families of cones*

As is explained in several places in this book, the spectral sensitivity of the visual system as a whole can be presented by one single curve. For photopic vision or day-time vision the curve is called the V_λ-curve. As mentioned earlier, it is assumed that in photopic vision only the cones are operational and the rods are inactive. Thus, the shape is essentially determined by the absorption characteristics of the visual pigments that are operational in the cones during daytime vision.

Now this sounds simple. However, as is explained in sec. 6.3.5b and 9.2.1, the normal adult human retina contains not one, but at least three families of cones. It is difficult to discern the three families of cones along anatomical characteristics (Boynton, 1979, p. 19). However, each family has its own spectral sensitivity curve. There are two ways to assess the spectral sensitivity:
1. To measure the absorption;
2. To study the response of colour blind observers.

The results are explained in Chapter 9, where colour vision is discussed. There, it is explained that the overall spectral sensitivity can be assessed by adding the separate spectral sensitivities of the three cone families, using a specific set of weighing factors (Smith & Pokorny, 2003, sec. 3.2.5, p. 117-120).

(b) *The V_λ-curve*
In the 1924 definition of the standard observer, a standard spectral luminous efficiency function is introduced. Commonly this is described as the V_λ-curve that has been mentioned already several times.

The 'Standard Visibility Curve' can be presented in different ways. The most easy to understand is a graphical representation. In a curve, the relative sensitivity is given versus the wavelength. The curve, called the 'V_λ-curve', this is shown in Figure 7.2.1.

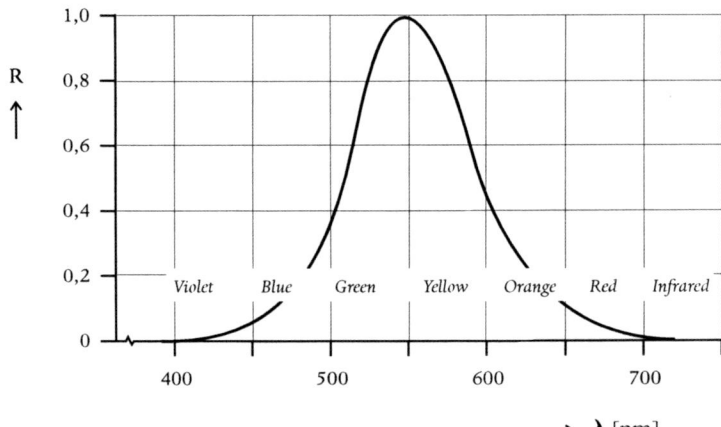

Figure 7.2.1 *The V_λ-curve. (After Narisada & Schreuder, 2004, Fig. 8.2.1. This figure has been given earlier as Figure 5.2.1.)*

The human observer; visual performance aspects 235

(c) *V_λ as a function*

In a symbolic way, the standard visibility curve can be written as a function: $V = f(\lambda)$. We may mention here that V_λ does not fulfill all mathematical requirements for a function in the mathematical sense (Bronstein et al., 1997; Daintith & Nelson, 1989; De Bruin, 1949). See also Wachter & Hoefer (2006, Appendix A4. This does not need to worry us, as it possible to derive a real mathematical function as an approximation of the points that represent an empirical relationship depicted as a curve. As long as the curve behaves well, that is when it does not includes infinitely large values, has no singularities, etc., such functions would also behave neatly. As regards V_λ, there are no problems.

An example of such a function is given by Baer, ed. (2006, p. 18). The function based on Mutzhas (1981). See Narisada & Schreuder (2004, sec. 8.2.2). It is very complex and not very accurate. Numerical approximations by means of modern high-power computers made that analytical representations did loose much of their interest. There might, however, still be one area of interest of such formulae. The integral of the function would give directly the conversion from lightwatts into lumens. The term is not very correct, but it gives some indication about the maximum efficacy that lamps theoretically might reach. Obviously, it is a dimensionless number. Its numerical value is, according to the modern definition of the lumen, 683. One might add 'lumen per watt' (Schreuder, 1998, p. 73). The value of 683 is sometimes called the radiometric equivalence. Incidently, for the scotopic observer, the corresponding value is 1699, as given by Wyszecki & Stiles (1982). This may not be expressed in lumen per watt, as the lumen is defined only for photopic vision.

(d) *Tabulated values of V_λ*

The CIE has produced a certified version of the table that gives the value of the sensitivity up to intervals of 1 nm (CIE, 1990). The data are summarised in Table 5.2.1 (After Narisada & Schreuder, 2004, table 8.2.1). It should be stressed that the presentations in graphical form and those in table-form represent exactly the same data. One is therefore completely free to use the one or the other. Furthermore, it cannot be stressed strongly enough that the Standard Visibility Curve is the central issue that links the photometric units to the other physical units. It is therefore no surprise that one is very reluctant to include any corrections in the visibility curve, such as would be needed to correct the underrating of deep blue colours. Such a correction would have its repercussions, not only in the lighting industry, but in the whole structure of physical measurements and nomenclature. To repeat it again, all photometry is based on the Standard Visibility Curve; it is therefore defined exclusively for photopic vision, for the standard observer, and for the 2° field of observation.

7.2.3 The scotopic spectral sensitivity curve

As is mentioned in sec. 6.3.5b, the chemical composition of the visual pigments in rods is very similar to that in the cones. It should therefore be no great surprise that the absorption functions for different wavelengths are very similar as well. As a rule of thumb, one may say that the spectral sensitivity curves for the different modes of observation – photopic, mesopic, and scotopic, show the same shape. This is the basis for the rule of Palmer that is explained in sec. 7.2.4, where mesopic vision is discussed. There is, of course, a difference. Otherwise there would be no 'different modes of observation'. The difference is the way the curves are shifted over the wavelength-scale. Thus, the spectral sensitivity curves of photopic vision and of scotopic vision are very similar in shape. In Figure 7.2.2 the two sensitivity curves are depicted together.

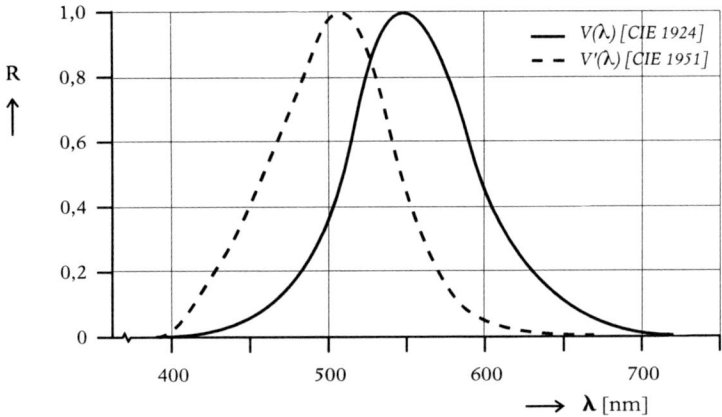

Figure 7.2.2 The spectral sensitivity curves for photopic and for scotopic vision, V_λ and V_λ' respectively. (After Narisada & Schreuder, 2004, fig. 8.2.6 Based on Schreuder, 1998, Figure 5.1.2, and on Baer, ed., 2006, fig 1.5.)

Notes to Figure 7.2.2:
- The two sensitivity curves are independently normalised to 100% for their maximum values;
- As is customary, the photopic curve represents measurements with a 2° measuring field, whereas the scotopic curve represents measurements with a 10° measuring field.

It might be added here that the spectral sensitivity curves of other visual functions, like e.g. the pupil reflex, and the non-image forming effects, all have a very similar shape. See Schreuder (2008) and Gall (2004, fig A4, D1).

7.2.4 Mesopic vision

(a) The limits of mesopic vision

There is not one single luminance value where photopic vision and scotopic vision meet. In contrary, there is a wide zone of transition zone between them. Because it is between photopic and scotopic vision, it usually called the zone of mesopic vision. The reason that the zone of mesopic vision does exist is because the activities of neither cones nor rods is simply switched 'on' or 'off'. There are reasons to believe that the cones and the rods both operate in all luminance conditions (Schreuder, 1991, 1991a; NSVV, 2003). However, their activity is not always apparent (Adrian, 1993; 1995; Schreuder, 1976; 1997).

The range of shared function varies with the wavelength but, as a rough generalization, rods and cones are active over a four-decade range of light levels (Buck, 2004, p. 863). This implies that there is no sharp borderline between the zones of photopic and mesopic vision, nor between the zones of mesopic and scotopic vision. Because there is no sharp criterion to determine whether one is in the zone of photopic or mesopic vision, or in the zone of mesopic or scotopic vision, the borders of the zone of mesopic vision that are used in literature, are somewhat arbitrary. It is often stated that the zone of scotopic vision begins at about 0,03 cd/m² (Baer, 1990, p. 41). Other data state that the zone of mesopic vision stretches from 10 cd/m² down to 10^{-3} cd/m² (Kokoschka, 1980; Hentschel, 1994, p. 32). At another place, it is stated that the zone of mesopic vision stretches from 3 cd/m² down to 0,01 cd/m² (Hentschel, 1994, p. 47). More alternatives are given in Narisada & Schreuder (2004, sec. 8.2.3).

The value of 0,01 cd/m² may seem low, but one may meet levels as low as these in road lighting, even if the lighting agrees with generally accepted standards or recommendations.

In traffic route lighting, the road surface luminance is between 2 and 0,5 cd/m² (CEN, 2002; CIE, 1992; 1995; NSVV, 1990, 2002). They fall in the high-mesopic region. When one looks at other areas of outdoor lighting, such as the lighting of streets in residential areas, the luminance values are much lower. As an example, the current recommendations for the Netherlands stipulate that in footpaths in quiet, dark, residential areas with a low crime rate, the average illuminance must be at least 2 lux with a uniformity defined as $E_{h,min} / E_{h,av}$ of at least 0,3 (NSVV, 1990, 2002). $E_{h,min}$ is the minimum horizontal illuminance and $E_{h,av}$ is the average horizontal illuminance. Thus, the illuminance on the footpath may be as low as 0,6 lux. With normal road surfaces this corresponds with about 0,03 cd/m². On the footpaths and sidewalks along such roads, the light level may easily be another factor 5 lower. So we may have to deal with luminance of about

0,01 to 0,02 cd/m². It is very likely that, in quite normal outdoor lighting practice, the luminance that have to be dealt with are near the lower end of the mesopic range and not near the higher end. The same seems to be true for roads without street lighting, for which vehicle low-beam lighting is supposed to provide adequate visibility (Schreuder & Lindeijer, 1987; Schoon & Schreuder, 1993; Narisada & Schreuder, 2004, sec. 12.2). The lighting levels on such roads fall in the range of mesopic vision.

(b) *The transition between photopic and scotopic vision*

In view of the considerations given ion the preceding part of this section, it is striking that for many decades many researchers apparently did assume that there was a sudden transition between rod and cone vision, and that there did not effectively exist any sort of mesopic range.

As an example, we will mention here the well-known Blackwell-data, the so-called Tiffany data that are the basis for the CIE Relative Contrast Sensitivity Reference Curve, or RSC-curve (Blackwell, 1946). See Figure 7.2.3. Actually, the Standard RSC-curve is only one of a whole family of curves. The RSC-curve is one of the basic tools in theoretical and practical illuminating engineering (CIE, 1981; De Boer & Fischer, 1981, p. 18; Narisada & Schreuder, 2004, p. 256). The RSC-curve is discussed in sec. 7.4.3h.

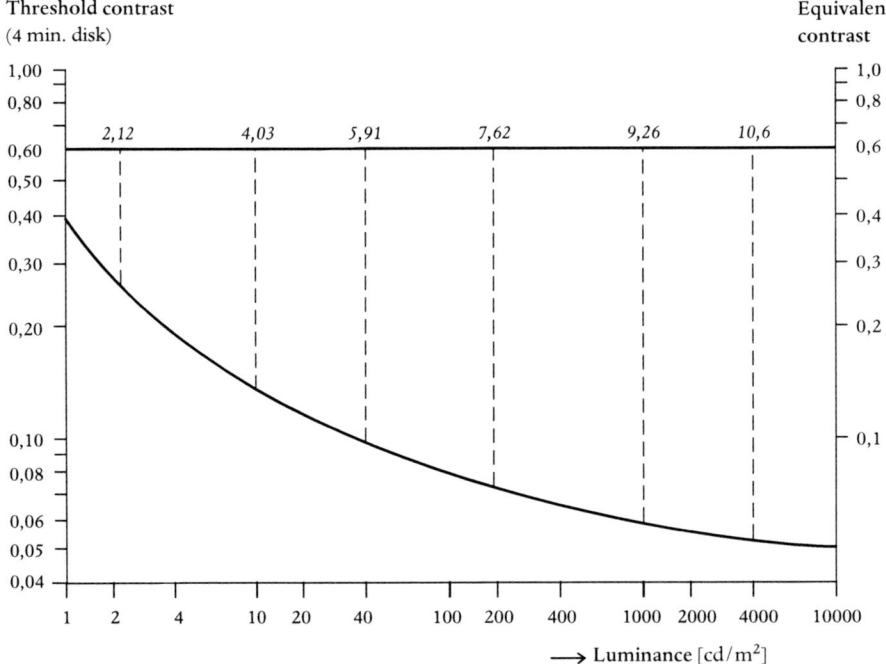

Figure 7.2.3 The relation between contrast sensitivity and adaptation: The CIE RSC-curve. (After Narisada & Schreuder, 2004. fig. 9.1.4.)

The human observer; visual performance aspects

It must be noted that the curve is smooth over the full range of adaptation luminances. However, in some representations of the Blackwell-data, a breach and not a smooth curve was given when the adaptation luminance is changed, e.g in Adrian's representation (Adrian, 1969). See Figure 7.2.4.

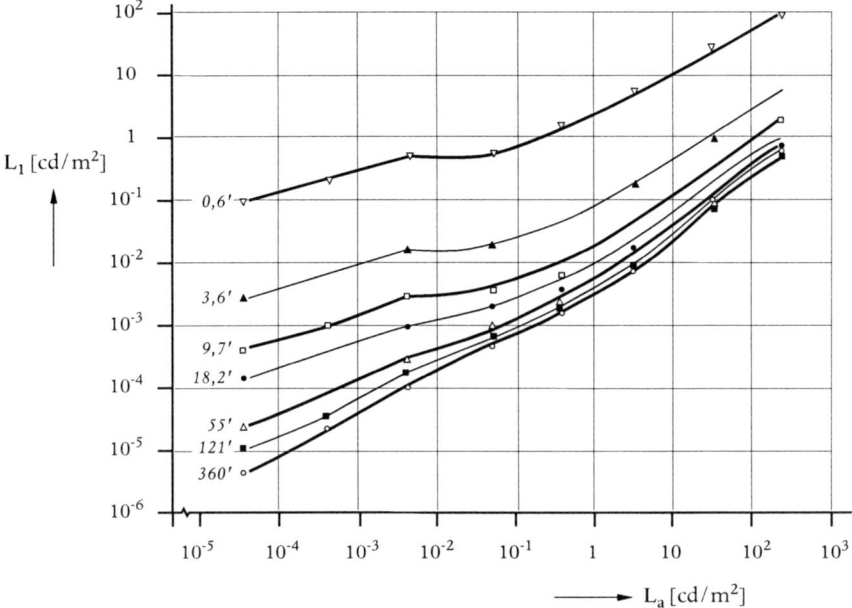

Figure 7.2.4 *The relation between the differential luminance threshold and the adaptation luminance for different sizes of the test objects and for a probability of observation of 50%, as recalculated by Adrian, 1969. (After Narisada & Schreuder, 2004, fig. 9.1.7. Based on Hentschel, ed., 2002, figure 3.7.)*

Most light adaptation studies are made in the fovea. It is likely that there will be no sudden breach in that region. When measurements are made in the peripheral area of the retina, the results seem to be different. However, the Adrian-representation shows a clear breach for small areas and a smooth curve for a large field of view. See also Reeves (2004, figure 54.7, p. 859). As all this is history, it is difficult to find out the reason for this. However, it seems justified to assume that the transition from scotopic vision toward photopic vision is smooth, without a conspicuous breach.

(c) *The high-mesopic region*

In order to avoid this confusion, Schreuder has introduced the concept of 'high mesopic vision' (Schreuder, 1976, p. 28; Narisada & Schreuder, 2004, p. 196). It has been found that for practical outdoor lighting applications the differences between mesopic vision and photopic vision can be neglected when the luminance is higher than 0,1 cd/m². In other words, for practical purposes, the lower limit of the zone of photopic vision may be put at 0,1 cd/m².

(d) The Purkinje-shift

The fact that the photopic vision and the scotopic vision are different, has considerable consequences for the human visual system and for visual perception (Narisada & Schreuder, 2004, sec. 8.3.4). There are four major differences:
1. Photopic vision is restricted mainly to the fovea and the near-periphery, whereas scotopic vision is mainly effective in the near, and even in the far periphery;
2. In photopic vision, the visual acuity, particularly in the foveal area, is much greater than in scotopic vision. Details are given in sec. 7.4.4;
3. Contrary to photopic vision, colour vision in not possible in scotopic vision. Details are given in sec. 9.2;
4. In photopic vision, the maximum of the spectral sensitivity is found at a much longer wavelength than in scotopic vision. Details are given the preceding part of this section.

As is explained in sec. 7.2.3, where scotopic vision is discussed, the shapes of the V_λ-curve and the V_λ'-curve are similar; however, the maximum of the V_λ-curve is at 555 nm and that of the V_λ'-curve is at 507 nm. The consequence is that the visual system in scotopic vision is much more sensitive to blue light, and much less sensitive to red and yellow light than in photopic vision. In other words, in scotopic vision, the visual system seems to be less effective for red and yellow light, but more effective for blue light compared to photopic vision. Thus, in scotopic vision, a blue flower looks white, and a red flower looks black. This was the original observation after which the 'Purkinje effect' was named (Moon, 1961, Appendix A). It should be noted that, because the visual observations were made in the scotopic region, no colours could be seen: flowers are either white or grey or black.

(e) Mesopic spectral sensitivity curves

As is explained in the preceding part of this section, there is not one single luminance value where photopic vision and scotopic vision meet. On the contrary, there is a wide zone of transition zone between them. Because it is between photopic and scotopic vision, it is usually called the zone of mesopic vision.

As has been indicated earlier, in the mesopic range both rods and cones are operational – or rather, where their activity is directly apparent. However, the 'mixture' of the two varies. It may be assumed that this mixture goes from 100% cones and 0% rods at the upper limit of the mesopic range (at about 3 cd/m²) to 0% cones and 100% rods at the lower limit of the mesopic range (at about 0,01 cd/m²). The simplest way to describe what happens between the two limits is to assume a linear relationship between these percentages and the (logarithm of the) luminance. This has been proposed by Cakir & Krochmann (1971). This simplified model is used by Schreuder (1967, 1998). In Figure 7.2.5, the graphical representation of this model is given.

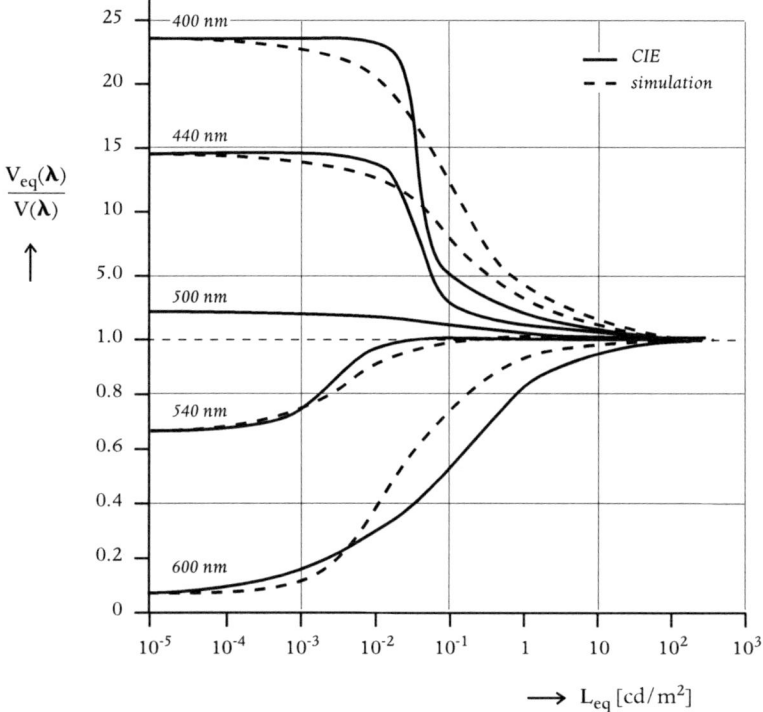

Figure 7.2.5 *The relative efficacy (V_{eq}/V) at various wavelengths as a function of the equivalent luminance (L_{eq}). (After Narisada & Schreuder, 2004, fig. 8.2.7 and Schreuder, 1976, figure 3. Based on Cakir & Krochmann, 1971, fig. 1.)*

Some authors state that this model is over-simplified (Kokoschka, 1971; Walters & Wright, 1943). Still it is often used. The implication of this model is that, for each luminance value between the upper and the lower limit of the mesopic range, a V_λ-curve may be found. Thus, a whole 'family' of V_λ-curves does exist, and not just one mesopic V_λ-curve. A further implication is that each V_λ-curve out of the family of V_λ-curves may be constructed by a linear interpolation between the photopic V_λ-curve and the scotopic V_λ'-curve. Details are given in CIE (1986; 1989).

(f) Mesopic photometry

In sec. 3.2.1c it is explained that the units of photometry are derived from the spectral sensitivity curves. Furthermore, the standard photometry is based on the photopic V_λ-curve. Also it is mentioned that, although a standard V_λ'-curve for scotopic vision has been defined by CIE, there are no serious proposals for a 'scotopic photometry'. Similarly, it would be possible to base, for each mesopic level, a whole set of photometric quantities and units, based on the V_λ-curve for that particular mesopic level. In CIE, suggestions for 'mesopic photometry' have been made (Eloholma et al., 2005; Halonen & Eloholma, 2005; Rea, 2005; Sagawa, 2005). It would seem that this is not a wise course to go. Not only because there would have to be many of them, one for each mesopic level, but mainly because it would also add to the confusion about how to proceed in lighting design, in calibration, and in data presentation.

(g) Mesopic metrics

There should be no misunderstanding. The criticism in the preceding part of this sections relates only to the proposals to establish a mesopic photometric standard. The use of a mesopic metric that allows to assess the spectral sensitivity at any level of mesopic vision is a completely different matter. Such a metric is exceedingly useful. Not for standardizing nor for calibration, but for the assessment of the performance of lighting installations, more in particular of outdoor lighting installations. In the recent CIE Symposium on mesopic vision, not only the proposals for a mesopic standard, but also proposals for a mesopic performance metric have been discussed in detail (CIE, 2005).

Over the years there have been made several proposals for such metrics. Most of them assume that the visual performance in the mesopic range is proportional to the ratio of the active rods and the active cones. One of the first proposals wad made by Opstelten in a non-published report (Opstelten, 1984; Narisada & Schreuder, 2004, sec. 8.3.5, p. 207-209). For this, the mixture-relation was used that had been proposed by Palmer (1971). This is sometimes called the Rule of Palmer, or the Rule of Opstelten. The relation is:

$$L_{mes} = \frac{L_v' + nL_v^2}{1 + nL_v} \qquad [7.2\text{-}1]$$

in which:
L_{mes} = the mesopic luminance (in cd/m^2);
L_v = the photopic luminance;
L_v' = the scotopic luminance.

On the basis of experiments with 16 observers, n was about 17 (Opstelten, 1984; Narisada & Schreuder, 2004, sec. 8.3.5, p. 206; eq. [8.3.1]).

In sec. 7.2.4a it is mentioned that it may be assumed that in the range of mesopic vision the mixture of operational rods and cones goes from 100% cones and 0% rods at the upper limit of the mesopic range (at about 3 cd/m^2) to 0% cones and 100% rods at the lower limit of the mesopic range (at about 0,01 cd/m^2). The simplest way to describe what happens between the two limits is to assume a linear relationship between these percentages and the (logarithm of the) luminance. As is mentioned earlier, this has been proposed by Cakir & Krochmann (1971).

The consequences of all this have been dealt with in considerable detail in Opstelten (1984). Some of the highlights from that – unpublished – report have been quoted in Narisada & Schreuder (2004, sec. 8.3.5)

As is mentioned earlier, one may define a number of different spectral sensitivity curves for the mesopic range, each describing the relation between brightness and radiant power for a particular adaptation level. This offers no problems for lighting design calculations and measurements, because all design methods and measuring apparatus are calibrated in units of photopic photometry (Akashi et al., 2005; Kostic et al., 2005). For the assessment of the effectiveness of lighting installations, this is, however, a factor of crucial importance. As is mentioned earlier there is a need for a general relationship between brightness and radiant power. This area is the subject of study of a recently established CIE Technical Committee CIE TC 1-58 (CIE, 2005). In the context of this TC, a European research consortium on Mesopic Optimisation of Visual Efficiency (MOVE) has been established. New research is under way (Varady et al., 2005; Eloholma et al., 2005; Akashi et al., 2005). It applies a performance-based approach for developing mesopic photometry, with emphasis on the problems of night-time driving (Halonen & Eloholma, 2005). The aim is to establish a data base for the assessment of the visual effectiveness of light sources with different spectral characteristics as lighting equipment in the mesopic range. In so far, it is an extension of the earlier work of Palmer and Opstelten that is mentioned earlier in this section (Halonen & Eloholma, 2005).

(h) Mesopic brightness impression

There is anecdotal evidence about the brightness impression of different light sources for road lighting. De Boer did find that low-pressure sodium street lighting seems to be brighter than that with high-pressure mercury lighting (De Boer, 1951, 1959, 1961, 1967; De Boer & Van Heemskerck Veeckens, 1955; De Boer et al., 1959; De Boer, ed., 1967). This result conflicts with other data, like those of Schreuder (Anon., 1987; Schreuder, 1989, 1989a). The reason might be that de Boer actually did measure visual performance, and not brightness impression. Performance is better with monochromatic light (Adrian, 1998).

In more recent times, in nearly all studies high-pressure sodium lamps have been compared to white light, usually to compact fluorescent lamps, sometimes to metal halide lamps. In most cases, white light was preferred. It is not always clear whether it was the brightness impression itself or rather the colour characteristics (Van Tilborg, 1991). Fotios & Cheal (2005) came to the same conclusion. It was found that white compact fluorescent lamps seem to be brighter than high-pressure sodium lamps. The trade-off against luminance would allow white light to be one S-class lower than SON. That would be a profit of about 30% (CEN, 2002). Contrary to this, Boyce and Bruno did not find any statistically significant difference between high-pressure sodium lamps and metal-halide lamps (Boyce & Bruno, 1999).

7.3   Visual performance

7.3.1   Human performance and visual performance

(a)   Ergonomic aspects

Ergonomics is the scientific discipline concerned with the understanding of interactions among humans and other elements in a system. Ergonomics contribute to the design and evaluation of tasks, jobs, products, environments, and systems in order to make them compatible with the needs, abilities, and limitations of people (Dirken, 2004). Ergonomics are in the last instance concerned with adapting tools and methods ('gadgets') to the needs of their human operators in order to maximize the efficiency (Daams, 1994). Ergonomics is also called human factors, although the two concepts are not identical (Boyce, 2003; Anon., 2007a; Forbes, 1972).

Usually, but not explicitly, one has a standard human operator in mind. Adaptations to the needs of individuals is not understood as an aspect of ergonomics, but concern for the needs of the elderly or the handicapped may be a subject of interest for the human factors approach (Schreuder, 1994, 2006; Wouters, 1991).

In the description of the IEA, there are three elements in the description of ergonomics that is given earlier:
1. the task;
2. the system;
3. the (human) operator.

In the past it was commonly thought that it was the human operator who had to adapt to the task and the system. This idea was ridiculed by Charlie Chaplin in his social protest movie 'Modern Times'. It was only during the Second World War that many began to understand that many 'human errors' could be avoided by adapting the system to the abilities and limitations of the operator. In the beginning, system was understood exclusively as the machine. Managerial systems were not included. This was ridiculed in the equally socially protesting novel of Joseph Heller 'Catch-22'. From this time dates the man-machine-system approach (Weir & McRuer, 1971). In more general terms, it became customary to put the human operator in a central position (Broadbent, 1958; Dirken, 2004; Foley, 1972; Wiener, 1954, 1966).

(b) *Task aspects*
A quite different approach is to look at the task. One may describe human performance as being derived from task performance, and task performance from visual performance (Boyce, 2003, figure 4.1, p. 124). Ergonomists will concentrate on human performance, trying to optimize it, whereas lighting engineers will concentrate on visual performance. In other words, the general task is for the human operator to perform, whereas the visual task is the subtask that involves the visual processes, and hence the lighting.

As an example, take motorized road traffic. The task aspects of car driving are discussed, in a different context, in sec. 1.3.4a. What follows here is in a way a repetition, in a way a clarification.

The general task is to reach the goal of the trip. Social aspects, trip generation, motivation, and economics play a role (Asmussen, 1972; Schreuder, 1974). From the point of view of the lighting engineer, the most important aspects is that the relevant decisions have been made at the moment the trip begins. Outdoor lighting plays hardly any role. The decisions are rational, and relate to the selection of the mode of transport, and to the route selection (Schreuder, 1991).

When the trip has started, and when the traffic participant / car driver is underway, a completely different set of decisions begin to be relevant. They refer to the operation of the vehicle, more in particular to maintain the correct lateral position within the driving lane, and the longitudinal control of position, or rather, of the speed. These two are the basic task elements. As in road traffic in the beginning of the 21st century, almost all information is gathered 'on line', and is visual in nature, it is visual performance that is crucial here.

The vehicle must be kept under control whilst negotiating curves, inclines, crossings, and road interchanges. An essential element is that all this must be done without being involved in accidents. The total is usually called the driving task analysis (Griep, 1971; Janssen, 1986).

Summarizing, ergonomics relate to the most efficient fulfillment of the overall task; the driving task analysis relates to the safe fulfillment of the driving task. And finally, the visual performance related to the visual aspects, more in particular to the input of visual information.

In sec. 11.1.3, when discussing the application of the visibility principles to road and street lighting, further details about the relation between the principles of outdoor lighting and the driving task are given. Amongst others, the visibility concept and the Small Target Visibility will be discussed.

7.3.2 Visual performance and Weber's Law

(a) *The concept of visual performance*

Although it may be difficult to define visual performance in a precise way, it is easy to describe it in general terms. As the word suggests, visual performance is the measure that describes and quantifies the degree in which the visual system can perform the requirements presented by the visual task. In this way we have given an operational definition of visual performance that agrees with the functional approach that is promoted throughout this book. It should be stressed that in this way both description and quantification are included in the definition.

There are several criteria that can be used to describe the visual performance. The most fundamental is the way the visual system responds to light. In operational terms, it relates to the increment threshold. The increment threshold is the minimal light 'power' is needed to evoke a sensation of 'light' for different values of the adaptation luminance. As usual, light power must be expressed in watts. Extensive research over several centuries has given us a clear picture about the nature of this relationship. This relationship is usually called Weber's Law (Krech et al., 1969, p. 113).

(b) *The law of Weber*

The relation between just noticeable differences (jnd's) to the size of the initial stimulus stands as a law of relativity in psychology: what must be added to produce a detectable difference is relative to what is already there. Weber proposed the principle that the ratio between the stimulus and the increment that must be added to it is invariant with intensity. And to a good approximation he was right. In vision and hearing the Weber fraction appears to be constant over more than 99,9 percent of the usable range of stimulus intensities (Stevens, 1951, p. 35-36). This is usually called Weber's Law after Weber, who postulated it around 1830. They were published in 1846 (Weber, 1846). According to Stevens (1951, p. 35) Weber's Law is usually written as:

The human observer; visual performance aspects

$$\Delta I = k \cdot I \qquad [7.3\text{-}1]$$

in which:
    I: the intensity of the stimulus;
    ΔI: the incremental threshold of the stimulus;
    k: a constant.

Small inconsistences in the law can be accounted for by the introduction of a correction factor (Miller, 1947). The equation [7.3-1] becomes, with the correction factor I_r:

$$\Delta I = k \cdot (I + I_r) \qquad [7.3\text{-}1a]$$

It is generally assumed that for practical lighting applications this correction may be disregarded.

An altogether matter is the range over which Weber's fraction is indeed constant. In practice the estimate of Stevens that is mentioned earlier seems to be too generous. This point will be discussed in a further part of this section. See Reeves (2004).

Fechner (1860) did try to extend Weber's Law to non-threshold increments. He postulated that all jnd's are subjectively equal. This would imply that a stimulus 20 jnd's above threshold should produce a sensation twice as big as a stimulus 10 jnd's above threshold. However, experiments did indicate that Fechner's postulate has to go. There is no Fechner's Law (Stevens, 1951, p. 36).

    (c) *Limits of Weber's Law*
        The light adaptation is discussed in more detail in the next part of this section. As the word suggests it is the way the visual system adjust its operation to increasing luminances in the surroundings. Small increments in the light stimulus are used to assess the light adaptation. The small increments are described in terms of the increment threshold. As has been explained in an earlier part of this section, Weber's Law is a model for the response to small increments.

In Figure 7.3.1, the relation between the increment threshold and the background is depicted over the full range of luminances to which the visual system may respond.

From Figure 7.3.1 it may be seen that Weber's Law seems to be valid throughout the middle range. The slope is 1,0 for cones and only slightly less that 1,0 for rods. At the upper end of the curve, in the saturation region, the increment threshold for rods increases steeply. Because bleaching protects the cones from saturation, it would seem that Weber's Law extends indefinitely for cones.

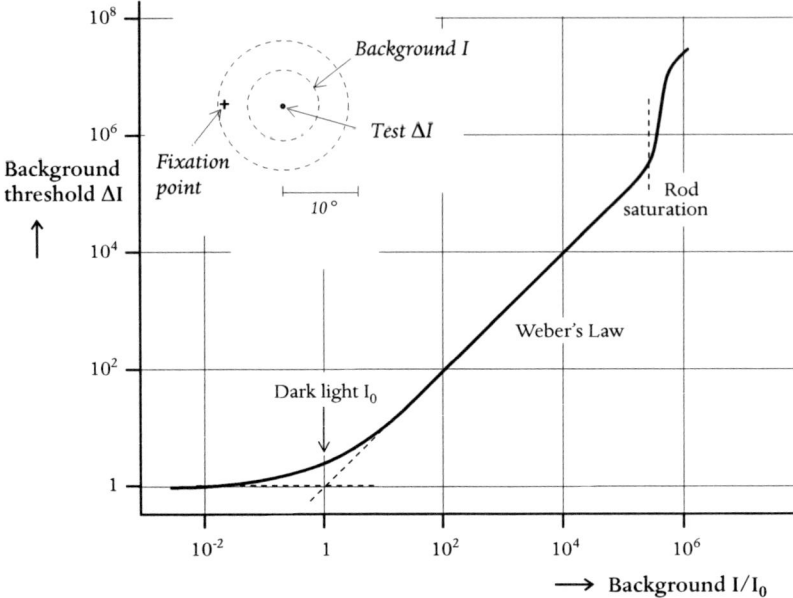

Figure 7.3.1 *The relation between the increment threshold and the background. (After Reeves, 2004, fig. 54.8, p. 859.)*

At the lower end the curve flattens out and becomes horizontal, thus marking the region where the background it too weak to have an effect. In Figure 7.3.1 it is explained that this value often is thought to be around 10^{-6} cd/m² (Hopkinson, 1969. See Table 7.4.1). This implies that it is difficult to allot quantitative values to the background luminance as given in Figure 7.3.1. It is probably best to consider that figure as a relative representation of the increment threshold. The graph being a relative representation allows to understand the remark of Reeves (2004, p. 859) that this curve is valid for cone vision as well as for rod vision, at least until the upper where saturation leads to a steep rise in the rod function.

Fechner did postulate a phenomenon that he called 'dark light'. This is supposed to be an activity of the visual system that does not provoke any sensation of light, but still influences the neural activities in such way that the increment threshold is increased. It may be this effect that is responsible for the curved section between the horizontal and the inclined parts of the graph in Figure 7.3.1, although it is not clear precisely how.

Reeves remarks that the dark light assumption is needed to explain dark adaptation, or, in his words, the recovery from light adaptation. The dark-light assumption is also used by Van Essen in postulating that blackness involves a visual activity even when there is no sensation of light (Van Essen, 1939; Schreuder, 2008).

(d) *Validity of Weber's Law*
Narisada and Schreuder have given an account about the history of Weber's Law. We refer for details to Narisada & Schreuder (2004, sec. 9.1.1). Also several of the possible pitfalls are explained. In the preceding part of this section, it is explained that Weber's Law is valid only for small differences in the luminance stimulus. The integration over greater differences as was proposed by Fechner does not agree with experiments.

Another pitfall is the fact that Weber's Law is an experimental relation. However, as it possible to represent it in graphical form, the graph may easily be mistaken as a representation of a mathematical function of the type $y = f(x)$, where x is called the independent variable or argument of the function, and y the dependent variable. This means that one may select an x-value and find, via the functional relationship, the corresponding y-value. In mathematics, when we have a 'real' function, one may in almost all practical cases exchange the dependent and the independent variable. This would lead to another function of the form of $x = f'(y)$ as well. Usually f and f' are closely related. Not always; when the function is not continuous, or when it contains singularities, there may be a problem in defining the relation between f and f'. There are other constraints as well. The concept of mathematical functions is described in Bronstein et al., (1997) and Daintith & Nelson (1989, p. 138-139).

A graphical representation of experimental, psycho-physical data may be regarded as a mathematical function. However, the data from the experiments themselves do not represent a mathematical function. They represent just the data, and nothing more. Exchanging the dependent and the independent variables is not justified for such data. If one would try to do so, one might end up with ridiculous results.

An example that did cause some excitement in the past was the recommendations given by Blackwell as regards the lighting of offices. Details are given in Narisada & Schreuder (2004, sec. 9.1.1).

Blackwell proposed to establish illuminating engineering on 'performance data' (Blackwell, 1959). Based on the RSC-curve that is discussed in sec. 7.4.3h, the required illuminance to be able to read an original type-written text. A field factor to incorporate non-ideal, day-to-day conditions was introduced. It was arbitrarily chosen as 15, a value well in agreement with modern measurements (Adrian, 1982, 1989; Bourdy et al., 1987; Schreuder, 1990, 1991a, 1997). The result was 75 lux. We may note here that this value agrees quite well with what is considered a the basic needs for adequate reading for young people (Schreuder, 2004). As was customary in that time, before photocopying was common practice, a number of carbon copies were made from the typewritten text. In each copy the contrast will decrease. When calculating the illuminance needed to read the consecutive copies, Blackwell unwittingly did exchange the dependent and independent variables in the graphical representation of the RSC-curve. Doing so, it was found that the first copy would required 1800 lux, the second 15 000 lux and the further copies would be unreadable because they would require more than 100 000 lux. Obviously, these results are absurd. The answer is, of course, in the RSC-curve itself: at high levels, the curve flattens and it approaches an asymptote that is parallel to the x-axis. This means that the visibility is independent on the light level – which is, of course, the case.

7.4 The primary visual functions

7.4.1 Introducing the primary visual functions

Because visual functions are interdependent, one has some freedom as to where to begin the discussion of them. It should be mentioned here that the distinction between the primary visual functions and the visual functions that are derived from them is not always clear.

The eye is an light-detecting organ, so it is natural to begin there. The physiology of the detection of light is discussed in sec. 6.1. The fundamental aspect of detection is the ability to judge whether the light is there or not; this is, in it simplest form, an important visual function with many practical implications like the detection of signalling lights, of stars etc. The visibility of point sources is discussed in detail in Narisada & Schreuder (2004, sec. 9.1.7).

As is mentioned already several times, the detection depends not only on the characteristics of the light, but also on the state of adaptation of the visual system. In sec. 7.4.2 the adaptation is explained in brief terms. Here we will begin the discussion of the primary visual functions with giving more details about the adaptation. The next function of general interest is the ability to detect differences of light, or, in other words, the discrimination of luminances. In practice, this means the ability to detect an object against its background. Visual objects can be detected only if they show a contrast to their immediate background that exceeds the threshold under the relevant conditions. For obvious reasons, the corresponding visual function is called the contrast sensitivity. In order to recognise the object, its shape or form must be discerned; this function is called the visual acuity. A number of other characteristics of the visual system that may be regarded as visual function are discussed in other parts of this book, and in more detail in Narisada & Schreuder (2004), and in Schreuder (2008).

7.4.2 Adaptation

As is mentioned earlier, the visual system works effectively over a vast range of light levels. The range is about 10^9 to one. In Table 7.4.1 an impression is given of what these different luminance values in outdoor scenes really mean.

Luminance log (cd/m^2)	Examples	Visual performance
5,0		dazzle
4,5	sun on snow	glare
4,0		
3,5	average daylight	
3,0		good perception
2,5		
2,0	interior lighting	
1,5		
1,0	dusk	
0,5	main road lighting	reading possible
0	candle light	reading difficult
-0,5	residential street lighting	
-1,0		colour perception impossible
-1.5	moonlight on snow	
-2,0		
-2,5	moonlit night	
-3,0		
-3,5	clear moonless night	only vague shapes
-4,0	starlight on snow	
-4,5	dark moonless night	
-5,0		only vague light impression
-5,5		
-6,0		perception threshold

Table 7.4.1 *Range of brightness. (After Narisada & Schreuder, 2004, Table 8.2.2. Based on Schreuder, 1998, Table 7.2.1, and Hopkinson, 1969.)*

At the lower end lies the absolute threshold of visual perception, the lowest luminance that still can be distinguished from 'absolute' black. This is about 10^{-6} cd/m² (Hopkinson, 1969). At the other extreme, the highest luminance where useful observations can be made near the condition of absolute glare, is about 10^4 cd/m² (Hentschel, 1994, p. 47; Schreuder, 1964). Light with much higher luminances may be observed, but the glare effects obstruct 'useful' observations, particularly because the after-images are predominant. In extreme cases, damage to the eye can follow (Augustin, 2007; Van Norren, 1995). It is interesting to note that, as is explained in sec. 6.3.5b, even at such high luminance values, where most people will use sun glasses, the amount of visual purple that is bleached out, is still small (Reeves, 2004, p. 853; Narisada & Schreuder, 2004, sec. 8.2.3a). This might suggest that the visual system would be capable of observing much higher luminances.

Although the visual system works effectively over a vast range of light levels, any specific moment in time, the span of adaptation is much smaller, ranging from some 3000:1 in nighttime surroundings to about 100:1 in bright daylight. This is depicted in Figure 7.4.1.

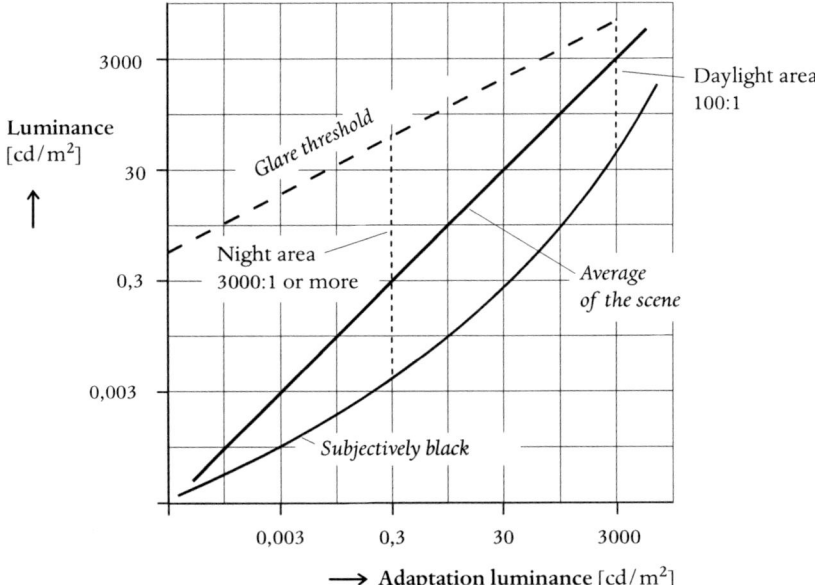

Figure 7.4.1 The scene and its limits of darkness and dazzle. (After Narisada & Schreuder, 2004, Figure 8.2.3. Based on Schreuder, 1998, Figure 7.2.5.)

These data agree with Noordzij et al. (1993). What is means is that only a small part of this range can be encoded by the nervous system at any time. Light with much higher

The human observer; visual performance aspects   253

luminances may be observed, but the glare effects obstruct 'useful' observations. Glare is discussed in Chapter 8. When the luminance is low, there is no sensation of light. Although the luminance may be quite far from zero, the scene looks black. This is sometimes called subjective black. The visual system must adjust its operating level to match the average ambient light level. This effect is called adaptation. Light adaptation means the adjustment of the operating level of the visual system to higher light levels; dark adaptation, to lower levels. The processes involved are explained in another part of this chapter, where the adaptation as a dynamic effect is discussed.

7.4.3   Luminance discrimination

(a)   The contrast

The ability to see small differences in the luminance seems to be the most fundamental visual capability, in the way that most other visual functions are either a corollary of this ability or can be derived from it. This ability is usually called luminance discrimination. In practice, the absolute difference between the sections of the field of view is not always relevant; this is much more the case for the relative difference or the contrast. The sensitivity to contrast is the measure for the smallest luminance difference which can still be perceived. It is therefore a threshold value. The most common way to designate this threshold is the contrast sensitivity. In illuminating engineering it is customary to define the contrast as:

$$C = \frac{L_0 - L_b}{L_b} \qquad [7.4\text{-}1]$$

in which:
  C: The contrast;
  L_0: The luminance of the object;
  L_b: The luminance of the background.

This definition is clearly inspired by Weber's Law that is discussed in an earlier part of this section. According to the definition, the contrast can have positive and negative values, running from -1 for $L_0 = 0$ to $+\infty$ for $L_b = 0$. The equation is asymmetric because both L_0 and L_b appear in the nominator and only L_b in the denominator. Other definitions of the contrast avoid this, but they are not commonly used in illuminating engineering, although they are common in establishing a diagnosis in case of ophthalmological disorders. An example is given in Schreuder (1990; 1991a; 1993). Still another way is given in a further part of this section (Fiorentini, 2004).

A contrast of zero corresponds with a situation where $L_0 = L_b$. The object is invisible. It was found that perception, at least for small contrasts, does not depend on the sign of the contrast according to [7.4-1]. An object can be seen equally well when it is light against a dark background (positive contrast) as when it is dark against a light background (negative contrast; Adrian, 1961; Blackwell, 1946; Schreuder, 1964).

For larger contrasts, this invariance does not apply because of the asymmetry of the way the contrast is defined.

(b) The minimum detectable contrast

In nature contrasts tend to be low. In most common scenes the retinal image is represented as peaks and troughs in intensity that differ from the local mean by about 20%. The contrast is defined in the common way as $\delta I/I$ with I some measure of stimulus intensity. Many fine structures exhibit far lower contrasts of only a few percent. This range is common in nature (Srinivasan et al., 1982). The visual threshold for a small stimulus, such as spanning a single ganglion cell dendritic field, corresponds to a contrast of about 3% (Watson et al., 1983; Dhingra et al., 2003).

These values do agree quite well with the practice of outdoor lighting. As an example, a contrast of 0,054 is used as the minimum in the practice of traffic tunnel lighting. This value is based on the minimum of 0,012 that is measured under laboratory conditions in combination with a practical field factor of 4,5: 0,054 = 4,5 · 0,012 (Schreuder, 1989b, 1990, 1991a; Alferdinck, 2000). It should be noted here that it is the main objective of illuminating engineers and of the designers of lighting installations to ensure that the contrasts that are met in real outdoor situations exceed, and preferably considerably exceed, these threshold values.

To create an optical image at low contrast, many photons are needed. Because light is quantified, a small difference from the mean of, say, 1% implies that the mean itself must contain at least 100 photons. Photons arrive at the retina random in time, and follow a Poisson distribution. The minimum detectable contrast is called δn.

In order to be detectable, δn must differ from the mean by at least one standard deviation. In a Poisson distribution, the standard deviation equals $\sqrt{n}$ (Moroney, 1990). This means that at least 10 000 photons are needed, because $\delta n/n > \sqrt{n}/n = 100/10\,000 = 1\%$. The root-mean-square function $\sqrt{n}$ is called the photon noise.

(c) Neural aspects of achromatic contrast phenomena

According to Fiorentini (2004), the objective contrast between two areas in the field of view with the luminances L_1 and L_2 is defined as the ratio between the luminance difference and the mean luminance:

$$K = \frac{L_1 - L_2}{(L_1 + L_2)/2} \qquad [7.4\text{-}2]$$

This differs from the definition that is normally used in illuminating engineering that is quoted earlier in this section, viz.:

$$C = \frac{L_0 - L_b}{L_b} \qquad [7.4\text{-}1]$$

with:
- L_o: the luminance of the object;
- L_b: the luminance of the background.

In practical illuminating engineering emphasis is usually on the contrast at the border between two adjacent areas in the field of view. This is especially relevant for the detection of objects on the road under public or vehicle lighting. This is called border contrast. Another important effects is area contrast (Fiorentini, 2004, p. 885). This effect is especially important for understanding visual illusions (Narisada & Schreuder, 2004, sec. 10.2.2; Schreuder, 2008).

A well-known area contrast phenomenon is that a grey square looks lighter when surrounded by a dark area than when it is surrounded by a light area. The area effect is that it looks uniformly light or dark. An example is given in Fiorentini (2004, fig. 56.1, p. 882). An example of border contrast effect is the existence of Mach bands at the border. They were described for the first time in 1865 by Ernst Mach (Fiorentini, 2004, p. 885).

Most contrast visibility effects can be satisfactorily described in neurophysiological terms. There is evidence that the ON and OFF systems that are discussed in sec. 6.2 are independently responsible for the supra-threshold perception of light increments and decrements. Further, many of the border contrast effects, such as Mach bands and some other visual illusions, can be described satisfactorily by means of two receptive fields of neurons with antagonistic excitary and inhibitory regions. This would imply that most contrast induced phenomena are of retinal origin, although some influence of actions in the visual cortex seem be involved as well. For other aspects of achromatic visual phenomena, such as lightness constancy, area contrast effects, and contrast induction, long range effects are involved. These long range effects are thought to involve lateral interactions by neurons in the visual cortex (Fiorentini, 2004, p. 890)

(d) *The laws of Ricco and Piper*

Luminance discrimination relates to the ability to discern two adjacent sections of the field of view if they differ in luminance. Here, it is assumed that the two areas as well as the adapting fields all have the same colour. In other words, we speak only of achromatic contrast phenomena (Fiorentini, 2004). If this is not the case, we have to do with the colour contrasts that are discussed in sec. 9.1. As has been indicated earlier, the related visual function is called the contrast sensitivity.

There is a direct relationship between the luminance to be found in the world outside the eye and the retinal luminance to be found within the eye. As is explained in an earlier part of this section, the relation depends on the state of adaption. The implication is that two adjoining sections of the field of view that have the same luminance will also look equally bright. It is customary to consider one of the areas as the test object and the other as the background.

The absolute threshold involves the concept that a constant number of retinal elements must be stimulated in order for a threshold response to occur (Bartley, 1951, p. 952). A general description of the absolute threshold is:

$$(A - n_t)^k I = C \qquad [7.4\text{-}3]$$

in which:
    A: the area of the test object;
    n_t: the threshold number of retinal elements;
    C: a constant;
    k: a constant.

In its simplest form the relation [7.4-3] is:

$$A \cdot I = C \qquad [7.4\text{-}4]$$

This is called Ricco's Law (Bartley, 1951, p. 953; Ricco, 1875). Ricco's Law pertains to the fovea. With $k = 0{,}5$, [7.4-3] becomes:

$$I \cdot \sqrt{A} = C \qquad [7.4\text{-}5]$$

This is called Piper's Law, that is valid for the periphery (Bartley, 1951, p. 953). Piper's Law is discussed in the next part of this section.

For small test objects, the two sections of the field of view look equally bright as long as the product of the diameter and the luminance is constant. This is called, as mentioned earlier, Ricco's Law. It is written in different forms. In its general form it is written as

$$\log C = k - \log A \qquad [7.4\text{-}6]$$

in which:
- C: the contrast, defined in the usual way;
- A: the area of the text object,
- K: a constant.

(Anon., 2007). Ricco's Law can also be written as:

$$L = \frac{r^2}{k_1} \qquad [7.4\text{-}7]$$

in which:
- L: the luminance of the test object;
- r: its radius;
- k_1: a constant.

(Moon, 1961, p. 404).

The upper limit of the validity of Ricco's Law is usually called the critical size. It should be noted that there is quite a difference of opinion about the critical size. According to Moon (1961, p. 404). the critical size is about 6° in diameter. Boyce mentions values between 0,5° and 5° for the critical size for the fovea, and of about 2° in the periphery at 35° of eccentricity (Boyce, 2003, p. 71; Hallet, 1963).

For the assessment of the design parameters for tunnel entrance lighting, the upper limit of Ricco's Law is stated as being only 10 minutes of arc (NSVV, 2003, p. 202). This value agrees to Bartley (1951, p. 953). For angles between 10 minutes of arc and 2°, an approximative relation is given:

$$K = 1 + 300/\varepsilon^2 \qquad [7.4\text{-}8]$$

(NSVV, 2003, p. 2003). This approximation is based on data from Phillips (1986) and Alferdinck (2000).

In a further part of this section, Piper's Law will be discussed. Sometimes, Ricco's Law is called the Law of Ricco-Piper (Hentschel, ed., 2002, p. 48). A value of 1° is quoted for the critical size (Hentschel, ed., 2002, p. 53)

It is not clear what is the reason for the large difference in the quoted values of the upper limit of Ricco's Law. It seems, however, that part of the differences arise from a certain confusions between the laws of Ricco and Piper. The one is pertains to the fovea; the other to the periphery. According to Bartley, Piper's Law is valid in the range of 2° to 7°. However, Bartley also mentions that the critical size of 10 minutes of arc for Piper's Law relates not only to the fovea but also to the periphery (Bartley, 1951, p. 953).

Above, we explained that Piper's Law is, just as Ricco's Law, a special case of the general description of the absolute threshold as given in [7.4-5] (Bartley, 1951, p. 952). It runs:

$$I \cdot \sqrt{A} = C \qquad [7.4\text{-}5]$$

According to Bartley, Piper's Law is valid in the range of 2° to 7°, and Ricco's Law under 10 minutes of arc.

(e) *The relation between size and threshold contrast*

It follows from the preceding part of this section that Ricco's Law, Piper's Law and Weber's Law all give information about the relation between the size of the test object and the minimum contrast needed for detection. It is must therefore be possible to bring these data together. An attempt to do this is depicted in Figure 7.4.2.

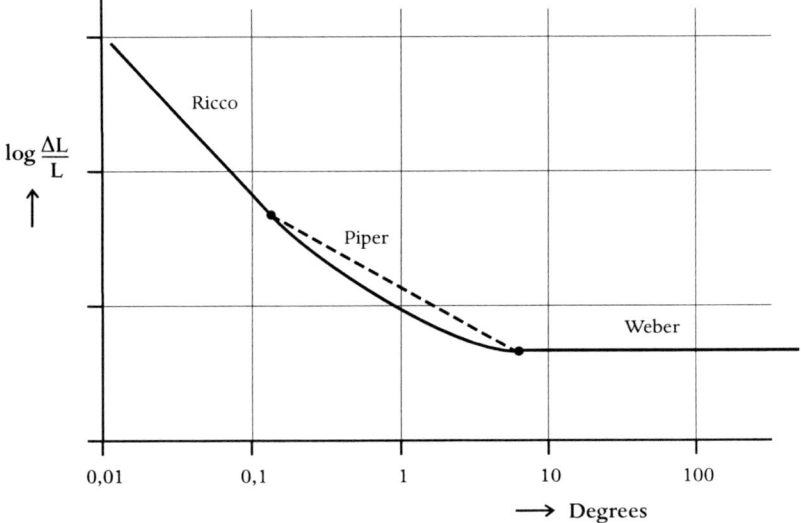

Figure 7.4.2 *A schematic picture of the relation between the size of the object and the threshold contrast.*

It is assumed that the adaptation is stable, and that the time of observation is unlimited. Both axes have logarithmic scales. The y-axis is relative. The x-axis contains actual values. The graph has three clear sections. For test objects smaller than about 10 minutes of arc, Ricco's Law is valid. In this graph, it means a straight line section under 45° down. For test objects over about 6°, Weber's Law is valid, meaning a horizontal line section in the graph. Between the two, Piper's Law is valid. This is depicted by a bend. This might be a better way to connect the two straight parts, rather than to use the relation [7.4-4] which would mean another straight line section.

It must be stressed that this graph is a schematic representation only. It is not verified by experiments. Also it is assumed that the distinction between the responses of cones and rods may be disregarded. Also, the detection of point sources like e.g. stars is omitted from the graph. The detection of point sources follows rather different rules. These rules are discussed in detail in Narisada & Schreuder (2004, sec. 9.1.7). It should not come as a surprise that this curve looks like the RSC-curve that is discussed in a further part of this section. The two are supposed to represent the same data.

(f) Lights with periodic brightness

It is mentioned in an earlier part of this section that two adjoining sections of the field of view that have the same luminance will also look equally bright. This is not always the case, however. Several important exceptions will be briefly discussed here. The first, and the one with the greatest importance for practical illuminating engineering relate to the Laws of Ricco and Piper that have been discussed in the preceding part of this section.

Apart from the Laws of Ricco and Piper, a second exception relates to the visibility of weak flashes of a duration shorter that 0,2 seconds (Moon, 1961, p. 404). It was found by Bloch in 1885; therefore the relation is called Bloch's Law:

$$E_{ret} \cdot \delta t = k_2 \qquad [7.4\text{-}9]$$

in which:
    E_{ret}: The retinal illuminance caused by the flash;
    δt: The duration of the flash;
    k_2: A constant.

The third exception is related to sections of the field of view where the luminance varies periodically in time. The related phenomena are explained in sec. 5.2.2g, where flicker effects are discussed in relation to flicker photometry, and where the laws of Blondel and Rey, and Talbot are explained.

(g) The sensitivity to changes in the contrast

The sensitivity to contrast is the measure for the smallest luminance difference which can still be perceived. In its most general form, contrast sensitivity can be written as:

$$CS = \frac{\delta L}{L} \qquad [7.4\text{-}1a]$$

which is, as mentioned earlier, clearly inspired by Weber's Law.

One often use its reciprocal value, the sensitivity to luminance differences. Often, the German term Unterschiedsempfindlichkeit is used in the international literature. It is written as:

$$LD = \frac{L}{\delta L} \qquad [7.4\text{-}1b]$$

Both CS and LD do not only depend on the adaptation level but also on the size of the test object, the time of exposure, the colours of objects and backgrounds and several other factors. Some are discussed in other parts of this section. Further details are given in Hentschel (1994, sec. 3.2) and in Narisada & Schreuder (2004, sec. 9.1.2).

(h) *The RSC-curve*

During the Second World War, a large number of tests were made in the USA. Often they are referred to as the Tiffany Studies. Historical notes about these studies are given in Blackwell (1946) and in Middleton (1952, p. 87-90). In his epoch-making studies, Blackwell did investigate the influence of many different parameters. A sample of Blackwell's data are depicted in Figure 7.4.3.

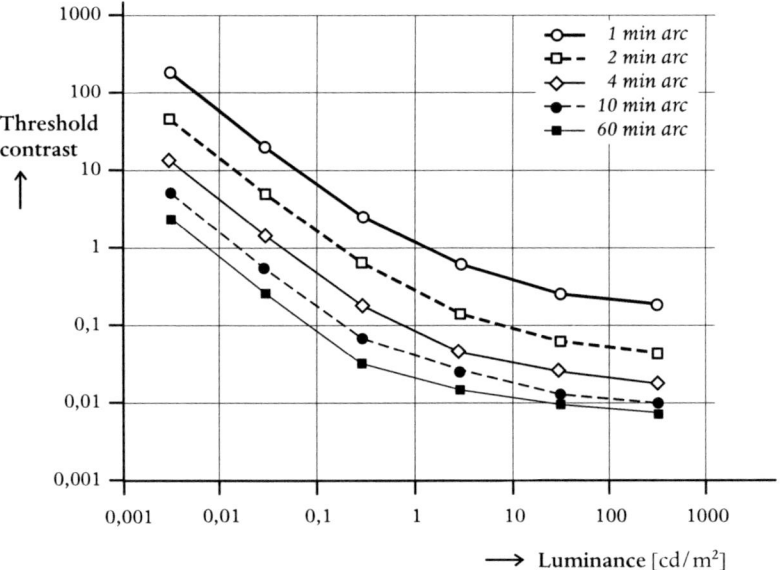

Figure 7.4.3 *A sample of the Blackwell data. (After Boyce, 2003, figure 2.15, p. 72. Based on data from Blackwell, 1959.)*

These studies are, in combination with other studies, the basis for the CIE Relative Contrast Sensitivity Reference Curve or RSC-curve (CIE, 1981). The Standard RSC-

The human observer; visual performance aspects

curve is depicted in Figure 7.4.4. Actually, the Standard RSC-curve is only one of a whole family of curves.

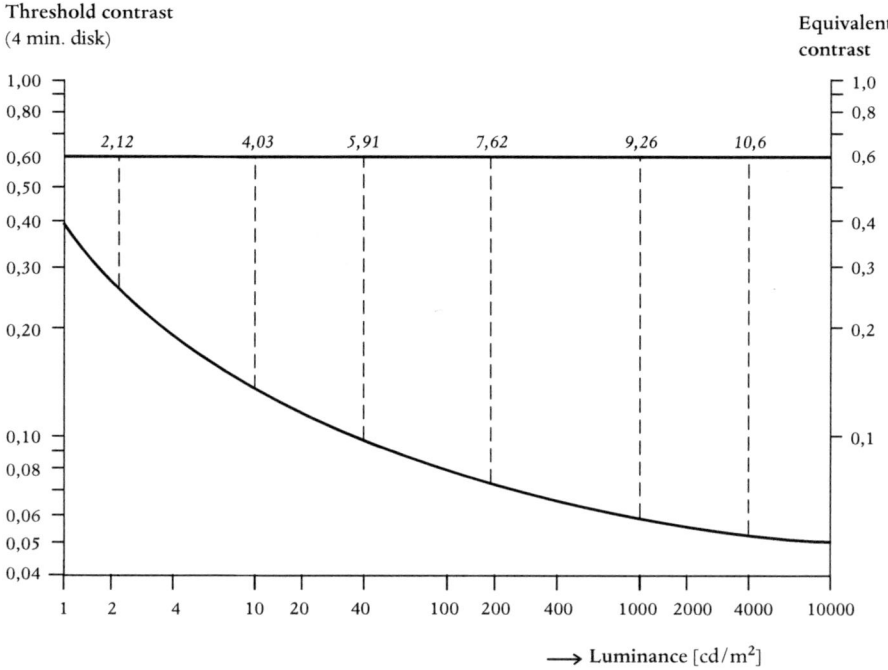

Figure 7.4.4 The relation between contrast sensitivity and adaptation: The CIE RSC-curve. (After Narisada & Schreuder, 2004. fig. 9.1.4.)

It should be noted that the RSC-curve covers the luminance area from 1 to 10 000 cd/m^2, whereas the Blackwell-data cover a range from $3 \cdot 10^{-4}$ to 300 cd/m^2, as is depicted in Figure 7.2.4 where the Adrian-representation of the Blackwell-data is given.

The RSC-curve is generally accepted as the standard for the contrast sensitivity. It is clear from the diagram that the contrast threshold is quite high for very low luminance levels. It decreases considerably with increasing luminances; the decrease seems to go on until the end of the range that is depicted in Figure 7.4.4. It should be pointed out that this decrease conflicts with Weber's Law. That law would require a constant value of $\delta L/L$. The threshold seems to level off for adaptation levels that correspond with average daylight conditions. Between 200 and 2000 cd/m^2, the contrast sensitivity is almost constant, in agreement with Weber's Law. When stating this, it should be realised that the curve is extrapolated towards its end. This is a major flaw in the RSC-curve.

Experiments that are extended to luminance levels up to 10 000 cd/m² or more, suggest that the threshold goes up again. In other words, the visibility level seems to decrease again (Schreuder, 1964; König & Brodhun 1889). This problem, as well as a great number of related problems, is discussed in detail in Narisada & Schreuder (2004, sec. 9.1.3c).

In spite of the discrepancies at very high adaptation levels, it seems to be justified to use the CIE RSC-curve, as depicted in Figure 7.4.4, as a standard for the contrast sensitivity at least for levels up to about 2000 cd/m². However, it should be noted that higher luminance values up to 10 000 cd/m² or more, are quite common in real life. This is important for the determination of the lighting requirements at the entrance of road traffic tunnels (CEN, 2003; CIE, 1990a; NSVV, 2003; Schreuder, 1964, 1998, Chapter 15).

The value of 10 000 cd/m² is easily exceeded in sunshine, particularly summer sunshine on the sand of beaches, and even more in sunshine on snow. The illuminance of the sun reaches easily 110 000 lux; as snow is an almost perfect diffuser, its luminance will be well over 35 000 cd/m² (110 000/π).

In the CIE 1981 publication, the concept of visibility level (or VL) was introduced. It was expected that this concept would replace the luminance concept as a metric for visibility in different conditions. In practice, however, its use was limited to road lighting (Adrian, 1993, 1995; Gallagher et al., 1975; Janoff, 1993; Narisada, 2007; Narisada & Karasawa, 2001; Narisada & Schreuder, 2004, secs. 9.1.3; 11.8.3; Narisada et al., 2003; Van Bommel & De Boer, 1980; TRB, 2007). It must be noted that there are doubts whether the Visibility Level is the most appropriate measure of visibility on road lighting.

In sec. 7.2.4a, the transition between photopic and scotopic vision is discussed. It is noted that as regards the contrast sensitivity several publications allow for a 'breach' between the two, whereas physiological considerations would prefer a gradual transition. The transition is, as is explained in sec. 7.2.4, the mesopic range.

7.4.4 The visual acuity

(a) Visus

If we would consider the contrast sensitivity as the most important visual function, the visual acuity would be the second. The most common description of the visual acuity is the reciprocal value of the smallest object or the smallest detail which still can be perceived. In optical instruments, this is usually designated as the resolving power. In medicine, often the term 'visus' is used. Visual acuity is usually defined as the angular measure of the smallest object that can be discerned. For healthy adults with good eye-sight or with good glasses, the smallest object usually is about half a minute of arc (Hubel, 1990, p. 46). The limit of one minute of arc is often indicated as a vision of 100/100 or as 100%. This applies to the fovea.

It should be mentioned that in order to acquire any degree at all of visual acuity, the eye must be able to move. Eye movements are essential for most visual processes. This is related to a special characteristic of the retina related to inhibitory processes (Steinman, 2004, p. 1339).

(b) *Measurement of the visual acuity*

In sec. 6.3.5d, when discussing the distribution of photoreceptors over the retina, it is explained that near the fovea the cone density is maximal, and that is decreases rapidly towards the periphery. This is depicted in Figure 6.3.11, where the relation is given between the visual acuity and the retinal location. See also Narisada & Schreuder (2004, Figure 8.2.4) and Bartley (1951).

The visual acuity depends heavily on the measuring method. One way is to measure visual acuity by means of parallel beams. An example of the differences of results one may encounter is depicted in Figure 7.4.5.

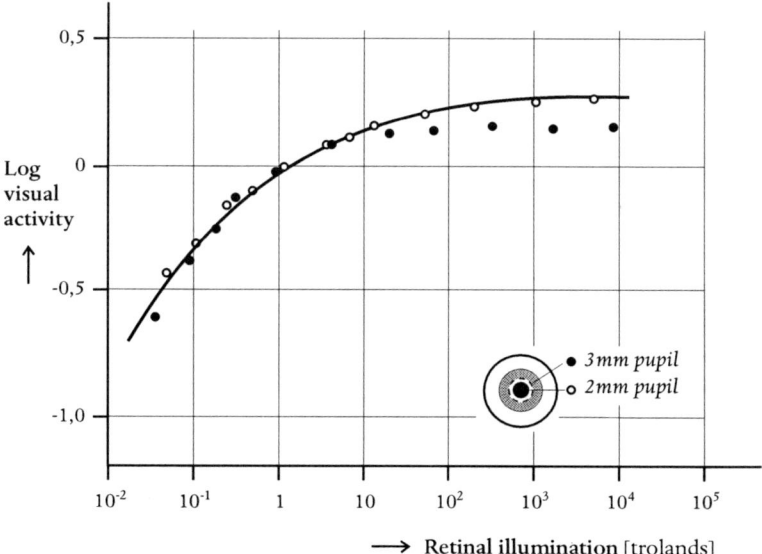

Figure 7.4.5 Visual acuity as measured by parallel bars in two ways: light bars on a dark background; and dark bars on a light background. (After Bartley, 1951, fig. 43, p. 961.)

In Narisada & Schreuder (2004, sec. 9.1.4) it is discussed in detail that there are many other ways to assess visual acuity. As a matter of fact, a different value may be found for almost every different test object. This implies that form perception is an essential element of visual acuity. This might be the reason that, as is explained in an earlier part of this section, there is some difference of opinion as to what is the upper limit of the validity of Ricco's Law, the reason being that the detection of objects is not only a matter of size, but also of shape and luminance contrast.

The most common test object is the Landolt Ring. The main reason is that it is very precisely defined. They look like the capital letter C; its gap is the actual test object detail to be detected. The shape is carefully defined (Schreuder, 1998, p. 101; De Boer & Fischer, 1981, p. 14-16).

The use of Landolt Rings as test objects is one of the reasons that the measurements made by Lythgoe (1932) have attained the status of a 'classic' in psycho-physics. They cover a range from 0,03 to 3000 cd/m² of adaptation luminance. The measurements were made in white light with a natural pupil and unlimited exposure time. The results are depicted in Figure 7.4.6.

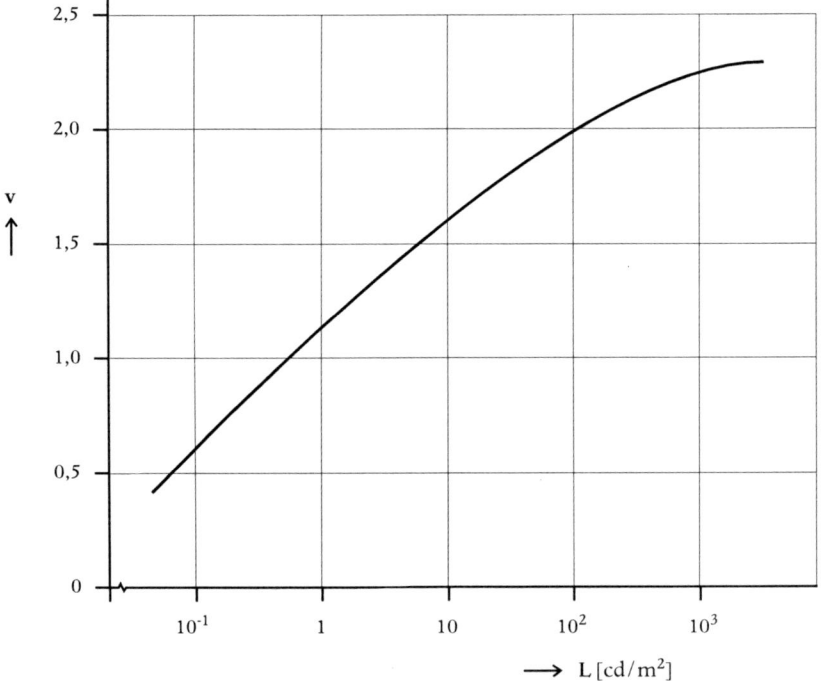

Figure 7.4.6 Relation between visual acuity v and luminance. (After Narisada & Schreuder, 2004, fig. 9.1.9. Based on Schreuder, 1976, figure 4, as represented by Walsh, 1958.)

We might mention here a special case, viz. the visual acuity for point sources. Point sources form an important class of test objects. They are for obvious reasons of special interest for astronomical observation. The detection and observation of point sources is discussed in detail in Narisada & Schreuder (2004, sec. 9.1.4c; sec. 9.1.7).

As has been explained by Schreuder (1976, sec. 3.3.2; 1988), the visual acuity is in reality not a good measure for visual performance. In spite of that, driving tests, that are very lenient in most countries as regards most physical or psychological factors, are very strict as regards visual performance (Vos & Legein, 1989). There are a number of reasons for this misconception:
1. Most real-life visual tasks are quite different from the detection of very small, colorless, high-contrast, stationary objects that are used in visus tests;
2. The visus is determined for the fovea, whereas a large part of observations in real life is done in the periphery;
3. Usually, the visual acuity is determined in white light, whereas the colour of the light has a considerable influence. This influence is discussed later in this section;
4. Most methods of assessing the visus are not adequate and cannot withstand scientific scrutiny as is explained by Vos (1969). More in particular, when normal letters are used, form recognition will play a role. Form recognition is discussed in an earlier part of this section.

(c) *Visual acuity in relation to colour*
Finally, the visual acuity depends significantly on the colour of the light. In the past, the colour of the light was considered an important aspect of road lighting and of vehicle lighting. As regards road lighting, this interest did stem from the time when only two 'families' of light sources were available for the lighting of traffic routes:
1. High-pressure mercury lamps (colour corrected or not);
2. Low-pressure sodium lamps.

In sec. 2.4.3c, the physical characteristics as well as the light-technical and the colour characteristics are discussed in detail.

The use of low-pressure sodium lamps was promoted for traffic routes, where the fact that their light is almost monochromatic is not considered as a disadvantage. Contrary to this, the higher visual acuity that is systematically reported was considered as an advantage, and so was the impression from studies that the light penetrated better through fog and haze, and caused less discomfort glare. Discomfort glare is discussed in Narisada & Schreuder (2004, sec. 9.2.3) and De Boer, ed. (1967). The different aspects of the colour of the light for road lighting are discussed in great detail in De Boer (1967); Schreuder (1967); Van Bommel & De Boer (1980). The aspects of the colour of the light for vehicle headlighting are discussed in great detail in Schreuder (1976).

7.5 Conclusions

In this chapter, visual performance aspects are discussed, concentrating on the image-forming aspects of the activities of the visual sense. A central issue is the duplicity theorem, stating that the main optical elements are the cones and the rods, each having their own characteristics. Many anatomical and neurological aspects have been discussed in the preceding chapter. It is concluded that the spectral sensitivity curves of photopic, mesopic, and scotopic vision are very similar in shape, but are shifted over a quite considerable distance in the wavelength dimension. The same is true for other spectral sensitivity curves like the ones that govern the pupil reaction or the biological clock.

Using a 2° field or a 10° field gives different results. All photometry is made with a 2° field and for photopic vision only. Mesopic and scotopic metrics are useful for research and for design but have no function in standards. Most outdoor lighting applications fall in the mesopic range, more in particular in the high-mesopic range.

Ergonomic aspects and task aspects are essential in human and visual performance. The primary visual functions are adaptation, luminance discrimination and visual acuity, of which the luminance discrimination or contrast sensitivity is the most important for the practice of outdoor lighting, its main tool being Relative Contrast Sensitivity Reference Curve, or RSC-curve.

References

Adrian, W. (1961). Der Einfluss störender Lichter auf die extrafoveale Wahrnehmung des menschligen Auges (The influence of disturbing light sources on the extrafoveal observation of the human eye). Lichttechnik. 13 (1961) 450-454; 508-511; 558-562.

Adrian, W. (1969). Die Unterschiedsempfindlichkeit des Auges und die Möglichkeit ihrer Berechnung (The contrast sensitivity of the eye and the possibilities for its calculation). Lichttechnik. 21(1969) no. 1, p. 2A-7A.

Adrian, W. (1982). Investigations on the required luminance in tunnel entrances. Lighting Res. Technol. 14 (1982) 151.

Adrian, W. (1989). A method for the design of tunnel entrance lighting. University of Waterloo, School of Optometry. Waterloo, Ontario, Canada, 1989.

Adrian, W. (1993). The physiological basis of the visibility concept. In: 'Visibility and luminance in roadway lighting'. LRI, 1993, p. 17-30.

Adrian, W. (1995). The visibility concept and its metric. In: Anon., 1995.

Adrian, W. (1998). The influence of spectral power distribution for equal visual performance in roadway lighting levels. In LRI, 1998.

Akashi, Y.; Morante, P. & Rea, M.S. (2005). An energy-efficient street lighting demonstration based upon the unified system of photometry. p. 38-43. In: CIE, 2005.

Alferdinck, J.W.A.M. (2000). Colour contrast in tunnels. Report No. TM-00-C009. Soesterberg, TNO/TM, 2000.

Anon. (1972). Abstract Guide XXth International Congress of Psychology, 13-18 Aug. 1972. Tokyo, Science council of Japan, 1972.

Anon. (1987). Jaarverslag 1986 (Annual report 1986). p. 6. Schiedam, Bewonersvereniging Schiedam-Zuid, 1987.

Anon. (1995). PAL: Progress in automobile lighting. Technical University Darmstadt, September 26/27, 1995. Darmstadt, Technical University, 1995.

Anon. (1995a). Symposium Openbare Verlichting, 22 februari 1995 (Symposium Public Lighting, 22 February 1995). Utrecht.

Anon. (2007). Ricco's Law. Internet, Wikipedia, 8 May 2007.

Anon. (2007a). What is ergonomics? International Ergonomic Association IEA Website. Internet, 9 May 2007.

Asmussen, E. (1972). Transportation research in general and travellers decision making in particular as a tool for transportation management. In: OECD (1972).

Augustin, A. (2007). Augenheilkunde. 3., komplett überarbeitete und erweiterte Auflage (Ophthalmology. 3rd completely revised and extended edition). Berlin, Springer, 2007.

Baer, R. (1990). Beleuchtungstechnik; Grundlagen. Berlin, VEB Verlag Technik, 1990.

Baer, R., ed. (2006). Beleuchtungstechnik; Grundlagen. 3., vollständig überarteite Auflage (Essentials of illuminating engineering, 3rd., completely new edition). Berlin, Huss-Media, GmbH, 2006.

Bean, A.R. & Simons, R.H. (1968). Lighting fittings performance and design. Oxford. Pergamon Press, 1968.

Blackwell, H.R. (1946). Contrast threshold of the human eye. Journ. Opt. Soc. Amer., 36 (1946) 624.

Bourdy, C.; Chiron, A.; Cottin, C. & Monot, A. (1987). Visibility at a tunnel entrance: Effect of temporal adaptation. Lighting Res. Technol. 19 (1987) 35-44.

Boyce, P.R. (2003). Human factors in lighting. 2nd edition. London. Taylor & Francis, 2003.

Boyce, P.R. & Bruno, L.D. (1999). An evaluation of high pressure sodium and metal halide light sources for parking lot lighting. Journ. Illum. Engng. Soc. 28 (1999) 16-32.

Boynton, R.M. (1979). Human color vision. New York, Holt, Rinehart and Winston, 1979.

Boynton, R.M & Kaiser, P.M. (1978). Temporal analog of the minimally-distinct border. Vision Research, 18 (1978) 111-113.

Broadbent, D.E. (1958). Perception and Communication. London, Pergamon Press, 1958.

Bronstein, I.N.; Semendjajew, K.A.; Musiol, G. & Mühlig, H. (1997). Taschenbuch der Mathematik. 3. Auflage (Manual of mathematics. 3rd edition). Frankfurt am Main, Verlag Harri Deutsch, 1997.

Buck, S.L. (2004). Rod-cone interactions in human vision. Chapter 55 p. 863-878. In: Chalupa & Werner, eds., 2004.

Budding, E. (1993). An introduction to astronomical photometry. Cambridge University Press, 1993.

Cakir, A. & Krochmann, J. (1971). A note on the equivalent luminance and spectral luminous efficiency of the human eye within the mesopic range. Lighting Res. Technol. 3 (1971) 152-157.

CEN (2002). Road lighting. European Standard. EN 13201-1..4. Brussels, Central Sectretariat CEN, 2002 (year estimated).

CEN (2003). Lighting applications – Tunnel lighting. CEN Report CR 14380. Brussels, CEN, 2003.

Chalupa, L.M. & Werner, J.S., eds. (2004). The visual neurosciences (Two volumes). Cambridge (Mass). MIT Press, 2004.

CIE (1924). Proceedings of the Commission Internationale de l'Eclairage, Geneva, 1924.

CIE (1959). Proceedings of the CIE Session 1959 in Brussels. Publications 4...7. Paris, CIE, 1959.

CIE (1971). Proceedings of the CIE Session 1971 in Barcelona (Vol. A, B, C). Publication No. 21. Paris, CIE, 1971.

CIE (1971a). Colorimetry. Publication No. 15. Paris, CIE, 1971.

CIE (1981). An analytical model for describing the influence of lighting parameters upon visual performance. Summary and application guidelines (two volumes). Publication No. 19/21 and 19/22. Paris, CIE, 1981.

CIE (1986). Colorimetry. Publication No. 15-2. Vienna, CIE, 1986.
CIE (1988). Spectral luminous efficiency functions based upon brightness matching for monochromatic point sources in 2° and 10° fields. Publication No. 75. Paris, CIE, 1988.
CIE (1989). Mesopic photometry: history, special problems and practical solutions. Publication No. 81. Paris, CIE, 1989.
CIE (1990). 2° Spectral luminous efficiency function for photopic vision. Publication No. 86. Vienna, CIE, 1990.
CIE (1990a). Guide for the lighting of road tunnels and underpasses. Publication No. 26/2. Vienna, CIE, 1990.
CIE (1992). Guide for the lighting of urban areas. Publication No. 92. Paris, CIE, 1992.
CIE (1995). Recommendations for the lighting of roads for motor and pedestrian traffic. Technical Report. Publication No. 115-1995. Vienna, CIE, 1995.
CIE (2003). Proceedings of the San Diego Session of CIE. 2 volumes. Publication no. 152. Vienna, CIE, 2003.
CIE (2005). Vision and lighting in mesopic conditions. Proceedings of the CIE Symposium '05. Leon, Spain, 21 May 2005. CIE X028. Vienna, CIE, 2005.
CIE (2005a). CIE 10 degree photopic photometric observer. Publication No. 165. Vienna, CIE, 2005.
CIE (2006). Colorimetry – Part 1: CIE Standard Colorimetric Observers. CIE Standard S 014-1/E:2006. Vienna, CIE, 2006.
CIE (2006a). TC 1-36 Fundamental chromaticity diagram with physiological axes. Part 1. Draft, 01-01-2006. Vienna, CIE, 2006
Daams, B.J. (1994). Human force exertion in user-product interaction. Physical Ergonomics Series, no. 2. Delft, Delft University Press, 1994.
Daintith, J. & Nelson, R.D. (1989). The Penguin Dictionary of mathematics. London. Penguin Books, 1989.
De Boer, J.B. (1951). Fundamental experiments of visibility and admissible glare in road lighting. Stockholm, CIE, 1951.
De Boer, J.B. (1959). La couleur de la lumière dans l'éclairage pour la circulation routière (The colour of the light in road traffic lighting). Lux. (1959). March – June.
De Boer, J.B. (1961). The application of sodium lamps to public lighting. Illum. Engng. 56 (1961) 293-312.
De Boer, J.B. (1967). Visual perception in road traffic and the field of vision of the motorist. Chapter 2. In: De Boer, ed., 1967.
De Boer, J.B.; Burghout, F. & Van Heemskerck Veeckens, J.F.T. (1959). Appraisal of the quality of public lighting based on road surface luminance and glare. In: CIE, 1959.
De Boer, J.B. & Fischer, D. (1981). Interior lighting (second revised edition). Deventer, Kluwer, 1981.
De Boer, J.B. & Van Heemskerck Veeckens, J.F.T. (1955). Observations on discomfort glare in street lighting. Influence of the colour of the light. Zürich, CIE, 1955.
De Boer, J.B., ed. (1967). Public lighting. Eindhoven, Centrex, 1967.
De Bruin, N.G. (1949). Beknopt leerboek der differentiaal- en integraalrekening (Short textbook on differential and integral calculus). Amsterdam, N.V. Noord-Hollandsche Uitgevers Maatschappij, 1949.
Dhingra, N.K.; Kao, Y.-H; Sterling, P. & Smith, R.G. (2003). Contrast threshold of a brisk-transient ganglion cell. Journ. Neurophysiol. (in press).
Dirken, J.M. (2004). Product-ergonomie; Ontwerpen voor gebruikers (Product ergonomics; Designing for users). Delft, VSSB, 2004.
Eloholma, M.; Ketomaki, J.; Orrevetlainen, P. & Halonen, L. (2005). Contrast threshold and reaction time experiments in developing a performance-based mesopic photometry system p. 5-9. In: CIE, 2005.

Fechner, G.T. (1860). Elemente der Psychophysik (Elements of Psychophysic). 2 volumes, 1860.

Fiorentini, A. (2004). Brightness and lightness. Chapter 56, p. 881-891. In: Chalupa & Werner, eds., 2004.

Foley, P.J. (1972). Design criteria in the human-centered man-machine system. L S 9-2. In: Anon., 1972.

Forbes, T.W. (1972). Human Factors in Highway Traffic Safety Research. New York, John Wiley & Sons, Inc., 1972.

Fotios, S. & Cheal, C. (2005). The white light illuminance trade-off permitted in UK design for pedestrian lighting: Can it be validated?. p. 50-55. In: CIE, 2005.

Gall, D. (2004). Grundlagen der Lichttechnik; Kompendium (A compendium on the basics of illuminating engineering). München, Richard Pflaum Verlag GmbH & Co KG, 2004.

Gallagher, V.P.; Koth, B.W. & Freedman, M. (1975). The specification of street lighting needs. FHWA-RD-76-17. Philadelphia, Franklin Institute, 1975.

Gleitman, H.; Fridlund, A.J. & Reisberg, D. (2004). Psychology. Sixth edition. New York, W.W. Norton, 2004.

Griep, D.J. (1971). Analyse van de rijtaak (analysis of the driving task). Verkeerstechniek 22 (1971) 303-306; 370-378; 423-427; 539-542.

Hallet. P.E. (1963). Spatial summation. Vision Research, 3 (1963) 9-24.

Halonen, L. & Eloholma, M. (2005). Development of mesopic photometry based on new findings on visual performance. p. 1-4. In: CIE, 2005.

Hentschel, H.-J. (1994). Licht und Beleuchtung; Theorie und Praxis der Lichttechnik; 4. Auflage (Light and illumination; Theory and practice of lighting engineering, 4th edition). Heidelberg, Hüthig, 1994.

Hentschel, H.-J. ed. (2002). Licht und Beleuchtung; Grundlagen und Anwendungen der Lichttechnik; 5. neu bearbeitete und erweiterte Auflage (Light and illumination; Theory and applications of lighting engineering; 5th new and extended edition). Heidelberg, Hüthig, 2002.

Hopkinson, R.G. (1969). Lighting and seeing. London, William Heinemann, 1969.

Janoff M.S. (1993). The relationship between small target visibility and a dynamic measure of driver visual performance. Journ. IES 22 (1993) no 1. p. 104-112.

Janssen, W.H. (1986). Modellen van de rijtaak; De state-of-the-art in 1986 (Models of the driving task; The state-of-the-art in 1986). IZF 1986 C-7. Soesterberg, IZF/TNO, 1986.

Keitz, H.A.E. (1967). Lichtmessungen und Lichtberechnungen. 2e. Auflage (The measurement and calculation of light. Second edition). Eindhoven, Philips Technische Bibliotheek, 1967.

Kokoschka, S. (1971). Spektrale Hellempfindlichkeit und äquivalente Leuchtdichte zentraler Gesichtsfelden im mesopischen Bereich (Spectral sensitivity and equivalent luminance in the mesopic range). In: CIE, 1971.

Kokoschka, S. (1980). Photometrie niedriger Leuchtdichten durch eine äquivalente Leuchtdichte des 10-° Feldes (Photometry of low luminance by means of an equivalent luminance of the 10-° Field). Licht-Forschung 2 (1980) nr. 1, p. 1-13.

König, A. & Brodhun, E. (1889). Experimentelle Untersuchungen über die psychophysischen Fundamentalformel in Bezug auf den Gesichtssinn (Experimental studies on the fundamental psychophysiological formulae regarding the visual sense). Sitz. Ber. Preuss. Akad. Wiss (1889) 641-644.

Kostic, M.; Djokic, L.; Pojatar, D. & Strbac-Hadzibegovic, N. (2005). Influence of the theory of mesopic vision on road lighting design. p. 44-49. In: CIE, 2005.

Krech, D.; Crutchfield, R.S. & Livson, N. (1969). Elements of psychology (second edition). New York, Alfred Knopf, 1969.

Le Grand, Y. (1956). Optique physiologique, Tome III (Physiological optics; volume III). Paris, Ed. Revue Optique, 1956.

LRI (1993). Visibility and luminance in roadway lighting. 2nd International Symposium. Orlando, Florida, October 26 – 27, 1993. New York, Lighting Research Institute LRI, 1993.

LRI (1998). Vision at low light levels. 4th International Lighting Research Symposium. Orlando, 1998. New York, Lighting Research Institute LRI, 1998.
Lythgoe, R.J. (1932). The measurement of visual acuity. Medical Council Special Report No. 175. London. H.M. Stationery Office 1932.
Middleton, W.E.K. (1952). Vision through the atmosphere. University of Toronto Press, 1952.
Miller, G.A. (1947). Sensitivity to changes in the intensity of white noise and its relation to masking and loudness. Journ. Acoust. Soc. Amer. 19 (1947) 609-619.
Moon, P. (1961). The scientific basis of illuminating engineering (revised edition). New York, Dover Publications, Inc., 1961
Moroney, M.J. (1990). Facts from figures (revised edition). Harmondsworth, Penguin Books, Pelican A236, 1990.
Mutzhas, M.F. (1981). The $2°$ spectral tristimulus value functions represented as exponential equations. Lichtforschung, 2 (1980) nr. 1. p. 15-21.
Narisada, K. (2007). Revealing power and road lighting design. In: TRB, 2007.
Narisada, K.; Karasawa, Y. & Shirao, K. (2003). Design parameters of road lighting and Revealing Power. In: CIE, 2003.
Narisada, K. & Schreuder, D.A. (2004). Light pollution handbook. Dordrecht, Springer, 2004.
Noordzij, P.C.; Hagenzieker, M.P. & Theeuwes, J. (1993). Visuele waarneming en verkeersveiligheid (Visual perception and road safety). R-93-12. Leidschendam, SWOV, 1993.
NSVV (1990). Aanbevelingen voor openbare verlichting (Recommendations for public lighting). Arnhem, NSVV, 1990.
NSVV (2002). Richtlijnen voor openbare verlichting; Deel 1: Prestatie-eisen. Nederlandse Praktijkrichtlijn 13201-1 (Guidelines for public lighting; Part 1: Performance requirements. Practical Guidelines for the Netherlands 13201-1). Arnhem, NSVV, 2002.
NSVV (2003). Aanbevelingen voor tunnelverlichting (Recommendations for tunnel lighting). Arnhem, NSVV, 2003.
OECD (1972). Symposium on road user perception and decision making. Rome, OECD, 1972.
Opstelten, J.J. (1984). Helderheid en luminantie in het overgangsgebied tussen fotopisch en scotopisch zien (Brightness and luminance in the transition zone between photopic and scotopic vision). Report LA 1003/84. Laboratory for lighting application and lighting engineering. Eindhoven, Philips Lighting, 1984 (not published).
Palmer, D.A. (1971). Table 2.3.2 $\bar{y}_{10}(\lambda)$. In: CIE, 1971.
Phillips, P.L. (1986). Minimum colour differences required to recognise small objects on a colour c.r.t. Journal of the Institution of Electronic and Radio Engineers. 56 (1986) no. 3, 123-129.
Rea, M.S. (2005). A model of mesopic vision: The bridge to a unified systen of photometry. p. 30-37. In: CIE, 2005.
Reeves, A. (2004). Visual adaptation. Chapter 50, p. 851-862. In: Chalupa & Werner, eds., 2004.
Ricco, A. (1875). Relazioni fra il minimo angolo visualo (The relation of the minimum visual angle). Ann. Ottalmol. 6 III (1875) 373-393.
Sagawa. K. (2005). Brightness in mesopic vision and the CIE supplementary system of photometry. p. 20-26. In: CIE, 2005.
Schober, H. (1960). Das Sehen; 2 Bände (Seeing; two volumes). Leipzig, Fachbuchverlag, 1958-1960.
Schoon, C.C. & Schreuder, D.A. (1993). HID car headlights and road safety; A state-of-the-art report on high-pressure gas-discharge lamps with an examination of the application of UV radiation and polarised light. R-93-70. Leidschendam, SWOV, 1993.
Schreuder, D.A. (1964). The lighting of vehicular traffic tunnels. Eindhoven, Centrex, 1964.
Schreuder, D.A. (1967). The theoretical basis for road lighting design. Chapter 3. In: De Boer, ed., 1967.
Schreuder, D.A. (1974). De rol van functionele eisen bij de wegverlichting (The rol of functional

requirements in road lighting). In: Wegontwerp en verlichting tegen de achtergrond van de verkeersveiligheid (Road design and lighting in view of road safety); Preadviezen Congresdag 1974, blz. 111 t/m 137. Vereniging Het Nederlandsche Wegencongres, 's-Gravenhage, 1974.

Schreuder, D.A. (1976). White or yellow light for vehicle head-lamps? Arguments in the discussion on the colour of vehicle head-lamps. Publication 1976-2E. Voorburg, SWOV, 1976.

Schreuder, D.A. (1988). Visual performance and road safety. R-88-46. Leidschendam, SWOV, 1988.

Schreuder, D.A. (1989). Enquête wijst uit: Straten zijn onveilig en licht is akelig (Enquiry shows: Streets are not safe and look ugly). De Gorzette, Verenigings- en informatieblad Bewoners Vereniging Schiedam-Zuid. 17 (1989) no 1. p. 23-25.

Schreuder, D.A. (1989a). Bewoners oordelen over straatverlichting (Residents judge street lighting). PT Elektronica-Elektrotechniek. 44 (1989) no. 5, p. 60-64.

Schreuder, D.A. (1989b). The field factor for the determination of tunnel entrance luminance levels. Proceedings, SLG/CIE Symposium, Lugano, 12 oct 1989.

Schreuder, D.A. (1990). De veldfactor bij de bepaling van de verlichtingsniveaus bij tunnelingangen; Verslag van experimenteel onderzoek (The field factor at the assessment of lighting levels in tunnel entrances; Report of experimental research). R-90-10. SWOV, Leidschendam, 1990.

Schreuder, D.A. (1991). Visibility aspects of the driving task: Foresight in driving. A theoretical note. R-91-71. Leidschendam, SWOV, 1991.

Schreuder, D.A. (1991a). De veldfactor bij de bepaling van de verlichtingsniveaus bij tunnelingangen; een nadere analyse (The field factor at the assessment of lighting levels in tunnel entrances; An additional analysis). R-91-65. SWOV, Leidschendam, 1991.

Schreuder, D.A. (1991b). Tegenstraalverlichting in tunnels; Een overzicht van de beschikbare literatuur (Counterbeam lighting in tunnels; A survey of available literature). R-91-96. Leidschendam, SWOV, 1991.

Schreuder, D.A. (1993). Contrastwaarnemingen in tunnels – Een meetmethode (The observation of contrasts in tunnels; A method for measuring) R-93-36. Leidschendam, SWOV, 1993.

Schreuder, D.A. (1994). Duurzame verkeersveiligheid voor ouderen en gehandicapten; Verslag van een pilot-studie ten behoeve van een onderzoek-opzet (Sustainable road safety for the elderly and the handicapped; Report of a pilot study for a research plan). Leidschendam, Duco Schreuder Consultancies, 1994.

Schreuder, D.A. (1997). The functional characteristics of road and tunnel lighting. Paper presented to the Israel National Committee on Illumination on Tuesday, 25 March 1997 at the Association of Engineers and Architects in Tel Aviv. Leidschendam, Duco Schreuder consultancies, 1997.

Schreuder, D.A. (1998). Road lighting for safety. London, Thomas Telford, 1998 (Translation of "Openbare verlichting voor verkeer en veiligheid", Deventer, Kluwer Techniek, 1996).

Schreuder, D.A. (2004). Verlichting thuis voor de allerarmsten (Home lighting for the very poor). NSVV Nationaal Lichtcongres 11 november 2004. Arnhem, NSVV, 2004.

Schreuder, D.A. (2006). Gerontopsychologische overwegingen bij het ontwerpen van binnenverlichting (Gerontopsycholgal considerations in interior lighting design). NSVV Nationaal Lichtcongres 2006, 23 november 2006. Ede, NSVV, 2006.

Schreuder, D.A. & Lindeijer, J.E. (1987). Verlichting en markering van voertuigen: Een state-of-the-art rapport (Lighting and marking of road vehicles: A state-of-the-art report). R-87-7. Leidschendam, SWOV, 1987.

Shevell, S.K., ed. (2003). The science of color. Second edition. OSA Optical Society of America. Amsterdam, Elsevier, 2003.

Smith, V. & Pokorny, J. (2003). Color matching and color discrimination. Chapter 3. In: Shevell, ed., 2003.

Srinivasan, M.V.; Laughlin, S.B. & Dubs, A. (1982). Predictive coding; A fresh view of inhibition in the retina. Proc. Roy. Soc. London. B,216 (1982) 427-459.

Steinman, P. (2004). Gaze control under natural conditions. Chapter 90, p. 1339-1356. In: Chalupa & Werner, eds., 2004.

Sterken, C. & Manfroid, J. (1992). Astronomical photometry. Dordrecht, Kluwer, 1992.

Stilma, J.S. & Voorn, Th. B., eds. (1995). Praktische oogheelkunde. Eerste druk, tweede oplage met correcties (Practical ophthalmology. First edition, second impression with corrections). Houten, Bohn, Stafleu, Van Loghum, 1995.

SWOV (1972). Psychological Aspects of Driver Behaviour. Symposium Noordwijkerhout, 2-6 August 1971. Voorburg, SWOV, 1972.

TRB (2007). Visibility symposium, 17-18 April 2007. College Station, Texas, 2007.

Van Essen, J. (1939). Etude psychophysiologique sur l'obscurité (Psychophysiological study on darkness). Archives Néerlandaises de physiologie et de phonétiqyue expérimentale (1939) p. 487-554 (Year estimated).

Van Norren, D. (1995). Lichtschade (Damage by light). Chapter 4. In: Stilma & Voorn, eds., 1995.

Van Tilborg, A.D.M. (1991). Evaluatie van de verlichtingsproeven in Utrecht (Evaluation of lighting experiments in Utrecht). Utrecht, Energiebedrijf, 1991 (not published; see Anon., 1995a).

Varady, G.; Szalai, A.; Bodrogi, P. & Schanda, J. (2005). Measuring mesopic visual performance: Contrast threshold experiments – effect of stimulus size and shape . p. 10-13. In: CIE, 2005.

Vos, J.J. & Legein, C. P. (1989). Oog en werk: een ergoftalmologische wegwijzer (Eye and work; Ergophthalmological guidance). Den Haag, SDU Uitgeverij. 1989.

Wachter, A. & Hoeber, H. (2006). Compendium of theoretical physics (Translated from the German edition). New York, Springer Science+Business Media, Inc., 2006.

Walsh, J.W.T. (1958). Photometry (3rd edition). London, Constable, 1958. Reprinted. New York, Dover, 1965.

Walters, H.V. & Wright, W.D. (1943). The spectral sensitivity of the fovea and the extra-fovea in the Purkinje range. Proc. R. Soc. B 131.

Watson, A.B.; Barlow, H.B. & Robson, J.G. (1983). What does the eye see best? Nature, 302 (1983) 419-422.

Weber, W.E. (1846). Das Tastsinn und das Gemeingefühl (Tactile sense and general sensation). In: Wagner, Handwörterbuch der Physiologie vol. III (Dictionary of physiologie), 1846.

Weigert, A. & Wendker, H.J. (1989). Astronomie und Astrophysik – ein Grundkurs, 2. Auflage (Astronomy and astrophysics – a primer, 2nd edition). VCH Verlagsgesellschaft, Weinheim (D), 1989.

Weir, D.H. & McRuer, D.T. (1971). Measurement and interpretation of driver-vehicle system dynamic response. In: SWOV, 1972.

Wiener, N. (1954). The human use of human beings (2nd ed). New York, Doubleday, 1954.

Wiener, N. (1966). God, mens en machine (translation of God and Golem, Inc., 1964). Rotterdam, Universitaire Pers, 1966.

Wouters, P.I.J. (1991). De veiligheid van oudere verkeersdeelnemers (Safety for elderly traffic participants). R-91-77. Leidschendam, SWOV, 1991.

Wyszecki, G. & Stiles, W. (1982). Color science: Concepts and methods, quantitative data and formulae. 2nd edition. First edition 1967. New York, Wiley, 1982.

8 The human observer; visual perception

In this chapter, the human observer is discussed, focussing on the visual perception aspects. In Chapter 6, the physical and anatomical aspects of vision are discussed. Chapter 7 concentrates on the visual performance aspects, and in Chapter 9 colour vision is discussed.

In this chapter, emphasis is on the derived visual functions like the field of view, flicker-effects, subjective brightness, and the detection of movement. The main subject of this chapter is glare, more in particular disability glare. Emphasis is placed on the stray light theory of disability glare. The subject with the greatest interest is the CIE Standard Glare Observer, and the influences of glare angle, age, and the pigmentation of the iris in it. The age effects of disability glare are highlighted.

Finally, discomfort glare is briefly discussed. In outdoor lighting, this aspect is mainly of historic interest.

8.1 Derived visual functions

8.1.1 Field of view

(a) *The field of vision*

As is explained in sec. 6.3.1, the optical elements of the visual system form an optical image of the outside world on the retina, the area at the rear end of the eye ball, that contains the photoreceptors. As a result of the shape of the eye ball and the lens, the overall field of view in human vision is very large as compared to common optical instruments that contain simple optics only. Even without moving the eye or the head, the overall field of view of human eyes has an elliptical shape with horizontal long axis. Usually, the field of vision is clinically assessed with a perimeter.

Often, the angular diameter of the longer axis is assumed to be about 200° and the shorter axis is about 140°. Usually, smaller values are used. The two eyes have optically the same field of view; however, functionally for each eye the nasal part is smaller than the temporal part. No wonder; the nose in is the way. This is depicted in Figure 8.1.1.

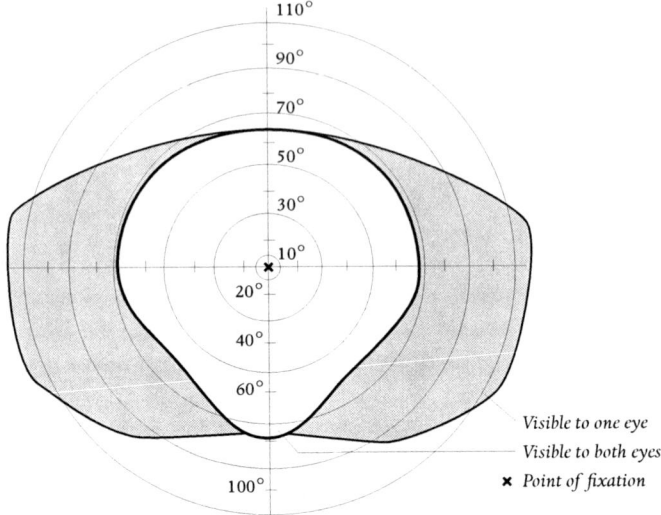

Figure 8.1.1 The binocular field of view in degrees in relation to the point of fixation. (After Boyce, 2003, fig. 2.1, p. 45.)

For a homogeneous field of view the standardised diameter is 140° (Fry, 1969). This is depicted in Figure 8.1.2.

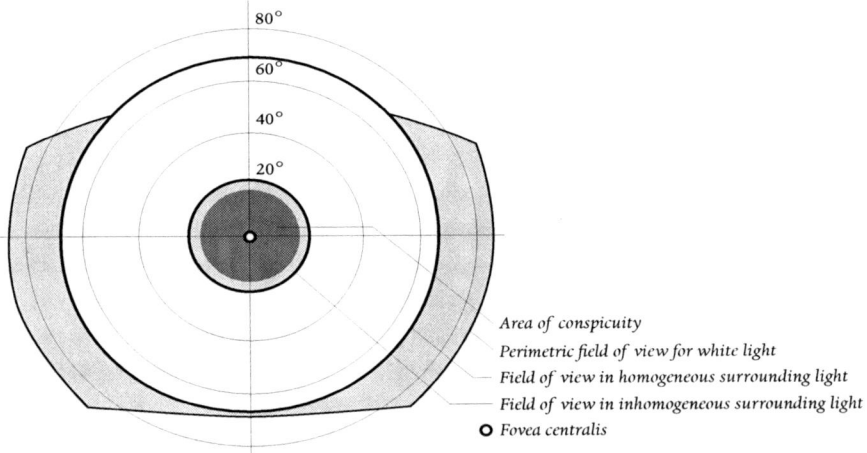

Figure 8.1.2 The binocular field of view. (After Baer ed., 2006, fig. 1.44, p. 53. Based on data of Fry, 1969.)

Figure 8.1.2 is valid for white light. The field of vision is much smaller for colored light. For yellow it is under 45°, for red under 40°, and for green under 30°. These are important facts for the visibility of signalling lights (Narisada & Schreuder; 2004, sec. 8.3.6d, p. 213; Middleton, 1952; Douglas & Booker, 1977; Schreuder, 1976).

As is explained in sec. 7.4.4, visual perception is concentrated in the fovea, although visual perception is possible toward the outer edges of the field of view, as is depicted in the Figures 8.1.1 and 8.1.2.

   (b) *The functional visual field*
       When the eyes and the head are kept still, the functionality of the perception decreases sharply towards these edges. In real life the stationary eye is not common. It is customary to introduce the functional visual field, where the eye and the head are allowed to move. This has been introduced by Bunt & Sanders (1973). See also Sanders (1967) and Schreuder (1998, chapters 6 and 8).

The functional visual field is subdivided as follows, expressed in degrees from the optical axis which is as usually considered as being the line from the lens centre toward the fovea:
1. For angles smaller than about 2°: foveal observation;
2. For angles between about 2° and about 25°, called the stationary field: near-periphery observation
3. For angles between about 25° and about 85°, called the eye-field: periphery observation
4. For angles over about 85°, called the head-field: far-periphery observation.

For foveal observation and for the stationary field, eye movements are not necessary, contrary to the eye-field. In the head-field also head movements are needed (Bunt & Sanders, 1973, p. 1a). However, practice will easily show that most people prefer to move their eyes and their already for much smaller angles. Maybe that is just a matter of convenience.

   (c) *Binocular vision*
       As is depicted in the Figures 8.1.1 and 8.1.2, there is a quite extensive overlap of the field of view of the two eyes of the human visual system. As is explained in a further part of this section, this overlap allows depth perception and distance estimation as a result of the stereopsis. The retinal images of the two eyes differ slightly. The processes that lead to the formation of a visually conscious image melts the two images into one, but one conscious image that contains depth information.

When discussing the estimation of distance one must discern between nearby and far-away objects. As will be explained further on, the limit for 'near by' is usually about 5 metres. The reason is that the just noticeable depth difference increases with the square of the distance (Walraven, 1989, p. 47). The distance estimation for objects farther away is mainly a cognitive affair, where familiarity with object plays an essential role. Also far away objects are usually coloured grey or blue from atmospheric scatter (Middleton, 1952). Far above sea level, and in an atmosphere that is free from water vapour and dust, this atmospheric perspective may be absent altogether, making it difficult to judge the distance of far-away objects like mountain ranges etc. (Schreuder, 2006). These atmospheric phenomena will not be discussed here.

Two factors are involved in depth perception and the related estimation of distance for near-by objects. Firstly, the accommodation must be adjusted in order to focus clearly on objects at various distances. This is a mono-ocular effect, similar to the well-know focussing of photo and video cameras (Longhurst, 1964, p. 367). Accommodation is explained in sec. 6.3.3b. More important is the stereopic effect (Longhurst, 1964, sec. 17-13, p. 365-367). When viewing a visual scene that contains objects with two eyes one may find an impression of the three dimensional solidity of the object.

  (d) Stereopsis
  When a three-dimensional objects is observed, the retinal images of the two eyes differ slightly. Not only there is a slight lateral shift, which requires an adjustment of the optical axes of the two eyes called convergence, but also the viewing angle to the object differs slightly. It seems as if the object is turned slightly around a vertical axis. A perspective view of the phenomena involved is depicted in Figure 8.1.3.

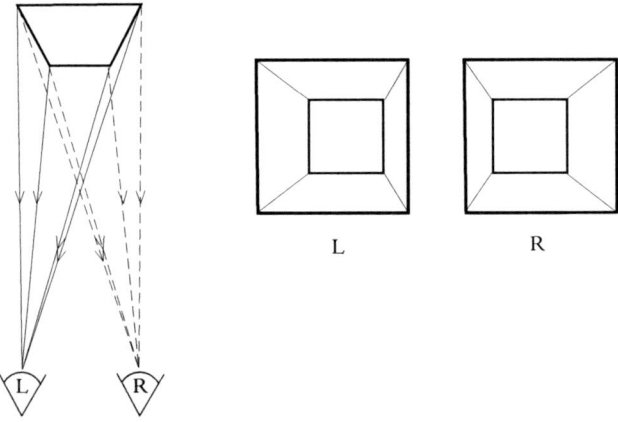

Figure 8.1.3 Perspective view of the left and right eyes. (After Longhurst, 1964, fig. 17-10, p. 365.)

The human observer; visual perception   277

Stereopsis is an extremely acute mechanism with one of the lowest visual thresholds, less than the width of a single retinal photoreceptor, which makes it one of the hyperacuities (Schor, 2004, p. 1300; Westheimer, 1979).

The processes that lead to the formation of a visually conscious image melt those two images into one, but then into a conscious image that contains depth information. Therefore the term cyclopean eye, that should be located halfway between the two human eyes is misleading, because such a cyclopean view would and could not contain stereoscopic depth information, precisely because Cyclops have only one eye! (Schor, p. 1300). Most of these processes are assumed to take place in the homologous areas in the retinal in the primary visual cortex (Schor, p. 1301).

(e) *The stereoscopic range*

The next aspect of distance estimation is usually called the stereoscopic range (Longhurst, 1964, sec, 17-14). The crucial phenomenon is the binocular parallax that is depicted in Figure 8.1.4.

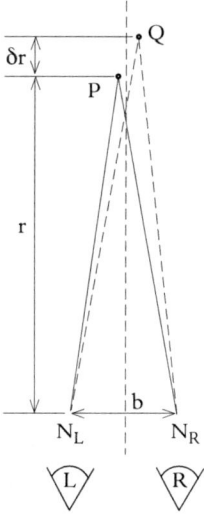

Figure 8.1.4 The binocular parallax (After Longhurst, 1964, fig. 17-13, p. 367).

It is assumed that the ability to detect the difference in distance between P and Q depends on the difference in size between the retinal images of P and Q. This is proportional to the difference-angle θ between the angles N_L-P-N_R and N_L-Q-N_R, N_L and N_R being the first nodal points of the left and right eye. The distance between N_L and N_R equals b in Figure 8.1.4. In practice this distance is about 65 mm in normal adult human beings (Longhurst, 1964, p. 365).

From Figure 8.1.4, it follows according to Longhurst (1964, p. 367) that

$$\delta\theta = \frac{b}{r} - \frac{b}{r+\delta r} \qquad [8.1\text{-}1]$$

Because δr is small, it follows that

$$\delta r = \frac{r^2 \delta\theta}{b} \qquad [8.1\text{-}2]$$

Relation [8.1-2] implies that the minimum detectable difference δr is determined by the minimum detectable difference $\delta\theta$. The latter is found to be around 0,5 minute of arc (Longhurst, 1964, p. 367).

8.1.2 The speed of observation; flicker-effects

(a) The discrimination in time

Many performance aspects of the visual system depend on the adaptation luminance. One of these aspects is the capability to discriminate time intervals. It seems that the integration time of stimuli increases when the adaptation luminance decreases. The integration time is the time interval where stimuli that enter the eye at different moments in time are observed as one single stimulus with an intensity greater than that of the individual stimuli. In this way, the system sensitivity can be increased (Gregory, 1965, p. 78). It might be considered as another adaptation effect. In this way, the retina behaves much like a CCD (Sterken & Manfroid, 1992, Chapter 13; Howell, 2001).

It has been found that under normal conditions and for high-mesopic and photopic conditions the integration time is in the order of 100 ms. In scotopic vision, a similar value is found, be it that there the integration is also spatial; and involves a number of rods. See secs. 6.4.2c and 8.1.2a.

In the visual cortex, the situation is much more complex. For different types of integration processes, values of about 50 ms are found (Albrecht et al., 2004, p. 759). It seems that, globally speaking, the integration times in the retina and in the cortex are of the same order of magnitude as the interval of natural fixation, i.e. about 200 ms (Albrecht et al., 2004, p. 758)

A consequence of this elongated integration time is that the response to a stimulus in low luminances is retarded compared to the response in higher luminances. This can be demonstrated very easily with what is commonly called the Pulfrich pendulum. This effect is described in Gregory (1965, fig. 6.3) and Narisada & Schreuder (2004, sec. 9.1.5a).

A similar effect can be observed when looking at a double star with two components of considerably different magnitude (Minnaert, 1942, Volume I, sec. 109, p. 150).

Two aspects of the discrimination in time are worth mentioning. First the speed of detection. This aspect is important for signalling lights, particularly as there are reasons to believe that there is a difference for different colours, implying that the different families of cones that are described in secs. 6.3.5c and 9.2.1 do not have the same integration time. This seems to be related to the fact that the regeneration in cone vision is not equally fast for the different visual pigments. This also results in the fact that after-images of different colours do not persists equally long (Minnaert, 1942, Volume I, sec. 94, p. 131-132). Furthermore, the flicker-fusion frequency for colours is lower than that for the brightness (Hentschel, ed., 2002, p. 73). Flicker effects are discussed in the next part of this section, as well as in sec. 5.2.2g, where flicker-photometers are discussed (Barrows, 1938, p. 123; Rood, 1893).

For details on the speed of detection for lights of different colours we refer to De Boer (1959); Dunbar (1939); Narisada & Schreuder (2004, sec. 9.1.5b); Reading (1966), and Schreuder (1976).

(b) Flicker effects

Lights with periodic brightness variations are often called flashing lights. As is mentioned in sec. 5.2.2, a periodic stimulus can be observed only if the stimulus is intense enough to be observed when the observation time is unlimited. This is called the Rule of Hoorweg. Also, a periodic stimulus can be observed only if the product of stimulus intensity and observation time reaches a specific minimum value. This is called the law of Blondel and Rey.

When the frequency of the light pulses is above a certain value, the pulses cannot be discerned separately any more. This pulse frequency is called the critical flicker-fusion frequency or CFF. The CFF depends on many variables, but primarily on the adaptation luminance.

Apart from the adaptation luminance, the detection depends on the wave-form (Jantzen, 1960; Schreuder, 1964, Schmidt-Clausen, 1968). In Figure 8.1.5, the relation between the CFF and the adaptation luminance is depicted for different wave-forms. GW is a factor that depends on the wave-form. GW = 0,637 for rectangular pulses; GW = 0,500 for a sinusoidal wave, and GW = 0,15 for a common fluorescent tube (Hentschel, ed., 2002, p. 63). Some points are discussed also in sec. 5.2.2.

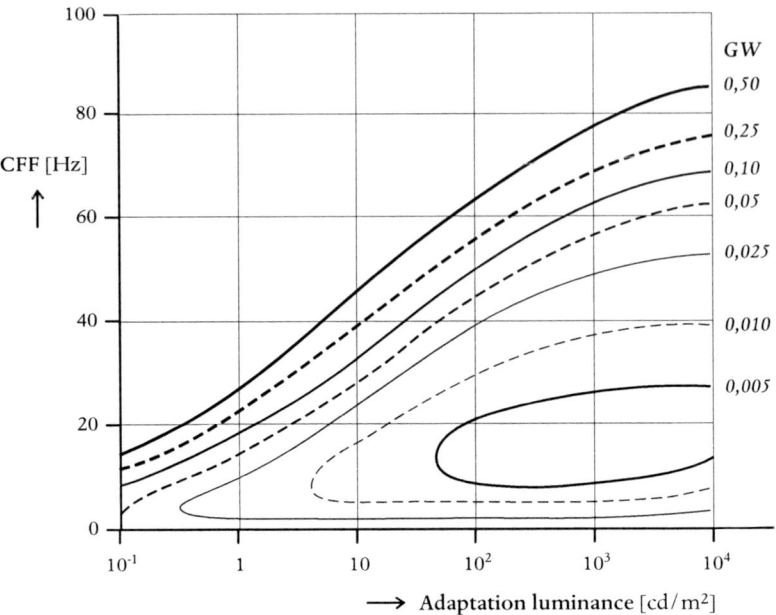

Figure 8.1.5 *The relation between CFF and adaptation luminance, with the wave form GW as a parameter. After Hentschel, ed., 2002, fig. 3.14, p. 64. Based on Kelly (1961-1962).*

Light appears to be steady when the frequency is higher than the flicker-fusion frequency. Talbot's Law states that the effective intensity equals the time-average of the instantaneous intensity.

Further details on the laws of Bloch, Blondel and Rey, and Talbot are given in Narisada & Schreuder (2004, sec. 9.1.2). See also sec. 5.2.2.

(c) *Discomfort by flicker effects*

When the frequency of the periodic light is below the flicker-fusion frequency, the flicker obviously can be perceived. The flicker, when perceived, may cause considerable discomfort. Problems may be encountered in practical lighting applications. In interior lighting, flicker from defective fluorescent lighting may cause the most nuisance when the frequency is close to the flicker-fusion frequency (Baer, ed., 2006, sec 1.3.2, p. 79; Collins & Hopkinson, 1957). Most discomfort can be avoided by using high-frequency ballasts, that are preferred for other reasons anyway (De Boer & Fischer, 1981, p. 161; Anon., 2007).

At lower frequencies, some other effects seem to play a role. It has been suggested that interference with the α-rhythm of the brain may be responsible for the discomfort.

The human observer; visual perception

See for the α-rhythm Greenfield (1997, p. 70); Gregory, ed., 2004, p. 97-98). The low-frequency flicker effects are particularly disturbing in the lighting of long tunnels for road traffic. It was found that the human visual system is the most sensitive for by flicker at frequencies between 5 and 9 c/sec (De Lange, 1957; Schreuder, 1964, p. 95; 1967; Saito & Narisada, 1968). In modern tunnel lighting recommendations it is required that this range must be avoided; it is often recommended to avoid a range between 3 and 12 c/sec (CEN, 2003; CIE, 1990; Narisada & Schreuder, 2004, 9.1.5c; NSVV 2003, and Schreuder, 1998, chapter 15).

8.1.3 Subjective brightness

It is well-known from experimental psychology that it is quite possible to scale impressions and sensations in a precise way (Steyer, 1997). However, not everyone seems to agree. "Sensations cannot be measured. There is no way of setting up a unit or of evaluating in terms of a unit. Any equation like the Weber-Fechner one is pure nonsense" (Moon, 1961, p. 421-422). In spite of such criticism, several proposals have been made in illuminating engineering to scale the visual impression, usually designated as subjective brightness or luminosity (Bodmann & Voit, 1962; De Boer & Fischer, 1981; Hopkinson, 1957; Stevens, 1969, Marsden, 1968).

During the Second World War, visibility of dim stimuli was important with respect to the wide-spread 'black-out'. Further on in this section we will have to say a little more about this. In 1941, a major study on the relation between the luminance stimulus and brightness response was published (Hopkinson et al., 1941; further details are given in Padgham & Saunders, 1966). A detailed discussion of these studies is given by Stevens (1969, p. 16-18). We will give here a summary of this discussion. More details are given in Narisada & Schreuder (2004, sec. 9.1.2c). See also Schreuder (2008).

The results of this study are depicted in Figure 8.1.6.
In this figure, three aspects are striking:
1. The curves for equal luminance bend off near the line of zero luminosity, causing a 'crowding' of the luminosities;
2. The curves for equal luminance are not equidistant. It is likely that this 'crowding' for low luminances is related to the fact that the contrast sensitivity decreases with decreasing luminance.
3. There does not seem a clear breach in the visibility process at the transitions from photopic to mesopic, and from mesopic to scotopic vision. The curves for equal luminance are smoothly grouped. This is to be expected because there is no sudden transition from photopic to scotopic vision as is explained in sec. 7.2.4. The whole mesopic range lies between them.

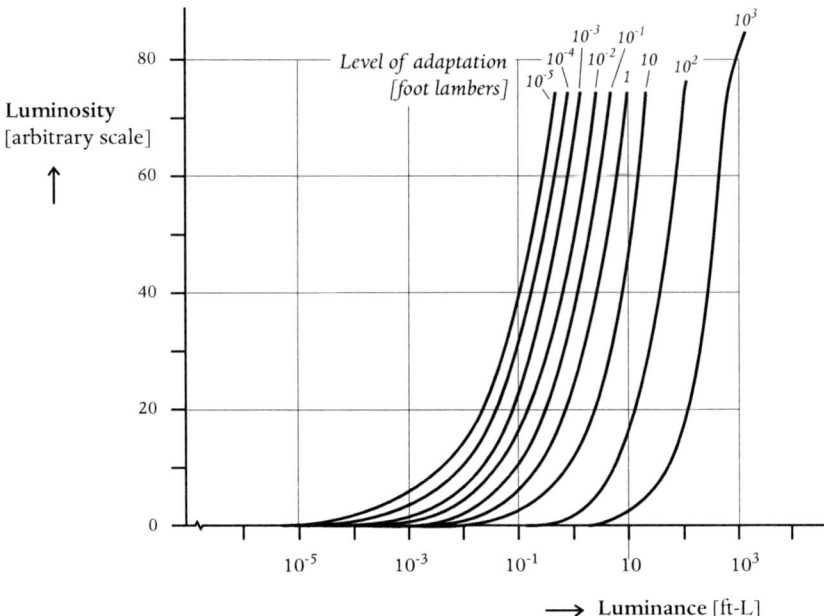

Figure 8.1.6 The relation between luminosity and luminance for different levels of adaptation (After Narisada & Schreuder, 2004, figure 9.1.2. Based on data of Stevens, 1969, fig. 1.7, and Hopkinson et al., 1941). Note: the original figure is reproduced. To convert the luminances into metric values, multiply the foot-Lambert-values by 3,43 to get cd/m^2-values.

From Figure 8.1.6 it can be seen that there is no linear relation between the subjective brightness and the luminance. The deviations from linearity are more pronounced at lower levels of adaptation. In moonlight, the adaptation level may be about 10^{-2} cd/m^2 (Middleton, 1952, p. 91).

The luminosity of a grey object at this luminance is very low; it is about 1 compared to zero for a black object. When looking at the graphs, it may be seen that the luminosities of all greyish objects are close ("crowded together"; Stevens, 1969, p. 17). For light coloured objects, like e.g. white painted road markings, however, the situation is different. Although the increase in luminosity is less than one might expect from the increase in luminance, they are highly conspicuous because they 'stick out'. This effect is clearly noticeable in dim surroundings, and it is a major contribution of white painted road markings to the safety and ease of pedestrians.

In Table 8.1.1, an example is given. We assume a street scene with an average scene reflection of 20% (Schreuder, 1964). In the scene we assume several objects:

The human observer; visual perception

1. a black-painted utility pole with a reflection of 5%;
2. a piece of asphalt road surface with a reflection of 10%;
3. a piece of cement concrete pavement surface with a reflection of 20%;
4. a white-painted road marking with a reflection of 70%.

Object	Reflection (%)	Luminosity for adaptation level	
		35 cd/m²	0,0035 cd/m²
1	5	26	0
2	10	42	1
3	20	63	3
4	70	95	7

Table 8.1.1 The luminosity of different objects at different adaptation luminances. Based on data from Figure 8.1.6.

In the 1980s, the available data were analyzed again by Haubner et al. (1980). A brief description of the procedure and of the main results is given by Hentschel, ed. (2002, p. 55). The following discussion is based on this description.

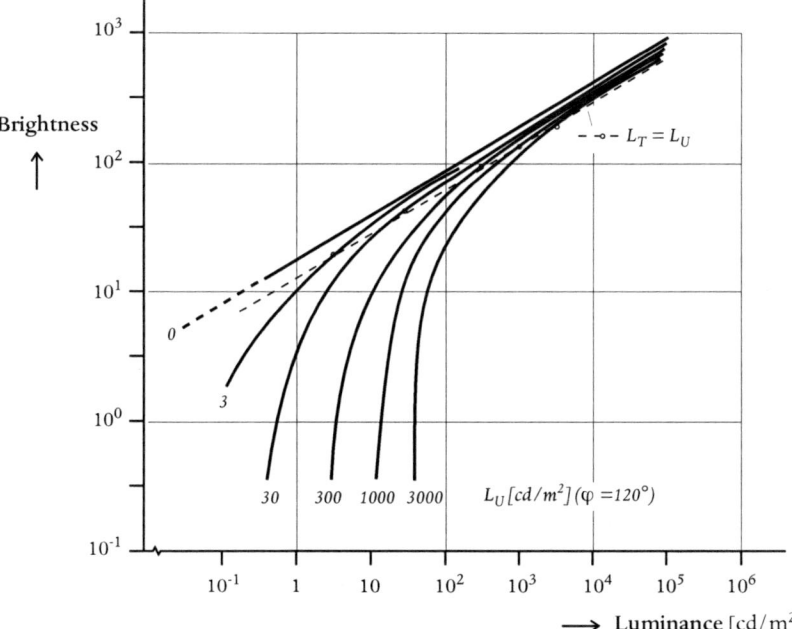

Figure 8.1.7 The brightness H of a test object of two degrees with luminance L_T for different surrounding luminances L_u (After Narisada & Schreuder, 2004, figure 9.1.3. Based on Hentschel, 1994, Figure 3.8 and Haubner et al., 1980).

The 'Haubner scale' that was introduced here, has a 'zero' point, contrary to the logarithmic scales that are used for the luminance (Haubner et al., 1980). The zero point corresponds with the impression 'black'. Its luminance depends on the adaptation luminance. However, it is more convenient to express the brightness as well in a logarithmic scale. The results established by Haubner et al. (1980) are depicted in Figure 8.1.7.

8.1.4 Detection of movement

(a) *Constancy*

Constancy is an essential element to understand the perception of movement (Narisada & Schreuder, 2004, sec. 9.1.6; Schreuder, 1998, sec. 6.3.2). Constancy is understood as the phenomenon that the object – or the whole world – is observed as being stationary even if the eye, or the observer as a whole, moves. Constancy is a poorly understood phenomenon. It seems that perception is remarkable in keeping things appearing much the same in varying conditions. Colours are hardly affected by changes of the ambient light, and sizes remain almost constant for different distances (Gregory, 1965, chapter 7; Gregory, ed., 2004, p. 186). Colour constancy is explained in sec. 9.3.2d. Constancy seems to be a typical 'Gestalt'-phenomenon (Schreuder, 2008).

As an example, we will consider a white piece of paper with black letters printed on it. Whatever the light level, we always will experience the letters as black and the paper as white, even if the luminances are quite different. We will assume that the reflection factor of the paper equals 90% and that of the ink of the letters 3%. In Table 8.1.2, the comparison is worked out for an office room with an illuminance of 1000 lux and a street with an illuminance of 10 lux. The luminances are calculated straightforwardly; the luminosities are derived from Figure 8.1.6.

Material	Reflection	Office 1000 lux		Street 10 lux	
		L	S.B.	L	S.B.
paper	0,9	286	60	2,86	22
ink	0,03	9,6	8	0,096	5

Table 8.1.2 *The luminance and the luminosity of paper and ink in different surroundings (L: Luminance; S.B.: Subjective brightness or luminosity).*

From Table 8.1.2, it follows that the luminance of the letters indoors is much greater than the luminance of the paper in the street. Still, we will never be in doubt that the letters are printed in black. Part of this may be explained by the fact that the luminosity of the letters indoors indeed is lower than the luminosity of the paper outdoors.

The human observer; visual perception 285

However, this explanation does not seem to be sufficient. In our example, we assumed that we knew beforehand that the paper was white and the letters were black. Another example is the brightness of the ceiling in a room with daylight entering at one side only, like e.g. an ordinary classroom. The luminance of the ceiling might easily differ from one point to another by a factor of 2 or even more. Observers will have no trouble at all to see this; still, they will assume that the colour and the reflection factor over the whole ceiling are the same, because that is normally the case in class-rooms and that is what the observers may expect.

This touches to one of the basic concepts of visual perception: observers usually will 'see' what they expect. The expectancy plays a crucial role here. The importance of the expectancy and its implications are discussed in Narisada & Schreuder (2004, sec. 10.3.2) and Schreuder (2008).

Another example is the Moon. We all know that the Moon shines with a 'silvery light'. It looks like silver because it is so much brighter than the night sky. The luminance of the night sky is about 10^{-4} cd/m² (Narisada & Schreuder, 2004, sec. 3.1). However, the Moon is illuminated by the Sun. The illuminance on the surface of the Moon – in vacuum! – is well over 150 000 lux. The reflectance is very low indeed, not more than about 3%. Assuming diffuse reflection, the luminance of the Moon is about $(0,03/\pi) \cdot 150\,000 = 1430$ cd/m² – many million times brighter than the night sky. No wonder it looks like silver. So, even in murky and misty conditions, or in twilight, we will see the Moon as a silver disk. This impression is not altered by the fact that, since the Apollo project, we know that the reflection factor of the Moon is about 3 to 5% – about the same as that of a cinder path! Even more so, a clever experiment, first reported by Sir John Herschel, allows us to see the fact that the Moon actually is fairly dark. All one needs to do is to look at the rising full Moon at the same time as looking at a wall that is illuminated by the setting Sun (Minnaert 1942, volume I, p. 86-87). A direct comparison between the Moon and the wall will show that the Moon is dark grey, not silver at all! And still we all persevere in this illusion.

(b) Movement detection

The perception of movement is complex and is only partly understood. The effects of constancy that are explained above play an important role. It is easy to understand how the movement of an object is detected when the eye does not move: the image of the object moves over the retina. However, if the eyes move as well, the visual system seems to be able to make a distinction between the movement of the retinal image due to the movement of the object itself and that of the eyes. It is clear that the muscles that govern the eye movement, play a role.

Modern studies have clarified some aspects of the role of the superior colliculus (SC) in controlling saccade duration, accommodation, fixation behaviour, smooth pursuit, vergence, coordinated movements of the eye and the head, and reaching movements of the arms (Ghandi & Sparks, 2004, p. 1462). The SC is part of the roof of the midbrain (Ghandi & Sparks, 2004, p. 1449). It seems that the role can be succinctly described as a control loop that governs muscles of the eye and other body parts on the basis of an error signal received by the SC. It is essentially a feed-back system. Still, although one may designate brain areas that relate to movement perception, the role of motion regions in the control of eye and body movements needs more work (Orban & Vanduffel, 2004, p. 1241).

In the literature, one may find a number of elegant, simple, but convincing test that may demonstrate the role of the eye muscles (Gregory, 1965, chapter 7). A slight pressure on the side of the eye may move the eye ball slightly. Objects in the surroundings are seen to move as well. Apparently, the mechanical movement of the eye ball circumvents the constancy effect, contrary to eye movements that are made by using the normal eye-movement muscles.

8.2 Blinding glare

Glare is related, as the word says, to phenomena that hamper or even obstruct visual observations. Glare has three aspects:
1. Absolute glare or blinding glare;
2. Disability glare;
3. Discomfort glare.

When the intensity of the light stimulus is over the maximum value where the visual system can process visual information, one speaks of blinding glare. The stimulus does not contribute to the visual information that is acquired from the surroundings, but it obstructs the processing of useful information. One is, to a certain extent, really 'blind' (Narisada & Schreuder, 2004, sec. 9.2.1; Schreuder, 1998, sec. 7.3.1). As is explained in sec. 7.4.2, where the range of useful luminances is discussed, the stimulus value where blinding glare takes place, depends on the adaptation luminance.

Blinding glare occurs rather frequently in daily life:
- when, in the dark, an opposing car does not switch back to low beam headlights;
- when leaving a road tunnel at daytime;
- when leaving a dark cinema hall in full sunshine;
- when driving on a wet, shiny asphalt road surface against a low sun;

- when walking (or skiing!) on snow in full sunshine;
- when sunbathing on a tropical beach in full sunshine.

It might be noted here that Vos proposed a further subdivision of 'sorts of glare' where he introduced a glare cube that contains eight different sorts of glare (Vos, 1999, p. 40-41, Fig. 5). Apart from the glare angle, the temporal effects have also been included so that the system could also cover flashing lights. Time must learn whether such a complicated system is an improvement for practical lighting engineers.

8.3 Disability glare

8.3.1 Sources of disability glare

(a) *Glare sources, the glare angle*

As is explained in sec. 8.1.1, the field of view in human vision is very large. It may reach up to an overall width of up to 200°, depending on its definition and its method of assessment. However, fine details can be perceived only in a very small area around the line of sight. It appears in practice that this line of sight usually coincides with the line of the highest interest. In general terms, the line of sight is along the tangent of the path that can be defined in the optical flow (Warren, 2004, p. 1254, footnote). In other words, in unrestrained vision, people usually look at the most interesting things in their surround. Sometimes the required eye adjustments are made consciously, but in most cases they are involuntarily. As is explained in sec. 6.3.1, the optical axis of the eye connects the centre of the eye lens, or rather the centre of the iris to the fovea. This automatically ensures the highest resolution of the visual perception. In most practical lighting applications the main visual task means the perception of small objects or of small elements of larger objects. The area of high resolution measures only two to three degrees in diameter (Bunt & Sanders, 1973).

In many cases there are bright sources of light that lay outside the line of sight, but may, as a result of their brightness, disturb the normal perception process. It is said that they cause disability glare. The seeing ability of the eyes at the line of sight is deteriorated. Under the circumstances such disturbing light sources are termed glare sources.

It seems to be generally accepted that the light scattered within the eye itself is the main cause for the disruption of the visual process. Of course, also light that is scattered outside the eye may cause trouble, but conventional wisdom has it that this is not included in what is called 'glare'. In many practical cases, the light scattered in car windscreens, in spectacle lenses, or in the atmosphere restricts vision much more severely that disability glare (Middleton, 1952; NSVV, 2003; Padmos & Alferdinck, 1983; 1983a; Schreuder, 1971, 1991).

It seems to be generally accepted, often tacitly, that all intra-ocular scatter is circle-symmetric. See e.g. Anon. (2007a); Franssen & Coppens (2007); Narisada & Schreuder (2004, sec. 9.2); Van Den Berg (1995); Vos (1963, 1999, 2003), and Vos et al. (2002). This is a very important statement, as it allows to describe all disability effects with one angular variable alone: the glare angle. The glare angle is the angle between the direction of the glare source and the line of sight. Conventionally it is indicated by θ. In astronomical terms: the elevation is enough, and the azimuth is not needed (Weigert & Wendker, 1989).

(b) Stray light in the eye
Disability glare is often called 'physiological glare' because its effects are of a physiological nature. It occurs when the field of view contains one or more glare sources. Part of the light from the glare source that enters the eye is scattered within the ocular media. It is termed the intra-ocular stray light (Van den Berg, 1995). Part of the light is scattered over large angles. One may assume that it gets 'everywhere' in the eye. It causes a light veil that seems to stretch over the complete field of view. One can define straylight in terms of the point spread function or PSF that is explained in sec. 6.3.3d. The PFS is known to spread over the full retinal surface, up to distances of 90° and more of the PFS-centre (Vos, 1984). It should be added that, as is explained in sec. 6.3.4b, the effectiveness of light to provoke a brightness sensation, desired or not desired, depends on the direction of light incidence upon the retina. This is called the Stiles-Crawford effect.

It is conventional wisdom that disability glare is exclusively the effect of intra-ocular stray light. "Already in the first half of the 20th century studies led to the now generally accepted view that disability glare can be fully understood on the basis of the optical phenomenon of light scattering in the eye, leading to stray light at the retina" (Anon., 2007b, p. 29). Since then, the words 'straylight' and 'disability glare' are used as synonyms. However, other studies suggest that there are neural effects involved as well (Schouten, 1937; Schreuder, 1964). In some cases, e.g. near the entrance of a dark traffic tunnel in a surrounding of snow in the full sun, discrepancies have been found, suggesting that the stray-light theory of glare is not sufficient to explain all glare phenomena (Schreuder, 1964, 1998). However, in practice it seems that the stray light components usually are more important than the neural effects (Van Den Berg, 1995, p. 52).

(c) *The equivalent veiling luminance*

Ocular stray light hits the retina in the same way as the image-forming light does. In other words, for the retina it is the same whether a number of photons originate from an object outside they eye or from the stray light inside the eye. In a further part of this section it is explained that most of the scattering particles are large in relation to the wavelength of the light, implying that the scatter is wavelength-independent and more or less diffuse. The light that hits the retina may be regarded as coming from a veil without structure. This in its turn implies that the stray light effect on the retina can be described by a – hypothetical – luminance veil outside the eye. The luminance of the veil must be regarded as being equivalent only. Hence the term equivalent veiling luminance or L_{seq}. In this way, L_{seq} can be expressed in cd/m². Further parts of this section are devoted to the quantitative assessment of the equivalent veiling luminance L_{seq}.

(d) *The nature of disability glare*

The type of the light scatter depends, of course, on the nature of the scattering particles. Mostly these particles are large in relation to the wave-length of the light, and they scatter light more than that they absorb it (Van Den Berg et al., 2007, p. 186). Most studies are made regarding irregularities in the eye lens. However, it is generally assumed that other particles in the eye, like e.g. cells in the anterior chamber, structures in the vitreous bodies, and deposits on the lens and the cornea would only add to the intensity of the scatter without changing its appearance (Van Den Berg et al., 2007, p. 192). This statement agrees with earlier findings that claim that for disability glare effects it is sufficient to consider the eye lens alone (Wolf & Gardiner, 1965; Schreuder, 1976, p. 20-21).

8.3.2 Characteristics of disability glare

(a) *The effect of the light veil*

The veil influences the contrast in the field of view. As is explained in sec. 7.4.3a, the luminance contrast is usually defined as:

$$C = \frac{L_o - L_b}{L_b} \qquad [8.3\text{-}1]$$

with:
  C: The contrast;
  L_o: The luminance of the object to be perceived;
  L_b: The luminance of the background of the object.

The veil can be expressed in luminance values in the way that is described in another part of this section. It is called the veiling luminance L_v. Because the veil extends over the complete field of view, all luminances in the field of view become larger. To all luminances, the veiling luminance has to be added. This implies that the contrast becomes smaller, as can be seen from [8.3-2]:

$$C' = \frac{(L_o + L_v) - (L_b + L_v)}{(L_b + L_v)} = \frac{(L_o - L_b)}{(L_b + L_v)} \qquad [8.3\text{-}2]$$

C' is always smaller than C: the nominator in [8.3-1] and [8.3-2] is the same, whereas the denominator is greater. Usually, C' is called the visible contrast and C the intrinsic contrast. The relations [8.3-1] and [8.3-2] describe what is the effect of disability glare: it is a reduction of the contrasts in the field of view, and hence a deterioration of the visual performance. This effect has been mentioned at several places throughout this book.

In practice, usually not only one, but a number of glare sources contribute simultaneously to the glare effect. Additionally, many glare sources are extended and do not represent point-sources. As long as we consider only the physical phenomena of light scatter, the effects of the different sources are additive, and that the effects of large sources may be integrated (CIE, 2002a).

(b) *Colour effects of disability glare*

Glare is the result of the scatter of light by particles in the eye. This type of phenomena are described in general optics by effects of diffraction. It is explained in sec. 10.1.1 that diffraction is wavelength dependent, the sort of dependency being related to the nature of the relevant particles. One would therefore expect that disability glare would depend as well on the wavelength of the light. In fact, this is correct (Coppens et al., 2007).

What happens depends on the size of the particles as compared to the wavelength of the light. For particles smaller than 0,1 of the wavelength of the light, the scatter of the light follows Rayleigh's Law:

$$s = K \cdot \frac{(n-1)^2}{N \cdot \lambda^4} \qquad [8.3\text{-}3]$$

in which:
    s: the scattering per unit of volume;
    K: a constant;
    n: the refraction index of the particles;
    N: the number of particles per cm^3;
    λ: the wavelength of the light.

This way to denote Rayleigh's Law is from Minnaert (1942, Vol. I, p. 240). See also Narisada & Schreuder (2004, sec. 8.3.7d, p. 218-219); Schreuder (1998, sec. 4.4., p. 30-

31); Van de Hulst (1981), and Douglas & Booker (1977). However, as is mentioned earlier in this section, the particles that cause ocular straylight are large compared to the wavelength of the light, so the strong colour dependency according to Rayleigh's Law is not to be expected (Vos, 1963, p. 77; Devaux, 1965).

(c) *Practical implications of the colour effects*
As mentioned in the preceding parts of this section, one would expect some sort of relation between the scatter and the wavelength of the light. Indeed, this is in general terms the case (Coppens et al., 2007).

In the past, people did not agree in how far one should take this relation into account for practical lighting applications, particularly in outdoor lighting situations. For indoor lighting this was never considered as a serious point, because in interior lighting, more in particular in functional interior lighting, all light sources were, within narrow limits, white (Jansen, 1946; De Boer & Fischer, 1981; Anon, 2007).

In outdoor lighting, this was different. In many cases, light sources of quite different colours were used, both in vehicle headlighting and in road lighting. First we will discuss the yellow headlamps that have been obligatory for many decades for motor vehicles in France. It was shown that intra-ocular stray light does not follow the λ^{-4} relation one would expect for Raleigh-type scatter for particles that are small in relation to the wavelength (Stiles, 1929). The Rayleigh scatter is described in the preceding part of this section. See also Van De Hulst (1981) and Middleton (1952). Still, many assumed that the lack of blue light would decrease ocular straylight (Le Grand, 1937). Also, less atmospheric scatter by water vapour, dust, and fog particles was suggested (Monnier & Mouton, 1939). The light transmission and light scatter in haze and in fog is discussed in Douglas & Booker (1977); Kocmond & Perchonok (1970); Middleton (1952); OECD (1976, 1976a, 1980), and Schreuder (1998, sec. 4.4, p. 32-33).

In the course of the discussion about any possible preference of yellow or white headlights for cars, many more aspects have been brought to the fore, many with elements of truth in them, and some that are just figments of imagination. In a very thorough study all possible arguments have been judged (Schreuder, 1976). The study included aspects of visual acuity, contrast sensibility, detection, speed of seeing, re-adaptation, age-effects, atmospheric scatter, and, most prominently, glare. The end conclusion was that there are some grounds for a slight preference of yellow headlights. However, this is cancelled out almost exactly by the light loss in the required yellow filters, so that the net result is almost exactly zero (Schreuder, 1976, p. 9-10).

Since then, the problem disappeared because the use of yellow headlamps is prohibited in Europe. This also holds for France, being a member of the European Union. The rationale is that vehicle lights have an important function as signalling lights indicating several vehicle aspects, like type and position of the vehicle, speed and direction of travel and its changes. The leading ideas are that position is indicated by the colour of the light: white for the front, yellow for the sides, and red for the rear. A further principle is that steady-states are indicated by steady lights and changes by intermittent – flashing – lights (Roszbach, 1972; Schreuder & Lindeijer, 1987). In this context, yellow headlamps simply do not fit.

The second aspect of possible colour effects of disability glare is related to the use of coloured light sources for public lighting (Narisada & Schreuder, 2004, sec. 11.3). In the first decades of road lighting, gas lamps, and later electric incandescent lamps were the rule. As is explained in sec. 2.3.1, incandescent lamps, electric or otherwise, are thermal radiators that show a continuous spectrum. Continuous spectra usually are whitish.

From the 1930s onward gas-discharge lamps took over, first the clear ones later the fluorescent high-pressure mercury lamps. Also sodium lamps were applied. Barrows, when discussing street lighting systems, mentions all three types: incandescent lamps, sodium-vapour lamps, and high-intensity mercury-vapour lamps (Barrows, 1938, p. 403-412). As is explained in sec. 9.4, the colour impression and the colour rendering of high-pressure mercury lamps, clear and fluorescent, are inferior to those of incandescent lamps. Still, their overall colour impression is also whitish – be it that whitish is interpreted rather widely. The concepts of colour impression and colour rendering are explained Chapter 9.

Also in the 1930s, another lamp type became conspicuously prominent in road lighting: the low-pressure sodium lamp. As is explained in sec. 2.4.3c, low-pressure sodium lamps emit their light in a very narrow band in the yellow part of the spectrum. For all practical considerations, they can be regarded as being monochromatic lamps. This implies that it is not possible to see any colours in the light of these lamps. As is explained sec. 11.1.3, this is considered as a grave disadvantage, particularly for the lighting of residential areas where amenity and social safety are the main quality criteria (Schreuder, 1998, sec. 7.4.4; Boyce, 2003, p. 416). CIE states that: "Monochromatic light sources should be avoided for areas where the crime rate is high, that are environmentally sensitive, or where pedestrian activities predominate" (CIE, 1995, Clause 9.3.4).

In the 1950s and 1960s it was a hotly debated issue whether low-pressure sodium lamps or high-pressure mercury lamps had to be preferred in road lighting, more in particular in traffic route lighting. The issue was never discussed in clear terms. There were two reasons for that. The first was that low-pressure sodium lamps were produced by a limited number of manufacturing companies, whereas high-pressure mercury lamps were produced by many companies all over the world – a point of market share. The other was that the driving task, and hence the vision task, of motor vehicle drivers had not yet been investigated in any depth. Almost all considerations regarding road lighting requirements were focussed on the detection of small, dark, diffuse objects on the road. So most research was focussed on this point as well. Surveys of the many studies can be found in De Boer, 1951, 1959, 1961, 1967; Narisada & Schreuder, 2004; Schreuder, 1998; Van Bommel & De Boer, 1980.

The fact that low-pressure sodium lamps emit monochromatic light, contrary to all other light sources that are commonly used in outdoor lighting applications, would imply a strong effect on disability glare, provided of course that such influences do exist in the first place. Curiously enough, this course of reasoning has been followed hardly, if at all. It seems reasonable to assume that the lack of insight in the driving task resulted in an approach where road luminance and glare were considered as unrelated subjects. It might be added here that, as is explained in sec. 11.2.1, the road surface luminance and the glare were the two quantitative photometric characteristics on which the quality requirements of road lighting were based. And so it could happen that the requirements for the road luminance were objective, whereas the requirements for glare were subjective. What 'subjective' means is explained in a following part of this section, where discomfort glare is discussed. The peculiar dichotomy, that cannot be explained rationally, is most clearly demonstrated in the work of De Boer and his collaborators (De Boer, 1951, 1959, 1961, 1967: De Boer et al., 1959; De Boer & Van Heemskerck Veeckens, 1955; De Boer & Schreuder, 1967). One of the consequences of this work was that for several decades road lighting recommendations and standards were based on discomfort glare, and not on disability glare as the main glare criterion (CIE, 1965, 1976, 1977, 1992, 1995; NSVV, 1957, 1974/1975, 1977, 1990). Discomfort glare is discussed in sec. 8.4.

After many years of debate, the discussions gradually petered out. There are three reasons for that:
1. Developments in lamp technology resulted in a great variety of lamps types that were used in road lighting. More in particular, in traffic route lighting, the whitish high-pressure sodium lamps almost completely took over from low-pressure sodium lamps and high-pressure mercury lamps;
2. In residential lighting the use of the white compact fluorescent lamps became ever more important;
3. The visibility approach became more prominent in road lighting theory, so that the dichotomy of ability and comfort disappeared (Adrian, 1993, 1995; Janoff, 1993; Keck, 1993; Narisada, 2007; Narisada & Karasawa, 2001; Narisada et al., 2003; Narisada & Schreuder, 2004, sec. 11.8);

4. The steep increase in the length of lit roads worldwide, and urge for cost-reduction caused to consider disability glare as something one must learn to live with, and disability glare as a luxury one could do just as well without.

(d) The four-component model

Recent improvements to the measuring methods allow more precise measurements concerning the colour effects of disability glare (Franssen et al., 2007). A more precise analysis did show that the light scatter is not a homogeneous phenomenon, but the combination of several more or less independent effects that each show a rather different dependence on the wavelength. The first is the light transmitted by the eye wall, or scattered by the retina. As these organs contain many blood vessels the stray light is highly red-dominated. It was found, however, that the red-dominance in its turn does depend heavily on the pigmentation of the eye, which can easily be observed from the colour of the iris. Secondly the scatter from the cornea and the eye lens are blue dominated, as is to be expected from the way the light is scattered by small particles, as is explained in an earlier part of this section (Coppens et al., 2007, p. 123; 124).

The studies led to the formulation of a four-component model (Coppens et al., 2007, p. 125):
1. Scatter in the cornea;
2. Scatter in the lens;
3. Translucency of the ocular wall;
4. Reflection of the fundus.

These different effects are described in Vos (1984) and in Van Den Berg et al. (1991).

From a further analysis two major effects became clear, effects that did not receive all too much attention in the past, viz. age and pigmentation. It was shown that the corneal contribution is independent of age (Van Den Berg & Tan, 1994) and also of pigmentation. The lenticular contribution depends on age and both other contributions depend on pigmentation (Van Den Berg et al., 1991; Van Den Berg & Tan, 1994).

So the four-component model can be rewritten as a three-component model:
1. Basal component;
2. Age-component;
3. Pigmentation-component.

The model is additive. The start is the 'base' value of young, healthy, and well-pigmented persons. This part follows closely the λ^{-4}-relation of Raleigh-like scatter (Coppens et al., 2007, p. 122). The age-component and the pigmentation-component can be

The human observer; visual perception

straightforwardly added, be it that they are weighted. Both are discussed in other parts of this section.

The end result is depicted in Figure 8.3.1.

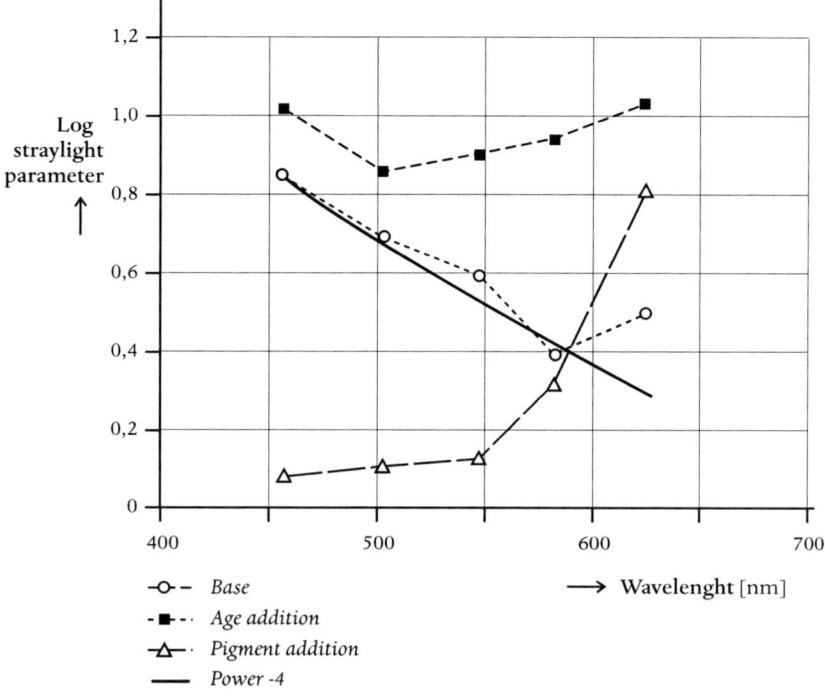

Figure 8.3.1 The three components of retinal scattering (After Coppens et al., 2007, fig. 4, p. 127).

(e) *The θ-dependence*

On the basis of the considerations of the preceding parts of this section it can be concluded that:

1. It is possible to define a Standard Glare Observer similar to the Standard Observers for Photopic and Scotopic Vision, as described in sec. 7.1.1a;
2. Disability effects are for practical lighting applications considered as caused exclusively by intra-ocular straylight, implying that time effects can be disregarded, and that glare effects from multiple glare sources or from extended glare sources can be assessed by summation or by integration respectively;
3. All glare effects show circular symmetry around the optical axis of the eye;
4. Colour effects of disability glare can be disregarded;

5. The only remaining variable for the assessment of all disability glare effects is the glare angle, with the proviso that correction factors for age and eye pigmentation must be taken into account.

So, by means of a process of elimination, the only variable in the assessment of the disability glare according to the definition of the CIE standard observer for disability glare is the glare angle θ. It is therefore sufficient to investigate the dependence of the glare effects in relation to this angle – with the proviso added earlier about age and eye pigmentation. And this is precisely what has been done over the last century.

The geometry that is used for studying the θ-dependence in glare assessments is depicted in Figure 8.3.2.

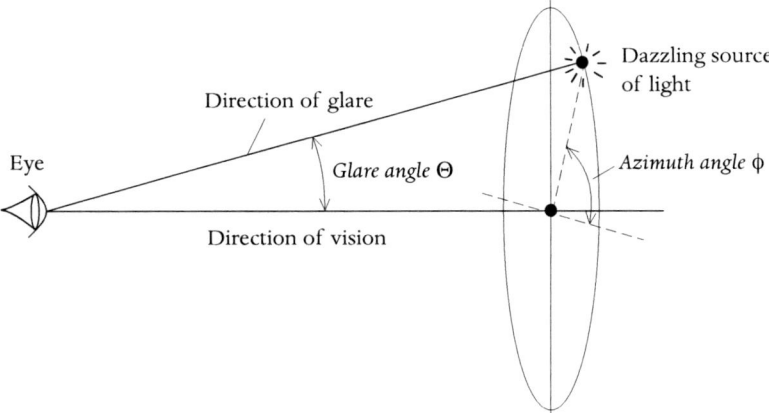

Figure 8.3.2 *The geometry used in glare assessments. θ: glare angle; φ: azimuth angle (After Narisada & Schreuder, 2004, fig. 9.2.1).*

(f) *Age effects of disability glare*

It is well-known that the eye media become turbid with increasing age (Schreuder, 1998, sec 7.5). Since disability glare is the result of scatter of light in the eye, the glare is bound to increase with age as well, the main reason of the steep increase of cataract with age (Mann & Pirie, 1950; Stilma, 1995). See also Augustin (2007, chapter 25). Cataract is the most common visual disturbance. In 1990, of the 15 million inhabitants of the Netherlands, 760 000 suffered from cataract in a disabling way (Schreuder, 1998, table 7.5.1.; Voorn, 1995, table 1-3).

It has been found that straylight, or disability glare, increases with age by a factor

$$1 + (A/D)^4 \qquad [8.3\text{-}4]$$

with
> A: age (in years);
> D: a factor (in years).

Usually, D is taken as being between about 62,5 and 70 (Anon., 2007b, p. 30-31).

As regards the θ-dependence, Fisher & Christie (1965) proposed to take account of the age dependence by exchanging the constant factor k in equation [8.3-5], the Stiles-Holladay relation by:

$$k = d + 0{,}2\,A$$

with A the age in years and d a factor depending on the geometry. According to Schreuder (1998, p. 111), d = 4,2. This means that if we take, according to Adrian (1963), k = 9,2 for 25 year old observers, k would be 14,2 for 50 year old people and 19,2 for 75 year old people (Schreuder, 1998, p. 111). Although is has not been mentioned anywhere at all, it seems safe to assume that the Fisher-Christie relation applies to healthy individuals, i.e. individuals who do not suffer from cataract in a disabling way.

> (g) *The limits for the θ-dependence*

As mentioned before, the θ-dependence has been the objects of many extensive studies over many decades. Surveys may be found in Anon. (2007b); CIE (2002); Narisada & Schreuder (2004, sec. 9.2.2); Schreuder (1964, 1981, 1998); Vos (1963, 1983, 1984, 1999, 2003); Vos & Padmos (1983); Vos & Van Den Berg (1999); Vos et al. (1976, 2002).

The first proposal that was applied on a large scale was that of Holladay (1927). This proposal is very simple: the glare depends on the illuminance that hits the eye and on the square of the glare angle. The relation is reproduced here in the form that was proposed by Holladay, Stiles and Crawford. It is usually called the 'Stiles-Holladay relation' (Stiles & Crawford, 1937; Adrian, 1961):

$$L_{seq} = k \cdot \frac{E_e}{\theta^2} \qquad [8.3\text{-}5]$$

with:
> L_{seq}: the equivalent veiling luminance (cd/m²);
> E_e: the illuminance on the plane of the eye pupil (lux);
> θ: the glare angle (degrees);
> k: a factor that depends on the age and on other parameters. It usually taken as 10.

It should be noted that 'time' is not an important parameter for disability glare, because the decrease in contrast takes place simultaneously with switching on the glare source.

It is not possible to get accustomed to disability glare, contrary to discomfort glare that is discussed in sec. 8.4.

There is, however, some difference of opinion about the θ-range over which it may be used. It seems that the difference of opinion results mainly from the fact that in different fields of application different degrees of accuracy are required. In other words, what is good enough for some applications may not be good enough for others. An estimate that seems to be at the safe side is given by CIE: "In the main practical glare angle domain, roughly between 3 degrees and 30 degrees, the description could be reliably used" (Vos, 1999, p. 38). In some field of application, lighting designers are more generous and allow a domain between 3° and 60° (Adrian, 1993) or even 2° and 60° (NSVV, 2003).

We may be brief about the upper limit. First, in practical outdoor lighting installation a glare angle of more than 30° is quite rare. And further, because the Stiles-Holladay relation as given in [8.3-5] contains the square of the glare angle in the denominator, the contribution of glare sources at large angles can be disregarded completely as $30^2 = 900$ – about 0,1 of a percent! So it is immaterial for practical applications to ascertain whether the glare formula can be used up to 30°, or 60°, or 100°.

Contrary to this, the lower limit of the Stiles-Holladay relation is of considerable practical importance. A circular area with a radius of 3° around the optical axis of the eye means an area with a diameter of 6° in the centre of the field of view. It should be realised that an area of 6° is really a quite large, particularly in road lighting. At a distance of 150 m, a reasonable distance in high speed motor traffic, 6 degrees will span more that 15 m, more than the width of a normal four-lane freeway, the median included. The fact that 150 m is a reasonable distance in high speed motor traffic is explained in Schreuder (1991, 1991a).

It must be stressed that in most road lighting applications, tunnel entrance lighting and glare assessment of opposing car headlamps included, the lower limit of about 2° to 3° of the Stiles-Holladay relation can be used without appreciable problems. Still, in some cases that are by no means exceptional, an area of well under 1° in diameter has to be considered. We will give, as an example, some explanation here.

Road tunnels are major constructions, made to overcome obstructions to traffic. They may present an extra safety risk for road traffic, particularly for motor vehicles. The major visual problem for motorized traffic when passing a tunnel is the day-time entrance, particularly in full sunshine. When the tunnel is not adequately lit, the entrance is dark, so that no details can be seen. This is often called the black-hole effect (Schreuder, 1964, 1999; CEN, 2003; CIE, 1973, 1990; NSVV, 2003). Basically, the black-hole effect is a special

The human observer; visual perception   299

case of disability glare. The bright surfaces near the tunnel portal act as a glare source. As an example, in Figure 8.3.3, the situation near a motorway tunnel in a flat country is depicted. This is often the worst situation as regards the black-hole effect (CEN, 2003).

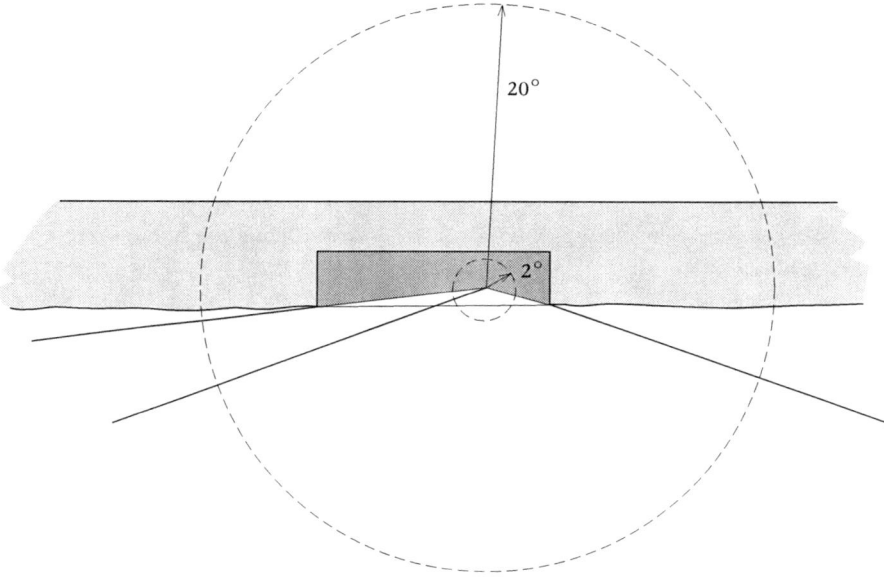

Figure 8.3.3 A tunnel entrance in a flat country.

From Figure 8.3.3 it can be seen that in such a case the brightest parts of the field of view are the sky over the tunnel portal, the tunnel facade right over the entrance itself, and the road surface right in front of the tunnel. For a tunnel portal of 5m high and 12 m wide, an eye height of 1,5 m and a distance of observation of 150 m, the tangent angles are about 2,3°, 1,3°, and 0,6°, as is depicted in Figure 8.3.3. So a lower limit of 3° as is given for the Stiles-Holladay relation is not small enough for tunnel lighting design (CEN, 2003; CIE, 1973, 1984, 1990; NSVV, 2003; Schreuder, 1998, chapter 15, 1999; Vos, 1983). An extension of the Stiles-Holladay relation for smaller angles is needed.

When considering the glare in vehicle headlighting this need is even greater. Car headlamps are aimed in such a way that the main luminous flux is emitted in a direction that is almost parallel to the road surface. Only in this way the required reach of headlamp beams can be realized. Traditionally, headlamp systems have two beams; the driving beam that is used when there is no opposing traffic, and the passing beam that is used when other traffic is met. Further details can be found in Narisada & Schreuder (2004, sec. 12.2), Schreuder & Carlquist (1969), and Schreuder & Lindeijer (1987, sec. 2.1).

In almost all cases, vehicle headlights are mounted lower than the eye height of drivers of normal cars and trucks to which the headlights are attached. This means that the maximum reach can be arrived at when the amount of light just under the horizon is as large as possible, because that can not cause direct glare for opposing drivers. The direct glare for opposing drivers is a minimum when the amount of light just above the horizon is as small as possible. What is needed, therefore, is a steep decrease in luminous intensity when crossing the horizon from below. Such steep decrease is called the coupure or cut-off (Anon., 2001, sec. 6.2.1, p. 10). This coupure is the most essential feature of passing-beam headlighting systems (Narisada & Schreuder, 2004, fig. 12.2.1). In order to give an idea of the angles involved, the standardized measuring screen that is used for measurements is depicted in Figure 8.3.4. See Schreuder & Lindeijer, 1987, Annex 1, Figure 1.

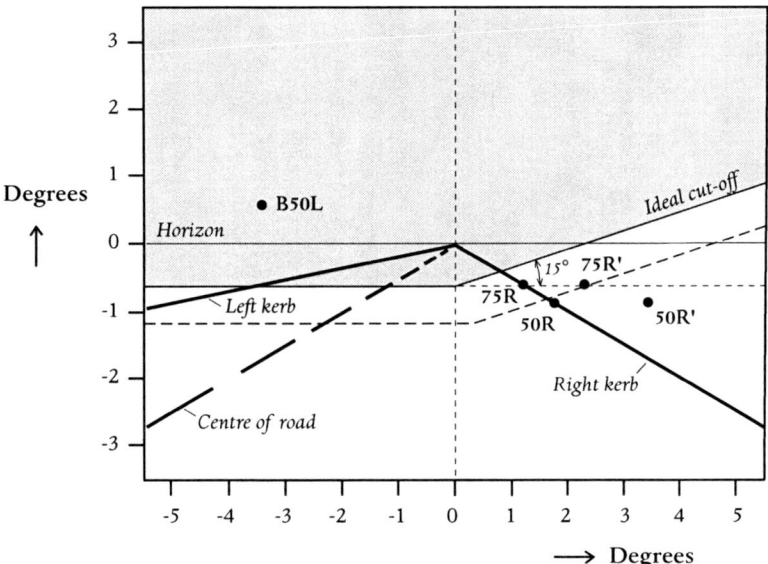

Figure 8.3.4 The standardized measuring screen for the measurement of vehicle headlamps (After Narisada & Schreuder, 2004, fig. 12.2.1a).

In the meeting situation on a two-way, two-lane rural road, the glare angle can be very small indeed. On a 7 m wide road – quite common – the near-side lamp of the one car and the eye of the driver of the other car are about 1,5 to 2 m apart. At a meeting distance of 50 m, the corresponding glare angle is less than 2° – the limiting value of validity of the Stiles-Holladay relation. At a more crucial distance between the two opposing vehicles of 200 m, the glare angle can be as small as 0,4°, or about 25 minutes of arc (Narisada & Schreuder, 2004, sec. 12.2.7c; Schreuder & Carlquist, 1969).

It may be concluded that, although the Stiles-Holladay relation is far from exact, it is still widely in use in its original form in many practical applications. However, glare angles below 3° will often occur.

Over the decades, many proposals have been made to define another value for the dependency from θ other than the square that is used in the Stiles-Holladay relation, particularly for smaller values of the glare angle. An overview of several proposals that have been made in the past is given in NSVV, 2003, Table 18-9, p. 218. Often, an exponent of θ of about 2,8 was found (see e.g. Hartmann & Moser, 1968; Hartmann & Ucke, 1974; Schreuder, 1981; Vos, 1963). The glare angle domain was extended to about 10 to 15 minutes of arc. Vos has proposed a new formula that was aimed in particular at the entrance lighting for long road traffic tunnels (Vos, 1983; Vos & Padmos, 1983). The formula is an extension of the Stiles-Holladay relation, as a term containing θ^{-3} is added.

$$\frac{L_{sec}}{E_e} = \frac{10}{\theta^2} + \frac{10}{\theta^3} \qquad [8.3\text{-}6]$$

It is recommended to use that formula in practical tunnel lighting design (NSVV, 2003, p. 219). It is assumed that this relation can be used down to glare angles of about 0,1°, or 6 minutes of arc; quite sufficient for almost all practical lighting applications.

CIE has defined three 'glare formulae' (CIE, 2002). The first is the CIE Age-adjusted Stiles-Holladay Disability Glare equation. Basically it is the traditional Stiles-Holladay relation to which an age-related factor is added. Just like the traditional Stiles-Holladay relation it is valid for values of θ between 1° and 30°. It may be noted that in this CIE report the lower limit of validity of the Stiles-Holladay relation is given as 1°, and not 3° as is given elsewhere. The second is the CIE Small Angle Disability Glare equation with an angular domain of 0,1° to 30°. The third is the CIE General Disability Glare equation with an angular domain of 0,1° to 100°. This general equation is based on the CIE Standard Glare Observer that is discussed in the next part of this section.

(h) The CIE Standard Glare Observer

In the late 1990s, CIE has defined a CIE standard observer for disability glare (Anon., 2007a,b; Vos, 1999, 2003; Vos & Van Den Berg, 1999; Vos et al., 2002). It must be stressed that the CIE standard observer for disability glare does not represent any living or dead person. It is a model in the sense that is adstructed in sec. 1.5.1, notwithstanding the usage of the term that might suggest otherwise. In this way, as in mentioned earlier, the Glare Observer is similar to the other Standard Observers that are introduced and defined by CIE. These standard observers are discussed in sec. 7.1.1a.

A long series of studies, researches, deliberations, and other considerations has lead in the end to the adoption by CIE of the Standard Glare Observer (Anon, 2007b, p. 36), or alternatively the General Glare Observer (Vos, 1999, eq. 4, p. 39). The two sources give equations that are different. As they are extremely complicated, it is difficult to judge whether the differences are material or not. So we decided to leave out the equations altogether. The data that are presented by Vos are depicted in Figure 8.3.5.

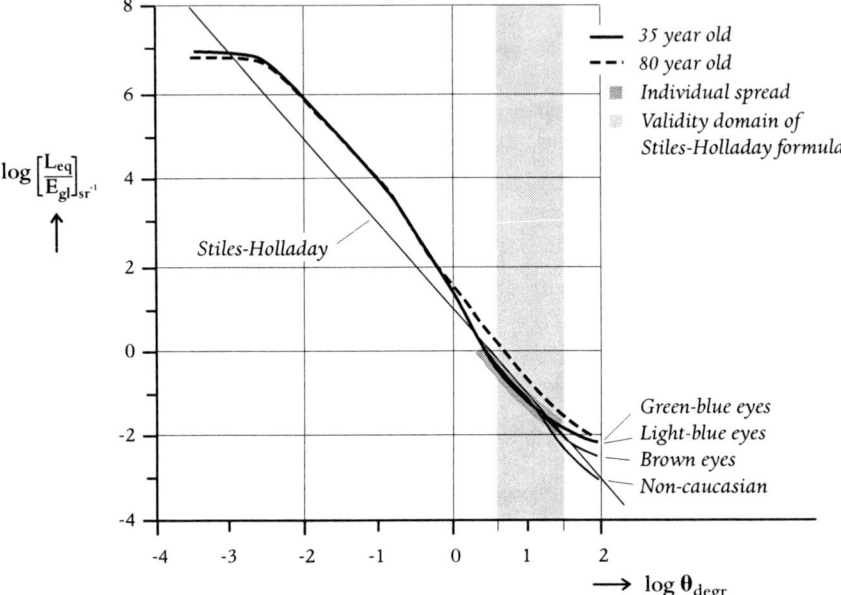

Figure 8.3.5 *Disability glare functions for four age/pigmentation conditions. The hatched zone indicates the individual spread in experimental data. The classic Stiles-Holladay relation is included, its realm of validity stretching from 1° to 30° (After Vos, 1999, fig. 1, p. 39).*

It is immediately clear from Figure 8.3.5 that the CIE Standard Glare Observer is valid over a much wider range of glare angles than the Stiles-Holladay relation. However, this wide range is not very relevant for most practical outdoor lighting installations. It is explained earlier in this section that in most practical cases the Stiles-Holladay relation is sufficient; in some cases it is recommended to extend it by adding the 'Vos term' that includes θ^{-3}.

One point must be added, however. The age dependence that is included in the CIE Standard Glare Observer seems to relate to the healthy elderly.

8.4 Discomfort glare

Discomfort glare is characterized by the fact that glare sources in the field of view cause disturbing effects and discomfort, although a decrease in the visual performance is usually not experienced. It is often called psychological glare because its effects are of a psychological nature, and because it is similar to 'real' or 'disability' glare. As is mentioned in an earlier part of this section, discomfort glare was in the past often considered as much more important than disability glare in outdoor lighting, notably in road and street lighting (De Boer, 1967; Schreuder, 1967a; De Boer, ed., 1967).

However, as has been explained earlier, in most modern recommendations the picture is different: disability glare is considered as the main or even as the exclusive glare aspect to be taken into account (CEN, 2003; NSVV, 2002). It was often argued that the glare of vehicle headlamps is so strong that it is futile to try to reduce the glare from road lighting installations. Furthermore, it has been argued that, if disability glare is absent, there will be no discomfort glare because the discomfort glare is dominant. Finally, reducing discomfort glare is often considered as a luxury on which no money ought to be spent. The recent national and international recommendations deal exclusively with disability glare and disregard discomfort glare.

Still, many road authorities and light designers are not fully satisfied by disregarding discomfort glare altogether. This is the reason to devote here a few words to the phenomenon. A full treatment of the subject van be found in the literature. We may refer to Boyce (2003, sec. 5.4.2.2); De Boer (1967); Narisada & Schreuder (2004, sec. 9.2.3b); Schreuder (1967a, 1998, sec. 7.3.3); Van Bommel & De Boer (1980, sec. 3.3).

Discomfort glare in road lighting has been a subject of intensive research for many years. The first studies were made in laboratory models (De Boer, 1951; De Boer & Van Heemskerck Veeckens, 1955; De Boer et al., 1959; De Boer & Schreuder, 1967; Schreuder, 1967; Adrian & Schreuder, 1968, 1971). The results have been validated in driving test in real streets (Cornwell, 1971; De Grijs, 1972). Based on these studies, a rather complicated formula is defined that relates the degree of discomfort glare to a number of geometric and photometric characteristics of a road lighting installation. The formula is created by Schreuder (1967; see also Adrian & Schreuder 1968; 1971; De Boer & Schreuder, 1967; Narisada & Schreuder, 2004, sec. 9.2.3). In 1976 it is accepted by CIE as the Glare Mark Formula (CIE 1976). Usually, G is called the 'Glare Control Mark' (Van Bommel & De Boer , 1980, p. 27; Baer, 1990, p. 76). The formula is incorporated in the CIE recommendations for road lighting, be it in an annex (CIE, 1995, Appendix B).

It proved to be difficult to use in practice. Several graphical methods have been proposed (Schreuder, 1967, p. 149-152; Adrian & Schreuder 1971; Baer, 1990, p. 77; Hentschel, 1994, p. 181-185). With the use of personal computers, such graphical methods are not needed any more.

Several simplifications have been proposed over the years like e.g. CIE (1994, formula 3.3); Einhorn (1991); Schreuder (1998, p. 114).

There is one more thing to mention. It seems that each area of light application has its own preferred method to assess discomfort glare. In the literature, at least four different methods can be found, each with its own glare formula to calculate the glare and with its own design methodology:
- Vehicle headlighting – glare by low beam headlamps;
- Road lighting – glare by fixed road lighting luminaires;
- The lighting of outdoor sports facilities – glare by lighting fixtures, both as regards the spectators and the sport participants;
- Indoor working areas. Here the glare in offices, factories and other facilities are again treated in different ways. Narisada & Schreuder (2004, sec. 9.2.3g) have proposed a general discomfort glare formula. This proposal is explained in Schreuder (2008).

It is interesting to consider the relation between discomfort glare and disability glare. In Figure 8.4.1, this relation is depicted for the luminance region that is relevant to most outdoor lighting.

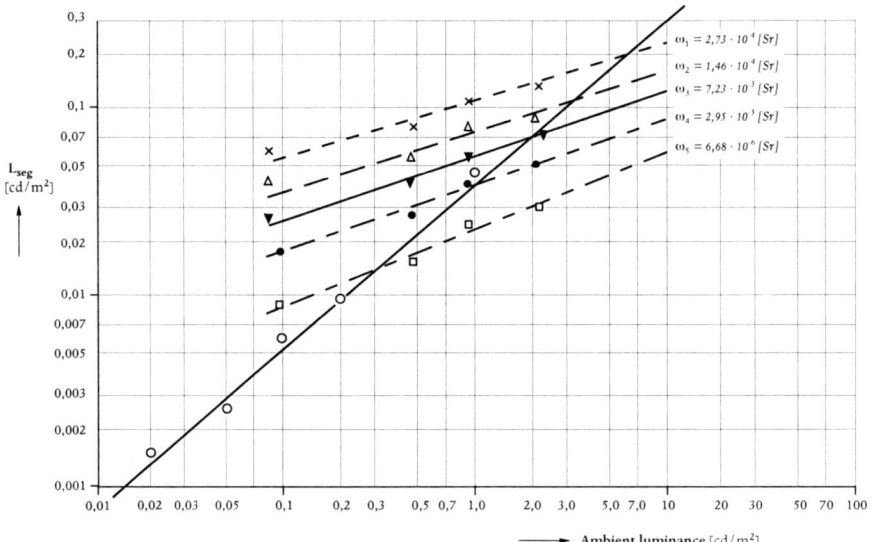

Figure 8.4.1 The relation between discomfort glare and disability glare (After Narisada & Schreuder, 2004, fig. 9.2.3, based on Adrian, 1966).

Figure 8.3.6 shows that there is some sort of relation between the two glare aspects; however, the slope of the two sets of curves is quite different, so the two are not directly proportional. As may be directly experienced by being outdoors, in very well-lit streets, discomfort glare is predominant even when disability glare is almost absent, whereas in poorly-lit streets the opposite is true. Disability glare can be considerable even when discomfort glare is hardly noticeable or absent. Thus, in poorly-lit streets, reducing discomfort glare to a degree that it is not noticeable any more, may cause hazardous situations for traffic, particularly for pedestrians who, contrary to cars, have to do without any lighting of their own. This aspect is often overlooked by those people who advocate low levels of street lighting on the grounds that high, or even mediocre, lighting is not only a waste of energy and a source of light pollution, but causes objectionable glare as well (Crawford, 1991, 1997, 1994; Mizon, 2002).

8.5 Conclusions

In this chapter, the human observer is discussed, focussing on visual perception. This discussion lead to a number of relevant conclusions. As regards the visual field, it measures for two eyes about 200° wide and 140° high. For critical vision, the field is much smaller. It measures not more than 2° to 3° around the optical axis of the eye. Flicker-effects are important in two ways. First, the flicker-fusion frequency depends on the adaptation level, the wave-form, and the colour of the light. In most cases it is well over 60 c/s, implying that normal gas-discharge lighting does not evoke flicker effects. Discomfort effects are important for tunnel lighting. Frequencies between 5 and 9 c/s should be avoided.

For all practical purposes disability glare can be considered as the result of intra-ocular stray light. This means that multiple glare sources can be added up, and that large glare sources can be integrated. The effects of disability glare are best described by the CIE Standard Glare Observer. In many outdoor lighting applications the Stiles-Holladay relation can be used, be it that sometimes this relation must be extended to smaller glare angles.

References

Adrian, W. (1961). Der Einfluss störender Lichter auf die extrafoveale Wahrnehmung des menschligen Auges (The influence of disturbing light sources on the extrafoveal observation of the human eye). Lichttechnik. 13 (1961) 450-454; 508-511; 558-562.

Adrian, W. (1966). Neuere Untersuchungen über die Blendung in der Strassenbeleuchtung (New research on glare in road lighting). In: Anon., 1966.

Adrian, W. (1993). The physiological basis of the visibility concept. In: 'Visibility and luminance in roadway lighting'. LRI, 1993, p. 17-30.

Adrian, W. (1995). The visibility concept and its metric. In: Anon., 1995.

Adrian, W. & Schreuder, D.A. (1968). The assessment of glare in street-lighting. Light and Lighting. 61 (1968) 12: 360-361.
Adrian, W. & Schreuder, D.A. (1971). A modification of the method for the appraisal of glare in street lighting. In: CIE, 1971.
Albrecht, D.G.; Geisler, W.S. & Crane, A.M. (2004). Nonlinear properties of visual cortex neurons: Temporal dynamics, stimulus selectivity, neural performance. Chapter 47, p. 747-764. In: Chalupa & Werner, eds., 2004.
Anon. (1966). The lighting of traffic routes. NSVV-Congress, Amsterdam, April 1966. Electrotechniek, 44 (1966) 509.
Anon. (1995). PAL: Progress in automobile lighting. Technical University Darmstadt, September 26/27, 1995. Darmstadt, Technical University, 1995.
Anon. (2001). Uniform provisions concerning the approval of motor vehicle headlamps emitting an asymmetrical passing beam or a driving beam or both and equipped with filament lamps. Addendum 111: Regulation No. 112. Date of entry into force: 21 September 2001. Geneva, United Nations, 2001.
Anon. (2007). Handboek verlichtingstechniek (Illuminating engineering handbook). Den Haag, SDU, 2007 (laatste aflevering; losbladig; last installment; loose-leaf).
Anon. (2007a). Introduction to retinal straylight. Chapter 2, p. 17-26. In: Franssen & Coppens, 2007.
Anon. (2007b). History of straylight measurement; A review. Chapter 3, p. 27-44. In: Franssen & Coppens, 2007.
Augustin, A. (2007). Augenheilkunde. 3., komplett überarbeitete und erweiterte Auflage (Ophthalmology. 3rd completely revised and extended edition). Berlin, Springer, 2007.
Baer, R. (1990). Beleuchtungstechnik; Grundlagen. Berlin, VEB Verlag Technik, 1990.
Baer, R., ed. (2006). Beleuchtungstechnik; Grundlagen. 3., vollständig überarteite Auflage (Essentials of illuminating engineering, 3rd., completely new edition). Berlin, Huss-Media, GmbH, 2006.
Barrows, W.E. (1938). Light, photometry and illuminating engineering. New York, McGraw-Hill Book Company, Inc., 1938.
Bodmann, H.W. & Voit, A.E. (1962). Versuche zur Beschreibung der Hellempfindung (Experiments to describe the impression of light). Lichttechnik. 14 (1962) no. 8, p. 394-400.
Boyce, P.R. (2003). Human factors in lighting. 2nd edition. London. Taylor & Francis, 2003.
Bunt, A.A. & Sanders, A.F. (1973). Informatieverwerking in het functionele gezichtsveld; Een overzicht van de literatuur (Information processing in the functional field of view; A survey of the literature). Rapport nr. IZF-1973 C-8. Soesterberg, IZF-TNO, 1973.
CEN (2003). Lighting applications – Tunnel lighting. CEN Report CR 14380. Brussels, CEN, 2003.
Chalupa, L.M. & Werner, J.S., eds. (2004). The visual neurosciences (Two volumes). Cambridge (Mass). MIT Press, 2004.
CIE (1959). Proceedings of the CIE Session 1959 in Brussels. Publications 4...7. Paris, CIE, 1959.
CIE (1965). International recommendations for the lighting of public thoroughfares. Publication No. 12. Paris, CIE, 1965.
CIE (1971). Proceedings of the CIE Session 1971 in Barcelona (Vol. A, B, C). Publication No. 21. Paris, CIE, 1971.
CIE (1973). International recommendations for tunnel lighting, Publication No. 26. CIE, Paris, 1973
CIE (1976). Glare and uniformity in road lighting installations. Publication No. 31, Paris, CIE, 1976.
CIE (1977). International recommendations for the lighting of roads for motorized traffic. Publication No. 12/2. Paris, CIE, 1977.
CIE (1983). Proceedings of the 20th Session of the CIE, Amsterdam, 1983. Publication No. 56. Paris, CIE, 1999.

CIE (1984). Tunnel entrance lighting: A survey of fundamentals for determining the luminance in the threshold zone. Publication No. 61. Paris, CIE, 1984.
CIE (1990). Guide for the lighting of road tunnels and underpasses. Publication No. 26/2. Vienna, CIE, 1990.
CIE (1992). Guide for the lighting of urban areas. Publication No. 92. Paris, CIE, 1992.
CIE (1994). Glare evaluation system for use within outdoor sports and area lighting. Publication No. 112. Vienna, CIE, 1994.
CIE (1995). Recommendations for the lighting of roads for motor and pedestrian traffic. Technical Report. Publication No. 115-1995. Vienna, CIE, 1995.
CIE (1999). Proceedings of the 24th Session of the CIE, Warsaw, June 24-30, 1999. Publication No. 133. Vienna, CIE, 1999.
CIE (2002). CIE equations for disability glare. Publication No. 146, p. 1-12. In: CIE Collection on glare. CIE, Vienna, 2002.
CIE (2002a). Glare from small, large and complex sources. Publication No. 147. In: CIE Collection on glare. CIE, Vienna, 2002.
CIE (2003). San Diego Session of CIE. 2 volumes. Publication No. 152. Vienna, CIE, 2003.
Collins, J.B. & Hopkinson, R.G. (1957). Intermittent light stimulation and flicker sensation. Ergonomics. 1 (1957) 61-76.
Coppens, J.E.; Franssen, L. & Van Den Berg, T.J.T.P. (2007). Wavelength dependence of intraocular straylight, Chapter 8, p. 121-130. In: Franssen & Coppens, 2007. Reprinted from Experimental Eye Research, 82 (2006) 688-692.
Cornwell, P.R. (1973). Appraisals of traffic route lighting installations. Lighting Res. & Technol. 5 (1973) 10-16.
De Boer, J.B. (1951). Fundamental experiments of visibility and admissible glare in road lighting. Stockholm, CIE, 1951.
De Boer, J.B. (1959). La couleur de la lumière dans l'éclairage pour la circulation routière (The colour of the light in road traffic lighting). Lux. (1959). March – June.
De Boer, J.B. (1961). The application of sodium lamps to public lighting. Illum. Engng. 56 (1961) 293-312
De Boer, J.B. (1967). Visual perception in road traffic and the field of vision of the motorist. Chapter 2. In: De Boer, ed., 1967.
De Boer, J.B. & Fischer, D. (1981). Interior lighting (second revised edition). Deventer, Kluwer, 1981.
De Boer, J.B.; Burghout, F. & Van Heemskerck Veeckens, J.F.T. (1959). Appraisal of the quality of public lighting based on road surface luminance and glare. In: CIE, 1959.
De Boer, J.B. & Schreuder, D.A. (1967). Glare as a criterion for quality in street lighting. Trans. Illum. Engn. Soc. (London). 32 (1967) 117-128.
De Boer, J.B. & Van Heemskerck Veeckens, J.F.T. (1955). Observations on discomfort glare in street lighting. Influence of the colour of the light. Zürich, CIE, 1955.
De Boer, J.B., ed. (1967). Public lighting. Eindhoven, Centrex, 1967.
De Grijs, J.C. (1972). Visuele beoordelingen van verlichtingscriteria in Den Haag en Amsterdam (Visual assessments of lighting criteria in The Hague and Amsterdam). Electrotechniek. 50 (1972) 515-521.
De Lange, H. (1957). Attenuation characteristics and phase-shift characteristics of the human fovea-cortex systems in relation to flicker fusion phenomena. Delft University, Doctoral Thesis, 1957.
Devaux, P. (1956). Unified European passing beam and yellow light. Int. Road Safety Traffic Rev. 4 (1956) 33.
Douglas, C.A. & Booker, R.L. (1977). Visual Range: Concepts, instrumental determination, and aviation applications. NBS Monograph 159. National Bureau of Standards, Washington D.C., 1977.
Dunbar, C. (1939). Visual efficiency in coloured light. Trans. Illum. Engng. Soc. (London). 4 (1939) 137-151.

Einhorn, H.D. (1991). Discomfort glare from small and large sources. Proceedigs First International Symposium on Glare. New York. Lighting Research Institute, 1991.

Franssen, L. & Coppens, J.E. (2007). Straylight in the eye; Scattered papers. Doctoral Thesis, University of Amsterdam. Amsterdam, 2007.

Franssen, L.; Coppens, J.E. & Van Den Berg, T.J.T.P (2007). Compensation comparison method for assessment of retinal straylight. Chapter 4, p. 45-63. In: Franssen & Coppens, 2007. Reprinted from Investigative Ophthalmology & Visual Science, 27 (2006) 768-776.

Fry, G.A. (1969). Limits of the field of view. Illum. Engn. 64 (1969) 5; 403-406.

Ghandi, N.J. & Sparks, D.L. (2004). Changing views of the role of superior colliculus in the control of gaze. Chapter 97, p. 1449-1565. In: Chalupa & Werner, eds., 2004.

Greenfield, S. (1997). The human brain; A guided tour. London, Weidefeld & Nicholson, 1997.

Gregory, R.L. (1965). Visuele waarneming; de psychologie van het zien (Visual perception; The psychology of seeing). Wereldakademie; de Haan/Meulenhoff, 1965.

Gregory, R.L., ed. (1987). The Oxford companion to the mind. Oxford, Oxford University Press, 1987.

Gregory, R.L., ed. (2004). The Oxford companion to the mind. Second edition. Oxford, Oxford University Press, 1987.

Hartmann, E.; Moser, E.A. (1968). Das Gesetz der physiologischen Blendung bei sehr kleinen Blendwinkeln (The laws of disability glare for very small glare angles). Lichttechnik. 20 (1968) 67A-69A.

Hartmann, E.; Ucke, C. (1974). Der Einfluss der Blendquellengröße auf die physiologische Blendung bei kleinen Blendwinkeln (The influence of the size of the glare source on disability glare at small glare angles). Lichttechnik. 26 (1974) 20-23.

Haubner, P.; Bodmann, H.W. & Marsden, A.M. (1980). A unified relationship between brightness and luminance. Siemens Forschung- und Entwicklungs Berichte. 9 (1980) nr. 6, p. 315-318.

Hentschel, H.-J. (1994). Licht und Beleuchtung; Theorie und Praxis der Lichttechnik; 4. Auflage (Light and illumination; Theory and practice of lighting engineering, 4th edition). Heidelberg, Hüthig, 1994.

Hentschel, H.-J., ed. (2002). Licht und Beleuchtung; Grundlagen und Anwendungen der Lichttechnik; 5. neu bearbeitete und erweiterte Auflage (Light and illumination; Theory and applications of lighting engineering; 5th new and extended edition). Heidelberg, Hüthig, 2002.

Holladay, L.L. (1927). Action of a light source in the field of view in lowering visibility. Journ. Opt. Soc. Amer. 14 (1927) 1.

Hopkinson, R.G. (1957). Assessment of brightness: What we see. Illum. Engng. 52 (1954) no 4. p. 211-222.

Hopkinson, R.G.; Stevens, W.R. & Waldram, J.M. (1941). Trans. Illum. Engng. Soc. (London). 6 (1941) 37.

Howell, S.B. (2001). Handbook of CCD astronomy, reprinted 2001. Cambridge (UK). Cambridge University Press, 2001.

Janoff M.S. (1993). The relationship between small target visibility and a dynamic measure of driver visual performance. Journ. IES 22 (1993) no 1. p. 104-112.

Jansen, J. (1946). Verlichtingstechniek (Illuminating engineering). Haarlem, H. Stam, 1946.

Jantzen, R. (1960). Flimmerwirkung der Verkehrsbeleuchtung (Flicker effects caused by road traffic lighting). Summary in Lichttechnik, 12 (1960) 211.

Keck, M.E. (1993). Optimization of lighting parameters for maximum object visibility and its economic implications. p. 43-52. In: LRI, 1993.

Kelly, D.H. (1961/1962). Visual response to time dependent stimuli. Journ. Opt. Soc. Amer. 51 (1961) 421-429; 747-754; 52 (1962) 8-95.

Kocmond, W.C. & Perchonok, K. (1970). Highway fog. NCHRP Report 95. Washington, DC, Highway Research Board, 1970.

Le Grand, Y. (1937). Recherches sur la diffusion de la lumière dans l'oeil humain (Studies on light scatter in the human eye). Rev. Opt. 16 (1937) 201-214-241-266.

Longhurst, R.S. (1964). Geometrical and physical optics (fifth impression). London, Longmans, 1964.
LRI (1993). Visibility and luminance in roadway lighting. 2nd International Symposium. Orlando, Florida, October 26 – 27, 1993. New York, Lighting Research Institute LRI, 1993.
Marsden, A.M. (1968). The relationship between brightness and luminance. Dissertation University Nottingham, 1968.
Monnier, A. & Mouton, M. (1939). La technique de l'éclairage des automobiles (Lighting technology for motor vehicles). Paris, Dunod, 1939.
Middleton, W.E.K. (1952). Vision through the atmosphere. University of Toronto Press, 1952.
Minnaert, M. (1942). De natuurkunde van 't vrije veld, derde druk (The physics of the open air, third edition). Zutphen, Thieme, 1942.
Moon, P. (1961). The scientific basis of illuminating engineering (revised edition). New York, Dover Publications, Inc., 1961.
Narisada, K. (2007). Revealing power and road lighting design. In: TRB, 2007.
Narisada, K. & Karasawa, Y. (2001). Re-consideration of the Revealing Power on the basis of Visibility Level. Proceedings of International Lighting Conference, Istanbul, Turkey, 2001, Vol. 2, p. 473-480. 2001.
Narisada, K.; Karasawa, Y. & Shirao, K. (2003). Design parameters of road lighting and Revealing Power. In: CIE, 2003.
Narisada, K. & Schreuder, D.A. (2004). Light pollution handbook. Dordrecht, Springer, 2004.
NSVV (1957). Aanbevelingen voor openbare verlichting (Recommendations for public lighting). Moormans Periodieke Pers, Den Haag, 1957 (year estimated).
NSVV (1974/1975). Richtlijnen en aanbevelingen voor openbare verlichting (Directives and recommendations for public lighting). Electrotechniek 52(1974)15; 53(1975) 2 en 5.
NSVV (1977). Het lichtniveau van de openbare verlichting in de bebouwde kom (The light level of urban public lighting). Electrotechniek 55 (1977) 90-91.
NSVV (1990). Aanbevelingen voor openbare verlichting (Recommendations for public lighting). Arnhem, NSVV, 1990.
NSVV (2002). Richtlijnen voor openbare verlichting; Deel 1: Prestatie-eisen. Nederlandse Praktijkrichtlijn 13201-1 (Guidelines for public lighting; Part 1: Performance requirements. Practical Guidelines for the Netherlands 13201-1). Arnhem, NSVV, 2002.
NSVV (2003). Aanbevelingen voor tunnelverlichting (Recommendations for tunnel lighting). Arnhem, NSVV, 2003.
OECD (1976). Polarized light for vehicle headlamps; Proposal for its public evaluation; the technical and behavioral problems involved. A report prepared by an OECD Road Research Group. Paris, OECD, 1976.
OECD (1976a). Adverse weather, reduced visibility and road safety. Paris, OECD, 1976
OECD (1980). Road safety at night. Paris, OECD, 1980.
Orban, G.A. & Vanduffel, W. (2004). Functional mapping of motion regions. Chapter 83, p. 1229-1246. In: Chalupa & Werner, eds., 2004.
Padgham, C.A. & Saunders, J.E. (1966). Trans. Illum. Engng. Soc. (London). 31 (1966) 122.
Padmos, P. & Alferdinck, J.W.A.M. (1983) Verblinding bij tunnelingangen II: De invloed van atmosferisch strooilicht (Glare at tunnel entrances II: The influence of atmospheric stray light). IZF 1983 C-9. Soesterberg, IZF-TNO, 1983.
Padmos, P. & Alferdinck, J.W.A.M. (1983a) Verblinding bij tunnelingangen III: De invloed van strooilicht van de autovoorruit (Glare at tunnel entrances III: The influence of windscreen stray light). IZF 1983 C-10. Soesterberg, IZF-TNO, 1983.
Reading, V. (1966). Yellow and white headlamp glare and age. Trans. Illum. Engn, Soc. (London) 31 (1966) 108-114.

Rood (1893). Am. Journ. Sci. 46 (1893) 173.
Roszbach, R. (1972). Verlichting en signalering aan de achterzijde van voertuigen (Lighting and signalling at the rear of vehicles). Voorburg, SWOV, 1972.
Saito, M. & Narisada, K. (1968). The effect of flickering light on visual comfort (In Japanese). National Technical Report. Vol. 14. no. 1. February 1968. English translation by Narisada. 1968.
Sanders, A.F. (1967). De psychologie van de informatieverwerking (The psychology of information processing). Arnhem, Van Loghum Slaterus, 1967.
Schmidt-Clausen, H.-J. (1968). Über das Wahrnehmen verschiedenartiger Lichtimpulse bei veränderlichen Umfeldleuchtdichten (On the observation of different light pulses by different levels of the surrounding luminance). Darmstadt, Dissertation TH, 1968.
Schor, C. (20045). Stereopsis. Chapter 87, p. 1300-1312. In: Chalupa & Werner, eds., 2004.
Schouten, J.F. (1937). Visueele meting van adaptatie en van de wederzijdse beïnvloeding van netvlieselementen (Visual determination of adaptation and the mutual influence of retinal elements). Doctoral thesis, University Utrecht, 1937.
Schreuder, D.A. (1964). The lighting of vehicular traffic tunnels. Eindhoven, Centrex, 1964.
Schreuder, D.A. (1967). Tunnel lighting. Chapter 4. In: De Boer, ed., 1967.
Schreuder, D.A. (1967a). The theoretical basis for road lighting design. Chapter 3. In: De Boer, ed., 1967.
Schreuder, D.A. (1971). Tunnel entrance lighting; A comparison of recommended practice. Lighting Res. Technol. 3 (1971) 274-278.
Schreuder, D.A. (1976). White or yellow light for vehicle head-lamps? Arguments in the discussion on the colour of vehicle head-lamps. Publication 1976-2E. Voorburg, SWOV, 1976.
Schreuder, D.A. (1981). De verlichting van tunnelingangen; Een probleemanalyse omtrent de verlichting van lange tunnels. Twee delen (The lighting of tunnel entrances; A problem analysis of the lighting of long traffic tunnels. Two volumes). R-81-26 I; II. Voorburg, SWOV, 1981.
Schreuder, D.A. (1991). Practical determination of tunnel entrance lighting needs. Paper presented at the TRB Annual Meeting. January 15, 1991, Washington DC.
Schreuder, D.A. (1991a). Tegenstraalverlichting in tunnels; Een overzicht van de beschikbare literatuur (Counterbeam lighting in tunnels; A survey of available literature). R-91-96. Leidschendam, SWOV, 1991.
Schreuder, D.A. (1998). Road lighting for safety. London, Thomas Telford, 1998 (Translation of "Openbare verlichting voor verkeer en veiligheid", Deventer, Kluwer Techniek, 1996).
Schreuder, D.A. (1999). The theory of tunnel lighting. LiTG-Sondertagung 'Aktuelles zur Tunnelbeleuchtung, 22-23 September 1999, Bergisch-Gladbach. Leidschendam, Duco Schreuder Consultancies, 1999.
Schreuder, D.A. (2006). Hoe uniek zijn de Hollandse wolkenluchten? (Are the skyscapes with clouds from the Dutch classical painters unique for Holland?). Zenit, 33 (2006) no 6, 419-421.
Schreuder, D.A. (2008). Looking and seeing; A holistic approach to vision. Dordrecht, Springer, 2008 (in preparation).
Schreuder, D.A. & Carlquist, J.C.A. (1969). Side lights and low-beam headlights in built-up areas. Report 1969-7. Voorburg, SWOV, 1969.
Schreuder, D.A. & Lindeijer, J.E. (1987). Verlichting en markering van voertuigen: Een state-of-the-art rapport (Lighting and marking of road vehicles: A state-of-the-art report). R-87-7. Leidschendam, SWOV, 1987.
Sterken, C. & Manfroid, J. (1992). Astronomical photometry. Dordrecht, Kluwer, 1992.
Stevens, W.R. (1969). Building physics: Lighting – seeing in the artificial environment. Oxford, Pergamon Press, 1969.
Steyer, R (1997). Forschungsmethoden; Quantitative Methoden (Research methods; quantitative methods). Chapter VII-1. In: Straub et al., eds., 1997.

Stiles, W.S. (1929). The effect of glare on the brightness difference threshold. Proc. Roy. Soc. 104B (1929) 322-355.
Stiles, W.S. & Crawford, B.H. (1937). The effect of a glaring light source on extrafoveal vision. Proc. Roy. Soc. 122b (1937). 255-280.
Straub, J.; Kempf, W. & Werbik, H., eds. (1997). Psychologie, Eine Einführung (Psychology; An introduction). DTV 2990. München, Deutsche Taschenbuch Verlag GmbH & Co. KG. 1997.
TRB (2007). Visibility symposium, 17-18 April 2007. College Station, Texas, 2007.
Van Bommel, W.J.M. & De Boer, J.B. (1980). Road lighting. Deventer, Kluwer, 1980.
Van De Hulst, H.C. (1981). Light scattering by small particles. New York, Dover, 1981.
Van Den Berg, T.J.T.P. (1995). Analysis of intraocular straylight, especially in relation to age. Optometry and Vision Science. 72 (1995) no. 2, p. 52-59.
Van Den Berg, T.J.T.P; Hanegouw, M.P.J. & Coppens, J.E. (2007). The ciliary corona; Physical model and simulation of the fine needles radiating from point light sources. Chapter 13, p. 183-194. In: Fransen & Coppens, 2007. Reprinted from Invest. Ophthalmol. & Visual Scie. 26 (2005) 2627-2632.
Van Den Berg, T.J.T.P.; IJspeert, J.K. & De Waard, P.W. (1991). Dependence of intraocular straylight on pigmentation and light transmission through the ocular wall. Vision Res. 31 (1991) 1361-1367.
Van Den Berg, T.J.T.P. & Tan, K.E. (1994). Light transmission of the human cornea from 320 to 700 nm for different ages. Vision Res. 34 (1994) 1453-1456.
Vos, J.J. (1963). On mechanisms of glare. Universiteit Utrecht, Dissertatie, 1963.
Vos, J.J. (1983). Verblinding bij tunnelingangen I: De invloed van strooilicht in het oog (Glare at the entrance of tunnels I; The influence of intra-ocular straylight). IZF 1983 C-8. IZF-TNO, Soesterberg, 1983.
Vos, J.J. (1984). Disability glare – A state of the art report. CIE Journal. 3 (1984) 39-53.
Vos, J.J. (1999). Glare today in historical perspective: Towards a new CIE glare observer and a new glare nomenclature. Paper No 182. P. 38-42; Volume 1 – Part 1. In: CIE, 1999.
Vos, J.J. (2003). Reflections on glare. Lighting Res. Technol. 35 (2003) 163-176.
Vos, J.J. & Legein, C. P. (1989). Oog en werk: een ergoftalmologische wegwijzer (Eye and work; Ergophthalmological guidance). Den Haag, SDU Uitgeverij. 1989.
Vos, J.J.; Cole, B.L.; Bodmann, H.-W; Colombo, E.; Takeuchi, T. & Van Den Berg, T.J.T.P. (2002). CIE Equations for disability glare. CIE Collection on glare. Vienna, CIE, 2002.
Vos, J.J. & Padmos, P. (1983). Straylight, contrast sensitivity and the critical object in relation to tunnel entrance lighting. In: CIE, 1983.
Vos, J.J. & Van Den Berg, T.J.T.P. (1999). Report on disability glare. Report 135/1. In: CIE Collection 135, Vienna, CIE, 1999.
Vos, J.J.; Walraven, J. & Van Meeteren, A. (1976). Light profiles of the foveal image of a point source. Vision Research 16 (1976) 215-219.
Wachter, A. & Hoeber, H. (2006). Compendium of theoretical physics (Translated from the German edition). New York, Springer Science+Business Media, Inc., 2006.
Walraven, J. (1989). Dieptezien met een en twee ogen (Depth perception with one and two eyes). Hoofdstuk 5, p. 46-54. In: Vos & Legein, 1989.
Warren, W.H. (2004). Optical flow. Chapter 84, p. 1247-1259. In: Chalupa & Werner, eds., 2004.
Weigert, A. & Wendker, H.J. (1989). Astronomie und Astrophysik -ein Grundkurs, 2. Auflage (Astronomy and astrophysics – a primer, 2nd edition). VCH Verlagsgesellschaft, Weinheim (D), 1989.
Westheimer, G. (1979). The spatial sense of the eye. Invest. Ophthalm. Vis. Sci. 18 (1979) 893-912.
Wolf, E. & Gardiner, J.S. (1965). Studies on the scatter of light in the dioptric media of the eye as a basis of visual glare. Arch. Ophthalm. 74 (1965) 338-345.

9 The human observer; colour vision

In the preceding chapters, other aspects of the human observer have been dealt with. In Chapter 6, the physical and anatomical aspects of vision are discussed. Chapter 7 concentrates on the visual performance aspects, and in Chapter 8 the different aspects of visual perception have been discussed. In this chapter, the human observer is discussed, focussing on aspects of colour vision. Also, the elements of colorimetry and the colour characteristics of light sources are discussed.

The three cone families are dealt with that differ mainly in the wavelength of their maximum sensitivity. The physiology and the related neural circuitry are briefly explained. The main theories of colour vision are discussed, including the related phenomena of colour blindness, or colour defective vision.

Next, colour metrics and colorimetry are discussed, including the additive and subtractive processes that are involved. The CIE system of colorimetry is discussed, based on the 1931 CIE Standard Chromaticity Diagram. The Munsell colour system for surface colours is briefly explained. Finally, the colour characteristics of light sources are explained, mainly the colour temperature and the colour rendition of light sources.

9.1 Colour aspects

9.1.1 A description of colour

Strange as it may appear, there is no definition of colour to which all people agree. Many allocate colour just like smell, taste, and sound to the physical world, whereas others regard colours as qualities of experience, projected onto external things. Still others maintain that colours are to be identified with complex neurological events (Blackburn, 1996, p. 68-69). There are several other definitions that combine these three. Still more strange is that most standard works on colour do not bother to give a precise definition of colour. It seems that the authors think that everybody understands what they are talking about. We will give a few examples of publications that are very useful in many other respects. Most of them have been referred to many times in this book: Barrows, 1938; Baer, ed., 2006; Bergmans, 1960; Boll & Dourgnon, 1956; Bouma, 1946;

Boyce, 2003; Boynton, 1979; Favié et al., 1962; Feynman et al., 1977; Gall, 2004; Graetz, 1910; Hentschel, ed., 2002; Moon, 1961; Reeb, 1962; Shevell, ed., 2003; Stevens, 1969; Walraven, ed., 1981; Weis, 1996.

For practical lighting application, the three descriptions given by Blackburn can be rewritten, and a fourth can be added:
1. Experience colours – a matter of psychology;
2. Seeing colours – a matter of physiology;
3. Creating colours – a matter of technology;
4. Measuring colours – a matter of colorimetry.

In the following parts of this section these subject will be discussed in some detail. The subjects are, however, very complicated and there is a tremendous amount of experimental material available. Discussing all this would lead the reader far away from the subject area of this book. For a more complete treatment see e.g.: Augustin (2007); Boyce (2003, sec. 2.2.7); Chalupa & Werner, eds. (2004, part VIII); Gleitman et al. (2004, p. 190-196), and Shevell, ed. (2003, parts 2,3, and 6). So, the following sections will be a summary only, based to a large extent on these works.

CIE and OSA both use a definition of colour like the one given by Wyszecki and Stiles in 1967. They did define colour as that aspect of visible radiant energy by which an observer may distinguish between two structure-free fields of view of the same size and shape, such as may be caused by differences in the spectral composition of the radiant energy concerned in the observation (Wyszecki & Stiles, 1982, after Boynton, 1979, p. 292). Although the sentence is somewhat crooked, some of the elements that are mentioned earlier can be tracked down. The most striking is that not colour, but only colour differences are described. An observer is introduced, and colour is considered as a physical quantity, linked to 'radiant energy', or rather to electromagnetic radiation.

9.1.2 The importance of colour

Colours are important in daily life. In many trades and professions, colours are used to code products and equipment. Coloured lights are quite common in sending messages, e.g. in road and rail traffic (Adrian, 1963; CEN, 2000; CIE, 1980, 1983). Colours are often considered to be the most important aspect of the surroundings. Surveys in lighted streets pointed out that many residents value a 'reasonable' colour impression above good visibility (Schreuder, 1989, 1989a, 2001, 2001a; Van Tilborg, 1991). It is difficult to think of architects or of fashion designers who would not put emphasis on the colour aspects of their creations. There is a large literature of colour vision, on detection of colours, and on colour coding, as well as on the physiological and psychological principles that are relevant in the detection of colours. So it would seem that life without colours is not feasible. However, not all people can see colours. It is interesting to note that colour defective people, even those who can see no colours at all, can have an almost normal life. It is for the sufferers, evidently, a severe inconvenience, and they may often feel that they 'miss' something valuable in their life. The restrictions in their professional life that was deemed necessary in the past, are, however, not justified. We will come back to this point later. So, missing the ability to see colours seems to have little or no consequences. However, those people who can see colours are very keen on using this ability – at least in the majority of cases. However, low-pressure sodium lighting, where no colours can be seen, is accepted in many outdoor lighting applications. And everybody, colour defective or not, has been perfectly happy with black-and-white photos and TV, at least until the mid-20th century. Black-and-white pictures and drawings are still popular amongst the aficionados. So it seems that the ability to see colours is not essential for human life, but if we have the ability we will use it with fervour (Narisada & Schreuder, 2004, sec. 8.3.1).

9.1.3 Experiencing colours

As mentioned earlier, experiencing colours is primarily a matter of psychology. Colours create their own experience; it is well-known that the colour scheme of a room influences the attitude and the mood of the people in that room. Many applications may be found in restaurants, hospitals, hotel lobbies, and homes for the elderly. This is a wide field, open to serious study, but also to pseudo-science. There a few architects that do not include the colour schemes in their design. In this respect, the rather new profession of light designer appeared. Although, as mentioned earlier, many people are still quite satisfied with black-and-white photographs, drawings and etches, most artists emphasise colour. And finally, colour plays an important role in psychotherapy.

In this book, that is devoted to topics of visual science and outdoor lighting engineering, we will not deal with the psychological aspects of experiencing colour. This subject is dealt with in great detail in a many excellent textbooks and treatises. We will refer here only to Schreuder, 2008, for a more detailed discussion of a number of the aspects involved. See also Narisada & Schreuder (2004, sec. 8.3).

9.2 Colour vision physiology

9.2.1 Three cone families

(a) *The relative spectral sensitivity*

As is explained in sec. 6.3.5c, the photoreceptors in the human retina can be classified in rods that are operational only at low values of the adaptation luminance, and cones that are operational only at high values of the adaptation luminance. Also it is mentioned that cones come in at least three different families, each family having its own visual pigment which results in a distinct spectral sensitivity curve for each family.

There are several ways to assess the spectral sensitivity of the three families. The two most common are to measure the absorption of the pigments, and to determine the overall spectral sensitivity of colour defective observers. The first might be considered as a physical, or chemical method. The second is essentially a psycho-physical method. As the two methods differ in a number of ways, one should not be surprised that the results of the two methods show some discrepancies.

The physical method, being 'in vitro', can only give relative results, whereas the psycho-physical method, being made in living creatures, can offer absolute values. It remains to be seen, however, in how far the two methods really measure the same thing!

In Figure 9.2.1a and b, the absorption curves of the three kinds of cones are given. In Figure 9.2.1a the absorption curves are given in an absolute scale. In Figure 9.2.1b they are given in a logarithmical scale; 'zero absorption' is not represented in the figure. The absorption for each kind of cones is normalised at 100%.

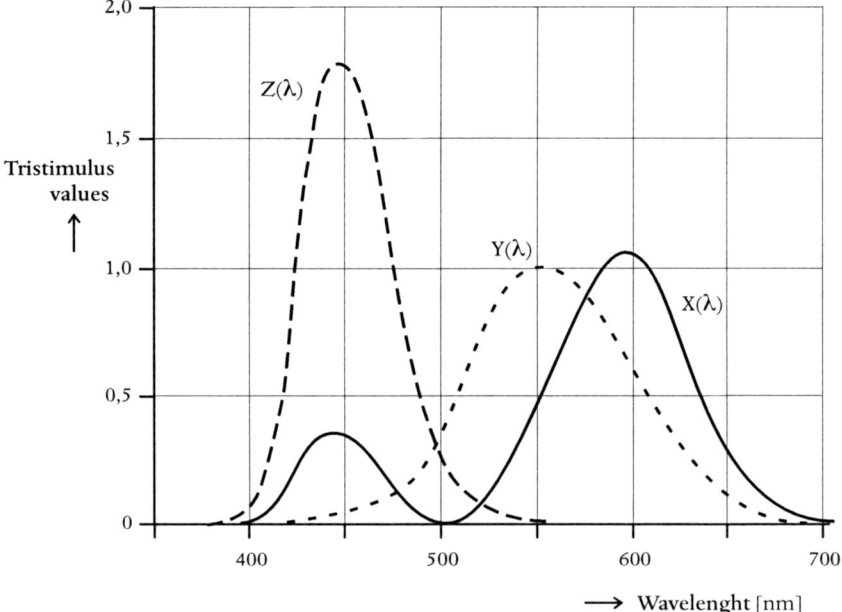

Figure 9.2.1a *The absorption spectra of the three kinds of cones in an absolute scale (After Smith & Pokorny, 2003, figure 3.4, p. 111).*

The human observer; colour vision

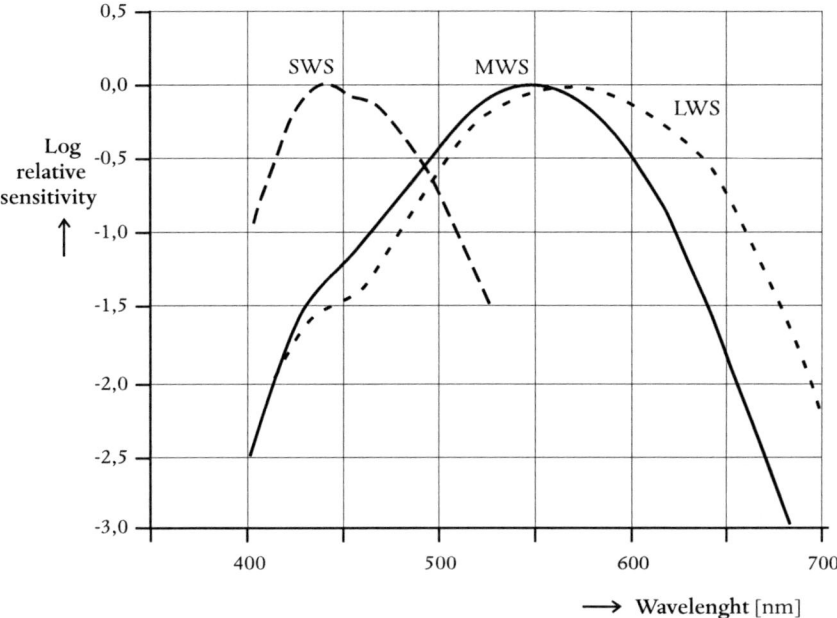

Figure 9.2.1b *The absorption spectra of the three kinds of cones in a relative scale (After Smith & Pokorny, 2003, figure 3.7, p. 119).*

Similar, but not identical, data are presented in the CIE report (CIE, 2006). The data are depicted in Figure 9.2.2. in this figure, the 10° –LWS, MWS, and SWS cone fundamentals are given. The meaning of these abbreviations is given further on.

In the past it was often assumed that the three curves were identical in shape (Hubel, 1990; Narisada & Schreuder, 2004, Figure 8.3.2). It is clear from Figure 9.2.1 that this is not completely true. Although their shape is somewhat similar, the three curves show considerable differences. More important is the shift along the wavelength-scale. The three cone families have their maximum sensitivity at wavelengths of about 415 nm, 530, and 560 nm (Neitz & Neitz, 2004, p. 974). Thus, they lay broadly in the blue, green and red regions of the spectrum. Consequently, the three families of cones were called 'blue', 'green', and 'red' cones. As all this is not completely true, the families are called nowadays 'short-wave-sensitive', 'medium-wave-sensitive', and 'long-wave-sensitive' cones; abbreviated to SWS, MWS, and LWS cones (Neitz & Neitz, 2004, p. 974). These indications are added in Figure 9.2.1 and 9.2.2.

The other approach is based on the fact that not everybody sees colours in the same way. Before this method is explained, it is better to discuss colour defective vision.

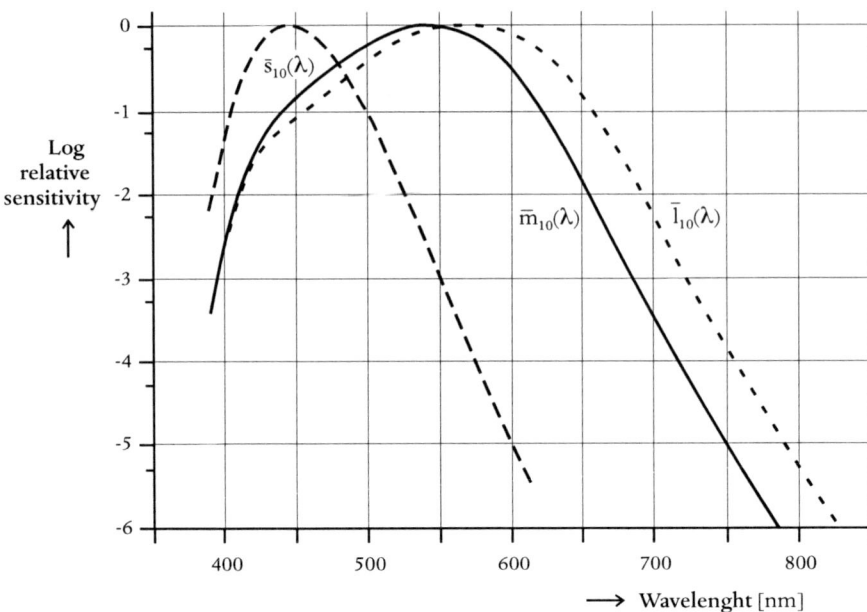

Figure 9.2.2 The 10° LWS, MWS, and SWS cone fundamentals, normalised at 1 at the maximum (After CIE, 2006, figure (2)3. Based on data of Stockman & Sharpe, 2000).

(b) *Colour defective vision*

As mentioned earlier, not everybody sees colours in the same way. Observers with what usually is called 'normal colour vision' can perceive four distinct – or unique – hues: red, yellow, green, and blue.

These four unique colours should not be confused with the three primary colours of the CIE system of colorimetry that is explained in sec. 9.3.2b. The colour information is extracted by neural circuits that compare the outputs of the cones. Red-green colour vision is mediated by the circuits that compare the outputs of the LWS and MWS cones. Blue-yellow colour vision is mediated by circuits that compare the output of the SWS cones to the summed outputs of the MWS and the LWS cones. These circuits are discussed in sec. 9.2.1c. Together, these two neural circuits provide the capacity to distinguish more that 100 different gradations of hue (Neitz & Neitz, 2004, p. 974). When further subleties are taken into account the number is far larger. The ICI Company developed an atlas with 27 850 different colours to be used in the textile industry (Walraven, ed., 1981, p. 134). The Guiness Book of Records suggests that there are many more than 100 000 different colours (Narisada & Schreuder, 2004, sec. 8.3.6d).

Colour defective vision – often called colour blindness – is the result of a malfunction of one or more kind of cones. Such malfunctions occur in 0,3% of females and 8% of males in Caucasians (Schreuder, 1998, sec. 7.5.4; Gleitman, 1995, p. 194; Neitz & Neitz, 2004, p. 975). It seems very likely that there is some ethnical component (Walraven, ed., 1980, p. 34; Sharpe et al., 1999). It seems to rather naive to state that there are no human races, as the politically correct want us to believe: "There is no line in nature between a 'white' or a 'black' race, or a 'Caucasoid' or 'Mongoloid' race" (Barnard & Spencer, eds., 2007, p. 464). Racism is a bloody stain on humanity, but there are differences in physiology.

In the past, it was felt that colour deficiencies would be a risk in many tasks, e.g in road traffic (Hopkinson, 1969; Mann & Pirie, 1950). Nowadays, it has been established that such excessive risks do not occur (De Jong, 1995; Boogaard & Vos, 1989). As there are three kinds of cones, and each may show different degrees of malfunction, there are many kinds of colour defective vision, each with its own name. People in which only one kind of cone function normally, are rare.

The terminology is rather arbitrary. A person who lacks the normal function of the photopigment in the LWS cones is called a protanope or just a protan. If the normal function of the photopigment in the MWS cones is lacking, such a person is called a deuteranope or a deutan. In the Caucasian population, about 2% of the males suffer from a protan defect and 6% from a deutan defect. About 1 in 230 females suffer from either protan or deutan defects. There is a third, more rare, defect, where the normal function of the photopigment in the SWS cones is lacking. This occurs in males and females equally frequent viz. about 1 in 10 000 people. These people are called tritanope or just tritan (Neitz & Neitz, 2004, p. 974-975). The terms are arbitrary; they just mean 'first', and 'second', and 'third'.

The three types of colour deficiencies have one thing in common, viz. the inclination to confuse colours with different hues. Protans and deutans may confuse red and green, whereas tritans may confuse blue and yellow (Boogaard & Vos, 1989, p. 42). In Figure 9.2.3 the lines where confusion may occur are depicted for the three types of colour deficiency.

There are of course people who lack the normal function of the photopigment in two or even in all three families of cones. Fortunately, these defects are rare. Additionally, those people usually suffer from other visual defects as well. They are not discussed here because these defects do not contribute to the establishment of the spectral sensitivity of the different cone families.

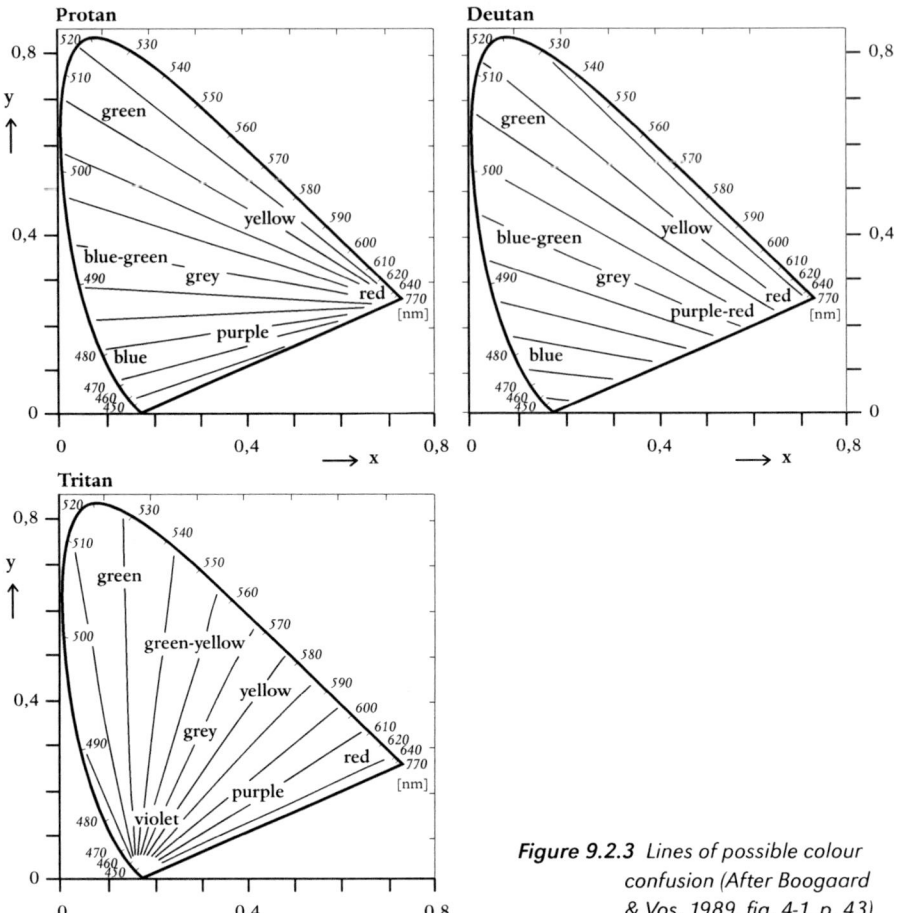

Figure 9.2.3 *Lines of possible colour confusion (After Boogaard & Vos, 1989, fig. 4-1, p. 43).*

(c) *The absolute spectral sensitivity*

There are two ways to assess the spectral sensitivity of the different cone families. The first, based on measuring the absorption of the photopigments, is discussed in an earlier part of this section. Being 'in vitro', it can only give relative results. The psycho-physical method, making use of colour defective observers, can offer absolute values.

As is mentioned earlier, there are three cone families, commonly labelled LWS, MWS, and SWS cones, meaning that the peak of their spectral sensitivity is at long, medium, and short wavelengths.

The human observer; colour vision

Now, a sizeable percentage of male Caucasians lack the normal function of the photopigment in the LWS and the MWS cones, and a small number lack it in the SWS cones. Making the normal measurements about the overall spectral sensitivity of the visual system, such as is used for the assessment of their V_λ-curve, discussed in sec. 7.2.2b, allows to find out the response of the three cone families separately. In Figure 9.2.4 the separate spectral sensitivity curve for each family is given. As mentioned above, they are determined by measuring the response of the total eye for people who have only two families of cones that function properly: the protanopes, the deuteranopes, and the tritanopes. The measurements are not recent, but it seems that they never have been repeated since the 1890s! The curves are similar to, but not identical with those in Figure 9.2.1 and 9.2.2, because a different procedure is used.

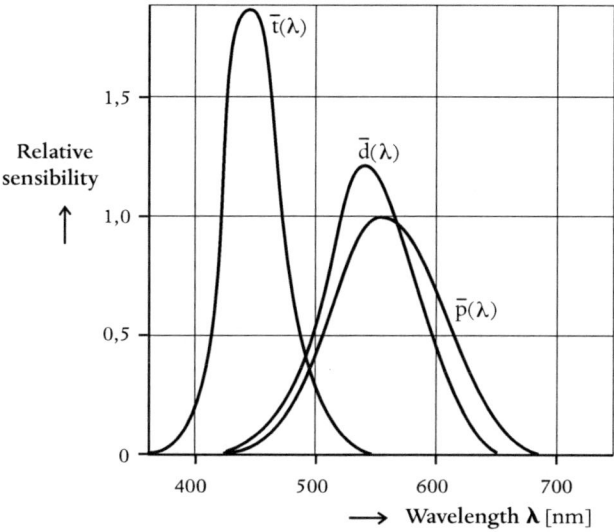

Figure 9.2.4 *The sensitivity curves for three types of colour defective vision; $\bar{p}(\lambda)$: protanopes; $\bar{d}(\lambda)$: deuteranopes; $\bar{t}(\lambda)$: tritanopes (After Narisada & Schreuder, 2004, figure 8.3.3. Based on Hentschel, ed., 2002, fig 4.46, and König & Dieterici, 1892).*

It seems that the three families of cones are evenly distributed over the full retina, taking into account the fact that the cones are predominantly present in the fovea and the near-periphery. The frequency of the three families is, however, quite different. It is stressed that there are 'tremendous' individual variations in the LWS:MWS cone ratio. On the average; the ratio is about 2:1 (Neitz & Neitz, 2004, p. 985; Cicerone & Nerger, 1989; Roorda & Williams, 1999). On the average there are about 7% SWS cones in the retina (Neitz & Neitz, 2004, p. 975).

These data lead to the following frequency distribution of the different cone families over the retina:
- LWS-cones: 62%;
- MWS cones: 31%;
- SWS cones: 7%.

This distribution is important for the construction of the V_λ-curve on the basis of the spectral sensitivity of the different cone families. This construction is explained in sec. 9.2.2d.

9.2.2 The neural circuity in cone vision

(a) Colour vision theories

Over the centuries, many different theories of colour vision have been proposed. A survey is given in Boynton (1979). See also Wright (1967). After the discoveries of Newton and Young, it is generally accepted that light is part of the electromagnetic spectrum and that each wavelength agrees with a specific colour. With this statement, all other colour theories, particularly those proposed by Goethe around 1810, were condemned to the history books (Walraven, ed., p. 84-85). It remains to be seen in how far this drastic steps are justified (Schreuder, 2008)

Later studies did show that all colours, white included, could be thought of as a combination of only three, well-chosen, colours. On those studies the theories of colour vision are based that are discussed in the further parts of this section. Also, colorimetry is based on the same principles. In sec. 9.2.3 colorimetry is discussed. It will be shown that there is not one colorimetry but two, one for lights and one for reflective surfaces. Attempts are made to unify these two colorimetries.

At first it was considered as a coincidence that in photopic vision, where colours can be observed, three kinds of light sensitive receptors are involved. As is explained in sec. 9.2.1, the cones can be subdivided in three classes, or families. As is mentioned earlier, they are similar in anatomy but the curves that represent their spectral sensitivity are different. The most important distinction in the wavelength where they are most sensitive. As mentioned earlier, they are called SWS, MWS, and LWS cones (Neitz & Neitz, 2004, p. 974).

It might seem logical to assume that, as the human retina contains three families of cones, colours are observed in the same way as colorimetry works. Three primaries ought to be sufficient to explain all colour vision aspects. These opinions were based on the classical experiments of Young and Helmholtz (Von Helmholtz, 1896). Non-colour defective observers can make all colours, white included, with three well-chosen primaries that can be adjusted as regards their light output. This is usually called the trichromatic theory. It should be stressed that many aspects of colour vision can adequately be explained by the trichromatic theories. The CIE Colorimetry System, that refers only to light sources and not to reflecting surfaces, is based on them (CIE, 1971, 1986, 1988, 2006).

In the second half of the 20th century, experimental methods were greatly refined. The processes of the response to light of individual retinal detectors could be measured in primates, apparently with little concern for our fellow-creatures. Also the related neuronal and cerebral processes could be measured in living human observers by means of non-aggressive methods. These studies did make it clear that there is not a one-to-one relationship between the cone families and the colour primaries, as would be required by the trichromatic theories. These considerations did lead to the opponent-process theory. Over the years it became clear that the opponent-process theory allows a better understanding of even many of the most complicated colour-vision effects. Nowadays, it seems that the opponent-process theory is almost universally accepted. This theory will be elaborated in the following parts of this section.

(b) *The opponent-process theory*

As mentioned earlier, the colour information from the retina is extracted by neural circuits that compare the outputs of the cones. Basically, there are two main circuits. The red-green colour vision is mediated by the circuits that compare the outputs of the LWS and MWS cones. Blue-yellow colour vision is mediated by circuits that compare the output of the SWS cones to the summed outputs of the MWS and the LWS cones. Further details are given in the next part of this section (Lennie, 2003).

The three families of cones that have been mentioned earlier are not connected directly to parts of the visual cortex. In fact, the way the visual signals travel from the retina towards the cortex is rather complicated.

As mentioned earlier, at present, the predominant theory about colour vision is the opponent-process theory (Gleitman et al., 2004, p. 194-195). Some alternative theories are mentioned in Feynman et al. (1977, vol. I, p. 36-22). The theory stems from Hering (1878) and is expanded by Hurvich and Jameson (Hurvich & Jameson, 1957). The basis thought is that the system is triggered by either excitation or inhibition. Excitation and inhibition are described in sec. 6.4.2e. Excitation and inhibition are the two 'opponent' actions that work on the system – hence the term.

As is explained in sec. sec 6.2.1b, excitation and inhibition processes find their origin in the neurotransmitter that may be released from their storage sites in the ends of the dendrites close to the synapse. After the encounter with a neurotransmitter receptor, the neuroreceptor shows a change in structure that shows a 'hole' or a 'passage' right through the neuroreceptor molecule, through which only a particular ion can pass. Their movement generates excitatory or inhibitory synaptic potentials.

The opponent-process theory finds its roots in the lateral geniculate nucleus or LGN, a brain part (Lennie, 2003, p. 221). LGN neurons and ganglion cells that drive them are very similar (Lennie, 2003, p. 231). There are M cells and P cells in the LGN. M cells are located in the magnocellular layer in the LGN (attached to many cells apparently, considering the name). P cells are located in parvocellural layers in the LGN (Lennie, 2003, p. 232). Both P cells and M cells have receptive fields organised into two concentric antagonistic regions: a centre (on- and off-) and a surrounding region of opposite sense (Lennie, 2003, p. 233). The M-cell's receptive field centre has a spectral sensitivity close to that of V_λ, therefore drawing its inputs from LWS and MWS cones, while the surround draws on some different mix of cone signals. P-cells fall into two chromatic classes, loosely red-green and yellow-blue (Lennie, 2003, p. 234). One class receives inputs from LWS and MWS cones only (De Valois et al., 1966; Wiesel & Hubel, 1966). A second class receives inputs from SWS cones opposed to some unspecified combination of signals from LWS and MWS cones (Derrington et al., 1984).

(c) *The opponent-process theory circuitry*

As mentioned in the preceding part of this section, the colour information from the retina is extracted by neural circuits that compare the outputs of the cones. Here, we will explain these circuits.

As mentioned earlier, the cones are not directly linked to parts of the optical nerve and thus also not to parts of the brain. Most of the circuitry in between is located in the retina, more in particular in the ganglion cells in the retina. In its most simple way, the circuits are sketched in Figure 9.2.5.

In this figure, the light comes from below. It hits the cones. Each cone family reacts in their own way. The cone output is delivered to the ganglion cells, part in the form of excitation, part in the form of inhibition. This information is processed by the ganglion cells, again each in their own way. Each type of ganglion cell creates pair of opponent signals: Blue-Yellow; Red-Green, and Black-White. From here the signals are propelled along the optical nerve for further processing in different brain parts, and lastly in the visual cortex.

The human observer; colour vision

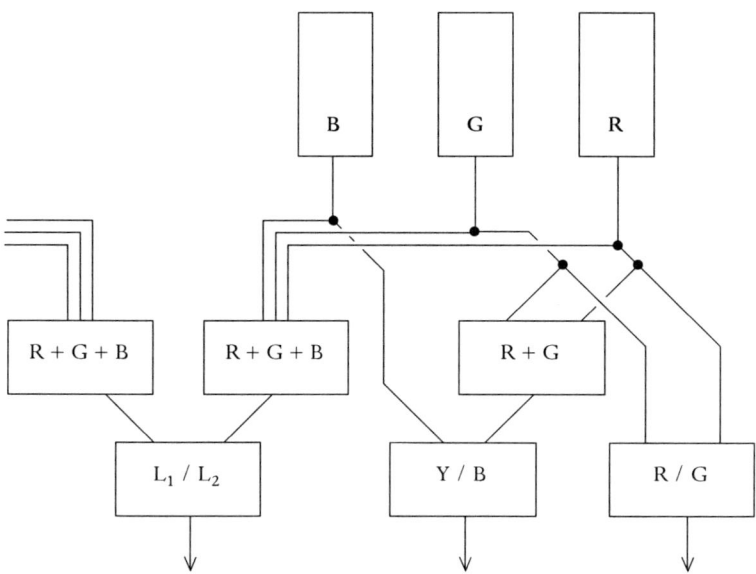

Figure 9.2.5 The information flow in the human retina (After Narisada & Schreuder, 2004, Figure 9.8.1b. Based on Walraven, ed., 1981, p. 34; Boogaard, 1990, fig. 7).

It should be noted that only the Blue-Yellow and the Red-Green channels contain colour information. The Black-White channel is therefore called the achromatic channel.

It should be added that in fact the circuitry is more complicated. A more detailed summary is given in Pokorny & Smith (2004, p. 918). They have presented a model of spectral processing based on current physiological knowledge of the retinal pathways. See Figure 9.2.6.

The model for the LWS and MWS cone discrimination (Fig 9.2.6A) is based on physiological data of the spectral PC pathway of primates (Derrington et al., 1984; Lee et al., 1990, 1994). The model for the SWS-cone discrimination (Fig 9.2.6B) is based on physiological data of the spectral opponency of the KC pathway (Zaidi et al., 1992)

Both models postulate an early stage of cone-specific multiplicative adaptation followed by a stage of spectral opponency between LWS and MWS cones for the L/M pathway or between SWS and L+M cones for the S pathway. Since gain control is not complete, the spectral opponent signal varies with wavelength and luminance. This signal is further controlled by subtractive feedback by the background. The signal then depends on the contrast between the background and test chromaticity. All ganglion cells respond to their preferred contrast with a negative accelerating function of contrast. Thus four components determine contrast detection and discrimination: the absolute threshold, the gain function, the subtraction at the surround chromaticity, and the saturation function.

Spatiotemporal contrast

a L/M pathway

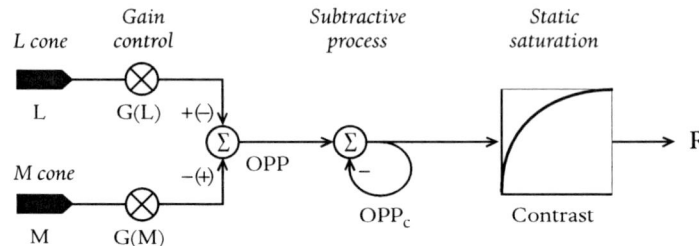

b S pathway

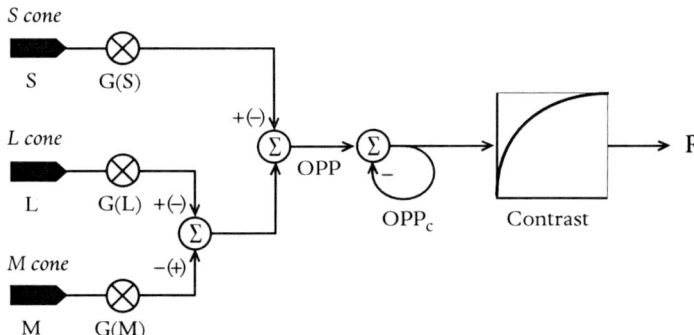

Figure 9.2.6 Schematic view of a retinal model for equiluminant chromatic discrimination. a) the L/M cone system; b) the S cone system (After Pokorny & Smith, 2004, figure 58.9, p. 918).

A given retinal ganglion cell responds best to a chromaticity change in its preferred direction (Kremers et al., 1993; Lee et al., 1994). Change in the nonpreferred direction will drive the cell below its resting level. The resting level is usually about 15% of the maximal response rate. This is an intrinsic nonlinearity that renders the cell response asymmetric: the cell behaves as if partially rectified.

For equiluminant chromatic pulses (+L–M) and (–M+L) give redundant information, responding positively to 'redward' changes from the adaptation point; similarly, (+M–L) and (–L+M) give redundant information, responding positively to 'greenward' changes from the adaptation point (Lee et al., 1994). To achieve a response for the entire

The human observer; colour vision

chromatic contrast range we require pairs of cells with opposite chromatic signature, such as (+L–M) and (+M–L).

(d) The construction of V_λ of self-luminous objects

A general definition of the luminance of self-luminous objects might run as follows:

$$L = \int R(I,\lambda, t,x,y...) \, V(I,\lambda, t,x,y...) \, d\lambda \qquad [9.2\text{-}1]$$

in which:
    R; the radiance of the light source;
    V: the spectral sensitivity of the visual system;
    I: the intensity of the light source;
    t, x, y... : other variables.

This representation is given by Crawford (2005, p. 58). It is interesting in a way that it shows that the intensity is involved in the fact that the spectral sensitivity of the visual system depends on the light level. The introduction of the terms of time and space (t, x, y...) serve only to show that "lighting, or vision, is complex".

It is explained in sec. 3.2.1c that all photometric definitions, all measurements, all design methods, and all data representation is exclusively based photopic vision. This allows to simplify the relation [9.2-1], so that the well-known definition of the luminance shows itself. Usually, it is written as follows:

$$L = K \int_{360\,\text{nm}}^{830\,\text{nm}} L_{e,\lambda} \cdot V_\lambda \cdot d\lambda \qquad [9.2\text{-}2]$$

in which:
    L: the luminance;
    $L_{e,\lambda}$: the spectral radiance;
    V_λ: the CIE Standard Spectral Sensitivity Curves for photopic vision;
    K: 683 lm/W.

It may be added that for scotopic vision, a similar definition is given. Also, for different levels of mesopic vision, descriptions are given that have the same structure. The only difference is that the appropriate spectral sensitivity curve of the visual system is used in stead of the CIE Standard Spectral Sensitivity Curve. Also the constant K may have a different value.

In the definition given here it is assumed that the luminance can be described completely when using the traditional 'achromatic' CIE Standard Spectral Sensitivity Curves for photopic vision. As is explained in sec. 7.2, this curve is essentially a composite of the spectral sensitivity curves of the different families of cones in the human retina.

As is explained in detail in sec. 7.2.2, there are three families of cones, each having its own spectral sensitivity curve. Although their shape is somewhat similar, the three curves show considerable differences. More important is the shift along the wavelength-scale. The three cone families have their maximum sensitivity at wavelengths of about 415, 530, and 560 nm. Usually, the families are called 'short-wave-sensitive', 'medium-wave-sensitive', and 'long-wave-sensitive' cones, abbreviated to SWS, MWS, and LWS cones (Neitz & Neitz, 2004, p. 974).

Here we consider the construction of V_λ. It is based on the fact, mentioned earlier, that each of the three cone families has its own spectral sensitivity curve, but that the three cone families are not equally frequent. The curves are depicted in Figure 9.2.1 and 9.2.2. It is generally assumed that their frequency distribution is as follows:
- LWS-cones: 62%;
- MWS cones: 31%;
- SWS cones: 7%.

Modern insight is that it is better to neglect the SWS-contribution altogether. This contribution had always been considered as minute. Also, the maximum sensitivity of the remaining two families is not the same. When the difference in sensitivity is taken into account, and when the 2:1 ratio of LWS and MWS-cones, the formula for the construction of the luminance would run:

$$kV_\lambda = 1{,}62\, R'(\lambda) + G'(\lambda) \qquad [9.2\text{-}3]$$

Using the notation that has been used earlier, the end result of the construction of the V_λ-curve is:

$$V_\lambda = 0{,}62\, \text{LWS}(\lambda) + 0{,}38\, \text{MWS}(\lambda) \qquad [9.2\text{-}4]$$

9.3 Colour metrics and colorimetry

9.3.1 Terminology

There are many different sets of terms in colorimetry, just as there are many different systems of colorimetry. Basically, they all include three concepts, that might be indicated in colloquial terms:

The human observer; colour vision 329

1. The actual colour (red, green, yellow etc.);
2. The saturation (strong or weak red; strong or weak green etc.);
3. The lightness (bright or dark red; bright or dark green etc.).

In the Munsell-system the corresponding terms are 'hue'; 'chroma', and value' (De Boer & Fischer, 1981, p. 90).

9.3.2 Colorimetry

(a) Additive and subtractive processes

As far as the colour impression goes, it is not important whether the light is coming from a 'coloured' light source directly, or whether it is reflected by a 'coloured' surface. For the metrics of the colour space, and consequently for colorimetry, it is essential (Narisada & Schreuder, 2004, sec. 8.3).

All colour considerations are based on the fact that by mixing colours, other colours emerge (Illingworth, ed., 1991, p. 74). Mixing coloured light sources is an additive process. Almost any colour can be produced or reproduced by mixing together lights of three colours, called the 'additive primary colours' (Illingworth, ed., 1991, p. 7). The most important characteristic of the additive colour mixing rules is that the three primary colours, when mixed in the right proportions, produce white light. This is the well-known Grassman's rule (Grassmann, 1853; Hentschel, ed., 2002, p. 115). As will be explained later on, usually red, green and blue are chosen as the additive primary colours. They form the basis of the CIE system of colorimetry. Mixing the colours of filters or pigments, however, follow completely different rules. With this process, again almost any colour can be produced or reproduced by mixing together the light of the three primary colours, being either filters or pigments. This process is called the subtractive process (Illingworth, ed., 1991, p. 466). It is characterised by the fact that the three primary colours, when mixed in the right proportions, produce black. As will be explained later on, usually yellow, magenta (purplish) and cyan (greenish blue) are chosen as the primary colours. They form the basis of the Munsell-system of colour representation. As an example, colour television and digital photo cameras make use of an additive colour process, whereas traditional – chemical – colour photography makes use of a subtractive process. The Munsell-system is explained briefly in sec. 9.3.2g. See also Narisada & Schreuder (2004, sec.8.3.7h).

(b) The CIE system of colorimetry

The CIE system of colorimetry, that is relevant for the assessment of the colour light sources, is based three primary colours. There is, in the use of three primaries, an analogy with the trichromatic theory of colour vision of Young and Helmholtz that is explained in sec. 9.2.2a. Here the analogy stops. The theory of Young and Helmholtz is based on three families of cones, each having its own spectral sensitivity curve, whereas the CIE system of colorimetry is based on the postulate that any – or at least almost any – set of three colours can be used as primaries. In other words, the theory of Young and Helmholtz is a physiological method, whereas the CIE system is a mathematical method. It may mentioned that recently CIE attempts to unify the two fundamentally different approaches (CIE, 2006).

In 1931, the CIE adopted a system of colour specification that is still in use to-day (Boynton, 1979, p. 19; CIE, 1932). It was based on the experimental studies of Maxwell (1855); Wright (1928), and Guild (1931). The major constituent of the CIE-system is a mathematically defined colour space. In the CIE XYZ colour space, the tristimulus values are not the SWS, MWS, and LWS cone spectral responses that are discussed in sec. 9.2.1, but rather a set of tristimulus values called X, Y, and Z, which are roughly red, green, and blue (Boynton, 1979, Appendix, Part I, The CIE System, p. 390-397; Anon., 2007). As mentioned earlier, X, Y, and Z themselves do not represent experimental, physiological data but a set of mathematically defined quantities.

The CIE system of colorimetry is rather complicated. Details can be found in the standard works of Baer, ed. (2006, sec. 1.2.4); Bouma (1946, 1971); Boynton (1979); De Boer & Fischer (1981, sec. 4.1); Judd & Wyszecki (1967); Richter (1976); Wright (1964), and Wyszecki & Stiles (1967). Here, we follow the summary given by Narisada & Schreuder (2004, sec. 8.3.6b).

As mentioned earlier, the CIE system of colorimetry is based on the postulate that any colour can be made by a combination of three – almost – arbitrary primaries. They are variables in the mathematical sense, termed X, Y and Z. The space they form together is called the CIE 1931 colour space. A transformation of these three can be defined in such a way that two variables represent the actual colour, whereas the third one represents the brightness. When the brightness is disregarded as not being a 'real' colour, only two variables are left over. This implies that colours can be graphically described in a two-dimensional diagram. Over the years, many proposals for such diagrams have been made (Bouma, 1946). One prevailed, the CIE xy chromaticity diagram, that is explained later on.

(c) *The 1931 CIE Standard Chromaticity Diagram*

As mentioned earlier, in 1931 the CIE adopted a system of colour specification called the CIE 1931 colour space based on the tristimulus values X, Y and Z (CIE, 1932). The tristimulus values X, Y and Z are transformed into the chromaticity coordinates x, y and z by using the following transformation formulae:

The human observer; colour vision

$$x = \frac{X}{X+Y+Z}; y = \frac{Y}{X+Y+Z}; z = \frac{Z}{X+Y+Z} \qquad [9.3\text{-}1]$$

Because, as is explained earlier, z represents the luminance, it is sufficient to use x and y. Also x and x are colours, real or hypothetical, so they can be characterised by normalized spectral values, or spectral tristimulus values (CIE, 1986). A graphical representation is given in Figure 9.3.1.

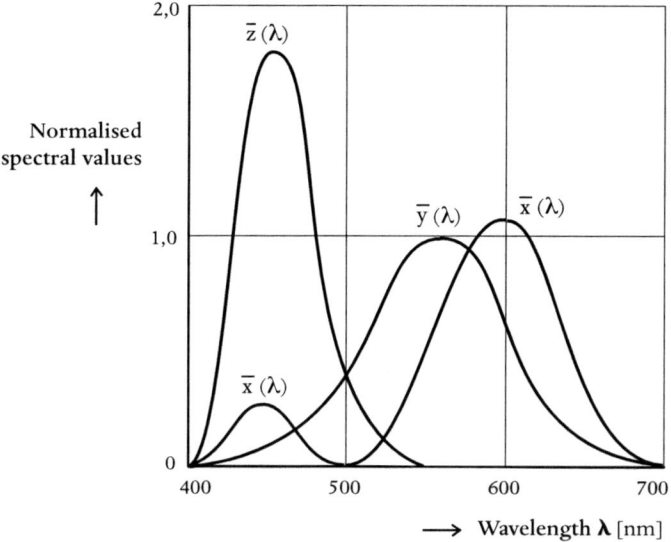

Figure 9.3.1 The standard curves for the CIE system of colorimetry (After Narisada & Schreuder, 2004, figure 8.3.6).

The variables of X, Y, and Z are chosen in such way that, after the transformation, the resulting two-dimensional x-y diagram looks like a triangle, be it somewhat blunted. This diagram is usually called the CIE colour triangle (CIE, 1932). It shows a number of specific characteristics:

1. The values for y = 1 and x = 1 fall outside of the area or real-world colours (De Boer & Fischer, 1981, p. 93);
2. Each point in the diagram represents one specific x-y combination, and hence one colour, real or hypothetical. This is called a colour point. Colour points are discussed in sec. 9.3.2f.
3. The spectral colours all lie on a curved contour. Red is at the bottom right extreme point, blue at the bottom left point;
4. If these two extreme points are connected by a straight line, it is found that all varieties of purple lie on this line. The resulting figure looks more or less like a triangle;

5. All real-world colours fall within this triangle-like contour;
6. Colours tend to become less saturated when one moves away from the contour of spectral colours toward the centre. In the middle, all 'near-white' colours are located.

As mentioned already, the resulting figure looks more or less like a triangle, be it with a blunted tip – hence the name of colour triangle. The triangular shape can easily be recognized in Figure 9.3.2, where a graphical representation of the 1931 CIE colour triangle is given.

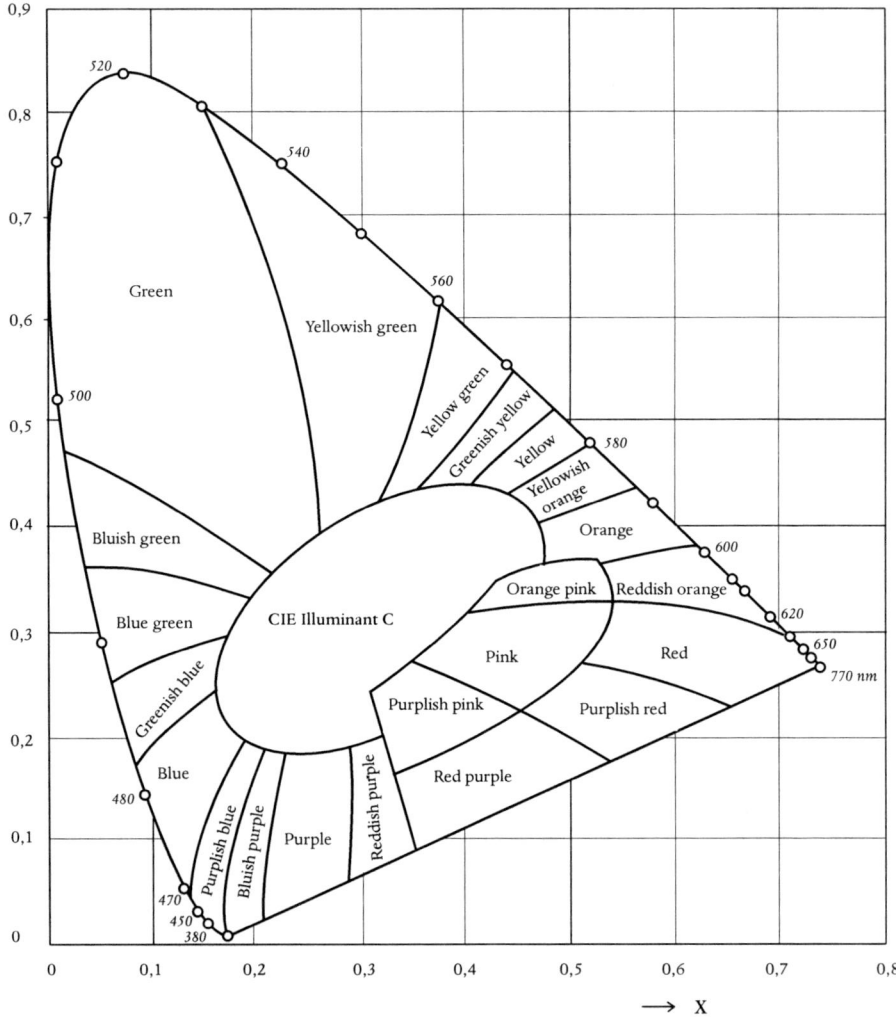

Figure 9.3.2 *The 1931 CIE Standard Chromaticity Diagram. The standardized colour names are included (After Narisada & Schreuder, 2004, figure 8.3.7).*

(d) Standard conditions for colour vision

The black-body radiator locus is important in two ways. The first was that it is linked to incandescent lamp lighting, which were for a long time the only useful light sources. More important is that it is linked to general illuminating engineering practice. It is generally accepted natural daylight represent the ideal lighting conditions (Moon, 1961, p. 2-3; Narisada & Schreuder, 2004, sec. 4.5.3). De Boer and Fischer, in their classic study on interior lighting, remark: "We compare the source with a familiar reference source. The best known and most widely used reference source is mid-day daylight." (De Boer & Fischer, 1981, p. 101). In the CIE-system of standard light sources, illuminant D65 is introduced to represent the mid-day daylight with an equivalent colour temperature of 6500 K. See Table 9.3.1

Above, CIE Illuminant D65 has been mentioned. This is one of the light sources – or spectral distributions – that have been standardized by CIE for the purpose of colour assessments. In Table 9.3.1, the CIE standard illuminants are listed. The standard illuminants are characterized by their colour temperature. The colour temperature is defined in sec. 9.4.2.

Illuminant	Colour temperature (K)	Representing
A	2856	incandescent lamp
B	4874	direct sunlight
C	6774	average daylight
D65	6500	mid-day daylight
E	infinite	white; equal-energy spectrum

Table 9.3.1 The CIE standard illuminants (After Narisada & Schreuder, 2004, Table 8.3.5).

Looking more closely at the preference for natural daylight, it is clear that not so much the colour temperature, but rather the way the colours in the real world appear. All practical lighting design considerations are based on the postulate that noon-time natural daylight renders colours 'natural'. Noon-time natural daylight can be represented by a black-body radiator of around 6500 – 7000K. Hence, CIE defined two standard illuminants in this region: C, with 6774K for average daylight, and D65, with 6500K, for mid-day daylight. See Table 9.3.1. Now, as is mentioned earlier, the spectral energy distribution of black-body radiators follow Raleigh-Jeans' Law. It is assumed that the Sun is also a Planckian radiator, so that its spectral energy distribution would also follow Raleigh-Jeans' Law. In Figure 9.3.3, the spectral energy distribution of the Sun is compared to that of a Planckian radiator. As can be seen the fit is not perfect (Budding, 1993. p. 36). It appears to be good enough for most practical lighting applications.

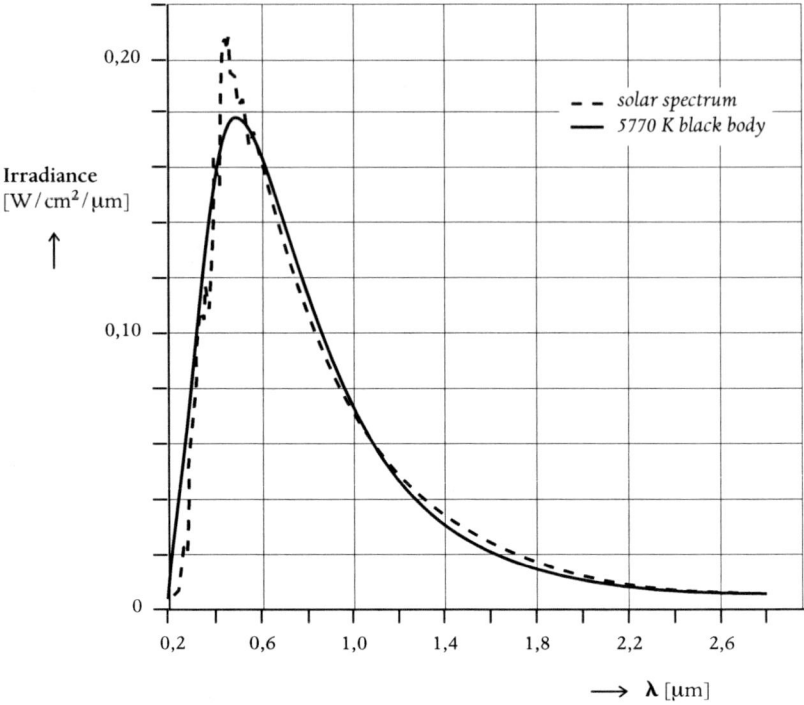

Figure 9.3.3 *The solar spectrum (dashed) and the spectrum of a 5770 K black body (full) (After Narisada & Schreuder, 2004, Figure 8.3.9. Based on Budding, 1993, fig. 3.3).*

More in particular it appears to be acceptable to disregard the many spectral absorption lines that can be observed in the Solar spectrum. Even for the much higher requirements of astronomical spectroscopy, the effect of absorption lines in stellar spectra can be eliminated by clever instrumentation (Budding, 1993, p. 69, 74). So, it can be stated that colours look 'natural' under a continuous spectrum that represents the for mid-day daylight.

When we want to know what happens to colours when observed under a continuous spectrum that represents another colour temperature, we have to introduce a curious characteristic of the visual system, viz. constancy. As is explained in sec. 8.4.1b, colour constancy means that objects more or less keep their colour as it is perceived in white light, in spite of the changes in the spectral composition of the light (Walraven, ed., 1981, p. 242). This implies that colours are observed as 'natural' as long as they are seen under a spectrum of a Planckian radiator, irrespective of its temperature.

The human observer; colour vision

(e) Colour names

Basically, each colour has its own name. Of course, there are many more shades of colour than there are suitable names. As mentioned earlier, estimates are up to more than 100 000 different colours. A system of colour names has been introduced by CIE with enough names – 23 different names in all – to suit most practical uses. The names and the corresponding areas they occupy in the CIE Colour plane are included in Figure 9.3.2.

It has been mentioned earlier that the CIE method is set up with 'self luminous' light sources in mind. This is clear from the way the colour triangle is depicted, e.g. in Figure 9.3.2. It is based on the rules of the additive colour mixing: Red plus Green plus Blue gives White! This is clear form the colour triangle itself. In the centre of Figure 9.3.2 the CIE Illuminant C is indicated. The CIE Illuminant C is defined very precisely. It is meant to represent tungsten-filament incandescent lamps in the time that there were no other light sources that could be used in practice. Also the white point is marked, defined as $x = y = 0{,}333$ (Hentschel, ed., 2002, p. 120). See Table 9.3.1.

(f) Colour points

Any colour can be produced by mixing, in the right proportion, three colours from the colour triangle. By using the x- and y-coordinates of the 1931 CIE colour plane, where z is assumed to be the luminance, this particular colour corresponds to one particular point in the colour plane. This point is called the colour point of that particular colour. As mentioned before, the CIE system is characterised by the spectral colours all located on the contour. In Figure 9.3.4, the colour triangle is depicted again.

The different light sources mentioned here are discussed in more detail in Narisada & Schreuder, 2004, sec. 11.3.

(g) The Munsell system

This is probably the best place to devote a few words to the Munsell system. The Munsell system is used in illuminating engineering to specify object colours under daylight conditions. Because the Munsell system is related to surface colours and thus to pigments, it is based on a subtractive colour mixing processes. For outdoor lighting applications, that seldom involve daylight conditions, the Munsell system is not important. A more detailed description is given in Narisada & Schreuder (2004, sec. 8.3.7h); De Boer & Fischer (1981, p. 90-91), and Walraven, ed. (1981).

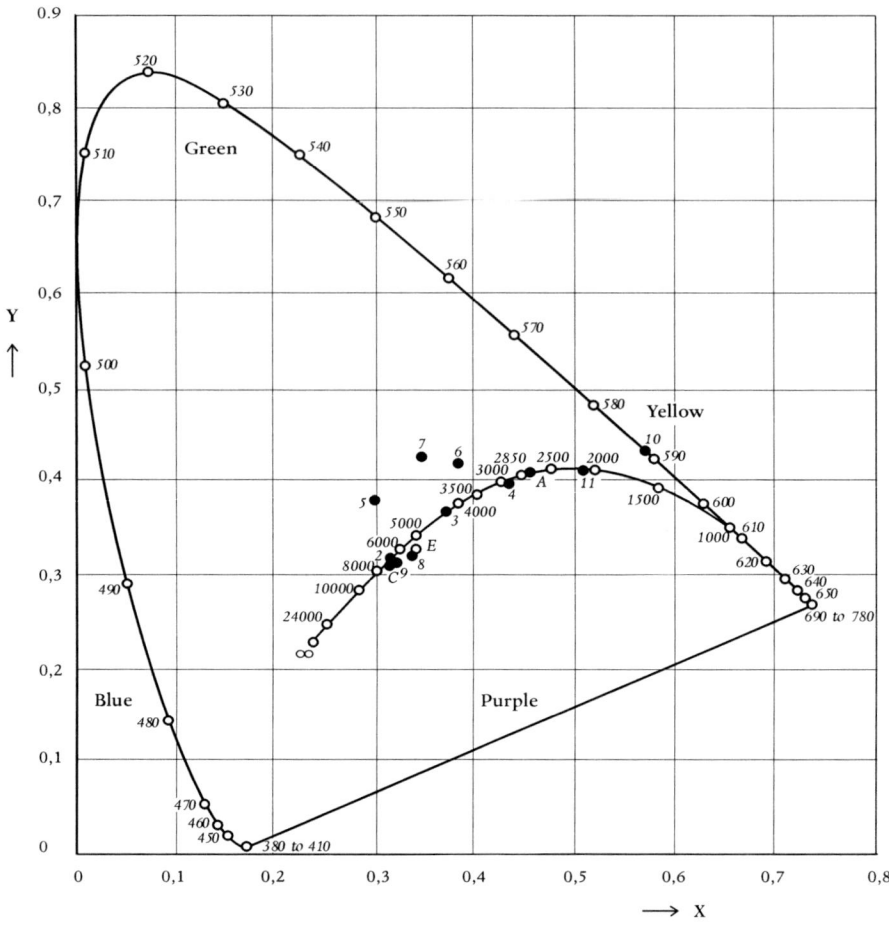

Figure 9.3.4 The 1931 CIE standard triangle (After Narisada & Schreuder, 2004, Figure 8.3.8).

Notes to Figure 9.3.4:
1. The locus of the colour points of all black-body radiators is included as a curve more or less through the centre of the triangle. The corresponding temperatures (in K) are added to the curve;
2. The point corresponding with 'white' is inserted in the figure as letter E. The conditions for white are: x = y = 0,333 (Hentschel, ed., 2002, p. 120);
3. For a number of specific light sources, the colour points are included as well:
    - A: CIE Standard for daylight;
    - C: CIE Standard for tungsten filament incandescent lamps;
    - 2: fluorescent tube colour 'daylight';
    - 3: fluorescent tube colour 'white';
    - 4: fluorescent tube colour 'warm-white';
    - 5: clear high-pressure mercury lamp;
    - 6 and 7: two types of fluorescent high-pressure mercury lamps;
    - 8 and 9: two types of high-pressure xenon lamps;
    - 10: low-pressure sodium lamp;
    - 11: high-pressure sodium lamp.

The human observer; colour vision

(h) Metamerism

As is explained in sec. 9.4.1a, the colour impression of coloured surfaces is created by the colour impression of the light that is reflected by the surface. This impression depends, apart from the chromatic adaptation that is not relevant here, on:
1. The spectral distribution of the incident light;
2. The spectral distribution of the reflection.

As is explained in sec. 9.3.2d, it is generally accepted that the colour impression of a surface can be considered is being 'normal' when the surface is illuminated by mid-day daylight. However, if a light source with a poor colour rendition is used, that means that the spectrum of the emitted light differs considerably from the mid-day daylight, the colours may look rather different. This effect is called metamerism (Boynton, 1979, p. 99-102; Pokorny & Smith, 2004, p. 910). A metameric pair is described as a pair of stimuli that have the same colour but different spectral composition (Judd, 1951, p. 863). The effect is illustrated in Figure 9.3.5. In that figure, as an example, only the spectral reflectance of two surface samples is given. The spectral emissivity is not included in the figure.

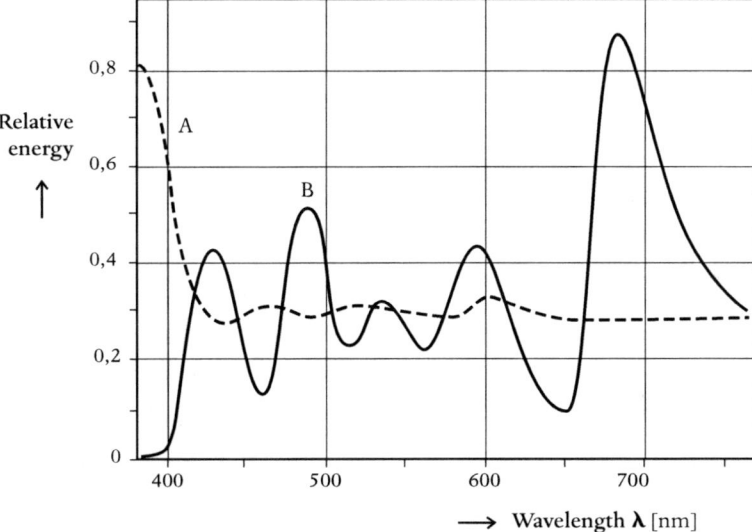

Figure 9.3.5 The spectral reflectance two surface samples that will show the same colour under standard lighting conditions but quite different colours when seen under light with a different spectral emissivity (After Boynton, 1979, figure 5.1, p. 100. Based on Wyszecki & Stiles, 1967 (1982), p. 351).

Also in practice, it often happens that two samples look the same under one light source but different under another light source. The effect is particularly striking in dresses or suits where the different parts stem from different rolls of fabric. The effects are most striking if both the sample and the two light sources show spectra that differ markedly from that of a Planckian radiator. There is only one solution to this: select the dress under lighting conditions that are the same as where it is going to be shown. Do not select a dress and shoes in a shop with poor fluorescent lighting when they are intended for a candle-light party!

9.4 The colour characteristics of light sources

9.4.1 Chromatic adaptation effects

(a) The colour impression

Coloured surfaces reflect incident light, but not to the same extent for different wavelengths of the light. Although the colour impression of a surface which does not emit light itself, is created by the colour impression of the light that is reflected by the surface, is it customary to speak of the colour impression of the surface. This impression depends on:
1. The spectral distribution of the incident light;
2. The spectral distribution of the reflection;
3. The chromatic adaptation of the visual system.

As is explained in sec. 9.3.2d, it is generally accepted that the colour impression of a surface can be considered is being 'normal' when the surface is illuminated by mid-day daylight.

If a light source is to fulfill the primary requirement of illuminating engineering, which means that all objects look like they do in mid-day daylight, the spectral distribution of the light emitted by the light source must be identical, or at least very similar, to that of the mid-day daylight. This is explained in sec. 9.3.2d.

(b) Chromatic adaptation

Chromatic adaptation can be described as follows: "It is the adjustment of the eye to the colour of the light in the surroundings. As an example, when the light contains much red, the nervous tracts that are activated by red light – to begin with the red sensitive cones – become less sensitive, The signal is reduced in strength. In this way, objects more or less keep their colour as it is perceived in white light, in spite of the changes in the spectral composition of the light" (Walraven, ed., 1981, p. 242).

The human observer; colour vision

This effect is often called the colour constancy. The reduction in strength of the signal is called inhibition. The adjustment is not instantaneous; usually, it takes a few seconds (Schreuder, 1964, p. 84-86). This persistence leads to after-images. It is recognised that the disappearance of after-images is a major factor to reckon with in tunnel entrance lighting. More in particular, the CIE Adaptation Curve that is mentioned in most CIE and CEN-documents is based on considerations of the disappearance of the after-images that remain after a car driver leaves the bright outside world when entering into a tunnel (CEN, 2003; CIE, 1990; Narisada, 1975; NSVV, 2003; Schreuder, 1964, 1999).

When the spectral composition of the light is changed suddenly, it takes some time for the inhibition to be effective or, conversely, to diminish. The result is a negative after-image. Both in brightness and in colour, the negative – or the opposite – of the object that was observed, seems to persist. A bright object seems to become dark (Schreuder, 1964, p. 8; Gregory, 1965, p. 49); a green object seems to become red (Walraven, ed., 1981, p. 43; p. 142). This is the reason that the phenomenon is also called the successive contrast (Walraven, ed., 1981, p. 40).

9.4.2 The colour temperature

(a) *The definition of the colour temperature*

As is explained in sec. 2.3.1a, the radiation of a black-body radiator is fully determined by its temperature, both as regards its maximum output (Stephan-Boltzmann Law), the wavelengths where the radiation is at its maximum (Wien's Law), and as regards the distribution of the radiation over different wavelengths (Raleigh-Jeans' Law). It is therefore possible to determine one specific colour point for each temperature of the black-body radiator, independent of the nature of this radiator. This temperature is the colour temperature of the source. Therefore, it is expressed in kelvin. When the points are connected, the result is the locus of the colour points of all black-body radiators. This black-body locus is included in Figure 9.3.4.

The colour temperature is a term used to describe the colour appearance of any light source by comparing its colour to the colour of the black-body radiator (De Boer & Fisher, 1981, p. 95). When the source is 'near white', its colour temperature gives a good approximation of the colour impression made by that source. The meaning of 'near white' is explained in sec. 9.4.2d. For light sources with saturated colours, the colour temperature is not a useful concept.

The way the colour temperature is determined is explained in sec. 9.4.2a.

(b) Incandescent light sources

As is explained in the next part of this section, the black-body radiator locus is important in two ways. The first is that it is linked to incandescent lamp lighting. Today, the most common incandescent lamp is the electric incandescent lamp. In an electric incandescent lamp, the actual light source is a tungsten filament that is made to glow by passing an electric current through it. Before the tungsten filament electric incandescent lamps were used, say before around 1900, other incandescent lamps common, like e.g. flames, candles, and gas lamps (Schreuder, 2000, 2001, 2004, 2005). In these lamps, tiny carbon particles are made to glow by the heat of the flame. Some characteristic data on these, and other, are given in Narisada & Schreuder, 2004, Table 8.3.6. See also Schreuder (2001); Mills (1999).

From the 1910s to the 1940s, the main lamp for almost all lighting applications was the tungsten filament electric incandescent lamp. The historical evolution of the different filament electric incandescent lamps is described in detail in Barrows (1938, p. 63-80). Although since 1930 gas discharge lamps and more recently also quantum lamps like lasers and LEDs are used on a large scale, the tungsten filament electric incandescent lamp is still by far the most common lamp type, in spite of its lower efficacy.

The low efficacy is recently a ground for trying to ban the 'light bulb' in favour of 'compact lamps'. This does not seem a well-considered policy, as incandescent lamps are needed in all cases where the lamp is used intermittently for short periods of time, like e.g. in toilets. More important, however, is that the environmental load of compact fluorescent lamps over the full cycle of manufacture, use, and waste disposal is at least as heavy as that of the incandescent lamps, mainly because all fluorescent lamps contain the very poisonous Mercury. Accurate data, however, are not available. In this respect, LEDs, although having a somewhat lower efficacy, seem to be a much better alternative to the incandescent lamp.

(c) The locus of the black-body radiators

The locus of the black-body radiators is included in Figure 9.3.4 as a curve that runs more or less through the centre of the triangle. The corresponding temperatures (in K) are added to the curve. The temperatures in the graph run from 1000 K to well over 24 000 K. It should be noted that the sign for 'infinity' is added. This is of course not a physical temperature. According to the kinetic theory of gasses, the mean kinetic energy of molecules in an ideal gas is:

$$E = 3/2 \cdot K \cdot T \qquad [9.4\text{-}1]$$

See Feynman et al. (1977, Volume I, p. 39-10) and Wachtel & Hoeber (2006, application no 57, 57, p. 403 406). A physical temperature of infinite height would mean an infinite kinetic energy – a physical impossibility. The sign for 'infinity' is just a symbol. It is supposed to indicate the colour point of the ideally blue sky. This follows from the way the light is scattered by small particles. According to Rayleigh's Law, for particles smaller than 0,1 of the wavelength, the scatter of the light can be determined from:

$$s = K \cdot \frac{(n-1)^2}{N \cdot \lambda^4} \qquad [9.4\text{-}2]$$

in which:
    s: the scattering per unit of volume;
    K: a constant;
    n: the refraction index of the particles;
    N: the number of particles per cm^3;
    λ: the wavelength of the light.

This way to denote Rayleigh's Law is from Minnaert (1942, Vol. I, p. 240). See also Schreuder (1998, sec. 4.4., p. 30-31); Van de Hulst (1981); Douglas & Booker (1977).

It should be added that tungsten filaments approach a black body only to a limited degree. The total emission of Tungsten is only between 0,46 for 1200K and 0,42 for 2800K (Hentschel, ed., 2002, p. 130; Narisada & Schreuder, 2004, sec. 11.1.1d, p. 435). The lamp efficacy for most common incandescent lamp types is between 7 and 20 lm/W (Anon., 1993, 1993a, 1997). Still, Tungsten is the best there is for electric incandescent lamps (Narisada & Schreuder, 2004, sec. 11.1.1e; 11.3.3).

    (d) Near-white light sources
    As is mentioned in the preceding part of this section, the black-body radiator locus is linked to general illuminating engineering practice. It is generally accepted natural daylight represent the ideal lighting conditions. Colours look 'natural' under a continuous spectrum that represents the mid-day daylight.

Now we will discuss near-white light. The graphical representation in Figure 9.3.4. includes the locus of the black-body radiators for temperatures from 1000 K to well over 24 000 K. In general lighting the light sources have to be regarded as 'whitish' by the casual observer. For this, the temperature range can be smaller. In Figure 9.4.1, a part of the CIE Chromaticity Diagram is given with the black-body locus between 2500 K and 8000 K. It also depicts the lines that represent those colour points that are similar to the corresponding black-body colour points. These lines are called lines of constant correlated colours (De Boer & Fischer, 1981, p. 96).

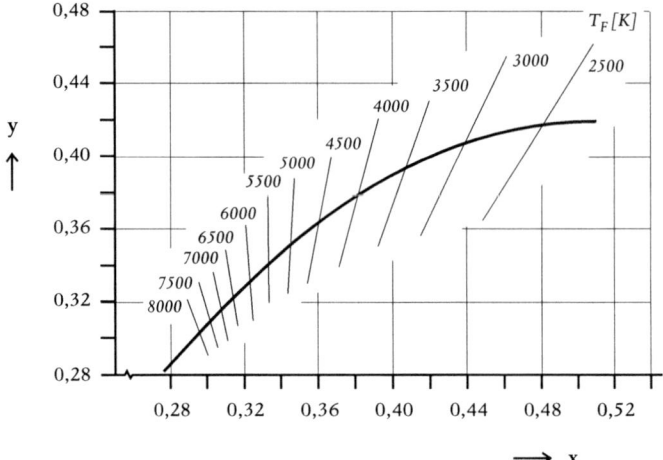

Figure 9.4.1 *Lines of constant correlated colours (After Narisada & Schreuder, 2004, figure 8.3.10).*

Each line in Figure 9.4.1 indicates the colour points of light sources that are equally 'whitish'. As the line intersect the black-body locus, it is possible to give a value of the colour temperature to each line. This value of the colour temperature belongs to the colour point of the intersection between the line of constant correlated colours and the black-body locus. This colour temperature can be allotted to each point of the line of constant correlated colours. In this way the equivalent colour temperature can be defined for each light source, independent of the spectral composition of its light, as long as its colour point is close to he black-body locus – as long as it is 'whitish'.

In summary, observers seem to prefer light that looks like mid-day daylight. A difference in light level is readily accepted as long as the spectrum is similar to that of a Planckian radiator, the levels stay within the area of photopic vision and as long as Weber's Law still holds. Deviations in the colour rendering of the light sources are, however, not easily tolerated.

(e) *Colour differences*

It has been explained earlier that it is possible to discern a very large number of different shades of colours. Although small changes in colour may be noticeable, they do not always deserve a new name. The name-giving procedures may be 'grainy'. Studies by MacAdam indicated how large the areas in the chromaticity diagram are that contain colours that are 'almost' the same. As these areas usually have a more or less elliptic shape, they are called 'MacAdam ellipses' (MacAdam, 1942; Schreuder, 1998,

p. 118; Smith & Pokorny, 2003, p. 135). The orientation and the size of the ellipses are determined by Stiles (1946). The MacAdam ellipses are depicted in Figure 9.4.2.

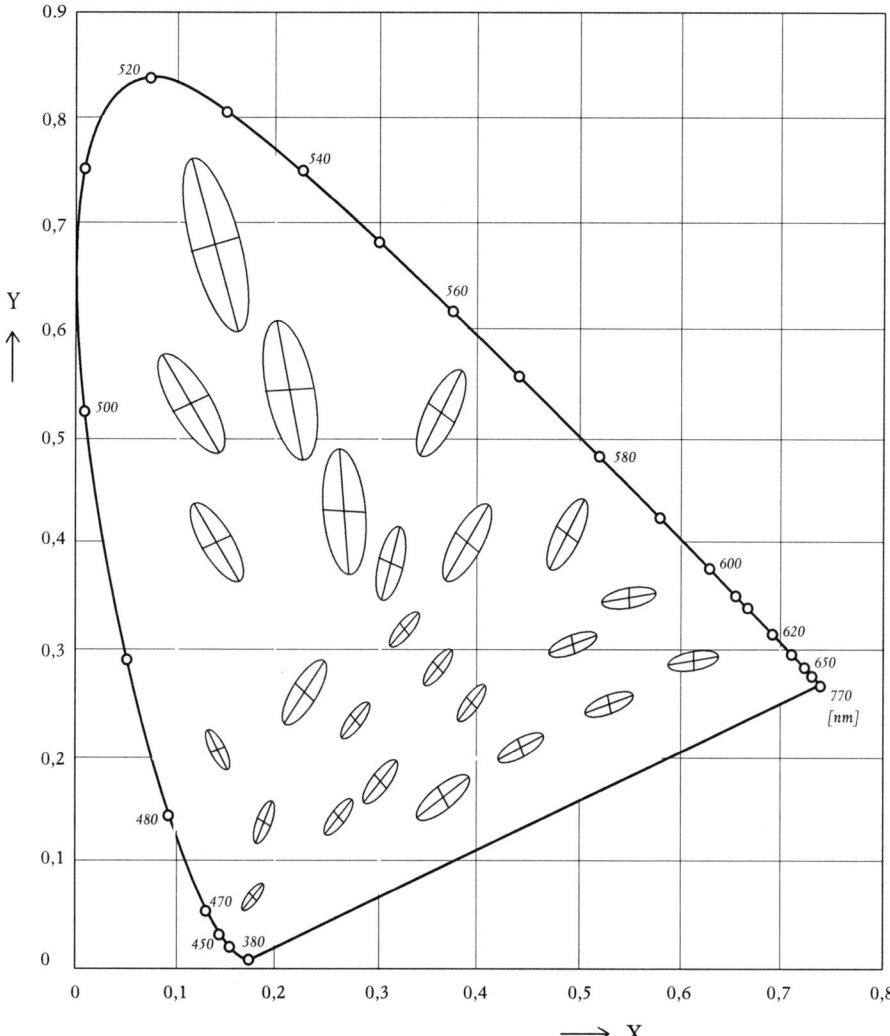

Figure 9.4.2 The MacAdam ellipses plotted in the CIE Chromaticity Diagram (After Schreuder, 1998, figure 7.4.5. Based on Smith & Pokorny, 2003, figure 3.17, p. 135).

A disadvantage of the MacAdam representation of Figure 9.4.2 is that the areas are not equally large. By means of an appropriate transformation of the coordinates, the ellipses can be made into circles with almost the same size. In 1960, CIE introduced the 'Uniform Chromaticity Scale Diagram' (the UCS-diagram). This diagram is depicted in Figure 9.4.3.

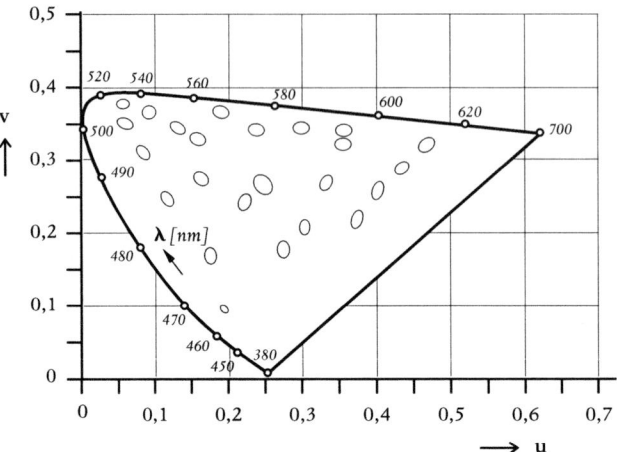

Figure 9.4.3 The CIE Uniform Chromaticity Scale Diagram (After Narisada & Schreuder, 2004, figure 8.3.12. Based on De Boer & Fischer, 1981, Figure 4.4).

9.4.3 The colour rendering

(a) *The colour rendering in illuminating engineering*

The primary goal of illuminating engineering is to make the visual surroundings look 'natural'. Light sources have to fulfill certain requirements as regards the way the colours of objects look when illuminated by them. In sec. 9.3.2d it is explained that 'natural' is defined as being observed under noon-time natural daylight. Noon-time natural daylight can be represented by a Planckian black-body radiator of around 6500 – 7000K. It is postulated rather arbitrary that 'natural' also means being seen under light of a Planckian radiator of different temperatures. This postulate agrees with most field observations. Earlier we mentioned the colour constancy that is supposed to play an important role here.

In many interior lighting applications like office rooms and factories, precise colour discrimination is a major aspect of the visual task. In many other interior lighting applications like staff offices and meeting rooms, aspects of environmental comfort have the highest priority. This is concentrated on the way the people look. Lighting where the complexion of people, man and women alike, is not perfect, are not acceptable (De Boer & Fischer, 1981; Visser, 1992). The same holds for other indoor lighting applications like hotel lobbies, restaurants, and residential living rooms (Visser, 1997). This is the main reason that subjects like colour experience and colour rendering have been studied primarily in the context of indoor lighting, more in particular in domestic architecture. However, colour aspects are also important in outdoor lighting. Cityscape lighting and city beautification put up high requirements for a good colour quality of the lighting (Narisada & Schreuder, 2004, secs. 2.4, 11.2.6; Schreuder, 1994, 1998, 2001, 2001a; Van Santen, 2005). In residential areas, good colour rendering is considered as a powerful crime prevention and crime reduction tool, as it helps to recognise criminals (Aelen & Van Oortmerssen, 1984; Atkins, et al., 1991; Farrington & Welsh, 2002; Narisada & Kawakami, 1998; Narisada & Schreuder, 2004, sec. 11.2.6; Painter & Farrington, 1999; Painter, 1999: Schreuder, 2000a). Also, as is mentioned in sec. 11.1.3g, amenity requires a good colour quality. And finally, even in traffic route lighting, colour quality is important. This is the reason that in a book that focussed on outdoor lighting, considerable attention is given to the colour aspects of illuminating engineering.

(b) The colour rendition

The requirement to the spectral distribution of the emitted light are called the colour rendering characteristics of the light source. It is quantified as the colour rendition of the source. It should be mentioned that usually the term 'colour rendering' is used as a synonym for the more correct term 'colour rendition'.

In principle, the procedure to assess the colour rendering a light source is simple. Take a number of standard surface colours and a standard light source. The samples are alteratively illuminated by the standard light source and by the light source that must be tested. In both cases, the spectral composition of the reflected light is measured by means of an objective colorimeter. The smaller the difference between the two, the better the colour rendering of the lamp. The actual procedure is more complex. It is described in detail in De Boer & Fischer (1981, sec. 4.3, p. 100-110).

The complexities relate to three aspects:
1. The selection of the standard colours;
2. The selection of the standard light source;
3. The way to take the chromatic adaptation into account.

(c) The standard colours

The standard colours were selected from a large number of test colours (Ouweltjes, 1960). Out of 19 test colours, 8 standard colours were selected (CIE, 1965; 1979; 1979a, 1986). Later, 6 more colours were added to these 8 to represent several strong colours, the complexion of the human face and of foliage (De Boer & Fisher, 1981, p. 104). It may be noted that only the Caucasian skin is included (De Boer &

Fischer, 1981, p. 104). In order to adapt the CIE system of standard colours to other parts of the world, notably to Asia, a Number 15 is added, that represents the Japanese female face complexion. The system, including the way the Number 15 has been selected, is described in more detail in Narisada & Schreuder (2004, sec. 8.3.8d). See also Kawakami (1955): Kaneko et al. (1979), and JIS (1967, 1990).

(d) The selection of the standard light source

As mentioned earlier, colours are by agreement understood to look 'natural' when viewed under daylight conditions. However, both the colour temperature and the spectral composition of daylight are far from constant (De Boer & Fischer, 1981, p. 101). The spectral energy distributions of 622 samples of daylight were studied by Judd et al. (1964). Based on these measurements, CIE selected a number of reference light sources with a colour temperature above 5000K. Additionally, a further number of reference light sources with a lower colour temperature were selected by CIE, using the black-body radiator as their basis (De Boer & Fischer, 1981, p. 101). It seems that such a large number of reference light sources were defined because that was, at the time, the only way to take the chromatic adaptation into account. The CIE Standard sources are listed in Table 9.3.1. The relative spectral energy distributions of a number of reference light sources are depicted in Narisada & Schreuder (2004, Figure 8.3.14). See also De Boer & Fischer (1981, fig. 4.9).

As mentioned earlier, the method to establish the colour rendering of a light source is based on the comparison of the chromaticities of a set of standard surface colours as seen under the light source in question and under the light of a standard light source.

The specific standard light source to be used depends on the colour temperature of the lamp to be tested. So this colour temperature must be found. The first step is to determine the chromaticity of the light source to be tested, that is its colour point in the Uniform Chromaticity Scale Diagram or UCS-diagram that is discussed in sec. 9.4.2e. This allows to determine the colour temperature of that lamp, assuming – as we always do in the considerations about colour rendering – that the light of the lamp is 'whitish'. After that, a reference light source is chosen with a colour temperature similar to, or even the same as that of the lamp to be tested. For each of the standard colours, a difference in chromaticity can be determined. This difference in chromaticity represents a colour shift. As is explained in sec. 9.4.2e, the main characteristic of the UCS-diagram is that areas that contain colours that are 'almost' the same, are circles of the same size. Thus, one may use the circle centre to establish the average colour shift. The average colour shift, corrected for the chromatic adaptation, is a measure of the colour rendering characteristic of the lamp to be tested. The smaller the shift, the better the colour rendering. The way to allow for chromatic adaptation is described in De Boer & Fischer (1981 p. 104-105) and in Narisada & Schreuder (2004, sec. 8.3.8g).

(e) *The colour rendering of light sources*
The colour shift for an individual standard colour is described as:

$$R_i = 100 - 4{,}6\, \delta E_{a,i} \qquad [9.4\text{-}3]$$

in which:
$\delta E_{a,i}$: the colour difference expressed values that correspond to the UCS-diagram as discussed in sec. 9.4.2e;
R_i: the Special Colour Rendering Index (SCRI) for an individual colour (After De Boer & Fischer, 1981, p. 105).

The colour rendering of a light source is expressed numerically in the 'General Colour Rendering Index' (GCRI) for the standard reference colours. The GCRI for the eight standard reference colours $R_{a,8}$ is defined, after Baer, ed. (2006, sec. 1.2.4.3, p. 51), as the average of the 8 different SCRI's, as follows:

$$R_{a,8} = 1/8 \cdot \sum_{1}^{i=8} SCRI_i \qquad [9.4\text{-}4]$$

The index $R_{a,14}$ for the fourteen standard reference colours is defined in a similar way. Details are given in De Boer & Fischer (1981, sec. 4.3, p. 100-108; see also Schreuder, 1998, sec. 7.4.4).

In Table 9.4.1, a lamp classification system based on $R_{a,8}$ is given.

Colour rendering class	CRI	Requirements
1A	$100 > R_{a,8} > 90$	very high
1B	$90 > R_{a,8} > 80$	very high
2A	$80 > R_{a,8} > 70$	high
2B	$70 > R_{a,8} > 60$	high
3	$60 > R_{a,8} > 40$	moderate
4	$40 > R_{a,8} > 20$	low

Table 9.4.1 Lamp classification based on $R_{a,8}$ (After Hentschel, 1994, Table 4.5).

In Table 9.4.2, the General Colour Rendering Index for the eight standard reference colours ($R_{a,8}$) is given for a number of light sources that are commonly in use in outdoor lighting. For convenience, the lamp coding system from Philips is used. Also, the luminous efficacy (lumen per watt) and the colour temperature are added.

Lamp type	Luminous efficacy (lm/W)	Colour temperature (K)	CRI $R_{a,8}$
incandescent lamp 100 W	14	2800	100
halogen lamp 500 W	19	3000	100
fluorescent tube 36/40 W:			
– colour 29	83	2900	51
– colour 33	83	4100	63
– colour 57	45	7300	94
– colour 93	64	3000	95
– colour 94	65	3800	96
– colour 95	65	5000	98
– SL 18 W	50	2700	85
sodium lamps:			
– SOX-E 131 W	2001	700	–
– SON-T 400 W	118	2000	23
high-pressure mercury lamps:			
– HP 250 W	47	6000	15
– HPL C 250 W	56	3300	52
– CSI 250 W	60	4200	80

Table 9.4.2 *Lamp characteristics of light sources commonly in use in outdoor lighting (After Schreuder, 1998, Table 7.4.2).*

Notes to Table 9.4.2:
- Incandescent lamp: normal General Lighting Service (GLS) lamp, mostly for indoor use;
- Halogen lamp 500 W: lamp specially used in floodlighting etc.;
- Fluorescent tube 36/40 W: the traditional TL-lamp with 26 mm diameter;
- SL 18 W: compact fluorescent lamp with integrated ballast;
- SOX-E 131 W: low-pressure sodium lamp;
- SON-T 400 W: high-pressure sodium lamp;
- HP 250 W: high-pressure mercury lamp without colour collection;
- HPL C 250 W: high-pressure mercury lamp with improved colour collection ('comfort');
- CSI 250 W: metal-halogenoid lamp for outdoor use.

Except when otherwise stated, ballasts not included.

9.5 Conclusions

For practical lighting application, four aspects of colour are of interest:
1. Experience colours – a matter of psychology;
2. Seeing colours – a matter of physiology;
3. Creating colours – a matter of technology;
4. Measuring colours – a matter of colorimetry.

The main aspect of colour vision physiology is the fact that rods are not involves. Colour vision is a matter of the cones in the retina. There are three cone families, each with their own characteristics. In view of their spectral sensitivity, they are called the 'short-wave-sensitive', 'medium-wave-sensitive', and 'long-wave-sensitive' cones; abbreviated to SWS, MWS, and LWS cones.

In colour defective vision, one or sometimes even more than one of the families of cones does not function properly. Although it may be very inconvenient, nowadays it is felt that there is no need for any professional restrictions of colour blind people.

It seems that the three families of cones are evenly distributed over the full retina, taking into account the fact that the cones are predominantly present in the fovea and the near-periphery. The frequency of the three families is, however, quite different. There are about 62% LWS-cones; 31% MWS cones, and 7% SWS cones.

Modern insight is that for colour vision the SWS-contribution can be neglected altogether. So the V_λ-curve can be constructed from the LWS and the MWS cones alone. So is the 1931 CIE Standard Chromaticity Diagram, commonly called the CIE colour triangle. These are the basis for the colorimetry of self-luminous sources.

The CIE system of colorimetry is based three primary colours. It is based on the 1931 CIE system of colour specification. The three arbitrary primaries are variables in the mathematical sense, termed X, Y and Z. A transformation of these three can be defined in such a way that two variables represent the actual colour, whereas the third one represents the brightness. When the brightness is disregarded as not being a 'real' colour, only two variables are left over. This implies that colours can be graphically described in a two-dimensional diagram.

Any colour corresponds to one particular point in the colour plane. This point is called the colour point of that particular colour. For light sources that have colour points close to the locus of the black-body radiator, the colour temperature is sufficient to characterise the colour. The colour rendition is a metric that indicates how different surface colours look under the light of that particular source. If its colour point lies on the locus of the black-body radiator, the colour rendition is per definition 100. In all other cases it is lower.

References

Adrian, W. (1963). Über die Sichtbarkeit von Strassenverkehrs-Signalen (On the visibility of road traffic control signals). Lichttechnik. 15 (1963) 115-118.

Aelen, J.D. & Van Oortmerssen, J.G.H. (1984). De effecten van openbare verlichting op criminaliteit; Een literatuurstudie (The effects of public lighting on crime; A survey of the literature). Interimrapport, Rijksuniversiteit, Leiden, 1984.

Anon. (1975). Symposium on tunnel lighting. Lighting Res. Technol. 7 (1975) 85-105.

Anon. (1993). Lighting manual. Fifth edition. LIDAC. Eindhoven, Philips, 1993.

Anon. (1993a). The comprehensive lighting catalogue. Edition 3. Borehamwood, Herts., Thorn Lighting Limited, 1993.

Anon. (1995). Symposium Openbare Verlichting, 22 februari 1995 (Symposium Public Lighting, 22 February 1995). Utrecht.

Anon. (1997). Philips lichtcatalogus 1997/1998 (Philips lighting catalogue 1997/1998). Eindhoven, Philips Lighting, 1997.

Anon. (2007). The CIE color space. Wikipedia. Internet, 21 May 2007.

Atkins, S.; Husain, S. & Storey, A. (1991). The influence of street lighting on crime and fear of crime. Crime Prevention Unit, Paper 28. London, Home Office, 1991.

Augustin, A. (2007). Augenheilkunde. 3., komplett überarbeitete und erweiterte Auflage (Ophthalmology. 3rd completely revised and extended edition). Berlin, Springer, 2007.

Baer, R. (1990). Beleuchtungstechnik; Grundlagen. Berlin, VEB Verlag Technik, 1990.

Baer, R., ed. (2006). Beleuchtungstechnik; Grundlagen. 3., vollständig überarteite Auflage (Essentials of illuminating engineering, 3rd., completely new edition). Berlin, Huss-Media, GmbH, 2006.

Barnard, A. & Spencer, J. eds. (2007). Encyclopedia of social and cultural anthropology. 2002 edition, reprinted 2007. London, Routledge, 2007.

Barrows, W.E. (1938). Light, photometry and illuminating engineering. New York, McGraw-Hill Book Company, Inc., 1938.

Bergmans, J. (1960). Seeing colours. Eindhoven, Philips' Technical Library, 1960.

Blackburn, S. (1996). The Oxford dictionary of philosophy. Oxford, Oxford University Press, 1996.

Boogaard, J. (1990). Samenvatting bij de kursus 'kleurenzien' voor optometristen (Summary of the course 'colour vision' for optometrists). Soesterberg, IZF, TNO, 1990 (Year estimated).

Boogaard, J. & Vos, J.J. (1989). Kleuren onderscheiden en herkennen (Colour discrimination and colour recognition). Chapter 4, p. 40-45. In: Vos & Legein, 1989.

Boll, M. & Dorgnon, J. (1956). Le secret des couleurs (The secret of colours). Que sais-je? no. 220. Paris, Presses Universitaires de France, 1956.

Bouma, P.J. (1946). Kleuren en kleurindrukken (Colours and colour impressions). Amsterdam, Meulenhoff & Co, 1946.

Boyce, P.R. (2003). Human factors in lighting. 2nd edition. London. Taylor & Francis, 2003.

Boynton, R.M. (1979). Human color vision. New York, Holt, Rinehart and Winston, 1979.

Budding, E. (1993). An introduction to astronomical photometry. Cambridge University Press, 1993.

CEN (2000). Traffic control equipment – Signal heads. European Standard EN 12368, January 2000, ICS 93.080.30. Brussels, CEN, 2000.

CEN (2003). Lighting applications – Tunnel lighting. CEN Report CR 14380. Brussels, CEN, 2003.

Chalupa, L.M. & Werner, J.S., eds. (2004). The visual neurosciences (Two volumes). Cambridge (Mass). MIT Press, 2004.

Cicerone, C.M. & Nerger, J.L (1989). The relative numbers of LWS and MWS cones in the human fovea centralis. Vision Research. 29 (1989) 115-128.

CIE (1932). Receuil des travaux et compte rendue des scéances, Huitième Session Cambridge – Septembre 1931. Cambridge, University Press, 1932.

CIE (1965). Method measuring and specifying colour rendering properties of light sources. Publication No. 13. Paris, CIE, 1965.

CIE (1971). Colorimetry. Publication No. 15. Paris, CIE, 1971.

CIE (1979). A review of publications on properties and reflection values of material reflection standards. Publication No. 46. Paris, CIE, 1979.

CIE (1979a). Absolute methods for reflection measurements. Publication No. 44. Paris, CIE, 1979.

CIE (1980). Light signals for road traffic control. Publication No. 48. 1980. Paris, CIE, 1980.

CIE (1983). Recommendations for surface colours for visual signalling. Publication No. 39/2. Paris, CIE, 1983.

CIE (1986). Colorimetry. Publication No. 15-2. Vienna, CIE, 1986.

CIE (1988). Spectral luminous efficiency functions based upon brightness matching for monochromatic point sources in 2° and 10° fields. Publication No. 75. Paris, CIE, 1988.

CIE (1990). Guide for the lighting of road tunnels and underpasses. Publication No. 26/2. Vienna, CIE, 1990.

CIE (2001). Criteria for road lighting. Proceedings of three CIE Workshops on Criteria for road lighting. Publication CIE-X019-2001. Vienna, CIE, 2001.

CIE (2005). Vision and lighting in mesopic conditions. Proceedings of the CIE Symposium '05. Leon, Spain, 21 May 2005. CIE X028. Vienna, CIE, 2005.

CIE (2006). TC 1-36 Fundamental chromaticity diagram with physiological axes. Part 1. Publication No. 170-1. Vienna, CIE, 2006.

Crawford, D. L. (2005). Mespopic? What is it, and what is its relationship to outdoor lighting? p. 56-60. In: CIE, 2005.

De Boer, J.B. & Fischer, D. (1981). Interior lighting (second revised edition). Deventer, Kluwer, 1981.

De Jong, P.T.V.M. (1995). Ergoftalmologie (Ergophthalmology). Chapter 20. In: Stilma & Voorn, eds., 1995.

Derrington, A.M.; Krauskopf, L. & Lennie, P. (1984). Chromatic mechanisms in lateral geniculate nucleus of macaque. Journ. of Physiology (London). 357 (1984) 241-265.

De Valois, R.J.; Abramov, L. & Jacobs, G.H. (1966). Analysis of response patterns of LGN cells. Journ. Opt. Soc. Amer. 56 (1966) 966-977.

Douglas, C.A. & Booker, R.L. (1977). Visual Range: Concepts, instrumental determination, and aviation applications. NBS Monograph 159. National Bureau of Standards, Washington D.C., 1977.

Farrington, D.P. & Welsh, B.C. (2002). Effects of improved street lighting on crime: A systematic review. Crime Reduction Research Series. Home Office Research Study 251, London, 2002.

Favié, J.W.; Damen, C.P.; Hietbrink, G. & Quaedflieg. N.J. (1962). Lighting. Eindhoven, Philips Technical Library, 1962 (original edition 1961).

Feynman, R.P.; Leighton, R.B. & Sands, M. (1977). The Feynman lectures on physics. Three volumes. 1963; 6th printing 1977. Reading (Mass.), Addison-Wesley Publishing Company, 1977.

Gall, D. (2004). Grundlagen der Lichttechnik; Kompendium (A compendium on the basics of illuminating engineering). München, Richard Pflaum Verlag GmbH & Co KG, 2004.

Gegenfurtner, K.R. & Sharpe, L.T. eds. (1999). Color vision: From genes to perception. New York, Cambridge University Press, 1999.

Gleitman, H. (1995). Psychology. Fourth edition. New York, W.W. Norton & Company, 1995.

Gleitman, H; Fridlund, A.J. & Reisberg, D. (2004). Psychology. Sixth edition. New York, W.W. Norton & Company, 2004.

Graetz, L. (1910). Das Licht und die Farbe; dritte Auflage (Light and colours, 3rd edition). Leipzig, B.G. Teubner, 1910.

Grassmann. H. (1853). Zur Theorie der Farbmischung (On the theory of colour mixtures). Pogg. Ann. der Physik. 89 (1853) 69-84.

Gregory, R.L. (1965). Visuele waarneming; de psychologie van het zien (Visual perception; The psychology of seeing). Wereldakademie; de Haan/Meulenhoff, 1965.

Gregory, R.L., ed. (1987). The Oxford companion to the mind. Oxford, Oxford University Press, 1987.

Gregory, R.L., ed. (2004). The Oxford companion to the mind. Second edition. Oxford, Oxford University Press, 1987.

Guild, J. (1931). The colorimetric properties of the spectrum. Philosophical Transactions of the Royal Society of London. A230 (1931) 149-187.

Hentschel, H.-J. (1994). Licht und Beleuchtung; Theorie und Praxis der Lichttechnik; 4. Auflage (Light and illumination; Theory and practice of lighting engineering, 4th edition). Heidelberg, Hüthig, 1994.

Hentschel, H.-J., ed. (2002). Licht und Beleuchtung; Grundlagen und Anwendungen der Lichttechnik; 5. neu bearbeitete und erweiterte Auflage (Light and illumination; Theory and applications of lighting engineering; 5th new and extended edition). Heidelberg, Hüthig, 2002.

Hering, K.E.K. (1878). Zur Lehre vom Lichtsinn (On the theory of sensibility to light). Vienna 1878.

Hopkinson, R.G. (1969). Lighting and seeing. London, William Heinemann, 1969.

Hubel, D.H. (1990). Visuele informatie; Schakelingen in onze hersenen (Visual information; The switchboard in our brain). Wetenschappelijke Bibliotheek, deel 21. Maastricht, Natuur en Techniek, 1990 (translation of: Eye, Brain and Vision. New York, The Scientific American Library, 1988).

Hurvich, L.N. & Jameson, D. (1957). An opponent-theory of color vision. Psychological Review. 64 (1957) 384-404.

Illingworth, V., ed. (1991). The Penguin Dictionary of Physics (second edition). London, Penguin Books, 1991.

Isobe, S. & Hirayama, T. eds. (1998). Preserving of the astronomical windows. Proceedings of Joint Discussion 5. XXIIIrd General Assembly International Astronomical Union, 18-30 August 1997, Kyoto, Japan. Astronomical Society of the Pacific, Conference Series, Volume 139. San Francisco, 1998.

JIS (1967). Japanese Industrial Standard, JIS Z 8726 "Method of measuring and specifying color rendering of light sources", 1967.

JIS (1990). Japanese Industrial Standard, JIS Z 8726 "Method of measuring and specifying color rendering of light sources". Revision, 1990.

Judd, D.B. (1951). Basic correlates of visual stimuli. Chapter 22. p. 811-867. In: Stevens, ed., 1951.

Judd, D.; MacAdam, D.L. & Wyszecki, G. (1964). Spectral distributions of typical daylight as a function of correlated colour temperature. Journ. Opt. Soc. Amer. 54 (1964) 1031.

Judd, D. & Wyszecki, G. (1967). Colour in business, science and industry. New York, John Wiley & Son, Inc. 1967.

Kaneko, S. et al. (1979). Skin colours and their synthesised spectral distribution of the reflection factor, Shyo-gi-shi, 13 (1979), 1.

Kawakami (1955). Colour sample of skin. Colour research, Vol.2 (1955) 1.

König, A. & Dieterici, C. (1892). Die Grundempfindungen im normalen und abnormalen Farbensysteme und ihre Intensitätsverteilung im Spektrum (The fundamental experiences in normal and abnormal colour systems and their intensity distribution in the spectrum). Z. Psychol. 4 (1892) 241-347.

Kremers, J.; Lee, B.B.; Pokorny, J.& Smith, V.C. (1993). Responses of macaque ganglion cells and human observers to compound periodic waveforms. Vision Research. 33 (1993) 1997-2011.

Lee, B.B.; Pokorny, J.; Smith, V.C.; Martin., P.R. & Valberg, A. (1990). Luminance and chromatic modulation sensitivity of macaque ganglion cells and human observers. Journ. Opt. Soc. Amer. A7 (1990) 2223-2236.

Lee, B.B.; Pokorny, J.; Smith, V.C. & Kremers, J. (1994). Responses to pulses and sinusoids in macaque ganglion cells. Vision Research, 34 (1994) 3081-3096.

Lennie, P. (2003). The physiology of color vision. Chapter 6, p. 217-246. In: Shevell et al., 2003.

MacAdam, D.L. (1942). Visual sensitivities to color differences. Journ. Opt. Soc. Amer., 32 (1942) 247-274.

Mann, I. & Pirie, A. (1950). The science of seeing. Harmondsworth, Penguin Books. Pelican A 157, 1950 (Revised edition).

Maxwell, J.C. (1855). Experiments on colour, as perceived by the eye, with remarks on colour-blindness. Trans. Royal Society (Edinburgh) 21 (1855) 275-298.

Mills, E. (1999). Fuel-based light: Large CO_2 source. IAEEL Newsletter 8 (1999) no 2 p 2-9.

Minnaert, M. (1942). De natuurkunde van 't vrije veld, derde druk (The physics of the open air, third edition). Zutphen, Thieme, 1942.

Moon, P. (1961). The scientific basis of illuminating engineering (revised edition). New York, Dover Publications, Inc., 1961.

Narisada, K. (1975). Applied research on tunnel lighting entrance lighting in Japan. In: Anon., 1975, p. 87-90.

Narisada, K. & Kawakami, K. (1998). Field survey on outdoor lighting in Japan. Summary of the IEIJ Report on a field survey on outdoor lighting in various areas in Japan. In: Isobe & Hirayama, eds., 1998, p. 201-242.

Narisada, K. & Schreuder, D.A. (2004). Light pollution handbook. Dordrecht, Springer, 2004.

Neitz, M. & Neitz, J. (2004). Molecular genetics of human color vision and color vision defects. Chapter 63, p. 974-988. In: Chalupa & Werner, eds., 2004.

Painter, K. (1999). Street lighting, crime and fear for crime; A summary of research. In: CIE, 2001.

Painter, K. & Farrington, D.P. (1999). Improved street lighting: Crime reducing effects and cost-benefit analyses. Security Journal, 12 (1999) 17-30.

Pokorny, J. & Smith, V.C. (2004). Chromatic discrimination. Chapter 58. p. 908-923. In: Chalupa & Werner, eds., 2004.

Reeb, O. (1962). Grundlagen der Photometrie (Fundaments of photometry). Karlsruhe, Verlag G. Braun, 1962.

Richter, M. (1976). Einführung in die Farbmetrik (Introduction in the metric of colours). Berlin, De Gruyter, 1976.

Roorda, A. & Williams, D.R. (1999). The arrangement of the three cone classes in the living human eye. Nature, 397 (1999) 520-522.

Schreuder, D.A. (1964). The lighting of vehicular traffic tunnels. Eindhoven, Centrex, 1964.

Schreuder, D.A. (1989). Enquête wijst uit: Straten zijn onveilig en licht is akelig (Enquiry shows: Streets are not safe and look ugly). De Gorzette, Verenigings- en informatieblad Bewoners Vereniging Schiedam-Zuid. 17 (1989) no 1. p. 23-25.

Schreuder, D.A. (1989a). Bewoners oordelen over straatverlichting (Residents judge street lighting). PT Elektronica-Elektrotechniek. 44 (1989) no. 5, p. 60-64.

Schreuder, D.A. (1994). Sick cities and legend analysis as therapy. International Workshop on Urban Design and Analysis of Legends. Kumamoto, Japan, July 26 – 28, 1994. Leidschendam, Duco Schreuder Consultancies, 1994.

Schreuder, D.A. (1998). Road lighting for safety. London, Thomas Telford, 1998 (Translation of "Openbare verlichting voor verkeer en veiligheid", Deventer, Kluwer Techniek, 1996).

Schreuder, D.A. (1999). The theory of tunnel lighting. LiTG-Sondertagung 'Aktuelles zur Tunnelbeleuchwtung, 22-23 September 1999, Bergisch-Gladbach. Leidschendam, Duco Schreuder Consultancies, 1999.

Schreuder, D.A. (2000). Energy efficient domestic lighting for developing countries. Paper presented at The "International Conference on Lighting Efficiency: Higher performance at Lower Costs" held on 19-21 January, 2001 in Dhaka (Bangladesh). Leidschendam, Duco Schreuder Consultancies, 2000.

Schreuder, D.A. (2000a). The role of public lighting in crime prevention. Paper presented at the workshop "The relation between public lighting and crime", held on 11 April 2000 at Universidade de Sao Paulo, Instituto de Eletrotecnica e Energia. Leidschendam, Duco Schreuder Consultancies, 2000.

Schreuder, D.A. (2001). Principles of Cityscape Lighting applied to Europe and Asia. Paper presented at International Lightscape Conference ICiL 2001, 13 – 14 November 2001, Shanghai, P.R. China. Leidschendam, Duco Schreuder Consultancies, 2001.

Schreuder, D.A. (2001a). Pollution free lighting for city beautification; This is my city and I am proud of it. Paper presented at the International Lighting Congress in Istanbul, Turkey, 6-12 September 2001. Leidschendam, Duco Schreuder Consultancies, 2001.

Schreuder, D.A. (2004). Verlichting thuis voor de allerarmsten (Home lighting for the very poor). NSVV Nationaal Lichtcongres 11 november 2004. Arnhem, NSVV, 2004.

Schreuder, D.A. (2005). Domestic lighting for developing countries. Prepared for publication in: UNESCO – A world of science, Paris, France. Leidschendam, Duco Schreuder Consultancies, 2005.

Schreuder, D.A. (2008). Looking and seeing; A holistic approach to vision. Dordrecht, Springer, 2008 (in preparation).

Sharpe, L.T.; Stockmann, A.; Jägle, H. & Nathans, J. (1999). Opsin genes, cone photopigments, color vision, and color blindness, p. 3-52. In: Gegenfurtner & Sharpe, eds., 1999.

Shevell, S.K., ed. (2003). The science of color. Second edition. OSA Optical Society of America. Amsterdam, Elsevier, 2003.

Smith, V. & Pokorny, J. (2003). Color matching and color discrimination. Chapter 3. In: Shevell, ed., 2003.

Stevens, S.S., ed. (1951). Handbook of experimental psychology. New York, John Wiley and Sons, Inc, 1951.

Stevens, W.R. (1969). Building physics: Lighting – seeing in the artificial environment. Oxford, Pergamon Press, 1969.

Stiles, W.S. (1946). A modified Helmholz line element in brightness colour space. Proc. Phys. Soc. London. 58 (1946) 41.

Stilma, J.S. & Voorn, Th. B., eds. (1995). Praktische oogheelkunde. Eerste druk, tweede oplage met correcties (Practical ophthalmology. First edition, second impression with corrections). Houten, Bohn, Stafleu, Van Loghum, 1995.

Stockman, A. & Sharpe, T.L. (2000). The spectral sensitivities of the middle- and long-wavelength-sensitive cones derived from measurements in observers of known genotype. Vision Research, 40 (2000) 1711-1737.

Van De Hulst, H.C. (1981). Light scattering by small particles. New York, Dover, 1981.

Van Santen, C. (2005). Light zone city: Light planning in the urban context. Basel, Birkhauser, 2005.

Van Tilborg, A.D.M. (1991). Evaluatie van de verlichtingsproeven in Utrecht (Evaluation of lighting experiments in Utrecht). Utrecht, Energiebedrijf, 1991 (not published; see Anon., 1995).

Visser, R. (1992). Verlichting & interieur (Lighting & interior). Amersfoort, Dekker/vd Bos & Partners, 1992

Visser, R. (1997). Spelen met licht in en om het huis (Playing with light in and around the house). Amersfoort, Dekker/vd Bos & Partners, 1997.

Von Helmholtz, H. (1896). Handbuch der physiologischen Optik, 2. Auflage (Handbook of physiological optics, 2nd edition). Hamburg, Voss, 1896.

Vos, J.J. & Legein, C. P. (1989). Oog en werk: een ergoftalmologische wegwijzer (Eye and work; Ergo-phthalmological guidance). Den Haag, SDU Uitgeverij. 1989.

Wachter, A. & Hoeber, H. (2006). Compendium of theoretical physics (Translated from the German edition). New York, Springer Science+Business Media, Inc., 2006.

Walraven, J., ed. (1981). Kleur (Colour). Ede, Zomer & Keuning. 1981. Original English edition: London, Marshall Editions Limited, 1980.

Weis, B. (1996). Beleuchtungstechnik (Illuminating engineering). München, Pflaum Verlag, 1996.

Wiesel. T.N. & Hubel, D.H. (1966). Spatial and chromatic interactions in the lateral geniculate body of the rhesus monkey. Journ. of Neurophysiology. 29 (1966) 1115-1156.

Wright, W.D. (1928). A re-determination of the trichromatic coefficients of the spectral colours. Transactions of the Optical Society. 30 (1928) 141-164.

Wright, W.D. (1967). The rays are not coloured. London, Adam Hilger, 1967.

Wyszecki, G. & Stiles, W. (1982). Color science: Concepts and methods, quantitative data and formulae. 2nd edition. First edition 1967. New York, Wiley, 1982.

Zaidi Q.; Shapiro, A. & Hood, D. (1992). The effect of adaptation on the differential sensitivity of the S-cone system. Vision Research. 32 (1992) 1297-1318.

10 Road lighting applications

In the preceding chapters of this book a number of fundamental aspects of lighting, more in particular of mathematical, physical, and physiological nature have been discussed. In this chapter it is studied in how far these fundamental aspects can be used in road lighting applications. In the next chapter several design aspects of road and tunnel lighting are discussed.

Road lighting luminaires have different functions some of which are described in this chapter. First the light control, which means to ensure that the light gets where we want it, and not where we do not want it. Light control is essentially applied optics. The principles of geometric optics are discussed, including the way they are applied in the optical design of road lighting luminaires. The most common systems of road lighting luminaire classification are described. Also, ingress protection is discussed. The consequences for low-maintenance luminaire design are described in the next chapter.

The next subject in this chapter is light pollution. Light pollution is closely related to the second aspect of light control: to ensure that the light does not gets to places or in directions where we do not want it. The reason to bother about the reduction of light pollution is not only a matter of energy conservation, but also to try to ensure that darkness allows human beings to enjoy the starry night, and to maintain the contact with the cosmos. These points are described in the first chapter of this book. Remedial measures are discussed. This section is based on the Light Pollution Handbook, published in 2004 by Kohei Narisada and Duco Schreuder.

The final subject of this chapter is the way that incident light is reflected by road surfaces. The importance is in the fact that for the detection of objects on the road, and for road safety more in general, the road surface luminance is the most important design criterion for road lighting installations. This is called the luminance technique in road lighting. The lighting design is helped by the documentation and the classification of road surface reflection characteristics. Finally, the measuring of the road reflection is discussed.

10.1 Geometric optics

10.1.1 Definitions of light

(a) Four models for the description of light

In sec. 2.1.1 a description is given about light and what it really is. Light is commonly understood as electromagnetic radiation that an observer perceives through visual sensation. As is explained in Chapter 3, where photometry is discussed, the observer is essential in the definition of light.

It is explained that there are four models to describe light as a physical phenomenon:
1. Light is a collection of light rays;
2. Light is an electromagnetic wave;
3. Light is a stream of rapidly moving particles, or photons;
4. Light is fluid of power (wattage).

(b) Light rays

The first model is used in geometric optics, more in particular in optical imaging. The light rays are straight as long as there is no reflection or refraction. They have no speed or propagation; they are just 'there'. The rays have no width. They are lines in the sense of geometry – hence the term geometric optics. Details can be found in Narisada & Schreuder (2004, sec. 11.1).

In every lighting design method and in every calculation that is used in illuminating engineering, the approach of geometric optics is used, both in luminaire design as in the design of lighting installations. Geometric optics are used in graphical design methods, as well as in computer-aided luminaire design. The same methods are used in the design of optical instruments, optical telescopes included.

Basically, geometric optics is based on the principle of Fermat (Feynman et al., 1977, Vol. I, p. 26-4). This principle states that a light ray always takes the shortest possible path; more precisely, when media of different refraction indexes are involved, the path that takes the least time. Here, the speed of light is apparently introduced, but it has no physical meaning. One of the consequences of Fermat's principle is that light rays can be reversed: all results of geometric optics are invariant with respect to the direction of the light. Fermat's principle is a special case of the Principle of Least Effort, one of the basic principles in Nature. The principle as such has a long history of at least 2000 years. See for this the section on the history of optics in Blüh & Elder (1955, p. 351-352).

Road lighting applications   359

Further on in this section it is explained that according to Huygens, light is to be considered as spherical waves that originate from the source and that propagate through space. Each wave has a wave front. At any moment in time, each point of the wave front acts as a secondary light source, emitting new elementary waves and creating a new wave front.

10.1.2   Design of optical devices

(a) *Principles of image-forming and lighting equipment*
The basic function of optical devices is to get the light where we want it. In this respect there are two large families that have quite different characteristics. Often, we are not primarily interested in the light as such but more in the information that is transferred by the light. One might call such equipment image-forming devices. Their function is in principle to make an image of an object in the object-space into the image-space. In this respect an optical image is essentially a model of the type that is discussed in sec. 1.5.1. The reason for wanting an image – or a model – of the original usually is that the object is too small or too weak for easy observation. Hence the need for microscopes and telescopes. Also, many object have only a fleeting existence; hence the need for cameras.

The difference between the two is that the image that is formed by a microscope or a telescope need to be observed by the eye. Therefore, the image must seem to be at a convenient distance from the eye. The implication is that one needs an virtual image. That is an image that, contrary to a real image cannot be caught up by a screen. As virtual images are not common in outdoor lighting, this matter will not be dealt with here. See Blüh & Elder (1955); Breuer (1994, vol. 1, p. 161); Longhurst (1964); Lorentz (1922); Van Heel (1950).

Microscopes and telescopes, being designed for direct visual observation, have the obvious drawback that they cannot save the image. For that, a camera is needed. A camera is a device where the image is formed on a surface that is treated in such a way that the image is kept, even after the light is gone. Cameras can, of course, be attached to microscopes and telescopes. This requires a real image. Evidently it also requires a surface that can keep the image, or, in other words, that can keep the effects of the energy from the light that did strike the surface.

At present, most cameras of amateurs and professionals alike use Charge Coupled Devices or CCDs (Illingworth ed., 1991, p. 57; Howell, 2001). The characteristics of CCDs are explained in some detail in sec. 5.4.1. Before reliable and affordable CCDs were available, photographic film was used. The processes involved are described in many outstanding books, like e.g. Folts et al. (2006). A summary is given in Illingworth ed. (1991, p. 348-349) and Reeb (1962, sec. 7.32, p. 166-167). Details on the photographic process, with emphasis on its use in astronomy, are given in Sterken & Manfroid (1992), Budding (1993), and Weigert & Wendker (1989).

(b) *Image-forming equipment*

As mentioned in the preceding part of this section, for imaging optical elements like lenses and mirrors are needed. First, we will consider the basic imaging function of lenses. A positive lens collects the light rays, whereas a negative lens disperses them (Kuchling, 1995; Longhurst, 1964). See Figure 10.1.1.

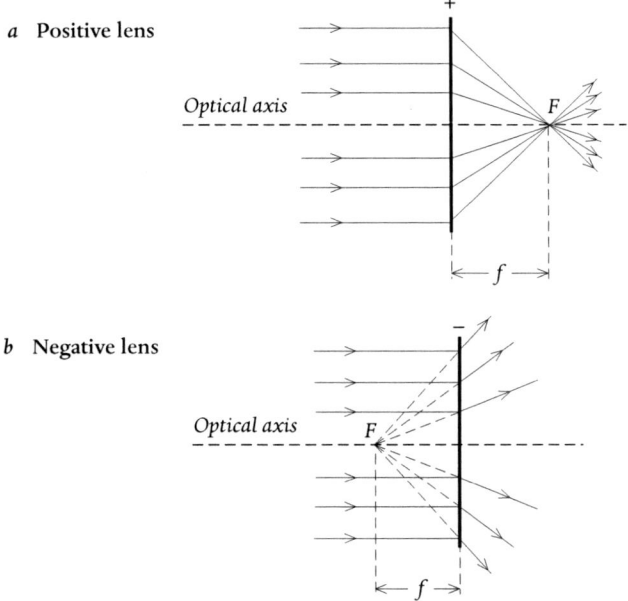

Figure 10.1.1 *The light rays in (a) a positive lens and (b) a negative lens.*

This means that it is not possible to make a real image with a negative lens. With a positive lens it may be done, provided the object is placed further away from the lens than the focal point. The real image is inverted. If the object is placed closer to the lens than the focal point, the image is virtual, but upright. See Figure 10.1.2.

Mirrors function in a very similar way, be it that, obviously, the direction of the light is always inverted. A concave mirror curves inward, it is hollow, whereas a convex mirror curves outward, like a Christmas tree ball. A concave mirror collects the light rays, whereas a convex mirror disperses them. See Figure 10.1.3.

Road lighting applications 361

a Real image

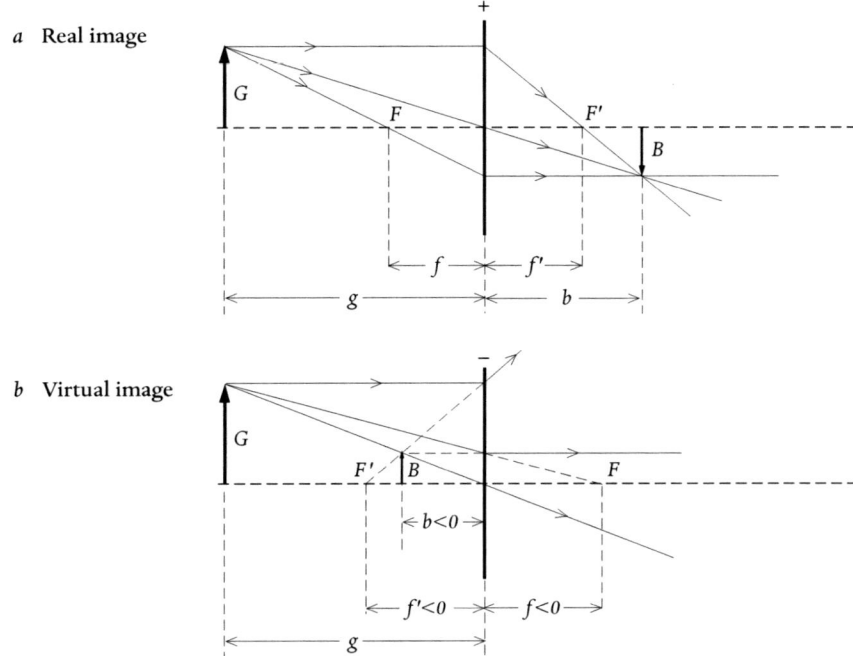

b Virtual image

Figure 10.1.2 The image-forming of a positive lens (a) real; (b) virtual.

a Concave mirror

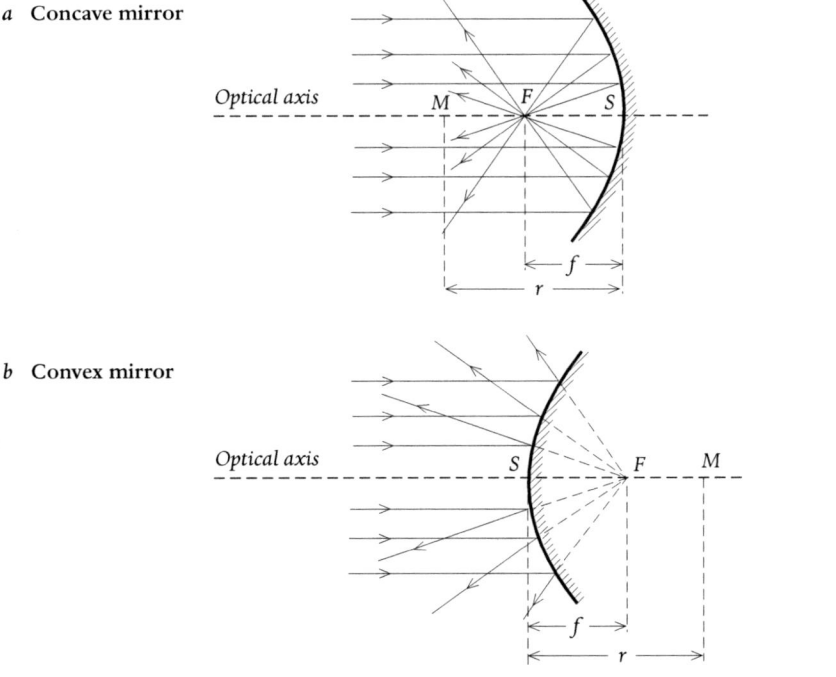

b Convex mirror

Figure 10.1.3 The light rays at (a) a concave mirror, (b) at a convex mirror.

The image forming is similar as with lenses. It is not possible to make a real image with a convex mirror. With a concave mirror it may be done, provided the object is placed further away from the mirror than the focal point. The real image is inverted. If the object is placed closer to the mirror than the focal point, the image is virtual, but upright. See Figure 10.1.4.

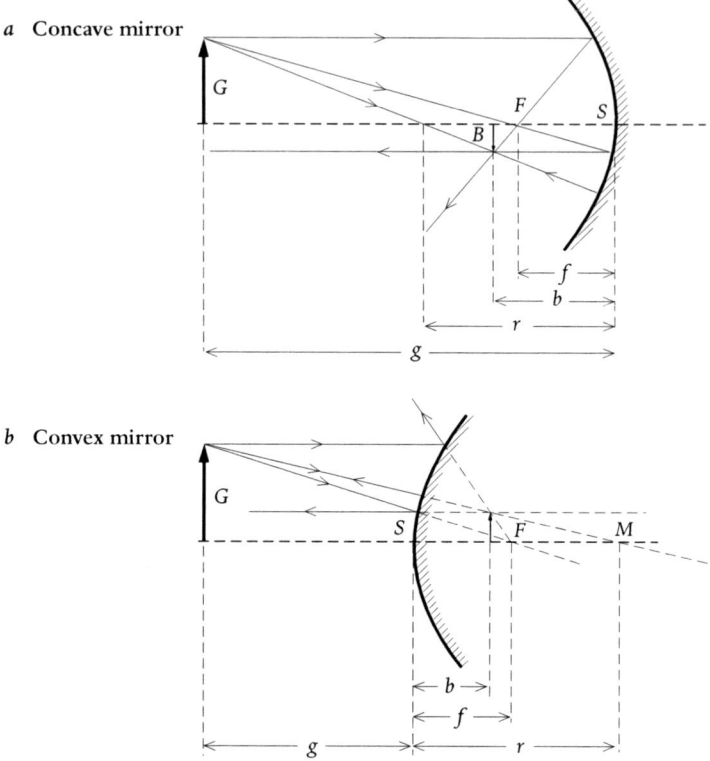

Figure 10.1.4 *The image-forming of a convex mirror (a) real; (b) virtual.*

For a camera it is sufficient to place the film or the CCD at the right point behind the rear focal point, at the location of the real image. When a lens is used as a magnifying device, e.g. as a reading glass, all one has to do is to place the object between the forward focal point and the lens. The upright virtual image is located behind the lens and can easily be viewed.

To improve the performance, it is possible to combine two lenses. The front lens or objective provides an enlarged, inverted, real image. This image, being real, can be viewed by using a second lens as a magnifier. This lens is, for obvious reasons, called the eye-piece. For visual observation, a virtual image is needed; for photography, we

Road lighting applications

need a real image. By choosing the right position, one may have at the end a real or a virtual image. Depending on the geometry, the device can be used to enlarge items that are close by, or objects that are far away. The first is the microscope, the other the telescope. Both were invented around 1600. As has been pointed out many times, they were the instruments that started the scientific revolution of the Renaissance by making accessible to humankind the previously invisible (Durant, 1962, p. 45).

(c) Non-image forming equipment

The function of non-image forming equipment is completely different from that of the image-forming equipment that is discussed in the preceding part of this section. Non-image forming equipment is the basic equipment for all applications in outdoor lighting, and, for that matter, also of most indoor lighting as well. In essence, it is a matter to bring light where we want it and to avoid light where we do no want it.

In sec. 1.3 the different functions of outdoor lighting are explained in detail. Basically it has to do with making those objects visible that are essential for the task at hand. Thus, all outdoor lighting is utilitarian. Non-image forming lighting has two main functions, viz.: signalling, and lighting or illumination. In this book we will deal only with the second function, the lighting or illumination.

This section is focussed on the application of geometric optics in the design of lighting equipment or luminaires. Other terms are used as well, such as lanterns, fittings, etc. Luminaires have several functions
1. Directing the light;
2. Attaching and supporting the lamps and other gear;
3. Protecting the gear against weather and vandalism;
4. Ensuring the correct lamps operating temperature, notably for fluorescent lamps.

In this book, some details regarding the optical design of luminaires will be given; the other aspects will be discussed only briefly. These functions and the related requirements are discussed in detail in many lighting engineering textbooks, like e.g. Boyce (2003); De Boer, ed. (1967); Narisada & Schreuder (2004, chapter 11); Ris (1992); Schreuder (1998); Van Bommel & De Boer (1980).

10.2 Luminaire design

10.2.1 Optical elements

(a) Lamp and road axis

As is explained in the preceding part of this section, the major function of luminaires is to direct the light to the places where we want it, and to prevent that light is aimed in other directions, so it gets at places where we do not want it. Directing the light in the desired directions is usually called light control.

On grounds of lamp life and lamp efficacy, most modern outdoor lighting, and road and street lighting in particular, is equipped with gas discharge lamps. Usually, gas discharge lamps are elongated or tubular in shape. This means that they show a circular symmetry around one of their axes – the axis of symmetry. Also the light emitted by the lamps shows this circular symmetry. The light distribution of the lamps consists of circles with the axis as centre. In a direction parallel to the lamp axis the light distribution is more irregular, for one thing as a result of the lamp cap. All lamps have lamp caps where they are connected to the electric supply, and which helps to support them. Tubular lamps have a cap at the end, and sometimes one at each end. The are also other constructional elements that cause the irregular length-axis light distribution. Appropriate light control requires optical elements to redirect some of the light into the desired direction.

These characteristics of the light distribution imply that light control around the lamp axis is usually rather simple, even with traditional low-pressure sodium lamps and many types of compact fluorescent lamps, that consist of two parallel, rather than one, gas-discharge tubes. Light control in a direction perpendicular to the lamp axis is mostly difficult, if possible at all. So, it is natural to position the lamp axis in a direction perpendicular to the lamp axis. Because roads, by their nature as traffic arteries, show a lengthwise character, the result is that in almost all cases of street lighting the lamp axis, and thus the luminaire axis, is perpendicular to the road axis. Obvious exceptions are on the one hand compact lamps and on the other hand squares, pedestrian precincts, and roundabouts. Additionally, as almost all roads carry two-directional traffic, the light distribution of road lighting luminaires usually is symmetric around their axis.

(b) Road lighting luminaire light distributions

As a result of the considerations given in the preceding part of this section, almost all road lighting luminaires show a light distribution that can be represented in one plane, the plane perpendicular to the luminaire axis. Further, in almost all cases the light distribution is symmetric around the vertical plane through the luminaire axis.

Road lighting applications

In sec. 10.4 it is explained that most road surfaces, particularly the almost universally applied asphalt road surfaces, show a marked specularity. For traffic routes, where the photometric requirements are expressed in terms of road surface luminances, it is preferred to have as much light in the direction where the light is reflected best by the road surface. That means that the light distribution of road lighting luminaires shows a marked peak in directions under the horizontal plane. According to CIE-specifications, the vertical angle of maximum luminous intensity is called throw (CIE, 1997; Van Bommel & De Boer, 1980, p. 102).

At the other hand, the need to restrict glare requires that little light is emitted in a direction close under the horizontal plane. In the CIE specifications glare control is termed control. It is characterised by SLI, the luminaire-dependent parts of the Glare Mark formula that is described in another sec. 8.4, where discomfort glare is discussed.

This, together with the symmetry mentioned earlier, makes that the light distribution of all road lighting luminaires for traffic routes are very similar in shape. An example of the light distribution in the plane parallel to the road axis is depicted in Figure 10.2.1.

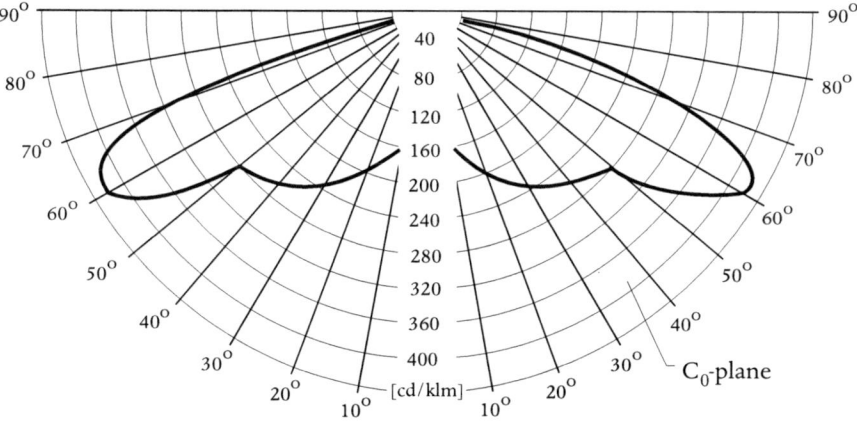

Figure 10.2.1 The light distribution of a typical traffic route lighting luminaire in the plane parallel to the road axis (After Schreuder, 1998, figure 5.4.3).

These considerations refer primarily to the light distribution in a plane parallel to the road axis. For luminaires that are designed for road side mounting on wide roads, an addition is made to this. By an adjustment of the optical system, extra light is thrown under an angle with the road axis, more or less in the direction of the opposing road side. The horizontal angle of maximum luminous intensity is called spread (CIE, 1997; Van Bommel & De Boer, 1980, p. 102). The spread may amount to some 10° to 30°. In some modern luminaires both throw and spread can be selected for the specific road conditions by adjusting the optical elements in the luminaire.

In Figure 10.2.2, the principles of the lay-out of the optical systems for road lighting luminaires are depicted.

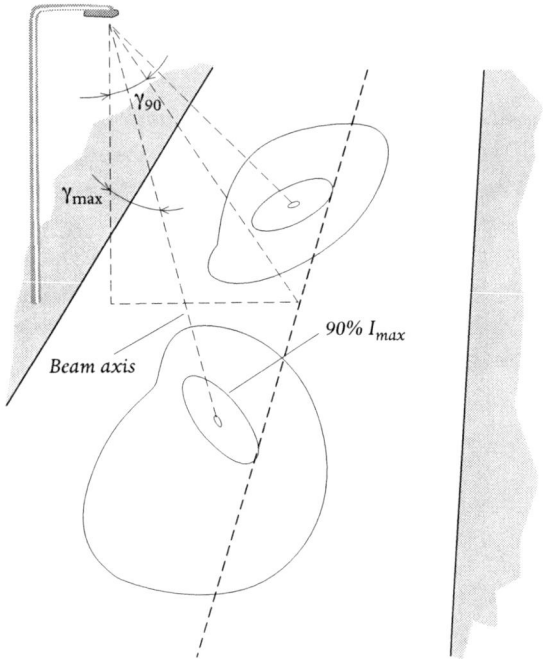

Figure 10.2.2 *The principles of road lighting luminaires (After Van Bommel & De Boer, 1980, Figure 7.3, p. 103).*

(c) Road lighting luminaire classification

As mentioned earlier, one of the most important characteristics of any outdoor lighting installation is the light distribution – or luminous intensity distribution – of the luminaires. A classification system of luminaires is helpful in the determination of the type of luminaire that can be applied in a specific lighting installation. For practical lighting design, it is useful to have a detailed luminaire classification system. The luminaire classes can easily be added to the product information of each luminaire type. Such classifications allow the designer to select, in an early stage of the design, an appropriate luminaire, or maybe several appropriate luminaires. The actual design calculations can be made with those luminaires, avoiding the extra work of doing the calculations with other luminaires that, in a later stage, prove to be not appropriate for the particular design.

In the preceding part of this section the CIE specifications of 1997 were mentioned. This classification system is based on the three concepts: Throw, Spread, and Control. For Throw, expressed in γ_{max}, three classes are introduced: Short, Intermediate and Long. The corresponding γ_{max}-values are 60°, 60° – 70°, and over 70°. For Spread, expressed in γ_{90}, also three classes are introduced: Narrow, Average, and Broad. The corresponding γ_{90}-values are given in Van Bommel & De Boer (1980, Table 7.2, p. 104). In spite of its merits, this classification is not used any longer. Modern standards refer primarily to complete lighting installations (CEN, 2002; NSVV, 2002). This trend makes it harder to use approximative ways for a preliminary road lighting design. The same fate befell the other classification systems that refer to the luminaires themselves (NSVV, 1957; CIE, 1965). After many years in use the CIE 1965 system was removed from the publications. However, as it is still used in lighting design practice, it is given in Table 10.2.1.

Luminaire type	Max. intensity (cd/1000 lm)	
	80°	90°
cut-off	30	10
semi-cut-off	100	50
non-cut-off	–	1000 cd total

Table 10.2.1 The CIE classification system of 1965 (After Knudsen, 1967, table 6.4, p. 225. Based on CIE, 1965).

A drawback of the CIE classification is that it refers only to the light distribution along the road axis. Also, most practical luminaires fall into the class 'semi-cut-off'. The cut-off-class is rather restrictive, whereas the non-cut-off class has no real restrictions.

The modern standards contain a luminaire classification of sorts. As the CEN-rapport states explicitly, the classification is aimed at glare restriction and stray-light reduction (NSVV, 2002, Annex A, p. 22). It may be mentioned again that NSVV, 2000, is a translation of CEN, 2002. This means that the use for general outdoor-lighting design is limited. The CEN-system is given in Table 10.2.2 and Table 10.2.3.

First the G-classes. They give a broad outline of the requirements for the luminous intensities of luminaires. As the requirements are expressed in relative values, the use for design is limited. It is not mentioned what is the experimental basis for the G-classes.

Class	Max intensity cd/1000 lm			Other requirements
	70°	80°	90°	
G1	–	200	50	–
G2	–	150	30	–
G3	–	100	20	–
G4	500	100	10	intensity over 95° zero
G5	350	100	10	intensity over 95° zero
G6	350	100	0	intensity over 95° zero

Table 10.2.2 The G-classes of CEN (After NSVV, 2003, Table A.1).

The D-classes refer to glare. They are based on one of the several ways to describe discomfort glare:

$$D = I \cdot A^{-0,5} \qquad [10.2\text{-}1]$$

The limits of D are given in Table 10.2.3.

Class	Maximum D
D1	7000
D2	5500
D3	4000
D4	2000
D5	1000
D6	500

Table 10.2.3 Maximum D-values of the CEN classes (After NSVV, 2002, Table A.2).

(d) *The proposal of Narisada and Schreuder*

A system that is adapted to the more recent requirements of road lighting, and with emphasis on stray light reduction is introduced by Narisada & Schreuder (2004, sec. 7.1.4b).

In Narisada & Schreuder (2004, sec. 7.1.4b, Table 7.1.4) a system of luminaire classification for road lighting is introduced with the following characteristics:
1. The upward light emission is divided in two angular areas:
   – from the zenith, downward to 15° above the horizon, designated as U_1;
   – from 15° above the horizon to the horizontal direction, designated as U_2.

Road lighting applications

2. The downward light emission is divided in four angular areas:
   - straight in the horizontal direction. Theoretically this is, of course, not an angular area. The direction will be designated as I_{90}, or, when appropriate, as D_1;
   - from the horizontal direction down to 15° below the horizon, designated as D_2. The area might, when appropriate, be designated as I_{80};
   - between the directions 15° below the horizon and 45° below the horizon, designated as D_3;
   - from the directions 45° below the horizon to the nadir (straight down under the luminaire), designated as D_4.
3. All values apply to all lateral angles around the luminaire.
4. All values are related to the rated (new) lamp lumens.
5. All recommendations apply equally to open or closed luminaires, independent of the way the closure is made.

The different angles and angle ranges are depicted in Figure 10.2.3.

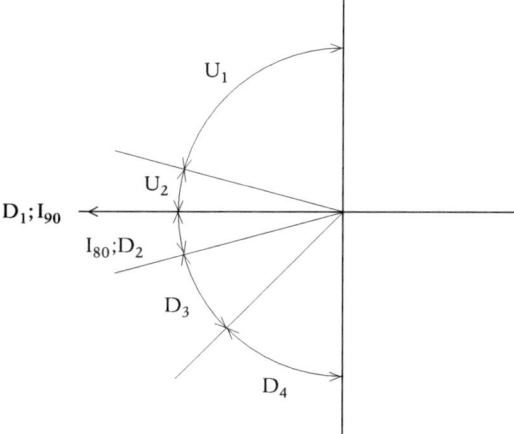

Figure 10.2.3 The angles and angle ranges of the Narisada-Schreuder proposal.

The Narisada-Schreuder proposal is based in part on earlier suggestions for luminaire classification. See e.g. CIE (1977); NSVV (1957, 2002); Pimenta (2002); Schreuder (1998, sec. 7.3) and Van Bommel & De Boer (1980, p. 103).

The system consists of six classes, designated as A to F. For each class, values of the different parameters are given. The recommended classification is given in Table 10.2.4. In this table, the CIE classification from 1977 is added for comparison as far as it is relevant (CIE, 1977).

Lum. class	Photometric requirements for different angular areas						
	U_1	U_2	I_{90} D_1	D_2 I_{80}	D_3	D_4	correspond to CIE class
A	0	0	0	10	30	100	n.r.
B	0	0	2	10	30	100	n.r.
C	0	2	10	30	100	n.r.	CO
D	0	5	30	50	n.r.	n.r.	SCO
E	10	10	50	100	n.r.	n.r.	SCO
F	20	20	100	n.r.	n.r.	n.r.	NCO

Table 10.2.4 The recommended classification of road lighting luminaires (After Narisada & Schreuder, 2004, Table 7.1.4). Note: n.r. means not relevant

10.2.2 The optics of road lighting luminaires

The accuracy of the light control is determined by the ratio between the lamp dimensions and the focal length of the optical system. In other words, the smaller the lamp, the better the light control, and consequently the better the luminaire efficiency. This is depicted in Figure 10.2.4.

In modern traffic route luminaires this ratio can be quite small for two reasons:
1. The discharge tube of modern high-powered gas-discharge lamps is small;
2. The luminaires cannot be made very small because of the thermal load caused by the lamps. So usually there is plenty of room in the luminaire for a fair-sized optical system.

The design of the optical systems of outdoor lighting luminaires is essentially based on the principles of geometric optics. The light rays are followed from the lamp toward the place where the light is needed – the road, or the sports field. For the light control either reflectors or refractors are used, sometimes in combination. In the past, diffuse reflectors with white paint or enamel were used. Modern reflectors generally are made of anodised aluminium having a total reflectance of 80-87% (Hentschel, ed., table 6.2, p. 187). Stainless steel is the second material that can be used for mirrors in luminaires. The reflectance of stainless steel in visible light is considerable lower than that of anodized aluminum, not much more than 60 % (Bean & Simons, 1968, p. 185, table 6.1). The reflection in the infrared region is very low. Stainless steel easily heats up. Furthermore, stainless steel is not easily 'workable'; it cannot easily be bent or folded in any shape. So it is not particularly suitable if precision in the light control is required. At the other hand, stainless steel is a sturdy material, so it is quite possible to give the housing the

Road lighting applications

right shape to act as a reflector. Usually, these luminaires have no cover – they are open luminaires. Open luminaires are discussed in a further part of this section. Refractors are usually made of clear plastics like metylacrylate. Reflectors are positioned inside the luminaire above the lamp, whereas refractors are incorporated, between the lamp and the road, in the luminaire cover.

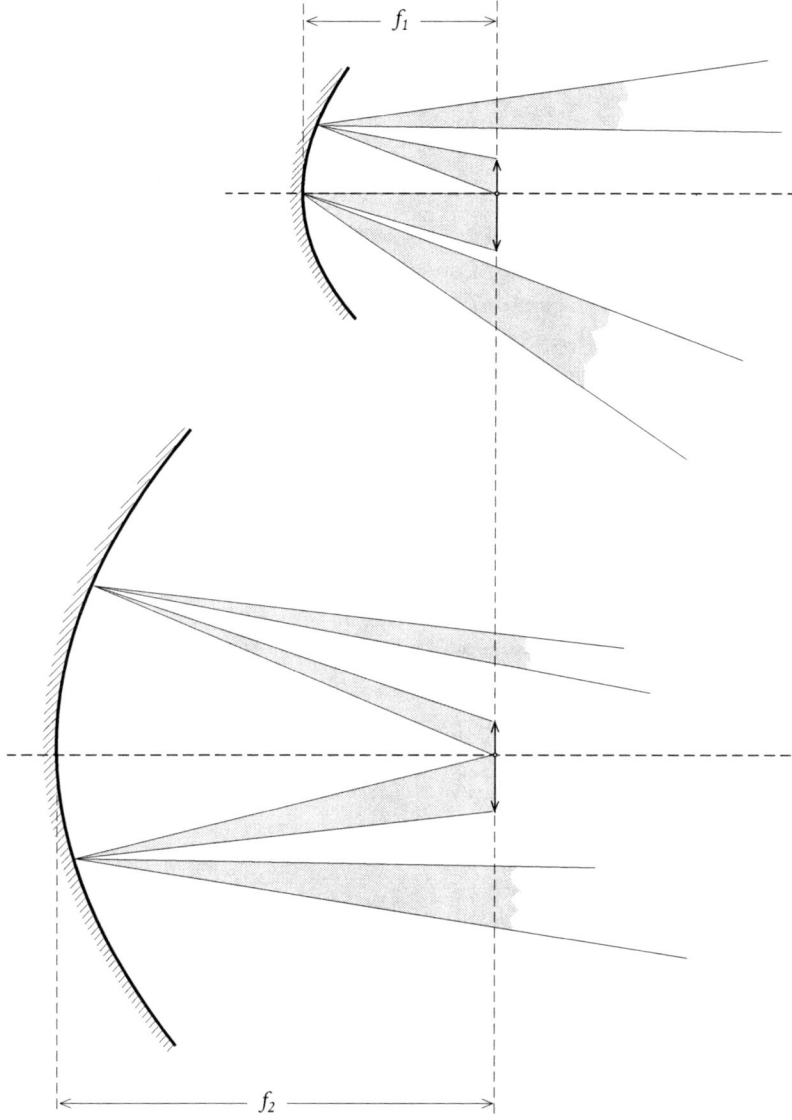

Figure 10.2.4 *The effect of lamp dimensions and the focal length on the light control.*

The luminaire efficiency, and hence the effective luminaire life are limited by soiling, corrosion, and general deterioration. It is taken for granted that luminaires must be closed. Close inspection of the available data shows, however, that in many places where maintenance is poor, open luminaires are a good alternative (Schreuder, 1999, 2001). We will come back to the potential merits of open luminaires in a further part of this section, where also the use of different reflector materials will be explained. It must be stressed that poor maintenance is not restricted to the developing world! Also, open luminaires are not contradictory to the IP-considerations that are discussed in a further part of this section.

10.2.3 Ingress protection

As has been mentioned earlier, one of the major functions of luminaires for outdoor lighting is to protect the lamps. This is termed ingress protection. The International Electrotechnical Commission (IEC) and the International Commission on Rules for the Approval of Electrical Equipment (CEE) have standardised the ingress protection by defining an IP-number for each type of luminaire. The IP-number consists of two digits. The first digit indicates the protection against touching electrically charged parts, as well as the protection against objects or dust entering the luminaire. The second digit describes the sealing against water. The two types of classes are summarised in Table 10.2.5 and Table 10.2.6. It should be stressed that the terminology in the Tables 10.2.5 and 10.2.6 is tentative only.

Digit	Protection against:	
	accidental touching	penetration
0	none	none
1	by hand	objects >50 mm
2	fingers	objects > 12 mm
3	wire > 2,5 mm	objects > 2,5 mm
4	wire > 1 mm	objects > 1 mm
5	impossible	no harmful dust
6	impossible	fully dust-proof

Table 10.2.5 IP, first digit (After Schreuder, 1998, Table 13.3.3, p. 225. Based on NSVV, 1997).

Road lighting applications

Digit	Protection against water	CEE-indication
0	none	–
1	drops falling straight down	–
2	drops max 15°	dripping waterproof
3	spray water max 60°	rainproof
4	splashing all directions	splash waterproof
5	spraying all directions	water spray-proof
6	waves, heavy seas	
7	submersion, limited	waterproof
8	submersion, unlimited	waterproof

Table 10.2.6 *IP, second digit (After Schreuder, 1998, Table 13.3.4, p. 225. Based on NSVV, 1997).*

Finally, the insulation of the luminaires is mentioned in Table 10.2.7. Also here, EEC-classes are relevant.

Class	Insulation
0	functional insulation, no earth facility
I	functional insulation, earthing facility for metal parts
II	double insulation. No touchable loaded metal parts
III	low voltage; maximum 42 V

Table 10.2.7 *EEC-classes for insulation (After Schreuder, 1998, Table 13.3.5, p. 226. Based on NSVV, 1997).*

10.3 Light pollution

10.3.1 Description of light pollution

(a) *Light pollution and sky glow*

Light pollution is one of the negative side-effects of artificial outdoor lighting. The term light pollution is an unhappy one, but as no better alternative seems to exist, it will be used here (Narisada & Schreuder, 2004). What we really want to express is the disturbance of the night itself, for which we have introduced the term 'starry night'. It goes without saying that there are many other ways to express the same effect.

As is explained already in several places in this book, the function of all outdoor lighting is to enhance the visibility or the aesthetics in the night-time environment. The light should come where it is needed. If not, it is 'spilled', causing economic losses as regards electric energy, and causing light pollution (Schreuder, 1995). It is not easy to estimate the amount of money nor the amount of energy that is involved in light pollution, but the amounts are considerable (Narisada & Schreuder, 2004, Chapter 2). Many view the visual disturbance and discomfort, as well as environmental damage as even more serious.

The overall effects are termed light pollution. A major form of light pollution is the glow extending over the night sky, usually called sky glow. Sky glow presents itself as a background luminance over the sky, against which the astronomical objects are to be observed. The interference of astronomical observations is caused by the resulting reduction of their luminance contrast. The effect is similar to the effects of disability glare, which is also caused by a veil, but then internally within the ocular media. Disability glare is discussed in sec. 8.3.

(b) *Victims of light pollution*

Light pollution has 'victims'. A victim matrix is given in Table 10.3.1, which includes both the victims as well as the major types of lighting installations that may cause the intrusive light. The table is used the recent Recommendations for the Netherlands (NSVV, 1999, 2003, 2003a,b).

Lighting for:	S	I	F	R	A	G
residents	✓	✓	✓	✓	✓	✓
astronomers	✓	✓	✓	✓		✓
life in nature	✓	✓	✓	✓	✓	✓
road users	✓		✓		✓	✓
shipping		✓	✓			✓

Table 10.3.1 *The victim matrix. The relevant areas are marked by ✓ (After Narisada & Schreuder, 2004, Table 3.2.1).*

Ad Table 10.3.1:
  S: sports lighting
  I: lighting for industry
  F: floodlighting
  R: road lighting
  A: lighting for advertizing
  G: greenhouse lighting

Road lighting applications 375

Further considerations did show that not all categories that are given in Table 10.3.1 are always relevant. Practice did suggest that several classes of lighting installations did need very similar requirements. Also, not all 'victim classes' were considered of equal importance. The result was that the subject of light pollution was split up in two main areas. These areas were committed to two CIE Divisions. Light intrusion, particularly in private premises, was studied by a Technical Committee in CIE Division 5. A report was produced (CIE, 2003). Overall light pollution, particularly sky glow, was studied by a Technical Committee in CIE Division 5 in close cooperation with IAU Commission 50 (Anon., 1978). Also here a report was produced (CIE, 1997). Proposals have been made to reunite the two areas again, because they have a lot in common. The complete area of light pollution is dealt with in great detail in the Light Pollution Handbook (Narisada & Schreuder, 2004).

(c) *Description of light pollution effects*

Light pollution shows itself in several different ways, the sky glow that is mentioned earlier being one of them:

1. Direct light entering the premises. Usually this is called light intrusion or light trespass. As mentioned earlier, residents suffer most when the light invades their private life. This is also called light immission. The effect are discussed in detail in Assmann et al., 1987; Hartmann, 1984; Hartmann et al., 1984. These studies form an essential part of the recommendations of the CIE (CIE, 2003). In astronomy direct light mainly refers to light shining directly into the telescope.
2. Sky glow. As is explained earlier, light pollution is the glow extending over the night sky. It presents itself as a background luminance over the sky, against which the astronomical objects are to be observed. Sky glow is the major obstruction for optical astronomy.
3. Horizon pollution. A major part of the light emitted by outdoor lighting is emitted more or less straight upward. This light is scattered in the atmosphere, and forms a haze over the light source. It is particularly conspicuous when the light stems from large, concentrated sources like city centres, airports, sports stadiums, or industrial premises. From a distance, the scattered light looks like a 'light blob' directly over the source, easily visible from a great distance – up to 200 km or more (Smith, 2001; Crawford, ed., 1991). As the blob is the result of the light emitted almost vertically upwards and scattered in the atmosphere, it can not be higher than the thickness of the atmosphere. If we take that as being about 8 km, above which the atmospheric pressure and the amount of aerosols are too low to cause much scatter, the elevation of the top of the blob will be at 50 km distance about 9° and about 4,5° at 100 km. Because few astronomical observations are made at smaller elevations than some 15°, this blob represents 'horizon pollution' but it is no threat to astronomical observations (Schreuder, 2000).

(d) *Conspicuity of point sources*

There are two further points that deserve some attention when discussing light pollution. The first is the visibility of point light sources, and more in particular the conspicuity of far-away lighting equipment. The second is what is sometimes called the flat-glass controversy. Both are rather technical; they will be summarised here. A full treatment is given in Narisada & Schreuder, 2004, sec. 9.1.7 and 12.4 respectively.

First the detection of point sources. The colloquial description of a point source is simply a source that is so small that no extension can be seen. Stars are point sources. However, the retinal image is not a point. As a result of diffraction that is explained in sec. 10.1, is measures about 0,002 mm, corresponding to about 0,2 minute of arc (Narisada & Schreuder, 2004, sec. 9.1.7).

Many people, astronomers and naturalists alike, complain that distant luminaires are conspicuous and that they cause horizon pollution. This effect is most noticeable at a distance between some 3 and 10 km. At such distances, the direct light from the outdoor lighting luminaires is clearly visible and may cause considerable disturbance (Mizon, 2002, fig. 1.62, p. 64). It must be noted that this relates to supra-threshold detection and not to the threshold of visibility. From the CIE data of 1981, it can deduced that the threshold of visibility is about 0,012 microlux (CIE, 1981; Narisada & Schreuder, 2004, p. 289). As an example, the supra-threshold values are given of a common CIE cut-off luminaire at different distances are given in Table 10.3.2. As is explained in sec. 10.2.1b, a common CIE Cut-off luminaire may not give more than 10 cd per 1000 lamp-lumens. With a high-pressure sodium lamp of 250 W of 28 000 lumen, a common street lighting lamps, this would mean 280 cd.

Distance (km)	Intensity (cd)	Threshold (microlux)	Supra-threshold ratio
1	280	280	23 333
3	280	31	2 583
10	280	2,8	233

Table 10.3.2 The illuminance at different distances and the supra-threshold ratio for a CIE cut-off luminaire (After Narisada & Schreuder, 2004, Table 9.1.14, p. 289).

From Table 10.3.2 is can be seen that a common street lighting luminaire is very conspicuous even at a distance of over 10 km.

(e) *Application of flat luminaire covers*

The first thing to do when reducing the light that is emitted upwards from outdoor lighting installations is to reduce the light emitted upwards by the luminaires.

Road lighting applications

As a measure of this emission CIE has introduced the Upwards Light Ratio ULR (CIE, 1997). Of course, the reflected light contributes also to the total upward flux. Here we will focus on the ULR.

As is explained in sec. 11.2.2, most outdoor luminaires are closed. In reflector luminaires, the cover has no optical function. It only serves for protection. In practice, several shapes of clear, transparent covers are used. We will use the following classification: flat glass, shallow bowl, deep bowl, and box. This classification is based on Laporte & Gillet (2003).

There is some controversy about the relative contribution of the reflected light to the upward flux of outdoor lighting installations. Most experts consider this as a minor point only (Narisada & Schreuder, 2004; Schreuder, 1996, 1998, 2000, Broglino et al., 2000). However, for different reasons this point became an important issue. Several comparisons have been made between luminaires with different cover shapes, both for the luminaires and for the complete road lighting installations. One set of data did show only marginal differences (Anon., 2002). The flat-glass version has a slightly better performance overall, but the reach is slightly shorter. Probably, the deep-bowl luminaire can be used at a slightly longer spacing for the same length-wise uniformity. However, the average luminance might be slightly lower. One may expect that the flat-glass version may be slightly better as regards the glare restriction. Details are given in Narisada & Schreuder, 2004, sec. 12.2.5, p. 669). One this is clear, however. Calculations or measurements regarding complete installations are needed in order to be certain about these points.

Studies regarding complete installations have been made in Italy (Broglino et al., 2000; Iacomessi et al., 1997) and in Belgium (Gillet & Rombauts, 2001, 2003). For this study, reflection measurements were made on different road surfaces. The study involved one member each of five 'families' of luminaires, each equipped with high-pressure sodium lamps. It followed that the high-efficiency refractor luminaire was the most efficient. The luminaire with the flat-glass cover with high-efficiency reflector performs considerably less well. As is to be expected, as regards the upward flux the roles are reversed.

As mentioned earlier, the studies suggested that, when luminaires with different cover shapes are compared directly, the differences are marginal. The overall performance of flat-glass covers is slightly higher, and the upward flux is slightly lower, that those of other cover shapes. At the other hand the reach of luminaires with flat-glass covers is slightly smaller than that of other cover shapes. All in all, however, it seems that these differences can be neglected in practice. Further details are given in Narisada & Schreuder (2004, sec. 12.4.6, p. 673-678).

10.3.2 Limits of light pollution

(a) The natural background radiation
Even under the most favorable conditions, the night sky is never completely black. Even from the interstellar space, electromagnetic radiation is coming toward us, light as well as other radiation. This is called the natural background radiation. It is defined as the radiation or luminance, resulting from the scatter of natural light by natural particles. For earth-bound optical astronomical observatories, the natural background radiation and the related natural background luminance is the absolute lower limit for observations.

According to Levasseur-Regourd (1992, 1994), the major contributions to the natural background luminance are:
1. The light from sub-liminal stars;
2. Interstellar dust (forming part of the galaxy);
3. Dust in the solar system (forming part of the solar system);
4. Air molecules;
5. Dust in the atmosphere;
6. Water vapour in the atmosphere.

Contrary to the contributions (1), (2) and (3), which come from outside the earth atmosphere, the contributions (4), (5) and (6) are from within the atmosphere. Obviously, their influence diminishes when the layer of air above the observatory is thinner and the air is cleaner. It should be mentioned that (5) may include a considerable fraction of man-made aerosols as well. This is the main reason that most major observatories are built on top of high mountains in desert areas – as far away as possible from any human activities.

A major contribution to the natural sky brightness comes from the influence of the Sun. As the solar radiation changes during the well-known 11-year period, the influence on the natural sky brightness varies as well. "The natural sky background at new moon near the zenith at high ecliptic and galactic latitudes varies by a factor of about 1,7 – i.e. by about 0,6 mag (21,3 < V < 21,9) per square arcsecond over the course of the 11-year solar activity cycle" (Smith, 2001, p. 40, quoting Krisciunas, 1997).

Under no earth-bound conditions is the natural background luminance a constant. The values given in rules-of-thumb for the natural background luminance must be regarded as an approximation only. There is not a single, simple value for 'the' natural background sky brightness. For that reason, it was suggested at the Vienna conference in 1999 to introduce a 'reference sky brightness' for the purpose of lighting design (Cohen & Sullivan, eds., 2001). The value should correspond to that of the solar minimum; it is

Road lighting applications 379

proposed to use the value of 21,6 magnitude per square arcsecond, which corresponds to $3,52 \cdot 10^{-4}$ cd/m². This value is proposed by Crawford (1997). There are other rules-of-thumb that are similar but not identical. See e.g. Narisada & Schreuder (2004, sec. 3.1. p. 62).

(b) *Artificial sky glow*

As is found from satellite observations as well as from earth-bound measurements, there are only very few places on Earth where the sky background brightness equals the natural background luminance (Beekman, 2001; Cinzano, 1997; Cinzano & Elvidge, 2003; Cinzano et al., 2001; Isobe, 1998; Isobe & Kosai, 1994; Isobe & Hamamura, 1998; Kosai & Isobe, 1988; Kosai et al., 1994; Narisada & Schreuder, 2004. sec. 5.2.1; Schreuder, 1999).

It is found that about two-thirds of the world population is subject to light pollution (Beekman, 2001). Here, light pollution is, in accordance with the usual IAU-definition, defined as having an airglow more than 10% of the natural background (Anon., 1978, 1984, 1985). We will come back to this value when we indicate that IAU recently amended their recommendations. In the USA and in Europe, for over 90% of the population it will never get darker than corresponds to a half-moon at 15° elevation. For 25% of the world population it will never get darker than the 'nautical twilight', corresponding to the sun being between 6^0 and 12^0 below the horizon. That means that for 25% of the world population, only the brighter stars can be seen, but not the Milky Way (Cinzano et al., 2001). For 85% of the population in the Netherlands, the artificial light exceeds that of the full moon, and in only 3% of the area of the country, the artificial brightness is less than the natural brightness (Schmidt, 2002, 2002a).

In spite of all efforts to avoid it, sky glow did increase considerably on a global scale over the last few years. At present, it seems safe to expect that the levels of light pollution at most locations in the world are increasing and that they will go on increasing for the foreseeable future. A rate of increase is difficult to guess but it might seem that 3% per year will not be too far off. It should be noted that 3% per year means a doubling in 23 years (Narisada & Schreuder, 2004, Chapter 4).

(c) *CIE Limits*

As mentioned earlier, in 1997 CIE published a report on the different aspects of sky glow (CIE, 1997). The recommendations are summarised in Narisada & Schreuder (2004, chapter 7).

The first general recommendation refers to zoning. The idea is that light pollution does not cause the same amount of disturbance at all places. More in particular, in brightly lit city centres or industrial estates, the requirements regarding the limits of sky glow can be lower that in a Natural Park or in a Protected Landscape. Zoning is as such a well-known method to discriminate between locations or regions. The British IES introduced in the 1990s a system of four zones (Pollard, 1993). This system is adopted by CIE (1997, 2003), and subsequently used by a number of national organizations (NSVV, 1999, 2003, 2003a,b). The zones are summarised in Table 10.3.3.

Zone	Surroundings	Lighting environment	Examples
E1	natural	intrinsically dark	national parks or protected sites
E2	rural	low district brightness	agricultural or residential rural areas
E3	suburban	medium district brightness	industrial or residential suburbs
E4	urban	high district brightness	town centres and commercial areas

Table 10.3.3 Description of the environmental zones, adapted from CIE, 2003, table 2.1 (After Narisada & Schreuder, 2004, Table 7.1.1).

In many cases a more detailed zoning system is needed. A more detailed zoning method is proposed by Narisada & Schreuder (2004, sec. 7.1.1), based on the CIE method at the one hand, and at the other hand on the ALCoR system proposed by Murdin (1997). The proposal made by Narisada and Schreuder is given in Table 10.3.4. See also Schreuder (1994, 1997).

Environmental zones	sub-zones	Examples
E1		Areas with intrinsically dark landscapes
	E1a	nature preserves
	E1b	national parks
	E1c	areas of outstanding natural beauty, protected landscapes
E2		Areas of low district brightness: rural agricultural areas, village residential areas
E3		Areas of middle district brightness
	E3a	sub-urban residential areas
	E3b	urban residential areas
E4		Areas of high district brightness
	E4a	urban areas having mixed residential, industrial and commercial land use with considerable nighttime activity
	E4b	city and metropolitan areas having mixed recreational and commercial land use with high nighttime activity

Table 10.3.4 Description of the environmental sub-zones (After Narisada & Schreuder, 2004, Table 7.1.2).

Road lighting applications

The second general recommendation refers to the fact that light pollution does not cause the same amount of disturbance at all times. A major part of the social activities that did begin in the day will continue in the evening. Therefore the strict requirements concerning the limits of sky glow that can be imposed in the middle of the night cannot be used in the evening. For this, the term curfew is often used, while introducing the evening regime and the night regime (CIE 1997, 2003). The moment when the evening regime ends and the night regime begins is mostly laid down in national or regional regulations (NSVV, 1999).

The third general recommendation refers to the upward light emission of outdoor lighting installations. It has been mentioned earlier that CIE introduced the Upward Light Ratio ULR (CIE, 1997). This ratio refers only to individual luminaires. Being the ratio between the upward flux and the lamp flux it is a relative value only. Therefore it does not give any information about what is the effect of a complete lighting installation. For lighting design purposes it has only a very limited significance. More recently, a number of suggestions have been made for methods to quantify the upward light emission of total outdoor lighting installations.

A suggestion, based on a limited number of field observations only, is given in Table 10.3.5. See also Narisada & Kawakami (1998).

Sub-zone	Maximum installed area luminous flux (lumen per square meter)	
	evening regime	night regime
E1a	0,02	0
E1b	0,06	0
E1c	0,18	0
E2	0,75	0,15
E3a	3	0,8
E3b	12	2
E4a	50	20
E4b	150	30

Table 10.3.5 Area luminous flux. Environmental sub-zones after Table 10.3.3. The 'area' includes the total surface of the area under consideration. Based on Schreuder, 2002.

The fourth general recommendation refers to the colour of the light. It is generally accepted that to use the monochromatic light of low-pressure sodium lamps is one of the most effective ways to reduce light pollution as regards astronomical observations (CIE, 1997, sec. 10). The reason is that it is not too difficult to filter out the one Sodium spectral line (Budding, 1993; Sterken & Manfroid, 1992). Furthermore, they are the most efficient light sources available at present (Schreuder, 2001a; Van Bommel & De Boer, 1980).

CIE has given a number of specific recommendations for the limitation of light pollution, light intrusion, and sky glow (CIE, 1997, 2003). We will not give any details here, primarily because most of the recommendations are still under discussion. Some of them are included in Narisada & Schreuder (2004, chapter 7). See also NSVV (1999, 2003, 2003a,b).

(d) IAU limits

In the 1970s, the IAU set up limiting values for the artificial sky glow for existing and new astronomical observatories. The sky luminance resulting from artificial lighting must not be more than 10% of the natural background (Anon., 1978, 1984, 1985). This value was difficult to maintain for two reasons:
1. The variations in the natural sky brightness resulting from variations in the solar radiation is far more than the 10%-rule. It is mentioned earlier in this section that over the 11-year period may amount to a factor of about 1,7 (Smith, 2001; Krisciunas, 1997).
2. There is hardly any place left on the whole Earth where the artificial sky glow is less than 10% of the natural background. In an earlier part of this section, some data are given.

Recently, IAU amended the requirements for the tolerable limits of sky brightness (IAU, 2006). Three classes of observatories have been introduced:
- Class A: large existing telescopes (over 6,5 meter diameter);
- Class B: future telescopes;
- Class C: telescopes over 2,5 meter diameter).

The new requirements are, for Classes A and B: "Undetected light pollution at zenith", for Class C: "Light pollution levels less than the natural variation in night-sky brightness associated with the 11 year solar cycle".

10.3.3 Remedial measures

(a) Limiting sky glow

Basically there are two ways to avoid or at least to reduce the light pollution, more in particular the disturbance for astronomical observations: avoid the light by switching and gated viewing, or reduce the disturbance by light control and spectral selection. The most important remedial measures will be indicated here briefly. More details are given in Narisada & Schreuder (2004, sec. 1.2.5). See also Schreuder (1987, 1991a).

(b) Switching off the lights

This obviously is the final solution as regards the astronomical observations. However, it cannot be done in full as other equally important aspects of our social life require night-time activities and consequently artificial light at night. Nevertheless, a fair number of light sources can be switched off without any damage; furthermore, in many cases light sources can be replaced by smaller sources emitting less light. Other, even more effective possibilities are discussed further on. An important way to improve the situation is to guarantee that most lights, particularly those that contribute most to light pollution like sports stadium lighting, are not used after a certain time. Many local by-laws and ordinances impose time restrictions in the use of such lighting installations. Some earlier ones are given in Anon (1981, 1982). As is explained in detail in Narisada & Schreuder (2004, secs. 15.5, 15.6) many local by-laws and ordinances impose time restrictions in the use of lighting installations. Modern electronic ballasts allow for a less drastic solution, as almost all gas-discharge lamps can be equipped with a dimmer.

(c) Gated viewing

Electric lamps emit light essentially only when electric current is passed through them. This means that (on a 50 c/s grid) lamps are extinguished 100 times per second. This gives the possibility to apply the principle of gated viewing: the shutter of the photo apparatus at the telescope, or of its electronic equivalent, is opened 100 times a second, and only during the periods that the lamp does not emit light. In this way, the observer will not notice the presence of the light (Schreuder, 1991, 1991a).

The principle of gated viewing is not new. However, in public lighting it has not been put into practice at any considerable scale. The first aspects that must be taken into account is the lamp type. Clearly, it is essential that during the 'off' period the lamp should emit no light for the method to have any appreciable effect. This automatically rules out incandescent lamps and all lamps that use fluorescence. Many types of gas-discharge lamps can be used, notably when they are equipped with dimmers that allow light emission only in a small part of the period of the alternating current. These dimmers are standard equipment, and can be used for most types of high-pressure sodium lamps. More convenient are, of course, Light Emitting Diodes or LEDs. As is explained in sec. 2.5.1, a LED will emit light only when the voltage is appropriate (Heinz, 2006, 2006a; Heinz & Wachtmann, 2001, 2002; Stath, 2006).

There are other factors to be taken into account. The whole outdoor lighting installation – probably a whole village – must be fed by a one-phase alternating current system, and not by the usual three-phase arrangement. Usually, this will require a separate electric generating plant. As in many cases the observatory is located in a dry, sunny, and isolated place, so is the village in question. This implies that stand-alone solar cells are probably the most convenient. This means also that the grid frequency can be selected at will. When the frequency is over 50 c/s, flicker effects will be avoided, as is explained in sec. 8.1.2.

The main problem for the application of the gated viewing system is that it has not been tested on any large scale. Most administrators are rather conservative. It will cost some money, particularly in the first stages of design and construction; it is not feasible in large cities with existing elaborate outdoor lighting installations; it is not feasible in places where a large part of the outdoor lighting is privately owned; and finally, it does not reduce the overall sky glow. It seems, however, to be a very promising system for small villages, that until now have no electric lighting, and are near to major astronomical observatories. In these cases, the financial burden is light, and it can easily be carried by those agencies that promote clear, dark skies.

(d) *Light control*

Light control means, in simple terms, preventing light being emitted above the horizon; more precisely, it means that the light is directed to the objects to be illuminated. In practical terms, this means primarily the selection of the appropriate light distribution and the best optical design of luminaires for outdoor lighting. The positioning of the luminaires needs to be taken into account because optimising the light distribution and the positioning are supplementary measures. As regards road lighting, the measures to reduce light trespass are similar to those that enhance the economic efficiency of the lighting installation (Schreuder, 1998).

(e) *Reduction of reflection*

Theoretically speaking, this remedy is just as important as the two that have been mentioned earlier. In practice, however, the possibilities to manipulate the reflective characteristics in the real world are very limited indeed. Furthermore, reducing the reflection of the road surface may be counterproductive, because in order to arrive at a certain luminance level for a 'dark' road surface, the luminous flux aimed at the road must be higher. A higher luminous flux means, however, a greater amount of stray light, and thus an increase in light pollution.

The effect of air pollution might be included here. Air pollution often creates haze, or even clouds of smoke and dust; the light from lighting installations partly is scattered in these clouds, and partly absorbed. The net result of air pollution on the sky glow, particularly that of far away observatories can therefore be positive or negative. When the scatter predominates, the sky glow may increase, whereas, when the absorption predominates, the net result may be a decrease of the sky glow (Anon, 1985).

(f) *Using monochromatic light*
As is mentioned earlier, the most effective way available is the use of quasi-monochromatic light sources, more in particular the use of low-pressure sodium lamps. These lamps emit a very narrow spectral band in the yellow part of the spectrum. Two advantages are obvious: no other spectral regions are involved, so that observations in other spectral regions – either photography or spectroscopy – are hardly affected. Secondly, as the yellow line is close to the maximum of the sensitivity of the eye, the luminous efficacy of low-pressure sodium lamps is high – they are far the most efficient light sources available at present (Sprengers & Peters, 1986). This is discussed in sec. 2.4.3c. Finally, they hardly emit any radiation in the non-visible parts of the spectrum. On these grounds, the application of low-pressure sodium lamps near sites of astronomical observatories is universally recommended. Low-pressure sodium lamps have several disadvantages as regards economic life, light control, and colour recognition. These aspects are discussed in sec. 11.1.3h.

(g) *Filtering the light*
A final countermeasure against the disturbance of light pollution is to narrow the spectral range of the emitted light. The idea is similar to using monochromatic light, but obviously it is less effective.

10.4 Reflection properties of road surfaces

10.4.1 Road reflection as a road lighting design characteristic

(a) *The luminance technique in road lighting*
As mentioned several times already, one of the main functions of road lighting is to make visible any object that may endanger traffic. The visibility of objects is determined by three factors:
1. The observer – age, visual performance, attention, vigilance, etc.;
2. The object – size, location, surface characteristics, contrast to background, etc.;
3. The level of adaptation.

In road lighting practice, the lighting engineer, nor the road authority, can do anything at all about the observer and very little about the object. The only variable that is available to the lighting designer to increase the chance of detecting the object, and thus to improve road safety, is the adaptation level. The obvious way is to improve the lighting level, or rather the luminance level on the road. As is well-known, the two are closely related, and often even directly proportional (Gregory, 1970; Le Grand, 1956; Schober, 1960; Schreuder, 1988a, 1998). This approach is sometimes called the luminance technique (Schreuder. 1964, 1970, 1998).

This leads directly to the most important characteristic of road lighting: the need to have a reasonable value of the average road surface luminance. Immediately, we have to make amends. This is only true for those cases where the main function of the lighting is to make objects visible. As is pointed out in sec. 1.4.3, when the driving task is taken into account, it is not so much the average road surface luminance, but rather the optical guidance as the most important quality criterion. And in areas where pedestrian traffic is predominant, e.g. in residential streets or shopping precincts, the road luminance has very little value at all (Schreuder, 1998, Chapter 10).

However, there is more. The road surface is not only the main part of the field of view, the part that contributes most to the adaptation level. The road is also the background for obstacles on the road that may endanger traffic. The relevance of this remark will become clear when three usual characteristics of traditional road lighting are mentioned:
1. Most road surfaces have a considerable, or even a high degree of specularity;
2. This implies that the counterbeam part of the light from the luminaire contributes most to the luminance;
3. Almost all objects on the road are dark and show a diffuse reflection.

Several of these characteristics will be explained in Chapter 11, where road lighting design is discussed. The net result of this is that in almost all practical cased, objects on the road are seen in negative contrast, that is to say as a dark objects against a bright background. From this, it can easily be seen that the contrast of the objects is greater, and thus the visibility better when the road surface luminance is higher.

Based on these considerations, in most national and international standards and recommendations for road lighting, the average road surface luminance is introduced as the most important quality requirement, at least for traffic routes and main roads (CEN, 2002; CIE, 1965, 1977b, 1992a, 1995b; NSVV, 1957; 1974/75; 1977; 1990; 1997).

(b) Reflection characteristics

The luminance of a non-emitting surface is determined by two quantities: the amount of incident light and the reflection characteristic of the surface. As is explained in sec. 4.4, both quantities have a vectorial character, showing both magnitude and direction – without being, mathematically speaking, 'true' vectors, as vector summation

Road lighting applications

usually is not valid. The direction of each vector can be described with two angles; this means four directional parameters in total, plus two scalars for the length of the vectors. As is explained in sec. 4.4, this implies a quantity with some of the characteristics of a tensor.

In many cases the directional dependency can be neglected, and the calculation of the luminance is reduced to a multiplication of two scalars: the illuminance and the reflectance. This is commonly the case in interior lighting design (Anon., 1993, sec. 4.8; Baer, ed., 2006, sec. 1.4.1; De Boer & Fischer, 1981; Hentschel, ed., 2002, sec. 8.2).

In road lighting, however, this is never allowed. The main reason is that driving a motor vehicle – the driving task – requires to see the road and what is on of near it, at a considerable distance in front of the vehicle (Schreuder, 1991). This means that the angle of observation, that is the angle between the line of sight and the surface of the road, is very small indeed; the angle is standardized to be one degree, but there are doubts whether this standard is always used. This glancing angle implies that also glancing angles of light incidence may be relevant. And finally, many surfaces are not isotropic – their reflection changes when the surface is rotated around a vertical axis. Fortunately, most asphalt surfaces are isotropic, and asphalt is the most common surface for traffic routes. So, when the angle of observation is kept constant, and the surface is considered as being isotropic, two angles are sufficient to describe the angular aspect.

The scalar aspect is simple: in order to find the luminance in the required direction, it is sufficient to multiply the scalar values of the light intensity and of the reflection characteristic, both in the relevant directions. And this is precisely what is done in the design of road lighting installations. As is explained in sec. 11.1.2, using I-tables and R-tables, the geometry of the installation, and the position of the observer, the luminance for each point on the road can be assessed. Repeating the process for all luminaires that contribute to the lighting of that point, and again for all points, the quality of the lighting installation can be determined. Obviously, it is necessary to know the light reflection of the road surface.

Each road surface has its own characteristic reflection distribution. As is explained later on, these distributions are called R-tables. An R-table can be measured by shining a light beam on the surface, and measure the resulting luminance under 1° down. The angles of incidence may have any value over the half-sphere over the surface. Practice indicated that not all angles are equally important. This is the way that hundreds of R-tables on samples have been measured in the 1950s, 1960, and 1970s in laboratories in Belgium, Canada, Denmark, England, Germany, and The Netherlands. The method is described in detail in Schreuder (1967, sec. 3.3.2; 1998, sec. 10.4).

(c) Documentation of reflection characteristics

The results of these measurements are collected in Erbay (1973); Erbay & Stolzenberg (1975). See also Kebschul (1968). As is mentioned earlier, they were the basis of all existing classification systems. Over the years, a number of major problems did arise:

1. The measurements were very expensive and time-consuming. The measuring equipment was large and required a lot of laboratory space. And finally, the funding of this type of expensive, long-term research became severely limited, both in industry and at universities. All this did lead to the almost complete end of road reflection research. The measuring set-up was decommissioned and finally scrapped.
2. Road authorities were increasingly reluctant to allow samples of almost a square metre to be take out of the road.
3. So, new measurements were not added to the atlases. These atlases reflected, however, the state of road building technology of the 1950s and 1960s. New developments were not taken into account. As a consequence, the R- and C-systems, including their standard R-Tables, did not represent the current state of affairs. Strangely enough, it seemed that nobody seemed to notice this. This is maybe a result of the fact that before about 2000, it was very difficult to do any luminance measurements in the field.

All this resulted in a new approach that focussed on portable measuring gear that was cheap to operate without damaging the road. Early attempts failed because the technology was not available. But with the fast progress of micro-electronics and the availability of simple portable computers like lap-tops, around 1995 the ideas could be put into practice. This is discussed in more detail in a further part of this section.

10.4.2 The classification of road surface reflection

All common dry road surfaces exhibit, when observed under a glancing angle, a mixture of specular and diffuse reflection. When wet, the specular component usually is the dominant; when viewed under a steeper angle – for instance as a pedestrian would do – the diffuse component dominates.

In order to characterise and classify road surfaces, systems are used that take these two components in account. The older one uses the Q_0 to quantify the diffuse component, and κ to quantify the specular component (Westermann, 1963). Because the two are not independent, or, in mathematical terms not orthogonal, and because they were very hard to measure, they are not used often any more. We will not discuss them here. Details can be found in Schreuder (1967, 1967a, 1998).

They have been replaced by another set of quantities: P(0;0), P(2;0) and P(1;90). This system has been proposed in the Netherlands (Burghout, 1977). It is subsequently adopted by CIE as an alternative of the Q_0-κ-system (CIE, 1976). Here, P means the luminance factor as commonly defined, but divided by π; the first digit is the tangent of the angle of incidence in the plane of observation, and the second digit is the angle (in degrees) in the plane perpendicular to the plane observation. The three reflection 'factors' can easily be measured in the laboratory, using samples cut out of the road. See e.g. SCW (1974, 1977, 1984); CIE (1976, 1984), and Schreuder (1967a, 1998).

Two remarks must be made.
1. Firstly, it has been proposed to replace Q_0 by the (theoretical) diffuse reflection Q_D, as is common in judging painted surfaces. See e.g. CEN (1995); CIE (1995); Sorensen (1995), Sorensen & Lundquist (1990) and Sorensen et al. (1991). The diffuse reflection Q_D may give important additional data for daytime situations, particularly for testing road marking materials. There is, however, some doubt in how far it may be considered as an improvement for assessing road surfaces and, consequently, in how far it is adequate for lighting design.
2. The second remark is that the measuring area must be large in comparison to the elements (the 'graininess') of the road surface. For traditional asphalt an area of 400 cm² did prove to be sufficient (SCW, 1974; 1984). There are, however, reasons to believe that for modern surfaces, the area must be considerably larger (Schreuder, 1998). It may be mentioned here that porous asphalt, or drainage asphalt as it is also known, is an extremely effective measure to reduce 'splash-and-spray' from wet road surfaces, and therefore reduce skidding accidents and improve visibility. In spite of the fact that part of these benefits are 'used up' by faster driving, the accident reduction is remarkable. Details can be found in PIARC (1990), Schreuder (1988, 1998), SCW (1977) and Tromp (1994). We will come back to porous asphalt is a further part of this section.

The C1-C2 system allows for the following steps:
1. The reflection of any road surface can be characterized by the values of the three factors;
2. The surface can be identified when the three values are identical (or close) to the three values of another surface (e.g. a surface that is designated as a 'standard' surface);
3. Road surfaces can be classified on the basis of P(2;0). A large study revealed that only two classes are needed to classify all 'traditional' surfaces, provided P(0;0) is used as a scaling factor. It should be noted that the C1-C2 classification is based on similarities of road lighting installations, and not, like the Q_0-κ-system on similarities in the reflection characteristics.

The C1-C2-system is applied for design purposes in a few steps.
1. First, the P(0;0) and the P(2;0) of the surface are determined, yielding the C-class;
2. Next, the standard surface of that class is selected, and the calculations of the luminance and the uniformity are made. Because the standards for C1 and C2 are normalised on P(0;0) = 1;, the result has to be multiplied by the value of P(0;0) of the actual surface.

This process seems to be simple; however, three major problems are not solved: the iterative process is tedious and time-consuming; the results cannot be measured accurately, and the reflection characteristics are usually not known.

10.4.3 Standard reflection tables

In the preceding section, two systems for the classification of the reflection properties of road surfaces are given:
1. The Q_0-κ-system (CIE, 1976);
2. The C1-C2-system (Burghout, 1977; CIE, 1984).

It may be noted here that in the past several other road surface classification systems have been used. They are not in use any more.

In each system a number of standard reflection tables is defined; four for the Q_0-κ-system, and two for the C1-C2-system. These tables are usually called R-tables. R-tables come from real roads; they are not constructed in any way. Each R-table is the best representative for the group of roads that are members of the class. That table is used as a standard that can represent all roads in that class in design procedures. An example is given in Figure 10.4.1.

10.4.4 Field measurements of the road reflection

As mentioned earlier, one of the problems of assessing the reflection properties of road surfaces is that at present measuring devices are rare. Only two or three are still available in the whole world. Furthermore, they require quite large road samples. All this is clumsy, time consuming and very expensive. In real life this means that the design of road lighting usually is based on estimated values of the reflection characteristics – or even, that the luminance concept is not used at all.

Numerous attempts have been made to design and operate mobile reflection meters that can be used on the road, without any damage to the surface. Most devices did not work fast and accurate enough, and have been abandoned. One of the problems had

Road lighting applications

tgγ	0°	2°	5°	10°	15°	20°	25°	30°	35°	40°	45°	60°	75°	90°	105°	120°	135°	150°	165°	180°
0,00	10000	10000	10000	10000	10000	10000	10000	10000	10000	10000	10000	10000	10000	10000	10000	10000	10000	10000	10000	10000
0,25	9222	9199	9139	9222	9246	9222	9199	9199	9187	9151	9115	9199	9067	9115	9151	9270	9199	9402	9342	9390
0,50	7608	7560	7632	7548	7458	7488	7404	7368	7333	7225	7117	7022	6902	7069	7093	7297	7356	7620	7548	7656
0,75	6077	6065	6041	5909	5933	5789	5586	5455	5323	5179	5060	4976	4844	4988	5084	5347	5443	5682	5694	5778
1,00	4904	4833	4844	4713	4510	4294	4079	3888	3708	3541	3421	3373	3254	3445	3612	3828	3959	4127	4199	4270
1,25	4007	3947	3959	3708	3505	3170	2835	2644	2512	2404	2321	2249	2249	2380	2524	2691	2907	3074	3086	3182
1,50	3349	3301	3266	2978	2644	2309	2033	1854	1734	1663	1615	1555	1555	1711	1818	2010	2117	2297	2321	2392
1,75	2623	2775	2667	2368	1986	1675	1435	1304	1232	1172	1124	1089	1148	1268	1340	1507	1603	1734	1782	1794
2,00	2440	2356	2261	1842	1507	1232	1041	945	897	837	801	825	825	933	1017	1148	1232	1364	1400	1411
2,50	1890	1770	1567	1172	861	694	598	538	502	478	467	467	502	574	646	718	778	861	897	921
3,00	1531	1400	1124	742	538	419	359	335	323	299	287	299	323	359	407	478	538	586	610	658
3,50	1256	1124	837	502	335	263	239	215	203	191	191	203	227	251	299	347	395	431	455	478
4,00	1041	897	646	371	227	179	167	156	144	144	144	144	167	191	227	251	287	335	347	371
4,50	909	754	478	275	167	132	120	108	108	108	108	120	132	156	179	203	227	263	275	287
5,00	778	658	371	191	120	96	69	84	72	72	84	96	96	120	132	156	179	215	227	239
5,50	682	526	299	156	96	72	72	72	60	48										
6,00	622	467	251	108	72	60	60	60	60											
6,50	574	419	215	96	72	60	60	60												
7,00	526	359	179	84	60	48	48	48												
7,50	478	335	156	72	48	36	36													
8,00	443	299	144	60	48	36	36													
8,50	419	275	120	60	48	36														
9,00	383	251	108	48	36	36														
9,50	347	227	96	48	36	36														
10,00	335	203	84	36	36	36														
10,50	323	203	84	36	24	12														
11,00	299	191	72	36	24	12														
11,50	287	179	72	36	24															
12,00	275	179	60	36	24															

Figure 10.4.1 An example of an R-table. Standard R table for class C1, multiplied by 10^4. $q_p = 0,084$; $q_0 = 0,109$ (After Schreuder, 1998, Figure 10.4.2).

to do with the overall dimensions of the devices. A small apparatus can have only a small measuring field. In order to make measurements over the required 400 cm² that is mentioned earlier, the measurements must be repeated very often. This requires a fast-working device. The history is described in Schreuder (1967, 1967a, 1998, p. 165); Van Bommel & De Boer (1980), and SCW (1974, 1984). See also Range (1972, 1973); Serres (1990). As has been mentioned earlier, only the rapid progress in micro-electronics made it possible to construct measuring gear that worked satisfactorily.

One of the first devices of the new generation is described by Schreuder (1991b, 1992, 1993, 2006). The measuring field measures about 50 cm², so that many measurements have to be made to have the result for one 'point' on the road. The reflection is measured in 8 directions. A complete R-table is made by interpolation. In order to allow these 8 measurements to be made in a parallel fashion, the direction of the light rays is inverted. Instead of many lamps and one luminance meter, one lamp was used and 8 luminance meters. The eight measurements were made by using an adapted digital photo camera. The measurements were made under the standard 1° downward angle with the horizontal. The construction is summarily described in De Kruijter et al. (2005). In one measuring session of about two hours, more that one hundred measurements could be made. Details are given in Tetteroo (2003) and Schreuder (2006). Results are given in De Kruijter et al. (2005) and Schreuder (2006).

One aspect of the results will be given here. As is mentioned earlier, asphalt road surfaces are inherently granular. That means that there is an inherent variability of the measurements of points on the road for the same geometry. A great number of measurements have been made on the type of asphalt that is used in the Netherlands for several decades as the standard road type for rural motorways and rural main roads. It is called DAB, meaning Coarse Closed Asphalt (SCW, 1974, 1984). Because of its long time of application, the roads of this type are very uniform as far as asphalt roads can be. A statistical analysis showed that the inherent spread of the measurements is about 4,3%, suggesting that very precise calculation systems may not be necessary nor relevant. Further research is under way. There are indications that drainage asphalt, even within the same type, shows a wider spread. To this comes that different types of drainage asphalt are quite different to begin with (De Kruijter et al., 2005).

In several countries, mobile measuring devices have been developed; many are fully operational. All employ different light sources and only one luminance meter, whereas in most cases the measuring is about 5° or more, instead of the standardized 1° down. For a full description and a summary of the results we refer to the literature (Blattner et al., 2006; Luisi, 2006; Frankinet et al., 2006; Delta, 2000, 2000a; Paumier, 2002; Goyat & Guillard, 2001). Further international harmonization and standardization is required.

10.5   Conclusions

This chapter deals with the way the fundamental aspects of lighting, more in particular of mathematical, physical, and physiological nature are applied in practical road lighting.

The optical design of road lighting luminaires is based on considerations of geometric optics. The most efficient light distributions with a minimum of glare can be reached by using reflectors in the luminaire instead of refractors at the outside. The use of flat covers does not bring much additional benefit.

Light pollution is a major concern for outdoor lighting, road lighting included. The tolerable maximum amount of emitted light is not equal for all times of the day and night, nor for all locations. Curfew and zoning are essential. It is concluded that in general the CIE Zoning system is adequate; however, in many special cases a more detailed subdivision of the environmental zones is required. It is possible to define lighting installation requirements for each environmental sub-zone. CIE and IAU have defined limits of light pollution based on the natural background radiation. These limits are applicable. However, there is an urgent need for an additional measure that indicates the overall requirements for complete lighting installations, like e.g a complete town.

When the luminance technique is applied in road lighting, the road surface luminance must be assessed. For this, it is essential to know the reflection characteristics of the road surface. The CIE C1-C2 classification system of road surface reflection characteristics can be used. The current standard R-tables need to be amended. For this, in situ-measurements are needed. Several measuring systems are available; international harmonization and standardization is required.

References

Anon. (1978). Report and recommendations of IAU Commission 50 (Identification and protection of existing and potential observatory sites) – published jointly by CIE and IAU in 1978. (Reproduced as Appendix 4.1. in McNally, ed., 1994, p. 162-166).

Anon. (1981). Relating to electricity, adapting the Tucson light pollution code. Ordinance No. 5338, adapted April 6, 1981. Tucson, 1981.

Anon. (1982). Real Decreto 2678/1982 de 15 de octubre (In Spanish). Boletin Oficial del Estado 28 de octubre. Madrid, 1982.

Anon. (1984). La protection des observatoires astronomiques et geophysiques (The protection of astronomical and geophysical observatories). Rapport du Groupe du Travail. Institut de France, Academie des Sciences, Grasse, 1984.

Anon. (1985). Identification and protection of existing and potential observing sites. (Draft). Report IAU Commission 50. International Astronomical Union, New Delhi, 1985.

Anon. (1993a). Lighting manual. Fifth edition. LIDAC. Eindhoven, Philips, 1993.

Anon. (1997). Control of light pollution – Measures, standards and practice. Conference organized by Commission 50 of the International Astronomical Union and Technical Committee 4.21 of the Commission Internationale de l'Eclairage. The Hague, 20 August 1994. The Observatory, 117 (1997) 10-36.

Anon. (2001). Luxjunior. 21-23 September 2001, Dörnfeld/Ilmenau. Proceedings. Ilmenau, University, 2001.

Anon. (2002). ARC Product Information. Cappelle aan de IJssel, Industria Technische Verlichting B.V., 2002 (year estimated).

Anon. (2006). Licht 2006, 17. Lichttechnische Gemeinschafttagung; Bern, September 2006.

Assmann, J.; Gamber, A. & Muller, H.M. (1987). Messung und Beurteilung von Lichtimmissionen (Measurement and assessment of light immissions). Licht 7 (1987) 509-515.

Baer, R. (1990). Beleuchtungstechnik; Grundlagen. Berlin, VEB Verlag Technik, 1990.

Baer, R., ed. (2006). Beleuchtungstechnik; Grundlagen. 3., vollständig überarbeitete Auflage (Essentials of illuminating engineering, 3rd., completely new edition). Berlin, Huss-Media, GmbH, 2006.

Batten, A.H. ed. (2001). Astronomy for developing countries. Proceedings of a Special Session at the XXIV General Assembly of the IAU, held in Manchester, UK, 14-16 August 2000. San Francisco, The Astronomical Society of the Pacific, 2001.

Bean, A.R. & Simons, R.H. (1968). Lighting fittings performance and design. Oxford. Pergamon Press, 1968.

Beekman, G. (2001). Snakken naar duisternis (Yearning for darkness). NRC Handelsblad, 25 August 2001, p. 37.

Blattner, P.; Dudli, H., & Schaffer, H. (2006). Mobiles Fahrbahnoberflächenreflektometer (Mobile reflection meter for road surfaces). In: Anon., 2006.

Blüh, O. & Elder, J.D. (1955). Principles and applications of physics. Edinburgh, Oliver & Boyd, 1955.

Boyce, P.R. (2003). Human factors in lighting. 2nd edition. London. Taylor & Francis, 2003.

Breuer, H. (1994). DTV-Atlas zur Physik, 4.Auflage (DTV atlas for physics. 4th edition). München, Deutsche Taschenbuchverlag DTV, 1994.

Broglino, M.; Iacomussi, P.; Rossi, G.; Soardo, P.; Fellin, L. & Medusa, C. (2000). Upward flux of public lighting: Two towns in Northern Italy. In: Cinzano, ed., 2000, p. 258-270.

Budding, E. (1993). An introduction to astronomical photometry. Cambridge University Press, 1993.

Burghout, F. (1977). Simple parameters significant of the reflection properties of dry road surfaces. In: CIE, 1977a.

CEN (1995). Road equipment – Horizontal signalization – Performance for road users. Draft prEN 1436. Brussels, CEN. 1995.

CEN (2002). Road lighting. European Standard. EN 13201-1..4. Brussels, Central Sectretariat CEN, 2002 (year estimated).

CIE (1965). International recommendations for the lighting of public thoroughfares. Publication No. 12. Paris, CIE, 1965.

CIE (1976). Calculation and measurement of luminance and illuminance in road lighting. Publication No. 30. Paris, CIE, 1976.

CIE (1977). Road lighting lanterns and installation data: Photometrics, classification and performance. Publication No. 34. Paris, CIE, 1977.

CIE (1977a). Measures of road lighting effectiveness. Symposium Karlsruhe, July 5-6, 1977. LiTG, Berlin, 1977.

CIE (1977b). Recommendations for the lighting of roads for motorized traffic. Publication No. 12/2. CIE, Paris, 1977.

CIE (1981). An analytical model for describing the influence of lighting parameters upon visual performance. Summary and application guidelines (two volumes). Publication No. 19/21 and 19/22. Paris, CIE, 1981.

CIE (1984). Road surfaces and lighting (joint Technical Report CIE/PIARC). Publication No. 66. Vienna, CIE, 1984.
CIE (1992). Proceedings 22th Session, Melbourne, Australia, July 1991. Publication No. 91. Vienna, CIE, 1992.
CIE (1992a). Guide for the lighting of urban areas. Publication No. 92. Paris, CIE, 1992.
CIE (1993). Urban sky glow, A worry for astronomy. Publication No. X008. Vienna, CIE, 1993.
CIE (1995). Road surface and road marking reflection characteristics. 1st draft, September 1995. Vienna, CIE, 1995.
CIE (1995a). CIE 23rd Session, New Delhi. Publication No. 119. Vienna, CIE, 1995.
CIE (1995b). Recommendations for the lighting of roads for motor and pedestrian traffic. Technical Report. Publication No. 115-1995. Vienna, CIE, 1995.
CIE (1997). Guidelines for minimizing sky glow. Publication No. 126. Vienna, CIE, 1997.
CIE (2003). Guide on the limitation of the effects of obtrusive light from outdoor lighting installations. Publication No. 150. Vienna, CIE, 2003.
CIE (2003a). 25th Session of the CIE, 25 June – 3 July 2003, San Diego, USA. Vienna, CIE, 2003.
Cinzano, P. (1997). Inquinamento luminoso e protezione del cielo notturno (Light pollution and the protection of the night sky). Venezia, Institutio Veneto di Scienze, Lettere ed Arti. Memorie, Classe di Scienze Fisiche, Matematiche e Naturali, Vol. XXXVIII, 1997.
Cinzano, P. & Elvidge, D.E. (2003). Modelling night sky brightness at sites from DMSP satellite data. In: Schwarz, ed., 2003, p. 29-35.
Cinzano, P.; Falchi, F. & Elvidge, C.D. (2001). The first world atlas of the artificial sky brightness. Mon. Not. R. Astron. Soc. 2001 (preprint).
Cinzano, P. ed. (2000). Measuring and modelling light pollution. Memorie della Società Astronomica Italiana (Journal of the Italian Astronomical Society). Vol 71, 2000, no. 1.
Cohen, R.J. & Sullivan, W.T., eds. (2001). Preserving the astronomical sky. Proceedings, IAU Symposium No. 196. Vienna, 12-16 July 1999. San Francisco. The Astronomical Society of the Pacific, 2001.
Crawford, D.L. (1997). Photometry: Terminology and units in the lighting and astronomical sciences. In: Anon., 1997, p. 14-18.
Crawford, D.L., ed. (1991). Light pollution, radio interference and space debris. Proceedings of the International Astronomical Union colloquium 112, held 13 to 16 August, 1989, Washington DC. Astronomical Society of the Pacific Conference Series Volume 17. San Francisco, 1991.
De Boer, J.B. & Fischer, D. (1981). Interior lighting (second revised edition). Deventer, Kluwer, 1981.
De Boer, J.B., ed. (1967). Public lighting. Eindhoven, Centrex, 1967.
De Kruijter, N.J.; Van Der Meij, F., & Schreuder, D.A. (2005). Lichtreflectieonderzoek van ZOAB wegdekken (Study of the reflection of drainage asphalt). Doorn, De Kruijter Openbare Verlichting, 2005.
Delta (2000). QD30 Reflectometer. Lyngby, DELTA Light & Optics, 2000 (Year estimated).
Delta (2000a). LTL2000S Reflectometer. Lyngby, DELTA Light & Optics, 2000 (Year estimated).
Durant, W. (1962). The story of philosophy; The lives and opinions of the greater philosophers. Third paperback printing. New York, Simon and Schuster, 1962.
Erbay, A. (1973). Verfahren zur Kennzeichnung der Refelxionseigenschaften von Fahrbahndecken (Assessment of the reflection of road surfaces). Dissertation Technische Universität Berlin, 1973.
Erbay, A. & Stolzenberg, K. (1975). Reflexionsdaten von allen praktisch vorkommenden trockenen Fahrbahnbelägen (Reflection data of all conventional dry road surfaces). Lichttechnik, 27 (1975) 58-61.
Feynman, R.P.; Leighton, R.B. & Sands, M. (1977). The Feynman lectures on physics. Three volumes. 1963; 6th printing 1977. Reading (Mass.), Addison-Wesley Publishing Company, 1977.
Folts, J.; Lovell, R. & Zwahlen, F. (2006). Handbook of photography. 6th edition. Thomson, Delmar Learning, 2006.

Frankinet, M.; Gillet, M.; Lang, V.; Longueville, J-L.; Maghe, L.; Marville, C.; Castellana, C.; Debergh, N., & Embrechts, J-J. (2006). Characterization of road surfaces using a mobile gonio-reflectometer. Liège, R-Tech, 2006 (year estimated).

Gillet, M. & Rombauts, P. (2001). Precise evaluation of upward flux from outdoor lighting installations (Applied in the case of roadway lighting). Paper presented at the ILE Light Trespass Symposium, held in London on 8 November 2001.

Gillet, M. & Rombauts, P. (2003). Precise evaluation of upward flux from outdoor lighting installations; The case of roadway lighting. In: Schwarz, ed., 2003, p. 155-167.

Goyat, Y. & Guillard, Y. (2001). Coloroute; coefficient de luminance des routes (Coloroute; luminance coefficient of roads). Lrcp, Cete de l'Est. Avancées techniques. Strassbourg, 2001 (year estimated).

Gregory, R.L. (1970). The intelligent eye. Weidenfeld & Nicholson, London, 1970.

Hartmann, E. (1984). Untersuchungen zur belästigende Wirkung von Lichtimmissionen (Studies on the disturbing effect of light immissions). LIS-Berichte, 51, 33-57, 1984.

Hartmann, E.; Schinke, M.; Wehmeyer, K. & Weske, H. (1984). Messung und Beurteilung von Lichtimmissionen künstlicher Lichtquellen (Measurement and assessment of light immisions from artificial light sources). München, Institut für medizinische Optik, 1984.

Heinz, R. (2006). Lichterzeugung mit organischen Werkstoffen; OLEDs für Displays und Allgemeinbeleuchtung (Light generation with organic materials: OLEDs for displays and general lighting). In: Anon., 2006.

Heinz, R. (2006a). Grundlagen der Lichterzeugung, 2. Auflage (Principles of light generation, 2nd edition). Highlight Verlag, 2006.

Heinz, R. & Wachtmann, K. (2001). Innovative Lichtquellen durch LED-Technologie (Innovative light sources by means of LED-technology). In: Anon., 2001, p. 199-207.

Heinz, R. & Wachtmann, K. (2002). LED-Leuchtmittel: moderne Halbleiterstrahlungsquellen im Visier (LED sources: Looking at modern semiconductor radiation sources). In: Welk, ed., 2002, p. 34-40.

Hentschel, H.-J. (1994). Licht und Beleuchtung; Theorie und Praxis der Lichttechnik; 4. Auflage (Light and illumination; Theory and practice of lighting engineering, 4th edition). Heidelberg, Hüthig, 1994.

Hentschel, H.-J., ed. (2002). Licht und Beleuchtung; Grundlagen und Anwendungen der Lichttechnik; 5. neu bearbeitete und erweiterte Auflage (Light and illumination; Theory and applications of lighting engineering; 5th new and extended edition). Heidelberg, Hüthig, 2002.

Howell, S.B. (2001). Handbook of CCD astronomy, reprinted 2001. Cambridge (UK). Cambridge University Press, 2001.

Iacomessi, P.; Rossi, G. & Soardo, P. (1997). La limitazione del flusso luminoso emesso verso l'alto (The limitation of the luminous flux emitted upwards). Atti del Convegno Inquinamento luminoso e risparmio energetico, Nove, 1997, p. 49-55.

Illingworth, V., ed. (1991). The Penguin Dictionary of Physics (second edition). London, Penguin Books, 1991.

Isobe, S. (1998). Light pollution situations of observatories. In: Isobe & Hirayama, eds., 1998, p. 185-189.

Isobe, S. & Kosai, H. (1994). A global network observation of night sky brightness in Japan – Method and some result. In: McNally, ed., 1994, p. 155-156.

Isobe, S. & Kosai, H. (1998). Star watching observations to measure night sky brightness. In: Isobe & Hirayama, eds., 1998, p. 175-184.

Isobe, S. & Hamamura, S. (1998). Ejected city light of Japan observed by a defence meteorological satellite program. In: Isobe & Hirayama, eds., 1998, p. 191-199.

Isobe, S. & Hirayama, T. eds. (1998). Preserving of the astronomical windows. Proceedings of Joint Discussion 5. XXIIIrd General Assembly International Astronomical Union, 18-30 August 1997, Kyoto, Japan. Astronomical Society of the Pacific, Conference Series, Volume 139. San Francisco, 1998.

Kebschull, W. (1968). Die Reflexion trockner und feuchter Strassenbeläge (The reflection of dry and wet road surfaces). Dissertation Technische Universität Berlin, 1968.
Knudsen, B. (1967). Lamps and lanterns. Chapter 6. In: De Boer, ed., 1967.
Kosai, H. & Isobe, S. (1988). Organized observations of night-sky brightness in Japan. National Astronomical Observatory, Mikata, Tokyo, Japan , 1988 (year estimated).
Kosai, H.; Isobe, S. & Nakayama, Y. (1994). A global network observation of night sky brightness in Japan – Method and some result. In: McNally, ed., 1994.
Krisciunas, K. (1997). P.A.S.P. 109, 1181.
Kuchling, H. (1995). Taschenbuch der Physik, 15. Auflage (Survey of physics, 15th edition). Leipzig-Köln, Fachbuchverlag, 1995.
Laporte, J-F. & Gillet, M. (2003). Meta analysis of upward flux from functional roadway lighting installations. In: CIE, 2003a.
LCPC (2003). Coluroute. Paris. Laboratoire Central des Ponts et Chaussées LCPC, 2003 (Year estimated).
Le Grand, Y. (1956). Optique physiologique, Tome III (Physiological optics, Vol. III). Ed. Revue Optique, Paris, 1956.
Levasseur-Regourd, A.C. (1992). Natural background radiation, the light from the night sky. Contribution to the conference held at UNESCO, Paris, 30 June – 3 July 1992.
Levasseur-Regourd, A.C. (1994). Natural background radiation, the light from the night sky. In: McNally, ed., 1994.
Longhurst, R.S. (1964). Geometrical and physical optics (fifth impression). London, Longmans, 1964.
Lorentz, H.A. (1922). Beginselen der natuurkunde. Achtste druk. Twee delen (Principles of physics, 8th edition, two volumes). Leiden, Boekhandel en Drukkerij voorheen E.J. Brill, 1922.
Luisi, F. (2006). Charakterisierung von Strassenbelägen mittels mobilem Gonio-Reflektometer (The chartacterisation of road surfaces with a mobile gonio-reflectometer). In: Anon., 2006.
McNally, D., ed., (1994). Adverse environmental impacts on astronomy: An exposition. An IAU/ICSU/ UNESCO Meeting, 30 June – 2 July, 1992, Paris. Proceedings. Cambridge University Press, 1994.
Mizon, B. (2002). Light pollution; Responses and remedies. Patric Moore's Practical Astronomy Series. London, Springer, 2002.
Murdin, P. (1997). ALCoRs: Astronomical lighting control regions for optical observations. In: Anon., 1997.
Narisada, K. & Schreuder, D.A. (2004). Light pollution handbook. Dordrecht, Springer, 2004.
NSVV (1957). Aanbevelingen voor openbare verlichting (Recommendations for public lighting). Moormans Periodieke Pers, Den Haag, 1957 (year estimated).
NSVV (1974/1975). Richtlijnen en aanbevelingen voor openbare verlichting (Guidelines and recommendations for public lighting). Electrotechniek, 52 (1974) 15; 53 (1975) 2, 5.
NSVV (1977). Het lichtniveau van de openbare verlichting in de bebouwde kom (The light level of urban public lighting). Electrotechniek, 55 (1977) 90-91.
NSVV (1990). Aanbevelingen voor openbare verlichting; Deel I (Recommendations for public lighting; Part I). Arnhem, NSVV, 1990.
NSVV (1997). Aanbevelingen voor openbare verlichting; Deel III, Richtlijnen voor het ontwerp van openbare verlichting (Recommendations for public lighting; Part III; Guidelines for the design of public lighting). Arnhem, NSVV, 1997.
NSVV (1999). Algemene richtlijnen voor lichthinder in de openbare ruimte. Deel 1, Lichthinder door sportverlichting (General directives for light intrusion in public areas. Part 1, Light intrusion by lighting of sports facilities). Arnhem, NSVV, 1999.
NSVV (2002). Richtlijnen voor openbare verlichting; Deel 1: Prestatie-eisen. Nederlandse Praktijkrichtlijn 13201-1 (Guidelines for public lighting; Part 1: Performance requirements. Practical Guidelines for the Netherlands 13201-1). Arnhem, NSVV, 2002.

NSVV (2003). Algemene richtlijnen voor lichthinder in de openbare ruimte. Deel 2, Terreinverlichting (General directives for light intrusion in public areas. Part 2, Area lighting). Arnhem, NSVV, 2003.

NSVV (2003a). Algemene richtlijnen voor lichthinder in de openbare ruimte. Deel 3, Aanstraling van gebouwen en objecten buiten (General directives for light intrusion in public areas. Part 2, Floodlighting of buildings and outdoor objects). Draft, September 2003. Arnhem, NSVV, 2003.

NSVV (2003b). Algemene richtlijnen voor lichthinder in de openbare ruimte. Deel 4, Reclameverlichting (General directives for light intrusion in public areas. Part 4, Lighting for advertizing). Draft, September 2003. Arnhem, NSVV, 2003.

NSVV (2003c). Richtlijnen voor openbare verlichting; Deel 3: Methoden voor het meten van de lichtprestaties van installaties (Guidelines for public lighting; Part 3: Measuring methods for the lighting quality of installations). Nederlandse Praktijkrichtlijn 13201-3. Arnhem, NSVV, 2003.

NSVV (2003d). Richtlijnen voor openbare verlichting; Deel 2: Prestatieberekeningen (Guidelines for public lighting; Part 2: Calculations of the quality). Nederlandse Praktijkrichtlijn 13201-2. Arnhem, NSVV, 2003 (To be published).

OTA (1970). Tenth International Study Week in Traffic and Safety Engineering. OTA, Rotterdam, 1970.

Paumier, J-L. (2002). Caractéristiques photométriques des surfaces routières. Compte-rendu établi pour le sous-groupe, 17 decembre 2002 (Photometric characteristics of road surfaces. Summary made for the sub-group, 17 December 2002). Lyon Cete – Lrpc Clermond-fd, 2002.

PIARC (1990). Final report. PIARC, Working Group on Pervious Coated Macadam, Draft, 1 October 1990. Paris, PIARC, 1990.

Pimenta, J.Z. (2002). A luz invadora (On light intrusion). Revisto Lumière. 50 (2002) June, p. 94-97.

Pollard, N. (1993). Sky glow conscious lighting design. In: CIE, 1993, Chapter 6.

Prins, J.A. (1945). Grondbeginselen van de hedendaagse natuurkunde, vierde druk (Fundaments of modern physics, 4th edition). Groningen, J.B. Wolters, 1945.

Range, H.D. (1972). Ein vereinfachtes Verfahren zur lichttechnische Kennzeichnung von Fahrbahnbelägen (A simplified method for the lighttechnical characterization of road surfaces). Lichttechnik. 24 (1972) 608.

Range, H.D. (1973). Strassenreflektometer zur vereinfachten Bestimmung lichttechnische Eigenschaften von Fahrbahnbelägen (Road reflectometer for the simplified assessment of the lighttechnical properties of road surfaces). Lichttechnik. 25 (1973) 389.

Reeb, O. (1962). Grundlagen der Photometrie (Fundaments of photometry). Karlsruhe, Verlag G. Braun, 1962.

Ris, H.R. (1992). Beleuchtungstechnik für Praktiker (Practical illuminating engineering). Berlin, Offenbach, VDE-Verlag GmbH, 1992.

Schmidt, W. (2002). Lichtvervuiling in kaart gebracht (Light pollution put on the map). Zenit, 29 (2002) no 4, p. 154-157.

Schmidt, W. (2002a). Landsdekkend karteren van nachtelijk kunstlicht (Nationwide cartography of nighttime artificial light). Utrecht, Sotto le Stelle, 2002.

Schober, H. (1960). Das Sehen; 2 Bände (Vision, 2 volumes). Fachbuchverlag, Leipzig, 1958-1960.

Schreuder, D.A. (1964). De luminantietechniek in de straatverlichting (The luminance technique in road lighting). De Ingenieur, 76 (1964) E89-E99.

Schreuder, D.A. (1967). The theoretical basis for road lighting design. Chapter 3 in: De Boer, ed., 1967.

Schreuder, D.A. (1967a). Measurements. Chapter 8 in: De Boer, ed., 1967.

Schreuder, D.A. (1970). A functional approach to lighting research. In: OTA (1970).

Schreuder, D.A. (1987). Road lighting and light trespass. Vistas in Astronomy 30 (1987) nr. 3/4, 185-195.

Schreuder, D.A. (1988). Zeer open asfaltbeton en de verkeersveiligheid (Porous asphalt and road safety). In: Wegbouwkundige Werkdagen 1988, Ede, 26 en 27 mei 1988, Deel 2; Stroom II: Kwaliteit in meervoud; Zitting II-1: Milieu en veiligheid; Bijdrage 27. Publikatie 8-II. Stichting C.R.O.W., Ede, 1988.

Schreuder, D.A. (1988a). Visual aspects of the driving task on lighted roads. CIE Journal 7(1988)1:15-20.
Schreuder, D.A. (1991). Visibility aspects of the driving task: Foresight in driving. A theoretical note. R-91-71. SWOV, Leidschendam, 1991.
Schreuder, D.A. (1991a). Lighting near astronomical observatories. In: CIE, 1992.
Schreuder, D.A. (1991b). A device to measure road reflection in situ. In: CIE, 1992.
Schreuder, D.A. (1992). Meting van de reflectie-eigenschappen van wegdekken ten dienste van het energetisch optimaliseren van openbare verlichting. Rapportage ten dienste van NOVEM. 26 december 1992 (Measurement of the reflection characteristics of road surfaces to be used for the energy-optimization of public lighting. Report for NOVEM, 26 December 1992). Duco Schreuder Consultancies, Leidschendam. 1992.
Schreuder, D.A. (1993). The in situ measurement of road reflection. ILE Journal. Duco Schreuder Consultancies, Leidschendam, the Netherlands. 1993.
Schreuder, D.A. (1994). Comments on CIE work on sky pollution. Paper presented at 1994 SANCI Congress, South African National Committee on Illumination, 7 – 9 November 1994, Capetown, South Africa.
Schreuder, D.A. (1995). The quantification of quality lighting – The need to cry over spilled milk. Paper, 3rd European Conference on Energy-Efficient Lighting, 18th-21st June 1995, Newcastle upon Tyne, England. Leidschendam, Duco Schreuder Consultancies, 1995.
Schreuder, D.A. (1997). Bilateral agreements on limits to outdoor lighting; The new CIE Recommendations, their origin and implications. In: Isobe & Hirayama, eds., 1998.
Schreuder, D.A. (1998). Road lighting for safety. London, Thomas Telford, 1998. (Translation of "Openbare verlichting voor verkeer en veiligheid", Kluwer Techniek, Deventer, 1996).
Schreuder, D.A. (1999). Road lighting in sparsely populated areas. Paper presented at the 4th SAARC Lighting Conference, held in Dhaka, Bangladesh, 29-31 January 1999. Leidschendam, Duco Schreuder Consultancies, 1999.
Schreuder, D.A. (1999a). De donkere nacht van 5 april 1997 (The 'dark night' of 5 April 1997). Zenit, 26 (1999), oktober, p. 444-446.
Schreuder, D.A. (2000). Obtrusive light audits: A method to assess light pollution. Paper presented at The 3rd National Lighting Congress Special session on "Light Pollution" held on 23-24 November 2000 at Taskisla-Istanbul Technical University, Istanbul, Turkey. Leidschendam, Duco Schreuder Consultancies, 2000; in Anon., 2000.
Schreuder, D.A. (2001). Pollution-free road lighting. Paper presented at the Special Session: "Astronomy for developing countries" held at the 24th General Assembly of the IAU, Manchester, UK, 7 – 16 August 2000. In: Batten, ed., 2000.
Schreuder, D.A. (2001a). Pollution free lighting for city beautification. Paper presented at the International Lighting Congress, Istanbul, Turkey, 6-12 September 2001. Leidschendam, Duco Schreuder Consultancies, 2001.
Schreuder, D.A. (2002). The area upward light emission of outdoor lighting installations. Proposal made for the Commission on Light Pollution of NSVV, 23 January 2002. Not published. Leidschendam, Duco Schreuder Consultancies, 2002.
Schreuder, D.A. (2006). In-situ Messung der Reflexionseigenschaften von Strassenbelägen (In-situ measurement of the reflection properties of road surfaces). In: Anon., 2006. Leidschendam, Duco Schreuder Consultancies, 2006.
Schwarz, H.E., ed. (2003). Light pollution: The global view. Proceedings of the International Conference on Light Pollution, La Serena, Chile. Held 5-7 March 2002. Astrophysics and Space Science Library, Volume 284. Dordrecht, Kluwer Academic Publishers, 2003.
SCW (1974). Wegverlichting en oppervlaktetextuur. Mededeling No. 34. SCW, Arnhem, 1974.
SCW (1977). Proceedings International Symposium on Porous asphalt. Amsterdam, 1976. Record No. 2. Arnhem, SCW, 1977.

SCW (1984). Lichtreflectie van wegdekken (Light reflection of road surfaces). Mededeling 53. Arnhem, SCW, 1984.

Serres, A.-M. (1990). Les images pour les études de visibilité de nuit (Images to investigate the night time visibility). Bull Liais. Labo. P et Ch. 165 (1990) jan-fév. 65-72.

Smith, M.G. (2001). Controlling light pollution in Chile: A status report. In: Cohen & Sullivan, eds., 2001, p. 40-48.

Sorensen, K. (1995). Road equipment – in situ measurement of reflection. In; CIE, 1995a.

Sorensen, K. & Lundquist, S.-O. (1990). A model for the specular reflection of road surfaces. Delta Light & Optics Note No. 2, 1990.

Sorensen, K.; Obro, P. & Rasmussen, B. (1991). A review of the suitability of the average luminance coefficient Q_0 for road surfaces and road markings – and proposal for a different parameter. Notat 8, 1 Feb 1991. Lyngby, Lys & Optik, 1991.

Sprengers, L.M. & Peters, J.I.C. (1986) SOX-E(conomy); A new generation of low-pressure sodium lamps with improved overall performance. The Lighting Journal. 51 (1986) 27-30.

Stath, N. (2006). Anorganische LEDs; Innovationen bei Halbleiter-Lichtquellen (Anorganic LEDs; Innovations in semiconductor light sources). In: Anon., 2006.

Sterken, C. & Manfroid, J. (1992). Astronomical photometry. Dordrecht, Kluwer, 1992.

Tetteroo, J. (2003). Wegdekreflecties. Onderzoek naar de noodzaak en mogelijkheid voor metingen van wegdekreflecties (Road reflection. Study of the need and possibility of measuring road reflection). 11 juli 2003. Utrecht, Bouwdienst Rijkwaterstaat, 2003.

Tromp, J.P.M. (1994). Road safety and drain asphalt. In: Road Safety in Europe and Strategic Highway Research Program (SHRP). Lille, France, 26-28 September 1994.

Van Bommel, W.J.M. & De Boer, J.B. (1980). Road lighting. Deventer, Kluwer, 1980.

Van Heel, A.C.S. (1950). Inleiding in de optica; derde druk (Introduction into optics, third edition). Den Haag, Martinus Nijhoff, 1950.

Weigert, A. & Wendker, H.J. (1989). Astronomie und Astrophysik – ein Grundkurs, 2. Auflage (Astronomy and astrophysics – a primer, 2nd edition). VCH Verlagsgesellschaft, Weinheim (D), 1989.

Welk, R., ed. (2002). Lichtlösung mit Leuchtdioden (Lighting solutions with light diodes). Licht Special 3. München, Richard Pflaum Verlag, 2002.

Westerman, H.O (1963). Reflexionskennwerte von Strassenbeläge (Reflection characteristics of road surfaces). Lichttechnik, 15 (1963) 507-510.

11 Road lighting design

This chapter deals with a limited number of subjects selected from the wide area of outdoor lighting applications. Emphasis is on the numerical design of road lighting installations, particularly on the luminance design. The CIE program LUCIE for computer-assisted luminance-design methods is discussed. Other subjects that are covered in this chapter are the road lighting for developing countries, the maintenance of lighting installations, and simplified design methods.

A number of important aspects of the design of outdoor lighting installations are not covered in this chapter. The main reason is that they do not involve many fundamental aspects of the type that are the core of this book, viz. the mathematical, physical, and physiological basis of vision and lighting. However, we will mention a few of these subjects here, and refer to the relevant literature for the benefit of those who may want further information. To begin with, we will list again several of the standard books on illuminating engineering where most subjects are treated in some detail: Baer 1990; Baer, ed., 2006; Boyce, 2003; De Boer, ed., 1967; Eckert, 1993; Forcolini, 1993; Hentschel, 1994; Hentschel ed., 2002; Narisada & Schreuder, 2004; Ris, 1992; Schreuder, 1998, 2001b; Van Bommel & De Boer, 1980.

- Vehicle lighting and marking (Alferdinck, 1987; Alferdinck & Padmos, 1986; Narisada & Schreuder, 2004, sec. 12.2; Roszbach, 1972; Schreuder, 1976; Schreuder & Lindeijer, 1987; Schmidt-Clausen & Bindels, 1974).
- Tunnel lighting (CEN, 2003; CIE, 1973, 1984, 1990a; NSVV, 1963, 1991, 2002; Schreuder, 1964, 1967; Vos & Padmos, 1983).
- General outdoor and sports lighting (CIE, 1994, 2002a; Forcolini, 1993).
- Floodlighting (Cohu, 1967; Forcolini, 1993; Schreuder (2001, 2001a).

11.1 Design methods for road lighting installations

11.1.1 Principles of lighting design

In sec 1.3.3 it is explained that the requirements of road lighting, particularly of traffic route lighting, can be viewed upon from different perspectives:

1. Promote the traffic through-put.
2. Arriving at the destination of the trip.
3. Avoiding accidents.

For residential streets a fourth has to be added:

4. Promote safety and well-being.

It is explained also that the first three are conventionally considered as the driving task. The driving task can be described as the collection of observations and decisions a car driver has to make in order to reach the goal of the trip. Decision making processes are not dealt with in detail in this book. See Narisada & Schreuder, 2004; Schreuder, 2008. The visual task is usually understood as being a part of the driving task. The goal of taking part in traffic is to reach the trip destination. As is explained earlier, this goal consists of three distinct sub-tasks. The three sub-tasks are:

1. Reaching the destination by selecting and maintaining the correct route;
2. Avoiding obstacles while under way towards the destination;
3. Coping with emergencies while performing the two other sub-tasks.

The selection of the correct route to reach the destination involves decisions that are made, for a large part, even before the beginning of the trip. Whilst on the road it is necessary to be able to see the road itself to maintain the correct route. As is explained in sec. 1.3.4, this is a matter of the road luminance. Its uniformity and road markings may play a role, and thus the question of whether the road is dry or wet.

The second sub-task only originates while driving. It refers to discontinuities in the run of the road, and to the presence of other traffic participants. It is necessary to see the road. Not only the road itself, but also its course – the 'run-of-the-road'. It is equally necessary to see the objects on or near the road. This is a matter of detection of objects, a matter of visibility.

In both cases the essential feature is that there is adequate time to acquire and process the necessary information, to make the decision, and to execute the manoeuvre. This is explained in sec. 1.3.4c, where foresight is discussed. See also Schreuder (1991, 1994, 1998).

In many instances, this is not the case. Unexpected and unwanted emergencies may arise that require a fast reaction of the driver in order to avoid collisions. This is the third sub-task – coping with emergencies. Usually, the information on which the decision must be made is inadequate, incomplete or even incorrect, the time for making the decision is very short, and the time to execute the manoeuvre is often simply not sufficient. Still, in most cases an experienced driver will be able to cope with the emergency, e.g.

by making an emergency stop or by swerving around the obstacle. It is more a matter of experience, vigilance, and alertness than a matter of photometric requirements. Vigilance and alertness are described in detail in Schreuder (1998). See also Schreuder (2008). In conclusion, considering the driving task leads to requirements regarding the road luminance and the visibility of objects. This in its turn requires different lighting design methods.

11.1.2   Design methods based on the road luminance

(a) Systems of road lighting quality assessment

As is explained in sec. 10.4.1a, where the luminance technique is discussed, modern road lighting design is based on the luminance concept. This concept implies that the road surface luminance is the most important criterion for the quality of the installation. As mentioned earlier, the non-uniformity of the luminance pattern and the (disability) glare are the two other criteria of quality. The luminance concept is the basis for most modern national and international codes and standards for road and street lighting, more in particular for the lighting of traffic routes.

The luminance concept is by no means the only system for the assessment and description of the quality of road lighting. Three more concepts are in wide use:
1. The visibility concept that reckons with the fact that the luminance is only an intermediate value, and that visibility is required for safe traffic. This approach is discussed in a further part of this section. A separate point is the visual or optical guidance. Usually it is included in the considerations of the luminance system (see e.g. De Boer, ed., 1967). As may be seen from discussions about the driving task, guidance is more like a medium-distance visual task (Griep, 1971; Schreuder, 1991).
2. The illuminance concept, stating that it is the light falling on the surface that determines the quality of the road lighting, rather than the light reflected by it. In the past, this concept was almost universally adopted, and today still is the most common concept for general outdoor lighting and for lighting of residential streets. This approach is discussed in a further part of this section, where simplified methods are discussed.
3. The comfort concept, that stresses the aspects of quality rather than quantity in road lighting. This concept is mainly relevant for shopping areas; usually, architectural aspects prevail. See Anon (2007); De Boer, ed. (1967); Forcolini (2003), Schreuder (2001, 2001a, 2008); Van Santen (2005); Visser (1997).

(b) The method using E-P diagrams

In 1952, De Boer published a method that allowed to determine the road surface luminance when the reflection characteristics of the road surface and the luminous intensity distribution of the luminaires to be used are known (De Boer, 1951; De Boer et al., 1952). This is the method of the E-P diagram. The name equivalent position diagram will be explained later on. The idea is both simple and elegant. The starting point was that the road surface and the luminaire to be used were known. So was the road width and the lamp post type, or rather the mounting height, as well as the luminaire arrangement. The first step is to draw on a transparency the reflection characteristics of the road surface. These characteristics are represented by lines of equal reflectivity multiplied by $\cos^3\gamma$ for reasons that will be explained later on. The numerical value related to each line represents the luminance that will be reached at the geometric origin of the graph when a one-candela light source is placed on a point on that line. This is true for any point of the line; the line is the – geometric – 'locus' of all points that deliver a particular luminance value for a one-candela light source on that line. All points are therefore equivalent; hence the term equivalent position diagram. Usually the mounting height is taken as the unit for the scale of the graph. An example is given in Figure 11.1.1.

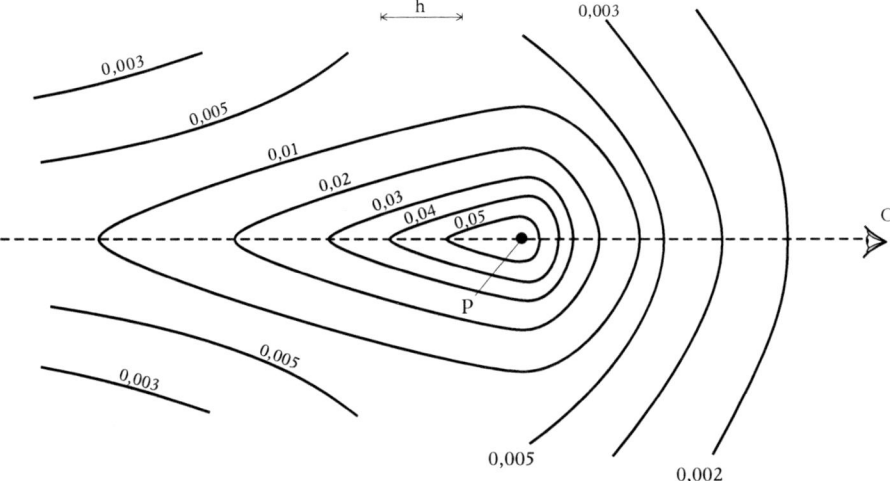

Figure 11.1.1 An example of an E-P diagram (After Schreuder, 1998, figure 13.1.1).

For normal road surfaces the graph is symmetric around the length-wise axis of the road, as is depicted in Figure 11.1.1. One might say that the shape of the graph represents what an observer would see under 1° down when the road is lit by a one-candela light source. It should be noted that the shape of the graph is the mirror image of that of the R-table that is described in sec. 10.4.1b.

Road lighting design

The second step is to make on a second transparency a graphical representation of the luminous intensity distribution – the light distribution – of the luminaire that is to be used. One might call this the I-diagram. The scale is the same is that of the E-P diagram.

Finally a plan on the same scale of the road, but now on normal paper, is needed. As the luminaire arrangement is known the position of the luminaires can be indicated on the plan.

The following procedure is used to determine the luminance of a particular point P on the road as observed from a particular point O. First, the transparency with the E-P diagram is placed on the plan of the road. The origin of the E-P diagram is placed on P. The transparency is turned in such a way that its axis falls on the line that connects P and O. In this position, the value of $q \cdot \cos^3 \gamma$ can be read off for any point on the road, also for any of the luminaire positions that are marked on the plan – if needed by interpolation between the lines of $q \cdot \cos^3 \gamma$. This step is depicted in Figure 11.1.2.

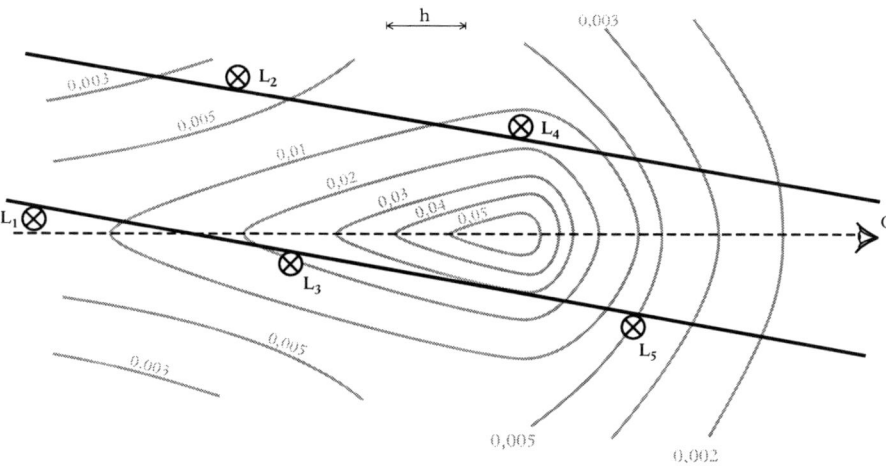

Figure 11.1.2 The use of the E-P diagram to determine the reflection for a specific point (After Schreuder, 1998, Figure 13.1.2).

The next step is to put the I-diagram over the plan and the E-P transparency. This is done to determine the luminance contribution to P as seen from O from one particular luminaire. The origin of the I- diagram is placed over the position of that particular luminaire, its axis over the line connecting the position of that luminaire and point P. So, the curves of the I-diagram represent the luminous intensity of that particular luminaire in the direction of P, again, if needed, by interpolation. Multiplying what one has found gives:

$$L = I \cdot q \cdot \cos^3\gamma \qquad [11.1\text{-}1]$$

This, of course, represents the contribution of that particular luminaire to the luminance in P as seen from O. The way of multiplication shows the need for transparencies; otherwise one could not see the required data simultaneously. As said earlier, the method is elegant indeed!

The only thing to do now is to repeat this procedure for all other luminaires that might contribute to the luminance in P, and repeat that again for all relevant points on the road. When finished, one has the complete luminance pattern of that particular road with that particular reflection, as lit by a number of specific luminaires in a specific arrangement. It goes without saying that this is a very time-consuming method. Further on, we will explain why we did describe it in such minute detail. But the process goes on.

From this pattern the average road surface luminance L_{av} as well the measures for the non-uniformity U_L and U_0 can be derived. These measures are the main criteria for the quality of road lighting installations. Standards and Recommendations contain numerical values of these measures as quality requirements. So, if the values that are derived by means of the E-P method as described above do not fulfill the requirements, the whole procedure has to be repeated for different values of the parameters of the installation, viz. the mounting height or the spacing, or maybe both. If the results are still not satisfactory, one may need to use a different luminaire. Clearly, this is an iterative process. The principle is explained by Schreuder (1967, sec. 3.3.3); the application is summarised by De Graaff (1967, sec. 7.4.1). A summary is given in Schreuder (1998, sec. 13.1.3).

There are two remarks that must be made. Their relevance will be made clear later on. The first is that although the method is cumbersome and time-consuming, one needs only one E-P graph for each road surface, and only one I-diagram for each luminaire type. The second remark is that the E-P method can be used for any conceivable road lay-out as long as it is flat, as well as for all possible luminaire positions and lighting installation geometries.

(c) *The basis of computer-assisted luminance-design methods*

It is stressed several times that the E-P method for the determination of the road luminance is cumbersome and time-consuming. Therefore its was hardly ever used

Road lighting design

in practice. The reason that we did explain the E-P diagram method in such minute detail is that, although the method as such is obsolete, the computer programs that are universally in use, all use exactly the same principles.

Current lighting design methods are based on the calculation system proposed by CIE (CIE, 1976, 1976a, 1990; NSVV, 1997). As will be explained further on, CIE standardised a computer program, called LUCIE. LUCIE is an acronym meaning Luminance Program of CIE. It is introduced, together with its predecessor STAN, in CIE (1976). The program uses the luminance concept. In essence it is based on the graphical design method that is developed in the 1940s by De Boer, and that is explained in the foregoing part of this section.

LUCIE has a long history. Already in 1960, the first steps were made to introduce computer-aided design instead of the graphical methods of the E-P graphs. As this took place in industry at a time that the development of computer software was a matter of company policy, hardly anything at all ever was published. The following survey stems from an unpublished report of October 1963.

The first attempts were not successful because the calculating and memory capacities of the computers of the time were too small for the job. In the early 1960s computers of newer generations could be used. Still, the principle was that of the E-P method, and so it remained to the present day. The new approach was characterised by the following aspects:
1. The E-P diagram was transformed from Cartesian into polar coordinates.
2. The interpolations were linear or quadratic, but also cubic.
3. The input data, viz. the R-tables and I-tables were presented either as paper tape or as punch cards, the usual media of the time. This implied little flexibility.
4. The capacity of the program was 7 I-tables of 2400 numbers each and 18 R-tables of 375 numbers each.
5. The use of polar coordinates simplified for computer calculations the rotation of the I-tables and the R-tables that was required by the E-P method. In this respect the principles of the E-P method were maintained.

(d) *LUCIE*

From the 1960s onward, several institutes started to set up computer programs for road lighting luminance calculations. They could be used for road lighting design, albeit not in an easy way (De Boer & Vermeulen, 1967; Vermeulen & Knudsen, 1967). The approach differs in a number of essential aspects from the programs that were developed for interior lighting:

1. In interior lighting, no attention is placed on the luminance. All calculations and all design methods refer only to the illuminance; primarily to the illuminance of the working plane.
2. In interior lighting all surfaces are considered to have diffuse reflection.
3. As interior lighting only refers to closed rooms, interreflection is essential.
4. In interior lighting, where comfort aspects prevail, much attention is given to the restriction of discomfort glare.

Shortly after, CIE stepped in and set up a standard, the program LUCIE that is mentioned earlier. The program is based on the considerations given in the preceding parts of this section: the E-P diagram method, and the first computer-aided design methods like e.g. the now obsolete program STAN that was mentioned earlier. As mentioned already several times, this implies that LUCIE is really rather ancient. Of course, over the years there have been made numerous proposals for improvements, but essentially they refer to the way the output is given, and in particular to ways and means to visualize the results – often in quite spectacular fashion. Visualizations are of course an important tool in selling lighting designs and lighting installations, but they do not help to overcome the inherent problems of LUCIE.

One thing must be mentioned here. In the preceding part of this section it is explained that the original E-P method allowed one to determine the luminance for any conceivable road lay-out as long as it is flat, as well as for all possible luminaire positions and lighting installation geometries. However, LUCIE only can be used for straight level roads, and for a small number of geometries only, that stretch regularly over the full length of the lighting installation under consideration. It is not possible to make with the help of LUCIE a lighting design for curves, slopes, intersections, roundabouts, and many other common traffic conditions.

(e) Shortcomings of LUCIE

LUCIE, and all subsequent programs that are derived from LUCIE, suffer from a number of shortcomings:

1. First, the program allows only the assessment of luminance values if and only if the installation is known. The luminance is calculated in each of a large number of grid-points for each relevant luminaire separately. The resulting luminance at that point is found by adding these separate luminance. From the geometry of the installation, and from the position of the observer, the relevant angles for each point on the road can be determined. The process is repeated for all points that are relevant for the description of the quality of the lighting installation – enough points to describe not only the average road surface luminance, but the non-uniformity as well. Obviously, the same data and the same system can be used to assess the

illuminances. When a convention regarding the glare formula to be used is included, glare can be assessed as well.

2. Secondly, LUCIE was designed in the earliest stages of computer technology, it is based on outdated programming methods. As a result of the lack of flexibility, of the slow calculation and of the limited computer memories available at that time, LUCIE is restricted to straight, horizontal and perfectly flat stretches of road of a considerable length (12 times the mounting height).

3. A third draw-back is directly clear form the description: when the road lighting installation is known, the program yields the required data immediately; in the design stage, however, the lighting installation is not know – to find the best one is precisely the purpose of the design process! In practice it implies that the designer selects one installation at random; if he or she is an expert in making road lighting designs, the 'gut feeling' is used. From that first trial installation, a number of time-consuming and cumbersome iterative steps must be made just as long as the designer is satisfied that the requirements are met 'close enough'. The design time is increased unduly because the programs do not cater for an automatic iterative process: each iterative step has to be inserted by hand. What obviously is needed, is a system in which the requirements (e.g. luminance level, uniformity and glare, or visibility, or illuminances) are inserted, and where the program assesses the 'best' installation. As there are usually many hundreds of possible installations that might meet the requirements, very extensive calculations are needed. It remains to be seen whether this is really a problem for modern PC's; if so, an expert system based on the 'gut feeling' selections must be made.

4. Another shortcoming of the current programs is that they are restricted to the light-technical parameters of the lighting design: reflection, intensity and geometry. In order to write a program that selects automatically the best solution, many more parameters must be included in the calculations. To name a few:
   – costs of equipment (masts, cables, transformers, luminaires, lamps, ballasts etc);
   – costs of installation;
   – interest and amortization rates for long term investments;
   – tariffs and rates for electricity;
   – maintenance costs (costs of wages, of keeping material in stock etc).
   When combining all these cost factors with the benefits from accident and crime reduction, the selection of the best installation can be based on cost-benefit considerations;

5. Finally, the programs cannot allow for inaccuracies in the input data. In other words, the programs assume that the R-tables as well as the I-tables do not show any errors at all. Obviously, this is not true, the tables being the result of measurements. Now, it is well-known that in practice photometric quantities cannot be measured better that about 3 to 20%, depending on the type of measurement and on the class of the

measuring apparatus (Ris, 1992, table 10.1 and 10.2). Worse still, the calculations are based on one R-table and one I-table only, assuming that all road surface points are identical as well as all individual luminaires. Even the most casual observation will show that these assumptions are not correct; however, quantitative data are almost completely lacking. This final point relates to the remark that the designer, at a certain moment in time, will stop trying to find an even better solution for the design requirements. Obviously, that 'certain moment in time' ought to be based on a sound assessment of tolerances in the result, and not on the work load of the designer, nor on the patience of his boss.

What happens in practice is that many lighting designers 'play for certain' by allowing a certain safety margin into their design. Such margins may result in an unnecessary increase of the costs and of the energy consumption of the lighting. Added to this is the fact that the requirements themselves contain some margins – that cannot be quantified either – based on considerations of visual impairment and/or driving comfort. Only a design based on sound cost-benefit considerations can avoid all these unnecessary expenditures.

11.1.3 Alternative design parameters

Over the years, many different parameters have been proposed that can be used for the design of road lighting installations. None of them, however, with the possible exception of the illuminance method, resulted in a system that could be used for the design of actual, practical lighting installations. Therefore, we will only mention them, without going into detail.

(a) Illuminance

In sec.1.3.4 it is explained that taking part in modern motorized traffic requires the acquisition of visual information. The way this may be done depends amongst many other variables on the state of adaptation. Also it is explained that the state of adaptation is determined primarily by the light level, and more in particular by the road luminance. This is the basis for the luminance technique that is discussed in sec. 10.4.1a, as well as for the CIE-adopted calculation program LUCIE. However, in most streets the luminance is hardy a good measure for the acquisition of the relevant visual information. One may think about the residential street, the homezone, residential yard, or woonerf, shopping malls, etc. (Narisada & Schreuder, 2004, sec. 11.2.6; Schreuder, 1998, chapter 9, 1994, 2000a). These consideration have lead to the inclusion of illuminance in almost all national and international standards and recommendations as the main quality criterion for areas where pedestrian traffic is predominant (CEN, 2002; CIE, 1965, 1977, 1992, 1995a, 2000a; NSVV, 1957, 1983, 1990, 2002).

Road lighting design

As regards the design based on the use of illuminance as the main quality parameter we can be short. It can be done directly with LUCIE. All one needs to do is to use $\cos^3\gamma$ for the R-table. This gives directly the illuminance values that are needed.

(b) Road lighting design methods based on revealing power

As has been stressed several times in this book, road lighting is installed for visibility, more in particular to allow road users to see the run of the road and to detect obstacles. Since nearly 70 years, attempts have been made to develop a method to design road lighting more directly on the basis of visibility. The first idea was the concept of 'revealing power' proposed by Waldram (1938). The revealing power was defined as the percentage of objects that can still be seen. The objects have precisely described characteristics (Narisada & Schreuder, 2004, sec. 11.8.3a). Since the time of Waldram, the concept have been expanded (Harris & Christie, 1951; Van Bommel & De Boer, 1980). In spite of the efforts by these investigators, however, practical applications of their results are still limited.

(c) Visibility Level

Another approach was to base road lighting design on a 'metric' for the visibility. This metric is the Visibility Level (LRI, 1993). The Visibility Level (VL) is defined as a ratio between two luminance differences (CIE, 1981). At the one hand ($\delta L / \delta L_{min}$), the actual luminance difference between the object (L_o) and its background (L_b), or $\delta L = L_o - L_b$. At the other hand, the luminance difference between the object just perceptible and its background associated with the adaptation condition in question. The concept of the Visibility Level stems originally from Holladay more than 75 years ago (Holladay, 1926). The direct use of the Visibility Level did not work out, because most parameters were not precisely defined. Furthermore they could not be measured accurately enough under practical road lighting conditions (Gallagher et al., 1975).

(d) Small Target Visibility

Above it is mentioned that the Visibility Level method did not prove to be useful because most parameters were not precisely defined, and could not be measured properly. As a further step in the Revealing Power approach, the objects were defined more precisely. This is called the Small Target Visibility or STV method. It is not clear why the objects were smaller than the generally accepted object of 20×20 cm^2. For each object, the Visibility Level is determined. The objects were weighted according to their potential danger. The weighted average of the Visibility Level is used as criterion for the quality of the road lighting (Anon., 2000). On the basis of this concept, and on studies by Janoff (1993), a method for the design of road lighting has been developed (Keck, 1993). There were a number of difficulties, however. The first was that there are no theoretical or practical reasons to assume that a small object is a good measure for road hazards. The 20 × 20 cm^2-object had been introduced as a measure for detection, and not as a measure of road hazards (Adrian, 1993; Padmos, 1984; Schreuder, 1964, 1967, 1984, 1987, 1984). The second problem was that, for experimental reasons, the measurements could be made only on specially prepared road segments, and not, as intended, on normal roads. This implied that it is not possible the validate the STV-method by means of traffic accident studies (Enzmann, 1993; Eslinger, 1993; Janoff, 1993a). The third problem was that the calculated and measured values did not agree well enough to base a design method on STV.

(e) The Narisada visibility-based design method for road lighting

Under practical traffic conditions, the distribution of the luminance in the field of view of a car driver is not uniform in time nor in direction. However, as is explained in sec. 7.4.2, the state of adaptation changes only slowly, the visual system of the driver is adapted to the dynamic average of the luminance (Narisada, 1999, 2000; Narisada & Karasawa, 2001, 2003; Narisada & Schreuder, 2004, sec. 11.8.4a). In practice this means that the driver is adapted to the average road surface luminance (Narisada et al., 1997). To this comes that drivers fix the line of attention only briefly on the same spot. In road traffic, the fixation time usually is not more than 0,1 or 0,2 seconds (Watanabe, 1965, Narisada & Yoshikawa, 1974; Narisada & Yoshimura, 1977). The implication is that the results of the experiments for uniform background cannot apply to non-uniform dynamic conditions, where the observer is not adapted to the luminance of the background (Narisada, 1995; Narisada & Yoshimura. 1977).

Subsequent experiments have shown that two independent relationships of the luminance difference threshold exist, one for the adaptation luminance and another for the background luminance. By adding the luminance difference thresholds for the adaptation luminance and that for the background luminance, the luminance difference threshold for any combination of the adaptation luminance and the background luminance for any non-uniform luminance field can be derived (Narisada, 1995; Takeuchi & Narisada, 1996). As a final step, a reference object is introduced (Narisada & Karasawa, 2001, 2003, Narisada et al., 2003). In Figure 11.1.3, experimental results are depicted about the relation between the luminance difference threshold and the adaptation luminance.

As is described in detail in Narisada & Karasawa (2003), Narisada et al. (2003), and Narisada & Schreuder (2004, sec. 11.8), the Visibility Level can be converted into Revealing Power. Finally, a procedure is established for deriving the distribution of the Revealing Power in road lighting installations (Narisada & Schreuder, 2004, sec. 11.8.6). For further details see the original papers (Narisada & Karasawa, 2001, 2003, Narisada et al., 1997, 2003).

Road lighting design

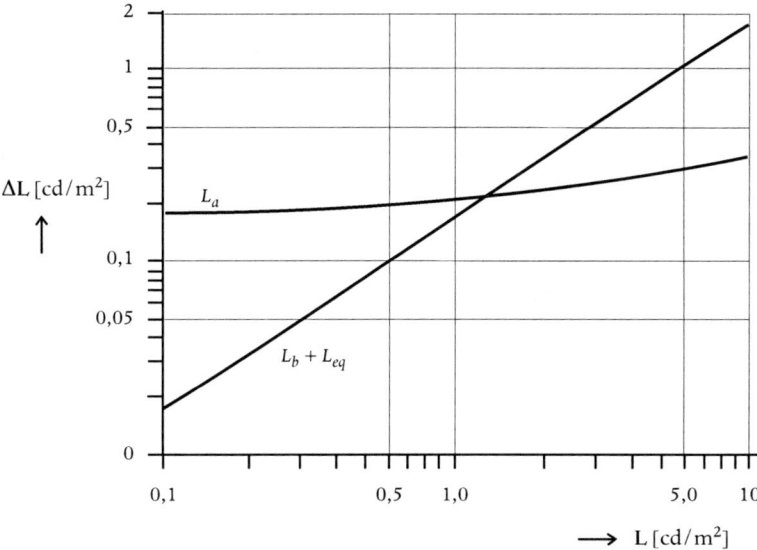

Figure 11.1.3 The relation between the luminance difference threshold and the adaptation luminance (After Narisada & Schreuder, 2004, Figure 11.8.2. Based on Narisada 1995).

1. By applying the widely used calculation method (CIE, 2000), calculate the luminance for a type of the luminance, at many points on the road surface for a type of road surface;
2. Calculate the average road surface luminance of the luminance at the many points;
3. Using the experimental results depicted in Figure 11.1.3 for a value of the average road surface luminance to which the driver's eyes are assumed to adapt, obtain the luminance difference threshold (δL_{min}) against the background road surface luminance at the many points on the whole road surface area. The location of the point of the road surface is tentatively assumed at 7 m behind the object. The point on the road surface 7 m behind the object 100 m is located just vertical centre of the Critical Object at 100 m ahead seen by a driver, whose eye height is about 1,5 m;
4. By applying the widely used calculation method (CIE, 2000), calculate the vertical illuminance on the visible surface of the object at many points on the road surface area;
5. Select a reflection factor of object, which corresponds to a Revealing Power (for example, a Revealing Power of 90%), of which distribution is to be calculated;
6. Calculate the luminance of the Reference Object, whose reflection factor, for instance, is 20%, corresponding to a Revealing Power of 90% at many points on the road surface;

7. Calculate the difference ($\delta L = L_b - L_o$) between the background luminance (L_b) and that of the object (L_o) at the many points on the road surface, on the basis of calculated results in (1) and (6);
8. Calculate the Visibility Level as a ratio ($\delta L / \delta L_{min}$) of the luminance difference (δL) to the luminance different threshold (δL_{min}) at all relevant points on the road surface. The difference in different polarities must not be confused. Only those in the same polarity are relevant;
9. Derive distribution curves for Visibility Level = 1, for the Reference Reflection Factor (For example a reflection factor of 20%);
10. The distribution curve for Visibility Level 1 shows the distribution curve of the corresponding Revealing Power (of 90% as the example);
11. By calculating for the percentage areal extent making up the whole road surface area, in which the Revealing Power is higher than, 90%, the Area Ratio can be obtained.

We have summarized this rather complicated discourse about the visibility-based design method for road lighting that is introduced by Narisada and his co-workers here, because it offers a theoretical way to apply the concept of Visibility in the design of road lighting installations. At present, however, there is no practically applicable design method available that can be used by the lighting engineer in the field.

(f) Guidance lighting

In secs. 1.4.3 and 10.4.1a, the importance of visual guidance, or optical guidance is stressed. Basically there are three ways to secure or improve of road traffic guidance:
1. Passive road markings;
2. Traditional road lighting;
3. Active road markings.

All three can be applied in daylight as well as at night (OECD, 1975, 1976; Narisada & Schreuder, 2004, sec. 7.1.4d). Passive road markings are the traditional means to improve guidance on the road. They are either road paints, retroreflective markings, and raised pavement markers, often called 'cat's eyes'. All rely on the vehicle headlighting to be effective (OECD, 1975; Schreuder, 1981).

One of functions of traditional road lighting is to support the optical guidance (De Boer, 1967, sec. 2.7; Schreuder, 1998, p. 163). For this, the luminaires must emit some light just above and just below the horizon. This requires a careful design of the luminaires. If the luminous intensity is too low, the guidance is not adequate; if the luminous intensity is too high, glare will be the result. Disability glare is discussed in sec. 8.3. In more

Road lighting design 415

recent times, lighting engineers seem to have lost interest in the subject of guidance (Van Bommel & De Boer, 1980, p. 148; Hentschel, 1994, p. 185; Hentschel ed., 2002). This might seem strange because from the driving task analysis point of view, optical guidance is the most important quality requirement of road lighting (Schreuder, 1991; 1998, p. 150; 1999a). The analysis of the driving task is discussed in sec.1.3.4.

The drawbacks of passive road markings and of traditional road lighting as a means to support optical guidance are clear: passive road markings require vehicle headlighting and therefore never can contribute to the guidance at a larger distance than the 'reach' of the low beams; traditional road lighting often causes too much glare and too much light pollution. Active road markings present a useful compromise. Several alternatives have been investigated (Jongenotter et al., 2000). The most promising seems to be to provide the raised pavement markers with Light Emitting Diodes or LEDs. LEDs are discussed in sec. 2.5. See also Narisada & Schreuder (2004, sec. 11.1.1h). The essence is that they emit enough light to be clearly visible over a considerable distance – say a few hundred meters – and still they are so dim that they do not cause any glare or light pollution. In fact, it was the restriction of light pollution that triggered this system (Anon, 1997; Schreuder, 1999). In the Netherlands, most research in this field is done in the Province of Noord-Holland (Jongenotter et al., 2000). On the basis of this research, Guidelines have been issued for the design of active markings (NSVV, 2004). However, again in this case, the recommendations do not present a design tool for the lighting engineer in the field.

  (g) *Visual comfort and city beatification*

Presently it is hard to imagine but in the 1950s and 1960s there did seem to be no end to desires about light levels. At the one hand, new lamp developments made easy, cheap, efficient, and long-life lamps common-place, whereas at the other hand, in the pre-OPEC times, energy, electric energy included, was very cheap indeed. It was the time of recommended value of 1000 and even 2000 lux in offices (Balder, 1957; Bodmann, 1962; CIE, 1975, 1982; De Boer & Fischer, 1981) and of well over 2,5 cd/m^2 in road lighting (de Boer, 1967, sec. 2.2). This allowed to emphasise ease of perception, because the minimal values for reliable perception were already met, and often even amply surpassed. See Schreuder (2005). One example was to use as a road lighting quality criterium discomfort glare instead of disability glare. This point is discussed in sec. 8.4.

It goes without saying that more recently, cost and environmental considerations did change a lot. For normal office lighting 200 lux is a minimum (Anon., 2007, sec. 2.2). This is of course still well over 40-50 lux that can be regarded as the minimum for reading (Schreuder, 2004, 2005). For road lighting something similar can be found.

Modern standards and recommendations give a wide range for the average road surface luminance for different road classes. Typical values for main roads are between 0,75 and 1 cd/m² (CEN, 2002; NSVV, 2002). This means to say, of course, that the emphasis is shifting from ease to reliability of visual perception.

A fully different aspect of visual comfort is city beautification. The basic idea is to make a city a place where it is pleasant to live, pleasant to be around, in short, a place one may be proud of, a matter of civic pride. The lighting requirements are higher, often much higher, than the minimum that is needed for safe traffic and for secure living. These requirements are discussed in some detail in Schreuder (2001a). Here they will be summarized.

A number of requirements must be fulfilled in order for any scheme for city beautification to be successful:
1. The scheme must represent the city. It must personify the love and pride the people have for their city.
2. It must be recognized and appreciated by the people for whom it is meant.
3. It must be functional; people must be able to 'do' something with it.
4. It must be a complete, homogeneous assembly of individual parts that fit together in an harmonious way.
5. It must be functional throughout the whole day and night, in rain, sunshine or frost.

The following design requirements for city beautification lighting can be defined:
1. The luminance of facades of buildings must be lower than 25 cd/m² (and preferably lower than 10 cd/m²).
2. The illuminance on windows of nearby residents lower than 25 lux in the evening and lower than 4 lux at night (preferably lower than 10 and 2 lux respectively).
3. The luminous intensity of fittings visible from windows of nearby residents lower than 7500 cd in the evening and lower than 2500 cd at night (preferably lower than 7500 and 1000 cd lux respectively).

Modern standards and recommendations give a wide range for the average road surface luminance for different road classes with the emphasis now on reliability of visual perception. This approach does not warrant separate design methods. City beautification does, but in spite of the fact that requirements for the lighting scheme, as well as for the design have been given, again in this case the recommendations do not present a design tool for the lighting engineer in the field.

(h) *Fear for crime and subjective safety*

As mentioned earlier, road lighting has a number of beneficial effects. Amongst them the subjective aspects of social safety of residents and pedestrians (fear for crime) and the amenity for residents are considered as being very important. In fact, in many countries, these aspects are the most important for the decision to light roads and streets, both urban and rural. Quantitative data, however, are very sparse. Some of the prevailing considerations are summarized in Schreuder (1989, 1998).

Over the years, many studies have been made regarding the relation between lighting and fear of crime. A detailed survey has been made in Narisada & Schreuder (2004, sec. 12.1.4) and Schreurs et al. (2007). As an example the following quote: "It does appear that street lighting reduces fear among specific population sub-groups. Favorable results have been shown in relation to various indicators of fear, anxiety, and worry about crime and disorder. The majority of projects provide strong evidence of fear reducing effect of lighting improvements, while others produce mixed results. There are fairly consistent indications that fears for personal safety and worries about crime are reduced following lighting improvements" (Painter, 1999).

As a conclusion of these studies it has been found that higher requirements, often considerably much higher, are needed than is customary in road and street lighting at present. In many cases this refers to the overall light level, but often also to the general lighting quality, including glare restriction and better uniformity (Schreurs et al., 2007).

A closer examination of the data suggests that one has to do with a rather complex phenomenon, incorporating a number of rather different aspects. In the 1980s and 1990s, these considerations have been studied in great detail in the Netherlands. The studies focussed on what was called 'petty crime'. Small crime is mainly street crime, such as hooliganism, street violence, vandalism, petty theft, and more in particular sexual harassment, not only of women. A comprehensive study was made (Anon., 1986, 1987, 1991, 1994). The studies included some theoretical investigations but were mainly based on 'in situ' inspection of 'weird places'. The studies resulted in the formulation of the 'Five criteria for social safety':
1. Presence of social control – supervision and surveillance. This point is rather obvious, and it includes supervision by casual bystanders or systematic surveillance.
2. Absence of potential criminals. Usually this means that areas with high crime rates or high crime menace are avoided. On private premises it is often possible to refuse admittance to suspect persons, but on public premises this is often difficult or impossible.

3. Visibility. The lighting must be adequate, but also it is important to avoid other restrictions to the visibility. Shrubs should be trimmed, pillars and recessed doors avoided etc. If one cannot see anything, surveillance is not effective.
4. The situation must be clear. One should be able to find the destination easily, and also have escape routes available if necessary.
5. The surround should look attractive. A clear, clean and well-maintained surround makes it clear that there are persons around who care, and suggest that they may be nearby. This is a reinforcement for the patrons of the accommodations, and a deterrent for the potential criminals.

(j) Cost-benefit considerations

Traffic at night requires artificial light that allows one to go around safely. In many cases, motor vehicles can use their own headlights. When traffic is dense, and in particular when slow traffic is a major contributor, stationary (overhead) road lighting is necessary. Before the design of a lighting installation is considered, one has to determine whether a particular road requires fixed road lighting. The decision to light a road or not can be made ultimately only on the basis of a cost-benefit analysis. We will not discuss cost-benefit analysis in this book, because it is a specialized field that would take us too far from our goal. Reference may be made to the literature, like e.g. COST, 1994; Flury, 1981, 1992; Schreuder, 1994a, 1995, 1998, 1998a; Stark, 1998. Furthermore, because in most cases the data to make this analysis are not available, one has to refer anyway to practical experience.

11.2 Road lighting for developing countries

11.2.1 Recommendations

(a) The visibility approach

There are two ways to look at road lighting for developing countries. Both are fully justified, although they lead to conflicting results. The first is to realise that the visual task in traffic is the same all the world over, in industrialised countries and in developing countries alike. The people have the same vision characteristics; their eye-sight is the same, and the objects to be seen are similar. This is the CIE-standpoint as regards recommendations for road lighting in developing countries. Basically they are the same as all other recommendations, the only difference being in the wording. It took a long time to set up recommendations. The first proposals were made at the CIE session in India in 1995. A separate workshop on the subject was organized. The Technical Committee TC 4-37 'Road transport lighting for developing countries' was established, and its Terms of Reference were defined (Chowdhury, 1995, 1997;

Chowdhury & Schreuder, 1998; Schreuder, 1998, sec. 13.4). In 2007, the final draft of the report is published (CIE, 2007).

(b) The traffic engineering approach

The second way to look at road lighting for developing countries is based on traffic engineering considerations. In most developing countries, the traffic characteristics are essentially different from those in industrialised counties:

1. The modal split over the different modes of transport is different. In industrialized counties almost all traffic is motorized, mainly cars, but with a fair percentage of lorries and buses. Vehicle lighting and marking is near-perfect. Motor and pedal bicycles, as well as pedestrians on the carriageway are rare, and other traffic like animal-drawn carts and push-carts are absent. The travel speed is high. Even in built-up areas the normal driving speed is some 50 km/h – barring traffic jams, of course. In most developing countries, the picture is nearly completely reversed. The predominant traffic participants are pedestrians, usually walking right on the carriageway. Pedal and motor bicycles are abundant, usually traveling unlit. Also, unlit animal-drawn carts and push-carts are common. And finally, the motor traffic consists mainly of buses and lorries with some cars in between. The travel speed of this mix is near a walking pace.

2. The infrastructure is different. In developing countries many roads are wide. That is to say that many roads have wide adjoining areas that form part of the road but that are not used for transport. They are used for small businesses etc., whereas the actual carriageway is narrow and usually unpaved. Another point is maintenance. Being usually mediocre or poor in industrialised countries, it is nearly absent in most developing countries. Still another is the electricity supply. In many developing countries, and more in particular in those with a steeply increasing GPN, the overall electricity supply on a national basis is often the limiting factor in a sustainable economical growth. In this, both power stations and power grids are in short supply. Consequently, often the decision is made to use the electric power for production purposes and not for road lighting. A road lighting power supply independent of the main grid is therefore a promising possibility. Often, stand-alone photovoltaic systems are suggested (De Gooijer & Looijs, 1997).

3. A major aspect to keep in mind is the difference in regional spread within countries. One might say that in the Central Business Districts of almost all capital cities in the world, the traffic situation is similar. Roads are usually overcrowded by cars and some lorries and buses; roads and road equipment is designed and constructed (be it often not maintained) in a similar way, and life in general is similar as well. In most industrialized countries the same 'life style' may be found over the whole territory of the nation – obvious situations like mountains and deserts excepted – whereas the life style in many developing countries changes rapidly when one leaves the CBD

of the capital. Usually, city outskirts consist of slums or 'shanty-towns', whereas villagers usually still live and let live like many centuries ago. The characteristics of urban developments in view of road traffic is described by Schreuder (1994b).

So these three points seem to warrant a different approach for road lighting recommendations for developing countries (Schreuder, 1989a, 1998b,c). This approach is, of course, not to allow poorer standards if living for the less-privileged members of the global human family.

In Table 11.2.1, an example is given how recommendations for road and street lighting in developing countries might look. The recommendations are expressed in illuminance values because the road surface is unknown, and may be gravel, sand, or mud. The speed (v) of the mode of transport is taken into account, and so is the related stopping distance (S.D). For the sake of the argument, lorries with poor brakes are included. Table 11.2.1 shows values that are sometimes, but not always, lower than those of current CIE-recommendations.

Primary transport mode	v (m/s)	S.D (m)	$E_{av,min}$ (lux)
pedestrians, carts	1	1	1
tricycles, carts	3	4	1 – 2
cycles	8	15	2 – 3
lorries, poor brakes	10	50	5 – 8
cars and lorries	20	130	10 – 15

Table 11.2.1 An example of possible recommendations for road and street lighting in developing countries. Based on data from Schreuder, 1995a.

11.2.2 Low-maintenance lighting installations

(a) *The maintenance of outdoor lighting installations*

In most countries, the budgets for installing and for maintaining lighting installations are managed by different authorities, resulting in the risk of poor maintenance. Poor maintenance leads to an inefficient lighting installation and may be considered as a waste of money, particularly a waste of invested capital. It should be stressed that these problems are by no means restricted to developing countries: in almost all industrialized countries, there are many installations where maintenance is poor or even absent. It seems that only the most prestigious installations (e.g. for high-volume rural motorways) are maintained on a regular base.

A lighting installation must follow certain requirements during its full lifespan. However, as with all technical equipment, the quality of street lighting deteriorates in time. Part of

Road lighting design

the effect is reversible such as soiling and break-down of the equipment. Other aspects are irreversible such as corrosion and surface deterioration. In order to keep up the quality, three different options are open:
1. Set the design criteria so high that, in spite of the deterioration, the requirements will be met at all times;
2. Design the lighting in such a way that deterioration is kept to a minimum;
3. Maintain the installation at regular intervals by relamping and by cleaning the equipment and replacing defective components.

All three options are expensive. The first and the second require high initial costs, and the third involves high labour costs. Calculations show that the cost optimum usually is found when combining the three options. A major problem is the fact that in most countries the initial costs and the running costs are dealt with in different ways – often by different agencies. In many developing countries the situation is confounded by the fact that expertise in equipment design usually is locally available or can be imported; installation and maintenance expertise is often low or even absent. For these reasons the second option – low maintenance design – seems to be the most promising for developing countries, in spite of the additional initial costs. Still, maintenance budgets usually are insufficient. This calls for two things: first to provide better information for the road authorities, and secondly to search for lighting installations that require a minimum of maintenance. The ideal is a 'no-maintenance' installation (Schreuder, 1996).

(b) Open luminaire options

In this section, an outline of a road lighting system conceived along these lines is presented, while illustrating some of the technical and economic aspects (Schreuder, 1996, 1998b). The focus is on an open luminaire with two lamps, that are used in sequence. That means that basically lamp 2 begins to operate only after lamp 1 has failed. This is a concept that was used in the past, and in many countries, amongst them the Netherlands, still is.

In most modern lighting installations, the practice of open luminaires has been replaced by the concept of the fully dustproof, waterproof, closed, one lamp luminaire with sophisticated optics. The required ingress protection is IP65 or better. The ingress protection is discussed in sec. 10.2.3. However, one should not expect that this high IP will be maintained after opening and closing repeatedly the luminaire, particularly if this is done by unskilled labour. Still, a high quality standard may be maintained at the cost of frequent cleaning and lamp replacement. Maintenance intervals usually are two years. However, in practice the luminaires are often not fully tight, even when new, and the maintenance interval is very long – if regular maintenance is done at all. The result is that usually dust and dirt will enter the luminaire and accumulate over the years, reducing dramatically the luminaire output. It has been stressed earlier that one may meet this state of affairs in developing countries and industrialised countries alike.

(c) Comparing open and closed luminaires

In Schreuder (1996) a detailed study of several alternative possibilities has been made. A low-maintenance two-lamp open luminaires with maintenance intervals of 12 years is compared to a high-quality one-lamp closed luminaire with a maintenance interval of 2 years and a low-quality one-lamp closed luminaire with a maintenance interval of 6 years. Here, only a summary of the results will be given. For more details see Schreuder (1996, 1998, 1998b).

The comparison is made for a 10 m wide road. Road shoulders may be present. The lighting is one-sided with 10 m mounting height, no overhang, and no luminaire tilt. We assume a climate where dust is a more common maintenance risk than mud or road salt. This implies that an open luminaire is to be preferred. The comparison is made for 3 lux as the minimum maintained value. As open luminaires are not common any more, the assessment is based on luminaire data from the 1960s (Knudsen, 1967; De Boer, ed., 1967). Only the average road illuminance is considered. Uniformity and glare are disregarded. The average illuminance is estimated by using the utilization factor method. This method is explained in a further part of this section. The average road illuminance E_{av} follows from:

$$E_{av} = \eta \cdot (\phi \cdot f)/(s \cdot w) \qquad [11.2\text{-}1]$$

in which:
    η: the luminaire utilization factor;
    ϕ: the initial (100h) lamp lumens;
    f: the maintenance factor;
    s: the spacing between lanterns;
    w: the effective road width.

Data for the maintenance factor over really long cleaning intervals of 12 years or over are not available. Based on available data, an estimate is made. Using the Philips HRD 10 with 2 · 80W lamps as an example, the luminaire utilization curve yields $\eta = 0{,}26$ for the carriageway. The depreciation factor or maintenance factor consists of two parts. First the lamp part. We assume that a SON-150W lamp is used. The 100-hour light output is 13 000 lumen (Anon., 1993, p. 6.2.11). When using a two-lamp sequential arrangement, one may assume that after 12 years about 82% of luminaires still will function with a lamp depreciation of 75% (Anon., 1993, p. 6.2.11; Schreuder, 1996, table 1). The second part is the luminaire part. Data for long cleaning intervals are not available. Studies in The Netherlands have suggested that luminaire depreciation may be severe, particularly

Road lighting design

for open, diffuse luminaires – the type we have suggested here (Knudsen, 1967, Table 6.5). From BS 5489 Part 2, 1992, it seems to follow that the deterioration of closed luminaires with poor closure is severe and will continues also after several years.

Based on these considerations it is assumed that the long term deterioration will amount to 0,45 for open diffuse luminaires. Based on some further extrapolation the maintenance factor f in eq. [11.2-1] for 12 years is taken as 0,277.

The spacing follows from eg. [11.2-1]. Inserting the values derived earlier, the maximum spacing for 12 years interval is $s_{12} = 0{,}26 \cdot 13\,000 \cdot 0{,}277 / 10 \cdot 3 = 31{,}2$ m. or 32,0 luminaires per km.

In new condition, the design value is:

$E_{av;12} = 0.26 \cdot 13000 / 10 \cdot 31.2 = 10{,}83$ lux

The low-maintenance installation is compared to a traditional installation. As an example the optical performance for the Philips luminaire HRP 012/1400 is used (Knudsen, 1967, p. 290-294). The reason to use this old luminaire is that the parameters are presented in the same way. For the high-quality installation, we will assume a more modern ingress protection.

The luminaire lamp utilization curve yields 0,31 for the carriageway, about 20% more than that of the open luminaire. Assuming a two-year cleaning and lamp replacement interval, an excellent ingress protection (IP 65) and one 150W SONCL-T lamp the depreciation is 0,88 and the lamp survival 0,93. The luminaire depreciation is 0,88 (Anon, 1993, p. 6.2.11) The deterioration is:

$f = 0{,}88 \cdot 0{,}93 \cdot 0{,}88 = 0{,}720.$

The maximum spacing is for a two year maintenance interval

$s_3 = 0{,}31 \cdot 13\,000 \cdot 0{,}72 / 10 \cdot 3 = 96{,}7$ m, corresponding to 10,3 luminaires per km.

In new condition (the design value) is:

$E_{av;3} = 0{,}31 \cdot 13\,000 / 10 \cdot 96{,}7 = 4{,}17$ lux.

The non-uniformity will be very pronounced under these conditions.

In Schreuder (1997) a detailed cost comparison is made. We will give here only the results. See Table 11.2.2. The costs are expressed in Hfl of 1992. If desired, the money-value van be converted into Euro or US$ of 2007, but as the values are used as an indication only, and because they are relative, there is no need to do this. As we consider only the relative costs, the absolute cost level is immaterial.

costs per column

system	low maintenance	traditional
interval (years)	12	2
installation costs		
column	1012	1012
luminaire	91	770
ballast	73	73
subtotal	1176	1855
amortisation (15 y; 8%: 0,1168)	137,36	216,66
maintenance per interval		
lamp	248,52	124,26
changing	48,38	24,19
cleaning	19,61	19,61
subtotal	316,51	168,06
per year	26,38	84,03
energy costs (per year, 4000h, 169W, 0,31/kWh)	209,56	209,56
running costs (per year)	235,94	293,59
total costs (per column per year)	373,30	510,25
columns (per km)	32,0	10,3
total costs (per km per year)	11945,60	5255,58
relative costs (per km per year)	227	100

Table 11.2.2 Cost comparison of low-maintenance and traditional installations (Based on Schreuder, 1997, Table 4).

From Table 11.2.2 is might be concluded that the traditional installation is cheaper than the low-maintenance installation. This is, of course, correct. However, when, as in many developing countries, there is no maintenance at all, the comparison might look different. Without more reliable data it cannot be indicated, however, how much different.

11.3 Simplified design methods

11.3.1 Characteristics of simplified design methods

(a) *The need for simplified design methods*

In the preceding parts of this chapter it is explained that the only solid design method for road lighting design that is based on the road luminance is LUCIE. It is also explained that LUCIE shows a number of shortcomings that make it difficult to use by non-expert designers. Furthermore it is not applicable for many situations where only a cursory impression of the results is needed, such as for feasibility studies, for a preliminary design, and more in particular for the use in developing countries where the relevant data in most cases are not available anyway. In the following part of this section, a simplified design method is described that may be used in general, but more in particular in developing countries. It should be noted that the method proposed here is based on the CIE recommendations, so it will essentially lead to roads that in general will obey the CIE requirements (CIE, 1995a). Another matter is the fact that there are many arguments to establish new recommendations specially for developing countries. This is discussed in another part of this section.

(b) *Road classification*

The method uses five classes regarding the traffic volume and also five classes in road width. These two are highly correlated so a 'matrix' can be established. See Table 11.3.1.

		traffic		
road width	high	med	low	very low
very wide	A	•		
wide	•	B	•	
medium		•	C	•
narrow			•	D

Table 11.3.1 Road classification.

The table is expressed in qualitative terms. In order to have an impression of the order of magnitude we are talking about, the following values might be attached to the concepts in the table. The traffic is expressed in the rush hour number of vehicles per lane per hour, or, more precise, in 'passenger car equivalents' per lane per hour for the yearly average rush hour. The road width is expressed in meters for the total road for both directions together. For dual carriageway roads, the two carriageways are added, irrespective of the width of the median. Shoulders are not included. For traffic, high means >1000; medium means around 100, low means 10, and very low means <2. For the road width, wide means 15 m or more; medium means 7 to 12 m and narrow means 5 m or less. These widths correspond to 4 lanes or more; 2 to 3 lanes, and 1 lane respectively.

In Table 11.3.1, the most common combination of traffic and road width are indicated by A, B, C, and D respectively. The design method is based on these combinations. The dots (•) mean combinations that may be found in practice, but should not be recommended. Usually, roads with predominantly a traffic function belong to the classes A or B; roads with a mixed function to all classes, and residential streets to classes C and D. The table is valid for rural and urban roads alike. The way the different road types are distributed is, obviously, not equal.

(c) *Lighted and unlit roads*
The first step is to assess whether the road under consideration requires lighting or not. From road traffic and road safety points of view, main roads of the class A almost always require road lighting. Rural roads of classes B and C often can remain unlit. Rural roads of class D require no lighting. Urban streets of all classes usually are lit on the grounds of public safety considerations (amenity, crime prevention). The question 'to light or not to light' refers only to rural roads of classes B and C.

A final answer to this question can only be given on the basis of cost-effectiveness considerations. These considerations must compare the all costs of the lighting, also cables and transformers included, and the benefits, usually understood as the monetary value of all accidents saved by the lighting. Usually these costs are expressed in money value per year per km.

The costs of the lighting do not depend on the number of vehicles that pass over the road. As regards the accident costs, one might say as a first approximation that they are directly proportional to the night time traffic. So for increasing traffic volume, the costs are equal and the benefits increase, giving a 'break-even point' where the two are equal. Lighting on roads with heavier traffic than this break-even point is cost-effective whereas the lighting is not cost-effective on roads with less traffic than the break-even point (Schreuder 1995, 1998).

11.3.2 A practical method for simplified lighting design

(a) *The design approach*
In this section an example is described of a simplified method that is based on the well-known utilization factor curves (Anon., 1993, sec. 7.1.5, p. 298; Van Bommel, 1978; Van Bommel & De Boer, 1980. Van Bommel & De Boer, 1980, sec. 7.2.2, p. 113).

Road lighting design

These curves depict the amount of light that actually falls on the part of the road under consideration. When the reflection characteristics of the road surface are taken into account, the utilization factor curves become the luminance yield curves or τ-curves (Anon., 1993, sec. 7.1.5, p. 304; Schreuder, 1967, sec. 3.4.2; De Graaff, 1967, sec 7.4). Examples of how to work with τ-curves are given in Schreuder (1967, sec. 3.4.3). The method that is explained here is described in more detail in Van der Lugt & Albers (1996) and Schreuder (1994, 1998b).

It should be stressed that in the simplified method that is presented here, the τ-curves are not used. It is essentially a method to assess the illuminance. When the luminance is needed, a 'rule-of-thumb' may be used. This rule is derived from the average reflection factor q_0 that is discussed in secs. 10.4.1 and 10.4.2. The rule-of-thumb states that to find the luminance if the illuminance is known, for a dark road the illuminance must be divided by 20; for an average, normal asphalt road by 15, and for a light road, like e.g. cement concrete, by 10. It must be stressed that this rule-of-thumb only gives the numerical values. The luminance is expressed in cd/m^2 and the illuminance in lux.

The method is built up in a number of separate steps; viz.:
1. The effective road width
2. The mounting height
3. The spacing
4. The light level
5. The utilization factor
6. The tabulated lighting design

    (b) The effective road width

    The first step in the design method is the assessment of the effective road width. Basically, this is the geometric road width minus the overhang of the luminaire, the overhang being the distance between the road edge and the projection of the luminaire. The overhang should not be confused with then outreach, the length of the horizontal part of the support. Common practice in most countries is that the overhang is about 20% of the road width for single-sided arrangements and 10% for double-sided arrangements. Dual carriageways are considered separately. It should be noted that some median-mounted installations on dual carriageway roads show a negative overhang, e.g. in catenary suspension installations. As these installations require an expert handling in design and construction, they are disregarded here. Summing up, we conclude that the effective road width is 80% of the geometric road width.

(c) The mounting height

The next step is the assessment of the mounting height of the luminaires. Based on practice, the mounting height of the luminaires is chosen in relation to the traffic and the effective road width. For major roads (e.g. class A) the mounting height is 1,3 times the effective road width. For roads with a mixed function (e.g. classes B and C) the mounting height is about 0,9 to 1,1 times the effective road width and for residential roads (e.g. class D) the mounting height is about 0,5 to 0,8 times the effective road width.

(d) The spacing

The spacing is understood as the distance between two consecutive light points, irrespective at which side of the road they are located. Increasing the spacing – with the same light level – reduces costs but decreases the effects of optical guidance and increases the non-uniformity of the lighting pattern, both luminance and illuminance. More in particular, the luminance pattern on wet roads deteriorates rapidly with increasing spacing. Practice in many European countries indicates that the conditions usually are still acceptable when the spacing on major roads (e.g. class A) is about 4 times the mounting height; on roads with a mixed function (e.g. class B and C) about 6 and on residential roads about 10. In some countries it is customary to use longer spacings.

(e) The light level

The lighting requirements for major roads with a traffic or a mixed function is expressed in luminance values, those for residential roads in illuminance values (CEN, 2002; CIE, 1992, 1995a; NSVV, 1990, 2002). This holds for the level as well as for the lighting pattern. As the different recommendations show, the actual level depends on a number of parameters. In the simplified design method, an average of the values given in the recommendations is used. For major roads with a traffic function (e.g. class A) an average road surface luminance of 1 cd/m^2 is proposed, for major roads with a mixed function (e.g. class B and C) an average road surface luminance of 0,5 - 0,7 cd/m^2 is proposed and for residential streets an average illuminance on the road of 2 - 4 lux is proposed. As has been mentioned earlier, it is often difficult to assess the road luminance in the design stage, because reflection data usually are lacking, the average illuminance is often used instead. It is proposed to assume that a rather light coloured road surface is used with $E_{av}/L_{av} = 15$, which corresponds approximatively with a q_0-value of about 0,8). This implies an illuminance of about 15 lux for class A and about 7 - 10 lux for class C.

(f) The utilization factor

As mentioned earlier, when the design is based on illuminances, the utilization-factor methods may be used In the past, utilization curves were used in lighting design; since personal computers are used, these curves are often considered obsolete and are

Road lighting design

not included in the manufacturer's documentation any more. For a simplified design method, however, they are still useful. This implies that in the examples given here only rather old data could be used.

The utilization curve is a graphical representation of the percentage of the lamp lumens that fall on the road, where the road width – for road side and kerb side separately – is a variable. The road width is expressed in the mounting height. For obvious reasons, the utilization increases when the roads are wider in relation to the mounting height – more light falls on the road. As a first approximation, the factors for lanterns using advanced optical systems are similar; so are those for lanterns with simple optics as well as those for diffuse reflectors. On might assume that the first will be used mainly on class A roads, the second on class B and C roads and the third on class D roads. In the simplified design method, for each road class a specific road width is suggested. This means that for each class al, relevant parameters have only one value.

(g) *The tabulated lighting design*

In the preceding subsections, all parameters that are relevant for the lighting design are discussed. The actual design can therefore be made in tabular form. See Table 11.3.2.

road class	A	B	C	ED
road width	15	12	10	5
eff. road width	12	10	8	4
height	15	11	8	3
spacing/height	4	5	6	10
spacing (m)	60	55	48	30
light level				
L_{av}	1	0.7	0.5	n.r.
E_{hor}	(15)	(10)	(7)	2
luminaire type	HRP	SRM	SRM	HRD
De Boer ed, p.	290	316	316	288
utilization	0,22	0,26	0,30	0,34
maint. factor	0,5	0,5	0,5	0,5
lumen needed	122700	50769	22400	1824
lamp type	son	son	son	pl
wattage	400	250	250	18
lamp lumen	48000	28000	28000	1000
lamps needed	2,68	1,81	0,86	1,82
min. number	3	2	1	2

Table 11.3.2 Simplified design method; results.

From Table 11.3.2 several conclusions may be drawn. The first conclusion is that this simplified design method is suitable for a first 'preliminary' design only, because the result is often far from optimum. A second iteration selecting a more suitable luminaire would improve the situation. Secondly, the result depends very much on the figure selected for the maintenance factor. And finally, only the light level is taken into account; the non-uniformity, the luminance level and the glare are disregarded.

11.4 Conclusions

In this chapter a number of subjects selected from the wide area of outdoor lighting applications have been discussed. Emphasis is on the numerical design of road lighting installations, particularly on the luminance design. The CIE program LUCIE for computer-assisted luminance-design methods is discussed. It is concluded that the program is cumbersome and time-consuming to use, and that it can be used only for a limited number of road layout types. Slopes, curves, changes in the cross-section or in the lay-out cannot be handled. There is a need for a more comprehensive design method, at least as long as the luminance approach is used. The simplified design methods that are discussed in this chapter are useful, but they suffer from the same shortcomings as LUCIE.

It is also concluded that there are no clear principles for the design of road lighting for developing countries. It is expected that the new CIE-document will fill at least part of that gap.

It is concluded that in industrialised countries as well as in developing countries, the maintenance of road lighting installations is essential, but that at the same time it is often insufficient or even completely absent. In some cases, open luminaires may be, on the long run, be a better proposition.

References

Adrian, W. (1993). The physiological basis of the visibility concept. In LRI, 1993, p. 17-30.

Alferdinck, J.W.A.M. (1987). Oorzaken van hoge verblindingslichtsterkten en lage bermlichtsterkten van autokoplampen (Causes of high glare intensities and low illumination intensities of car headlamps). Report TNO-TM 1987, C 19. Soesterberg, TNO, 1987.

Alferdinck, J.W.A.M. & Padmos, P. (1986). Car headlamps: Influence of dirt, age and poor aim on the glare and illumination intensities. Lighting Res. Technol. 20(1988)195-198.

Anon. (1986). Eindrapport Commissie Kleine Criminaliteit (Final report Commission Petty Crime). Den Haag, SDU, 1986.

Anon. (1987). Juryrapportage Buiten Gewoon Veilig Prijs 1987 (Report of Jury: Outdoor Safety Price 1987). Rotterdam, Stichting Vrouwen Bouwen Wonen, 1987.

Anon. (1991). Scoren met sociale veiligheid; Handleiding sociale veiligheid in en om sportaccomodaties

Road lighting design 431

(Scoring with social safety; Manual social safety in and at sports facilities). Rijswijk, Ministerie van WVC, 1991.

Anon. (1993). Lighting manual. Fifth edition. LIDAC. Eindhoven, Philips, 1993.

Anon. (1994). Zien en gezien worden; Voorbeeldprojecten 'sociale veiligheid' (See and be seen; Example projects 'social safety'). Rijswijk, Ministerie van WVC, 1994.

Anon. (1997). Richtlijn openbare verlichting in natuurgebieden (Guideline for road lighting in nature reserves). Publicatie 112. Ede, CROW/NSVV, 1997.

Anon. (2000). American National Standard Practice for Roadway Lighting, ANSI/IESNA RP-8-00. The Illuminating Engineering Society of North America, 2000.

Anon. (2007). Handboek verlichtingstechniek (Lighting engineering handbook). Loose-leaf edition, 2007 issue. Den Haag, SDU, 2007.

Baer, R. (1990). Beleuchtungstechnik; Grundlagen (Fundaments of illuminating engineering). Berlin, VEB Verlag Technik, 1990.

Baer, R., ed. (2006). Beleuchtungstechnik; Grundlagen. 3., vollständig überarbeitete Auflage (Essentials of illuminating engineering, 3rd., completely new edition). Berlin, Huss-Media, GmbH, 2006.

Balder, J.J. (1957). Erwünschte Leuchttdichten in Büroräumen (Preferred luminance in offices). Lichttechnik. 6 (1957) 455.

Bodmann, H.W. (1962). Illumination levels and visual performance. International Lighting Rev. (Amsterdam), 13 (1962) 14.

Boyce, P.R. (2003). Human factors in lighting. 2nd edition. London. Taylor & Francis, 2003.

CEN (2002). Road lighting. European Standard. EN 13201-1..4. Brussels, Central Sectretariat CEN, 2002 (year estimated).

CEN (2003). Lighting applications – Tunnel lighting. CEN Report CR 14380. Brussels, CEN, 2003.

Chowdhury, R.S. (1995). Technology appropriate to developing countries. Special Workshop. In: CIE, 1995.

Chowdhury, R.S. (1997). The working programme of CIE TC 4-37 and preliminary results of pilot study. In: SANCI, 1997.

Chowdhury, S.R. & Schreuder, D.A. (1998). Road lighting in developing countries; the task of CIE TC 4-37. Workshop 'Warrants for road lighting'. 24 October 1998. Bath, UK., CIE, 1998.

CIE (1965). International recommendations for the lighting of public thoroughfares. Publication No. 12. Paris, CIE, 1965.

CIE (1967). Proceedings of the CIE Session 1967 in Washington, DC. (Vol. A, B). Publication No. 14. Paris, CIE, 1967.

CIE (1973). International recommendations for tunnel lighting. Publ. No. 26. CIE Paris, 1973.

CIE (1975). Guide on interior lighting. Publication No. 29. Paris, CIE, 1975.

CIE (1976). Calculation and measurement of luminance and illuminance in road lighting. Publication No. 30. CIE, Paris, 1976.

CIE (1976a). Glare and uniformity in road lighting installations. Publication No. 31. CIE, Paris, 1976.

CIE (1977). International recommendations for the lighting of roads for motorized traffic. Publication 12/2. Paris, CIE, 1977.

CIE (1980). Proceedings of the CIE Session 1979 in Kyoto. Publication No. 50. Paris, CIE, 1980.

CIE (1981). An analytic model for describing the influence of lighting parameters upon visual performance. Volume I: Technical foundations. Publication No. 19/2. Paris, CIE, 1981.

CIE (1982). Guide on interior lighting. Publ. No. 29-2. Paris, CIE, 1982.

CIE (1984). Tunnel entrance lighting. Publ. No. 61. Paris, CIE, 1984.

CIE (1990). Calculation and measurement of luminance and illuminance in road lighting. Publication No. 30/2. CIE, Paris, 1982 (reprinted 1990).

CIE (1990a). Guide for the lighting of road tunnels and underpasses. Publication No. 88, 1990.

CIE (1992). Guide for the lighting of urban areas. Publication No. 92. Vienna, CIE, 1992.
CIE (1994). Glare evaluation system for use within outdoor sports and area lighting. Publication No. 112. Vienna, CIE, 1994.
CIE (1995). 23nd Session of the CIE, 1-8 November 1995, New Delhi, India. Publ No. 119. Vienna, CIE, 1995.
CIE (1995a). Recommendations for the lighting of roads for motor and pedestrian traffic. Technical Report. Publication No. 115-1995. Vienna, CIE, 1995.
CIE (1999). 24th Session of the CIE, Warsaw – June 24-30, 1999. Proceedings; Two volumes. Publication No. 133. Vienna, CIE, 1999.
CIE (2000). Road lighting calculations. Publ. No. 140. Vienna, CIE, 2000.
CIE (2000a). Guide to the lighting of urban areas. Publ. No. 136. Vienna, CIE, 2000.
CIE (2001). Criteria for road lighting. Proceedings of three CIE Workshops on Criteria for road lighting. Publication No. CIE-X019-2001. Vienna, CIE, 2001.
CIE (2002). Visibility design for roadway lighting. A report of CIE TC 4-36. First draft. CIE, 2002.
CIE (2002a) Lighting of work places – outdoor work places. Draft Standard. DS015/E;2002. Vienna, CIE, 2002.
CIE (2003). 25th Session of the CIE, 25 June – 3 July 2003, San Diego, USA. Vienna, CIE, 2003.
CIE (2007). Road transport lighting for developing countries. Publication 180. Vienna, CIE, 2007.
Cohu, M. (1967). Floodlighting of buildings and monuments. Chapter 10. In: De Boer, ed., 1967.
COST (1994). Socio-economic cost of road accidents. COST 313. EUR 15464 EN. Commission of the European Communities, Brussels, 1994.
De Boer, J.B. (1951). Fundamental experiments of visibility and admissible glare in road lighting. Stockholm, CIE, 1951.
De Boer, J.B. (1967). Visual perception in road traffic and the field of vision of the motorist. Chapter 2. In: De Boer, ed., 1967.
De Boer, J.B. & Fischer, D. (1981). Interior lighting (second revised edition). Deventer, Kluwer, 1981.
De Boer, J.B.; Onate, V. & Oostrijck, A. (1952). Practical methods for measuring and calculating the luminance of road surfaces. Philips Research Records, 7 (1952) 45-76.
De Boer & Vermeulen, J. (1967). Simple luminance calculations based on road surface classification. In: CIE, 1967.
De Boer, J.B. ed. (1967). Public Lighting. Centrex, Eindhoven, 1967.
De Gooijer, H. & Looijs, G. (1997a). Photovoltaics for public lighting. In: SANCI, 1997.
De Graaff, A.B. (1967). Practical lighting design. Chapter 7. In: De Boer, ed., 1967.
Eckert, M. (1993). Lichttechnik und optische Wahrnehmungssicherheit im Strassenverkehr (Lighting engineering and security of visual perception in road traffic). Berlin – München, Verlag Technik GmbH., 1993.
Enzmann, J. (1993). Development and principles of the luminance and visibility calculations. In LRI, 1993, p. 1-4.
Eslinger, G.A. (1993). Practical aspects of the application of VL in roadway design. In: LRI, 1993, p. 149-154.
Flury, F.C. (1981). Cost/Effectiveness aspects of road lighting. Lichtforschung, 3 (1981) no. 1, p. 37-41.
Flury, F.C. (1992). De kosten van de verkeersonveiligheid; Een interimrapport (The cost of road accidents; An interim report). A-92-31. Leidschendam, SWOV, 1992.
Folles, E. & Poort, W.J.J.M. (1997). Processturing bij dynamische verlichting (Process control for dynamic lighting). In: NSVV, 1997a.
Forcolini, G. (1993). Illuminazione di esterni (Exterior lighting). Milano, Editore Ulrico Hoepli, 1993.
Gallagher, V.P.; Koth, B.W. & Freedman, M. (1975). The specification of street lighting needs. FHWA-RD-76-17. Philadelphia, Franklin Institute, 1975.

Griep, D.J. (1971). Analyse van de rijtaak (Analysis of the driving task). Verkeerstechniek, 22 (1971) 303-306; 370-378; 423-427; 539-542.

Harris, A. J. & Christie, M. A. (1951). The revealing power of street lighting installations and its calculation. Trans. Illum. Eng. Soc. (London) 16 (1951) p. 120.

Hentschel, H.-J. (1994). Licht und Beleuchtung; Theorie und Praxis der Lichttechnik, 4. Auflage (Light and illumination; Theory and practice of lighting engineering, 4th edition). Heidelberg, Hüthig, 1994.

Hentschel, H.-J., ed. (2002). Licht und Beleuchtung; Grundlagen und Anwendungen der Lichttechnik; 5. neu bearbeitete und erweiterte Auflage (Light and illumination; Theory and applications of lighting engineering; 5th new and extended edition). Heidelberg, Hüthig, 2002.

Holladay, L. L. (1926). The fundamentals of glare and visibility. J. Opt. Soc. Amer. 12 (1926) p. 271.

IES (1988). Annual Conference of the Illuminating Engineering Society of North America. Augst 7-11, 1988. Minneapolis, 1988.

Janoff, M. S. (1993). Visibility vs response distance; A comparison of two experiments and the implications of their results, J. IES 22, 1993, no. 1, p. 3-9.

Janoff, M.S. (1993a). The relationship between small target visibility and a dynamic measure of driver visual performance. Journ. IES, 22 (1993) nr. 1. p. 104-112.

Jongenotter, E.; Buijn, H.R.; Rutte, P.J. & Schreuder, D.A. (2000). Nieuwe richting voor wegverlichting (New directions in road lighting). Verkeerskunde, 51 (2000) no 1, January, p. 32-36.

Keck, M.E. (1993). Optimization of lighting parameters for maximum object visibility and its economic implications. In: LRI, 1993, p. 43-52.

Knudsen, B. (1967). Lamps and lanterns. In: De Boer, ed., 1967, Chapter 6.

LITG (1977). Measures of Road Lighting Effectiveness. 3rd International Symposium, Karlsruhe, 5th and 6th July 1977. Transactions. Berlin, Lichttechnische Gesellschaft e.V., 1978.

LRI (1993). Visibility and luminance in roadway lighting. 2nd International Symposium. Orlando, Florida, October 26 – 27, 1993. New York, Lighting Research Institute LRI, 1993.

Muhlrad, N., ed. (1989). Proceedings of the second European Workshop on Recent developments in road safety research, Paris, 25-27 January 1989. Actes INRETS No. 17, p. 261-269. Paris, INRETS, 1989.

Narisada, K. (1995). Perception in complex fields under road lighting conditions. Lighting Res. Technol. 27 (1995) p. 123-131.

Narisada, K. (1999). Balance between energy, environment and visual performance. In: CIE., 1999, Vol. I. p. 17-22.

Narisada, K. (2000). A method to balance between energy, environment and visual performance. Lighting & Engineering. 8 (2000). No. 1, p. 1.

Narisada, K. (2001). Visibility under motorway lighting conditions (in Japanese). Report submitted to the Express Highway Research Foundation, Japan, 2001.

Narisada, K. (2002). Distribution of revealing power and road lighting design (draft). Chapter 8. In: CIE, 2002.

Narisada, K. & Inoue, T, (1973). Uniformity in road lighting installations. Paper presented at Karlsruhe University on 15 October, 1973.

Narisada, K. & Inoue, T. (1981). Full scale driving experiments – Uniformity and perception under road lighting conditions. Journ. of Light & Visual Environment. 5 (1981) no. 2, p. 30-37.

Narisada, K. & Karasawa, Y. (2001). Reconsideration of the revealing power on the basis of visibility level. Proceedings of International Lighting Congress, Istanbul, 2, p. 473-480, 2001.

Narisada, K. & Karasawa, Y. (2003). Revealing power and visibility level. Light & Engineering. 11 (2003) no. 3, p. 24-31.

Narisada, K.; Karasawa, Y. & Shirao (2003). Design parameters of road lighting and revealing power. In: CIE, 2003, Vol. 2, D4, p. 10-14.

Narisada, K.; Saito, T. & Karasawa, Y. (1997) Perception and Road Lighting Design. In SANCI, 1997, p. 83-86.

Narisada, K. & Schreuder, D.A. (2004). Light pollution handbook. Dordrecht, Springer, 2004.
Narisada K. & Yoshikawa, K. (1974). Tunnel entrance lighting, Effect of fixation point and other factors on the determination of requirements, Lighting Res. Technol. 6 (1974), No. 1, p. 9-18.
Narisada, K. & Yoshimura, Y. (1977). Adaptation luminance of driver's eyes at the entrance of tunnel – An objective method. In: LITG, 1977.
NSVV (1957). Aanbevelingen voor openbare verlichting (Recommendations for public lighting). Moormans Periodieke Pers, Den Haag, 1957 (year estimated).
NSVV (1963). Aanbevelingen voor tunnelverlichting (Recommendations for tunnel lighting). Eletrotechniek. 41 (1963), 23; 46.
NSVV (1983). Fietspadverlichting: Een studie van de Commissie voor Openbare Verlichting van de NSVV (Cycle track lighting; a study of the commission for Public Lighting of the NSVV). Elektrotechniek 61 (1983) 233-245.
NSVV (1990). Aanbevelingen voor openbare verlichting (Recommendations for public lighting). Arnhem, NSVV, 1990.
NSVV (1991). Aanbevelingen voor de verlichting van lange tunnels voor het gemotoriseerde verkeer (Recommendations for the lighting of long tunnels for motor traffic. Arnhem, NSVV, 1991.
NSVV (1997). Aanbevelingen voor openbare verlichting; Deel III, Ontwerpen (Recommendations for public lighting; Part III, design). Arnhem, NSVV, 1997.
NSVV (1997a). Het licht in de toekomst (Light in the future). Conference, 20 March 1997. Arnhem, NSVV, 1997.
NSVV (2002). Richtlijnen voor openbare verlichting; Deel 1: Prestatie-eisen. Nederlandse Praktijkrichtlijn 13201-1 (Guidelines for public lighting; Part 1: Performance requirements. Practical Guidelines for the Netherlands 13201-1). Arnhem, NSVV, 2002.
NSVV (2004). Aanbeveling active markering (Recommendation active road marking). Arnhem, NSVV, 2004.
OECD (1975). Road marking and delineation. Paris, OECD, 1975.
OECD (1976). Adverse weather, reduced visibility and road safety. Paris, OECD, 1976.
Oudhaarlem, G.H. (1997). Telemanagement voor de openbare verlichting (Telemanaging public lighting). In: NSVV, 1997a.
Padmos, P. (1984). Visually critical elements in night time driving in relation to public lighting. In: TRB (1984).
Painter, K. (1999). Street lighting, crime and fear for crime; A summary of research. In: CIE, 2001.
Ris, H.R. (1992). Beleuchtungstechnik für Praktiker (Practical illuminating engineering). Berlin, Offenbach, VDE-Verlag GmbH, 1992.
Roszbach, R. (1972). Verlichting en signalering aan de achterzijde van voertuigen (Lighting and signalling at the rear of vehicles). Voorburg, SWOV, 1972.
SANCI (1997). SANCI-CIE International Conference on 'Lighting for Developing Countries'. Durban, South Africa, 1-3 September 1997.
Schmidt-Clausen, H.-J. & Bindels, J.T.H. (1974). Assessment of discomfort glare in motor vehicle lighting. Lighting Research & Technology. 5 (1974) 79-88.
Schreuder, D.A. (1964). The lighting of vehicular traffic tunnels. Eindhoven, Centrex, 1964.
Schreuder, D.A. (1967). The theoretical basis for road lighting design. Chapter 3. In : De Boer, ed., 1967.
Schreuder, D.A. (1967a). Tunnel lighting. Chapter 4. In : De Boer, ed., 1967.
Schreuder, D.A. (1976). White or yellow light for vehicle head-lamps? Arguments in the discussion on the colour of vehicle head-lamps. Publication 1976-2E. Voorburg, SWOV, 1976.
Schreuder, D.A. (1979). The lighting of residential yards. In: CIE (1979). In: CIE, 1980.
Schreuder, D.A. (1981). Visibility of road markings on wet road surfaces; A literature study. Arnhem, SCW, 1981.

Schreuder, D.A. (1984). Visibility aspects of road lighting. In: TRB, 1984.
Schreuder, D.A. (1987). Visual aspects of the driving task on lighted roads. CIE Journal. 7 (1988) 1 : 15-20.
Schreuder, D.A. (1989). Bewoners oordelen over straatverlichting (Residents judge the lighting of their streets). PT Elektronica- Elektrotechniek 44 (1989) nr. 5, p. 60-64.
Schreuder, D.A. (1989a). Road safety in developing countries: problems and research. In: Muhlrad, N., ed., 1989.
Schreuder, D.A. (1991). Visibility aspects of the driving task: Foresight in driving. A theoretical note. R-91-71. SWOV, Leidschendam, 1991.
Schreuder, D.A. (1994). New developments in urban street lighting. Paper presented at Technion, Haifa, Israel, 14 december 1994. Leidschendam, Duco Schreuder Consultancies, 1994.
Schreuder, D.A. (1994a). Kosten-Nutzen Ueberlegungen für Straßenbeleuchtung (Cost-benefit considerations in road lighting). Paper presented at LICHT94, Interlaken, Switzerland, September 1994. Leidschendam, Duco Schreuder Consultancies, 1994.
Schreuder, D.A. (1994b). Sick cities and legend analysis as therapy. Conference contribution. International Workshop on Urban Design and the Analysis of Legends, to be held on July 26 – 28, in Oguni-town, Kumamoto Prefecture, Kyushu Island, Japan. Duco Schreuder Consultancies, Leidschendam, the Netherlands. 1994.
Schreuder, D.A. (1995). The cost/benefit aspects of the road lighting level. Paper PS 161. In: CIE, 1995; Leidschendam, Duco Schreuder Consultancies, 1995.
Schreuder, D.A. (1995a). Road lighting in developing countries. In: CIE, 1995.
Schreuder, D.A. (1996). Street lighting maintenance; The need and the possibilities. Paper presented at the 3rd SAARC Lighting Conference in Kathmandu, Nepal, 21 – 23 November 1996. Leidschendam, Duco Schreuder Consultancies, 1996.
Schreuder, D.A. (1997). Theory and background for road lighting in developing countries. In: SANCI, 1997.
Schreuder, D.A. (1998). Road lighting for safety. London, Thomas Telford, 1998 (Translation of "Openbare verlichting voor verkeer en veiligheid", Deventer, Kluwer Techniek, 1996).
Schreuder, D.A. (1998a). Cost effectiveness considerations. In: CIE, 2001.
Schreuder, D.A. (1998b). Road lighting in developing countries; design and maintenance. Paper presented at the 14th Bi-annual Symposium on Visibility, April 20-21 Washington, DC, USA. Leidschendam, the Netherlands, Duco Schreuder Consultancies, 1998.
Schreuder, D.A. (1998c). Introduction of the draft recommendations for road lighting in developing countries. Workshop 'Warrants for road lighting'. 24 October 1998. Bath, UK., CIE, 1998.
Schreuder, D.A. (1999). Environmental-friendly lighting design. Paper for the Conference on Light Pollution, held in Athens, Greece, 7 – 9 May 1999. Leidschendam, Duco Schreuder Consultancies, 1999.
Schreuder, D.A. (1999a). Bogen en bochten; Enige rijtheoretische aspecten (Curves and bends; Some aspects from the theory of driving). Leidschendam, Duco Schreuder Consultancies, 1999.
Schreuder, D.A. (2001). Principles of Cityscape Lighting applied to Europe and Asia. Paper presented at International Lightscape Conference ICiL 2001, 13 – 14 November 2001, Shanghai, P.R. China. Leidschendam, Duco Schreuder Consultancies, 2001.
Schreuder, D.A. (2001a). Pollution free lighting for city beautification; This is my city and I am proud of it. Paper presented at the International Lighting Congress in Istanbul, Turkey, 6-12 September 2001. Leidschendam, Duco Schreuder Consultancies, 2001.
Schreuder, D.A. (2001b). Straßenbeleuchtung für Sicherheit und Verkehr (Street lighting for safety and transport). Aachen, Shaker Verlag GmbH, 2001.
Schreuder, D.A. (2004). Verlichting thuis voor de allerarmsten (Home lighting for the very poor). NSVV Nationaal Lichtcongres 11 november 2004. Arnhem, NSVV, 2004.

Schreuder, D.A. (2005). Domestic lighting for developing countries. Prepared for publication in: UNESCO – A world of science, Paris, France. Leidschendam, Duco Schreuder Consultancies, 2005.

Schreuder, D.A. (2008). Looking and seeing. A holistic approach to vision. Dordrecht, Springer, 2008 (in preparation).

Schreuder, D.A. & Lindeijer, J.E. (1987). Verlichting en markering van voertuigen: Een state-of-the-art rapport (Lighting and marking of road vehicles: A state-of-the-art report). R-87-7. Leidschendam, SWOV, 1987.

Schreurs, J.P.G.; De Vries, N.E. & Schreuder, D.A. (2007). Openbare verlichting & veiligheidsbeleving; Een experimentele studie (Public lighting and the sensation of security; An experimental study). Utrecht, Gemeente Utrecht, Dienst stadwerken, 2007.

Stark, R.E. (1998). Warrants for road lighting, Experience in the USA. In: CIE, 2001.

Takeuchi, T. & Narisada, K. (1996). Additivity of the luminance difference thresholds for foveal adaptation luminance and for the veiling luminance. J. Illum. Inst. Japan. 80 (1996). no. 8A, p. 14-18.

TRB (1984). Providing visibility and visual guidance to the road user. Symposium July 30 – August 1, 1984. Washington, D.C. Transportation Research Board, 1984.

Van Bommel, W.J.M. (1978). Optimization of road lighting installations by the use of performance sheets. Lighting Res. Technol. 10 (1978) 189.

Van Bommel, W.J.M. & De Boer, J.B. (1980). Road lighting. Deventer, Kluwer, 1980.

Van der Lugt, D.B. & Albers, H. (1996). The art of lighting. Arnhem, NSVV 1996 (year estimated).

Van Santen, C. (2005). Light zone city: Light planning in the urban context. Basel, Birkhauser, 2005.

Vermeulen, J. & Knudsen, B. (1968). Het ontwerpen van een verlichting van voorgeschreven luminantie en gelijkmatigheid (Road lighting design with prescribed luminance and uniformity). Philips Tech. Tijdschr. (168)29.

Visser, R. (1997). Spelen met licht in en om het huis (Playing with light in and around the house). Amersfoort, Dekker/vd Bos & Partners, 1997

Vos, J.J. & Padmos, P. (1983). Straylight, contrast sensitivity and the critical object in relation to tunnel entrance lighting. CIE, Amsterdam, 1983.

Waldram, J. M. (1938). The Revealing power of street lighting installations. Trans. Illum. Eng. Soc. (London). 3 (1938) 173-186.

Watanabe, S. (1965). NHK Gijyutsu Kenkyu, 17, 1965.

Index

24/7 economy 1
τ-curves 427

A

aberrations 200
    heterochromatic ~ 200
    monochromatic ~ 200
absolute glare 252
absolute threshold 252, 256
absorbance 93, 132
absorption 30
acceptors 69
accommodation 197, 198, 276
achromatic channel 325
action
    qualitative ~ 13
    quantitative ~ 13
action potential 210
active road markings 415
acuity
    visual ~ 121, 209, 240, 262
adaptation 208, 410
    chromatic ~ 338
    dark ~ 253
    light ~ 253
Adaptation Curve
    CIE ~ 339
adaptation level 132, 145, 203
adaption
    state of ~ 256
additive process 329
aerosols 378
aether 21
after-images 252, 339
age 297
Airy-disk 200

ALCoR 380
alkali metal 43
aluminium
    anodised ~ 370
amacrine cells 212, 218
amenity 2, 292, 345, 417
anatomy 192
angle
    solid ~ 89
anode 45, 47
anodised aluminium 370
anorganic LED 70
approach
    functional ~ 3
aqueous humour 194, 196
area contrast 255
artificial light 1
asphalt
    drainage ~ 389
    porous ~ 389
atmospheric perspective 276
attribution 144
autonomous nervous system 191
axiom 30
axon 188

B

background luminance
    natural ~ 378
background radiation
    natural ~ 378
ballast 42, 47
band
    conduction ~ 52
    conductivity ~ 67
    valence ~ 52, 67

437

baryons 42
basic formula of photometry 118
Bernoulli 129
    theorem of ~ 119
binning 179
binocular parallax 277
biological clock 206, 211, 220
bipolar cells 212, 218
black-body locus 339
black-body radiator 147, 339
black-hole effect 298
black body 38
blackening 39
bleeding 178
blinding glare 286
blind spot 194
Bloch's Law 259
Blondel-Rey's Law 157
body
    black ~ 38
    grey ~ 34
Bohr 66
bolometer 146, 149, 230
border contrast 255
bosons 42
brain 191
    large ~ 221
brain stem 220
brightness 103, 133
    reference sky ~ 378
    subjective ~ 281
Bunsen-Roscoe's Law 157
burner 48, 63

C

camera 359
candela 34, 91, 104, 146, 148
candle
    international ~ 147
cat's eyes 414
cataract 196, 296
cathode 47
CCD 68
CCDs 177
    line scan ~ 178
CEE 372
cells
    amacrine ~ 212, 218
    bipolar ~ 212, 218
    ganglion ~ 203, 204, 206, 212, 218, 324

nerve ~ 193
off-cells 213
solar ~ 384
Celsius 12
central nervous system 191
cerebellum 219
cerebrum 220, 221
CFF 158, 160, 279
Charge-Coupled Devices 177
chiasma
    optical ~ 214, 222
chromatic adaptation 338
chromaticity coordinates 330
CIE 1931 colour space 330
CIE Adaptation Curve 339
CIE colour triangle 332
    CIE ~ 331
CIE Illuminant C 335
city beautification 345, 416
civic pride 416
closed loop 162
coiled 38
colorimetry 322, 328, 329
colour 31, 313
    constant correlated ~ 341
    primary ~ 329
    spectral ~ 331, 335
colour blind 234
colour blindness 319
colour constancy 339
colour defective vision 319
colour point 331, 335
colour recognition 385
colour rendering 342
colour rendering characteristics 345
colour rendition 42, 337, 345
colour shift 346
colour space 329
    CIE 1931 ~ 330
colour temperature 339
    equivalent ~ 342
colour triangle 331
    CIE ~ 332
communication engineering 164
compact fluorescent lamp 62
concave mirror 360
conduction band 52
conductivity band 67
conductor
    electric ~ 44

Index

semiconductor 44, 66
cones 151, 207, 230, 233, 316, 322, 328
cone vision 150
conflicts 6
consciousness 187
constancy 284, 334
constant correlated colours 341
continuity
    equation of ~ 120
continuity principle 118
continuum 202
contrast 156, 253
    area ~ 255
    border ~ 255
    intrinsic ~ 290
    negative ~ 386
    successive ~ 339
    visible ~ 290
contrast method 156
contrast sensitivity 156, 253
control 365
    light ~ 364, 384
    pro-active ~ 163
    reactive ~ 163
control loop 162
convergence 276
convex mirror 360
cornea 194
cortex 220
    visual ~ 193, 206, 221
cosine correction 176
cosine law 101, 176
cosine to the third law 102
cost-benefit analysis 418
counterbeam 386
coupure 300
crime
    fear for ~ 417
    petty ~ 417
critical flicker-fusion frequency 160
critical size 257
curfew 381
curve
    τ-curve 427
cut-off 300
cycle
    halogen ~ 41
cyclopean eye 277

D

dark adaptation 253
dark light 248
dark noise 215
daylight vision 87, 230
day vision 146
decision
    pro-active ~ 164
    reactive ~ 164
dendrites 188
dependent variable 151, 249
depreciation 39
descriptive models 10
design voltage 35
detection 250
detector 162
determinism 10
deutan 319
deuteranope 319
differences
    just noticeable ~ 246
diffraction 22, 121, 122, 290, 376
diffraction limited 24
diffuser
    perfect ~ 114, 135
    quasi-perfect ~ 135
dimmer 383
dimming 39
diode 67
    Light Emitting Diode 70
    OLED 75
    semiconductor ~ 67
diopter 193
disability glare 287, 288, 374, 415
discomfort glare 12, 303, 415
discomfort glare scale 144
distribution
    light ~ 364, 366, 384, 405
donors 69
doping 67
doping material 52
drainage asphalt 389
driving task 6, 386, 387, 402, 415
driving task analysis 245
duplicity theory 233
dynode 169

E

E-P diagram 404
earths
  rare ~ 57
economic life 39
efficacy 28, 29
  luminous ~ 92
electric conductor 44
electro-luminescence 69
electromagnetic field 112
electromagnetic wave 18
electron shell 43
element
  optical ~ 193
  refractive ~ 195
ellips
  MacAdam ~ 342
emission spectrum 202
emissivity 30
enamel 370
energy 2
energy level 43
envelope 19
epitaxy 70
equation of continuity 120
equation of hydrodynamics 120
equivalent colour temperature 342
equivalent position diagram 404
equivalent veiling luminance 289
ergonomics 244
evening regime 381
excitation 45, 323
exclusion principle 43
expectancy 285
expert system 409
exposure 92
external photo-effect 167
extrinsic semiconductor 67
eye-ball 193
eye-piece 362

F

fear for crime 417
Fechner's Law 149, 247
feed-back loop 163
feedback 162
Fermat 19, 358
Fermi-level 70
Fermi gas 66

fermions 42
field 19, 21, 110, 127
  functional visual ~ 275
  light ~ 112
  morphic ~ 111
  photic ~ 112
field intensity 110
field line 110, 127
field map 127
field strength 110
filament 35, 38
filling-in process 195
filter
  neutral ~ 155
first law 133
fixation time 412
flat-glass controversy 376
flicker 160
flicker-fusion frequency 158, 217, 279
  critical ~ 160
flicker effects 259, 384
flicker phenomena 157
flicker photometry 154, 156, 157, 160, 259
flood lighting 4
flu-powder 57
fluorescence 51, 61
fluorescent tube 53, 61, 202
flux
  geometric ~ 127, 129, 136
  luminous ~ 91
  radiation ~ 136
flux density 94
flux line 110
focal length 200
forbidden zone 165
force 110
force line 110
foresight 6, 402
form perception 264
form recognition 265
Fourier analysis 199
Fourier transform 199
fovea 232, 287
fovea centralis 209
friction 119, 121
frontal lobes 221, 223
full radiator 33
function 151, 249
functional approach 3
functionality 4

Index

functional PSF 200
functional visual field 275

G

ganglion cells 203, 204, 206, 212, 218, 324
gas
    Fermi ~ 66
    noble ~ 44
gas-filled lamp 39
gas discharge 48
gated viewing 383
Gauss's theorem 120
GCRI 347
General Glare Observer 302
geometric flux 117, 127, 129, 136
geometric optics 19, 112, 358, 370
Gestalt 9
glare 286, 414
    absolute ~ 252
    blinding ~ 286
    disability ~ 287, 288, 374, 415
    discomfort ~ 12, 303, 415
    discomfort ~ scale 144
    physiological ~ 288
    psychological ~ 303
glare angle 199, 288, 296
Glare Control Mark 303
glare cube 287
glare sources 287
glaucoma 196
Globe at night 149
glow
    sky ~ 149, 374
GLS 33
goal 163
Grassman's rule 329
grey body 34
grey matter 220
guidance
    optical ~ 386, 414
    visual ~ 414
gut-feeling 156

H

halogen 41
halogen cycle 41
halogen lamp 40, 41
headlamps, vehicle ~ 100
heat sink 71
Hefner lamp 146
Heisenberg 27
heterochromatic 154, 160, 232
heterochromatic aberrations 200
heuristic 164
high mesopic vision 239
hippocampus 220
holes 52, 67
holism 9
holon 111
homezone 410
Hoorweg 279
Hoorwegs' Rule 157
horizon pollution 375
horizontal cells 212
human factors 244
humour
    aqueous ~ 194, 196
    vitreous ~ 194
Huygens Principle 19
hydrodynamics
    equation of ~ 120
hyperacuities 277
hypermetropy 197, 201
hypothalamus 220

I

IEC 372
illuminance 94, 133, 134, 137, 170
    semicylindrical ~ 95
Illuminant C
    CIE ~ 335
illusions
    visual ~ 195, 255
image 359
    real ~ 359
    virtual ~ 359
image-forming devices 359
image forming
    non-~ 211
immission
    light ~ 375
incandescent 34
increment threshold 246, 247
independent variable 151, 249
index
    refractive ~ 200
infinity 113
ingress protection 372, 421
inhibition 323, 339
inner plexiform layer 218

insulator 44
integration time 278
intensity
    field ~ 110
    luminous ~ 89, 93
interference 21
internal photo-effect 167
international candle 147
interval scale 12, 144, 149
intra-ocular stray light 288
intrinsic contrast 290
intrinsic semiconductor 66
intrusion
    light ~ 375
inverse square law 155
ion 44, 45, 46
ionization 46
IP-number 372
iris 194, 202, 287, 294
irradiance 134
isochromatic photometry 155
isotropic 387
Istwert 163
iterative process 155, 406

J
Judd modification 231
just noticeable differences 246

K
Kelvin 13
Kirchhoff's law 30

L
Lambertian 100
Lambertian radiator 114
lamp
    compact fluorescent ~ 62
    gas-filled ~ 39
    halogen ~ 40, 41
    Hefner ~ 146
lamp efficacy 151
Landolt Ring 264
large brain 221
law
    Bloch's Law 259
    Blondel-Rey's Law 157
    Bunsen-Roscoe's Law 158
    cosine ~ 101, 176
    cosine to the third ~ 102
    Fechner's Law 149, 247
    first ~ 133
    first ~ of thermodynamics 30
    inverse square ~ 155
    Kirchhoff's ~ 31
    Piper's Law 256, 258
    Raleigh-Jeans' Law 333
    Rayleigh's Law 290, 341
    Ricco-Piper's Law 157
    Ricco's Law 256
    Stefan-Boltzmann's ~ 31
    Stokes' ~ 53
    Talbot's Law 158
    Weber's Law 246, 342
    Wien's ~ 31
LED 70
    anorganic ~ 70
LEDs 415
lens 194
    negative ~ 360
    positive ~ 360
LEP 76
life 39, 59
    economic ~ 39
light 17, 358
    artificial ~ 1
    dark ~ 248
    duplicity of ~ 20
    intra-ocular stray ~ 288
    intraocular stray ~ 205
    monochromatic ~ 201, 381
    near-white ~ 341
    phenomenological approach to ~ 18
    speed of ~ 113
light adaptation 253
light blob 375
light control 364, 384
light designer 315
light distribution 364, 366, 384, 405
Light Emitting Diode 70
light field 18, 95, 112
light immission 375
lighting
    flood ~ 4
    low-beam ~ 238
    public ~ 4
lighting designers 5
lighting engineers 5
light intrusion 375
light pencil 128

Index

light pollution 2, 374
light rays 18, 112, 119, 358
light trespass 375
light tube 127, 203
lightwatts 235
limbic system 220
line
    field ~ 110, 127
    flux ~ 110
    force ~ 110
    resonance ~ 46
    spectral ~ 46
line scan CCDs 178
lobes 221
    frontal ~ 221, 223
    occipital ~ 221, 222
    parietal ~ 221
    temporal ~ 221
loop
    closed ~ 162
    control ~ 162
    feed-back ~ 163
    open-loop systems 162
low-beam lighting 238
LUCIE 407
lumen 92
lumens 235
luminaire efficiency 370
luminaires 363
    open ~ 371
luminance 103, 125, 133, 137
    equivalent veiling ~ 289
    natural background ~ 378
    road ~ 402
    veiling ~ 289
luminance discrimination 253
luminance factor 105, 134
luminance level 386
luminance technique 386, 403, 410
luminescence 51
    electro-~ 69
    photoluminescence 68
luminosity 281
luminous efficacy 92
luminous flux 91, 137
luminous intensity 89, 93
luxmeter 96

M

MacAdam ellipses 342
Mach bands 255
magnitudes 141
maintenance factor 422
Maxwell equations 118
measurement 143
measuring grid 96
memory 163
Menedeleyev 43
mercury 46, 48, 340
mesopic metric 242
mesopic range 262
mesopic vision 92, 209, 216, 236, 237, 240
    high ~ 239
message 162
metal 43, 44
    alkali ~ 43
metal halide lamps 202
metameric pair 337
metamerism 337
metric
    mesopic ~ 242
metric scale 13, 144
metylacrylate 371
microscope 363
midbrain 219
minimum separable 24, 123
mirror
    concave ~ 360
    convex ~ 360
mirror neurons 223
modal split 419
model 301, 359
    Stimulus-Response ~ 163
models 10
    descriptive ~ 10
    predictive ~ 10
momentum 113
monochromatic 62
monochromatic aberrations 200
monochromatic light 201, 381
morphic field 111
morphic unit 111
mounting height 404
Munsell system 336
myopia 197
myopic 201

N

natural background luminance 378
natural background radiation 378

Natural Park 380
nautical twilight 379
near-white light 341
negative contrast 386
negative lens 360
neocortex 220
nerve
    optical ~ 206
nerve cells 193
nerve tract 193
    visual ~ 204
nervous system 191
    autonomous ~ 191
    central ~ 191
    peripheral ~ 191
nervus opticus 213
neuron 188
neurons 193
    mirror ~ 223
neurotransmitter 190, 324
neutral filter 155
night
    starry ~ 2
night regime 381
noble gasses 44
noise 164
    dark ~ 215
    photon ~ 164, 254
    readout ~ 179
    shot ~ 164
    signal-to-noise ratio 164
nominal scale 12, 143
non-image forming 211
nucleus geniculatus lateralis 221
number 143
Nyquist limit 216

O

objective 362
observer
    standard ~ 150
occipital lobes 221
Ockham's razor 10
off-cells 213
OLED 75
olfactory regions 220
ontogenesis 219
open-loop systems 162
open luminaires 371
operating voltage 35

opponent-process theory 323
Opstelten
    Rule of ~ 242
optical chiasma 214, 222
optical elements 193
optical guidance 386, 414
optical nerve 206
optics
    geometric ~ 19, 112, 358, 370
ordinal scale 12, 144
organogenesis 219
orientation 5
Orion 149

P

p-n junction 67, 69
Palmer 236
    Rule of ~ 242
paradigm 10
parallax
    binocular ~ 277
parietal lobes 221
pattern recognition 181
perception
    form ~ 264
perfect diffuser 114, 135
performance
    visual ~ 201, 245, 246
peripheral nervous system 191
periphery 209, 210
perspective
    atmospheric ~ 276
petty crime 417
phosphorescence 51
phosphors 57, 63
photic field 112
photo-effect
    external ~ 167
    internal ~ 167
photo-electric effect 68
photoluminescence 68
photometer bench 155
photometric threshold distance 100
photometry 86, 150, 229
    flicker ~ 154, 156, 157, 160, 259
    heterochromatic ~ 232
    isochromatic ~ 155
    split-field ~ 156
    subjective ~ 154
    visual ~ 154, 231

Index

photomultipliers 146, 168
photon noise 164, 254
photons 18, 26, 42, 91, 109, 119, 134, 150, 162
photopic mode 87
photopic vision 36, 146, 151, 230, 327
photopigments 206
photoreceptors 193, 206, 233, 315
photosynthesis 211
phylogenesis 219
physiological glare 288
Piper's Law 256, 258
pituitary gland 220
pixels 178
Planck-constant 26
Planck's law 31
Planckian radiator 333
plexiform layer
    inner ~ 218
point sources 259
point spread function 199, 288
pollution
    horizon ~ 375
    light ~ 2, 374
porous asphalt 389
positive lens 360
positivism 10
potential 110
power 39, 162, 229
predictive models 10
presbyopia 198
pride
    civic ~ 416
primaries 322
primary colours 329
primary standard 147
pro-active control 163
pro-active decisions 164
process
    additive ~ 329
    filling-in ~ 195
    iterative ~ 155, 406
    subtractive ~ 329
protan 319
protanope 319
Protected Landscape 380
PSF 199, 288
    functional ~ 200
psychological glare 303
psychophysical experiments 232
public lighting 4

Pulfrich 278
pupil 193, 202
Purkinje effect 240
Purkinje shift 209
purple
    visual ~ 252

Q

qualia 223
qualitative action 13
quantitative action 13
quantum 146, 206
quantum yield 55
quasi-perfect diffuser 135

R

R-table 390, 404
radiation
    natural background ~ 378
    thermal ~ 30
radiation density 136
radiation flux 136
radiator
    black-body ~ 147
    full ~ 33
    Lambertian ~ 114
    thermal ~ 147
radiometric equivalence 235
radiometry 18, 86, 145, 149, 229
Raleigh-Jeans' Law 333, 339
range
    mesopic ~ 262
    stereoscopic ~ 277
rare earths 57
Rayleigh's Law 290, 341
rays
    light ~ 358
reach 299, 377, 415
reactive control 163
reactive decisions 164
readout noise 179
real image 359
recognition
    colour ~ 385
    form ~ 265
    pattern ~ 181
recombination 206
reductionism 9
reference object 412
reference sky brightness 378

reflectance 92, 132
reflection 18, 19, 358
reflection formula 135
reflex 163
refraction 18, 19, 358
refraction errors 197
refractive element 195
refractive index 200
regime
    evening ~ 381
    night ~ 381
rendering
    colour ~ 342
    colour ~ characteristics 345
rendition
    colour ~ 345
residential street 410
resistance 44
resolving power 262
resonance line 46
response
    spectral ~ 145
rest mass 27
reticular formation 219
retina 150, 193, 194, 204
revealing power 411
rhodopsin 206
Ricco-Piper's Law 157
Ricco's Law 256
road luminance 402
road marking
    active ~ 415
rods 207, 233, 315
rod vision 150
RSC-curve 261
Rule of Opstelten 242
Rule of Palmer 242
run-of-the-road 402
run-up 63

S

S/N ratio 164
saccades 192
safety
    social ~ 292
scalar 134, 387
scale
    discomfort glare ~ 144
    interval ~ 12, 144, 149
    metric ~ 13, 144
    nominal ~ 12, 143
    ordinal ~ 12, 144
sclera 204
scotoma 194
scotopic vision 36
SCRI 347
selective radiators 37
semiconductor 44, 66, 165
    extrinsic ~ 67
    intrinsic ~ 66
semiconductor diode 67
semicylindrical illuminance 95
senses 187
sensitivity
    spectral ~ 29
sensor 162
set 11
shift
    colour ~ 346
shot noise 164
SI 13
si-unit 90
signal 162, 164
signal-to-noise ratio 164
signalling lights 292
size
    critical ~ 257
skull 193
sky glow 149, 374
SLI 365
Small Target Visibility 246, 411
smell 220
social safety 292
sodium 46, 48
solar cells 384
solar panel 68
solid angle 89
solidity 276
Sollwert 163
spectral colours 331, 335
spectral line 46
spectral response 145
spectral sensitivity 29
spectral tristimulus values 331
spectrum
    emission ~ 202
speed of light 113
sphere
    Ulbricht ~ 172
    unit ~ 89

Index

spikes 206
spin 43
spinal chord 191
splash-and-spray 389
split
    modal ~ 419
split-field photometry 156
spot
    yellow ~ 209
spread 365
stainless steel 370
standard 146
    primary ~ 147
Standard Glare Observer 295, 302
standard observer 150
starry night 2, 373
state
    steady ~ 230
state of adaptation 230, 256
steady state 230
steel
    stainless ~ 370
Stefan-Boltzmann's Law 31
steradian 89
stereopic effect 276
Stereopsis 277
stereoscopic range 277
Stiles-Crawford effect 200, 204, 208, 288
Stiles-Holladay relation 297
stimulus 206, 247
Stimulus-Response model 163
Stokes' law 53
strategy 164
stray light
    intra-ocular ~ 288
    intraocular ~ 205
STV 411
subjective brightness 281
subjective photometry 154
subtractive process 329
successive contrast 339
supra-threshold detection 376
synapse 189, 324
synesthesia 222

T

Talbot's Law 158
task
    driving ~ 6, 386, 387, 402, 415
    driving ~ analysis 245

    visual ~ 6, 246, 287, 402
telescope 363
temperature
    colour ~ 339
    equivalent colour ~ 342
    thermodynamic ~ 13
temporal lobes 221
tensor 137, 387
thalamus 220
thermal radiation 30
thermal radiator 147
thermodynamic temperature 13
threshold
    absolute ~ 252, 256
    increment ~ 246, 247
    supra-threshold detection 376
throughput 129, 130, 131
throw 365
Tiffany Studies 260
transduction
    visual ~ 206
transmittance 93, 132
trichromatic theory 323
tristimulus values 330
    spectral ~ 331
tritan 319
tritanope 319
tube
    fluorescent ~ 53, 61
    light ~ 127
tungsten 35, 38
tungsten filament 340
twilight
    nautical ~ 379

U

Ulbricht sphere 172
ULR 377, 381
uncertainty principle 27, 113
Uniform Chromaticity Scale Diagram 344
unit
    morphic ~ 111
    si-~ 90
unit sphere 89
universal gravitational constant 110
Unterschiedsempfindlichkeit 260
utilization factor 422, 426

V

vacuum lamp 39

valence band 52, 67
variable
    dependent ~ 151, 249
    independent ~ 151, 249
vector 109, 134, 137, 387
vehicle headlamps 100
veil 288
veiling luminance 289
victim matrix 374
viewing
    gated ~ 383
virtual image 359
visibility 402
    Small Target ~ 246, 411
visibility concept 246
Visibility Level 411
visibility level 262
visible contrast 290
vision
    cone ~ 150
    day ~ 146
    daylight ~ 87, 230
    high mesopic ~ 239
    mesopic ~ 92, 209, 216, 236, 237, 240
    photopic ~ 36, 146, 151, 230, 327
    rod ~ 150
    scotopic ~ 36
visual acuity 121, 209, 240, 262
visual cortex 193, 206, 221
visual field
    functional ~ 275
visual guidance 414
visual illusions 195, 255
visual nerve tract 204
visual performance 201, 245, 246

visual photometry 154, 231
visual purple 252
visual task 6, 246, 287, 402
visual transduction 206
visus 262
vitreous humour 194
voltage
    design ~ 35
    operating ~ 35
V_λ-curve 152, 232, 234

W

warming-up 63
wattage 39
wave
    electromagnetic ~ 18
wave front 19, 359
wave mechanics 26
wWeber's Law 246, 342
Weber fraction 149, 246
white matter 220
white point 335
Wien's Law 31
Wolfram 35
woonerf 410
work 29, 134, 146

Y

yellow spot 209

Z

zone
    forbidden ~ 165
zoning 379

Printed in the United States
125036LV00002B/98/P